Combinatorial Game Theory

GRADUATE STUDIES IN MATHEMATICS 146

Combinatorial Game Theory

Aaron N. Siegel

AMERICAN MATHEMATICAL SOCIETY
Providence, Rhode Island

2010 *Mathematics Subject Classification.* Primary 91A46.

For additional information and updates on this book, visit
www.ams.org/bookpages/gsm-146

Library of Congress Cataloging-in-Publication Data

Siegel, Aaron N., 1977–
Combinatorial game theory / Aaron N. Siegel.
pages cm. — (Graduate studies in mathematics ; volume 146)
Includes bibliographical references and index.
ISBN 978-0-8218-5190-6 (alk. paper)| ISBN 978-1-4704-7568-0 (Softcover)
1. Game theory. 2. Combinatorial analysis. I. Title.

QA269.S5735 2013
519.3—dc23

2012043675

Printed in the United States of America.
Reprinted by the American Mathematical Society, 2023.

∞ The paper used in this book is acid-free and falls within the guidelines
established to ensure permanence and durability.
Visit the AMS home page at http://www.ams.org/

11 10 9 8 7 6 5 4 3 2 28 27 26 25 24 23

To Elwyn Berlekamp

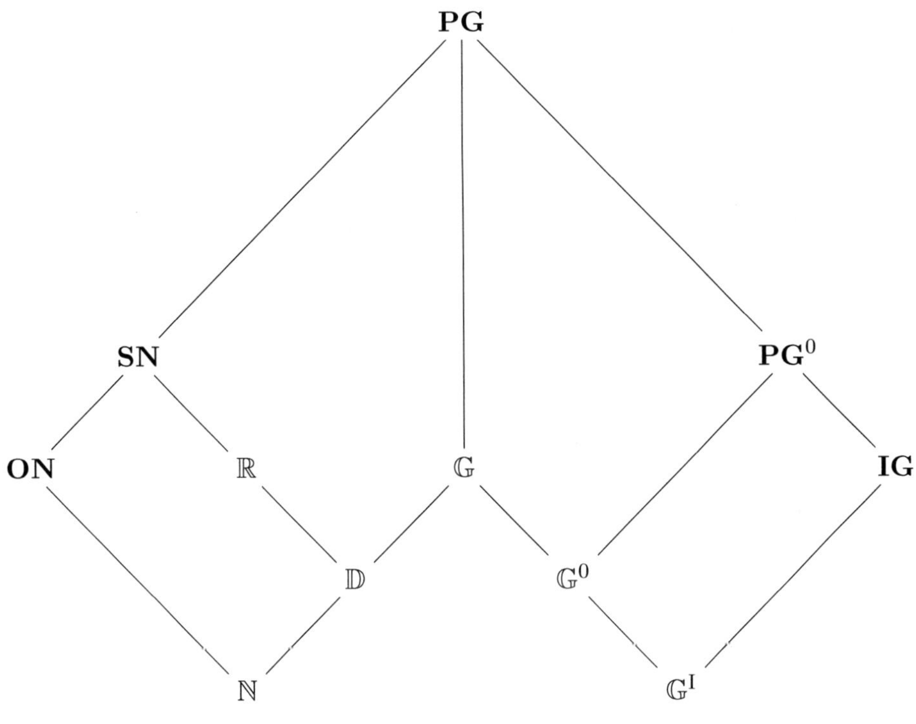

Figure. Various systems of numbers and games.

Contents

Preface

The disjunctive theory of combinatorial games can trace its roots to the work of Sprague and Grundy in the 1930s, but its modern form was born with the arrival of Conway's *On Numbers and Games* in 1976 and the classic *Winning Ways for Your Mathematical Plays* by Berlekamp, Conway, and Guy in 1982. In the ensuing three decades, combinatorial game theory has blossomed into a serious and active branch of combinatorics, with connections to coding theory, computational complexity, and commutative algebra.

This book is intended as a second course on combinatorial games, at the first- or second-year graduate level, and most readers will benefit from some prior exposure to the subject. *Winning Ways* is a fine introduction; in addition, an excellent new textbook by Albert, Nowakowski, and Wolfe, titled *Lessons in Play: An Introduction to Combinatorial Game Theory*, has recently appeared. Either (or both) of these references should serve as adequate preparation for this volume.

Nonetheless, this book is completely self-contained and traces the development of the theory from first principles and examples through many of its most recent advances. It should serve those who have read *Winning Ways* and crave a more rigorous development of the theory, as well as professionals seeking a cohesive reference for the many new ideas that have emerged in recent years. Among those advances appearing for the first time in textbook form (as far as I know) are Berlekamp's generalized temperature theory, Thane Plambeck's elegant theory of misère quotients, David Moews' results on the group structure of $\mathbb{G}$, and the construction of misère canonical forms for partizan games.

A great many people contributed insights, support, and suggestions. Elwyn Berlekamp has been a continued source of inspiration and encouragement throughout the past decade. Others whose input has influenced more than one page include Michael Albert, Dean Allemang, John Conway, Aviezri Fraenkel, J.P. Grossman, Richard Guy, Bob Hearn, Neil McKay, Richard Nowakowski, Thane Plambeck, Bill Spight, and David Wolfe. My wife, Olya, and daughter, Tali, have remained heroically supportive throughout endless years of writing and revising. Finally, I wish to thank my acquisitions editor, Ina Mette, and the AMS editorial committee for their seemingly infinite patience and unwavering support of this project.

Combinatorial game theory, in its modern form, is still a young field; many areas of the subject are still poorly understood and many questions remain unanswered. Open problems and conjectures are sprinkled throughout the text and are summarized in Appendix A. If this book provides a reservoir of tools and inspiration to attack these problems, then it will be a success.

Aaron Siegel
San Francisco

Chapter I

Combinatorial Games

1. Introduction

Combinatorial games are two-player games with no hidden information and no chance elements. They include child's play such as TIC-TAC-TOE and DOTS AND BOXES; mathematical abstractions "played" on arbitrary graphs or grids or posets; and some of the deepest and best-known board games in the world, such as GO and CHESS.

The mathematical theory of combinatorial games pursues several interrelated objectives, including:

- exact solutions to particular games, usually in the form of an algebraic description of their outcomes;
- an understanding of the general combinatorial structure of games; and
- hardness results, suggesting that for certain games, or in certain situations, no concise solution exists.

In all of the games considered in this book, there will be just two players, **Left** and **Right**, who play alternately and whose moves affect the **position** in a manner defined by the rules of the game. Both players will have complete knowledge of the game state at all times ("no hidden information"), and the effect of each move will be entirely known before the move is made ("no chance elements"). We'll defer a formal definition until after we've introduced some examples.

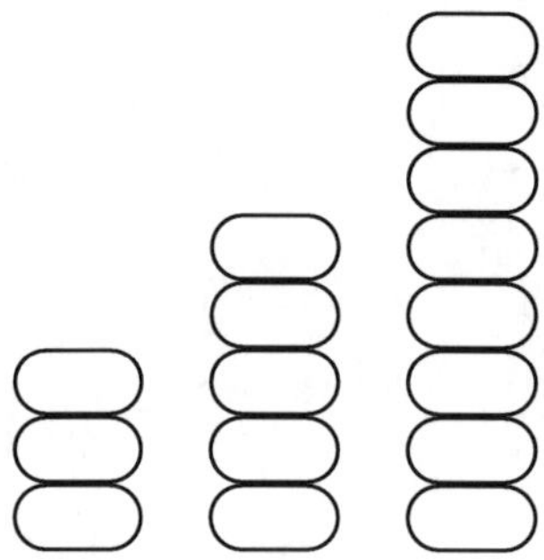

Figure 1.1. A typical NIM position, with heaps of 3, 5, and 8 tokens.

NIM

NIM was among the earliest games to be subjected to a complete combinatorial analysis. It is played with several heaps of tokens; a typical position is shown in Figure 1.1. On her turn a player may remove any number of tokens (but necessarily at least one) from any *one* heap. The removed tokens are discarded and play continues. Whoever removes the last token wins.

Obviously any position containing just one (nonempty) heap is a first-player win, since whoever moves first will simply discard the entire heap. What about two-heap positions? If both heaps contain the *same number* of tokens, then the position is easily seen to be a second-player win: second player can simply respond to each of his opponent's moves with an identical move on the other heap, and (by induction) the game must eventually end. This is known as the **mirror-image strategy**. Conversely, a position containing exactly two *unequal* heaps is a first-player win, since first player can equalize the heaps and thereafter follow the mirror-image strategy.

A complete strategy for the game was discovered by the Harvard geometer Charles Bouton in 1902. Bouton's strategy is sufficiently simple to describe that we might as well do so now. Given two nonnegative integers a and b, the **nim-sum** $a \oplus b$ is given by the following computation: first write a and b in binary; then "add without carrying." For example, here's the calculation of $29 \oplus 21 \oplus 11 = 3$:

$$\begin{array}{r} 11101 \\ \oplus\ 10101 \\ \oplus\ \ \ 1011 \\ \hline =00011 \end{array}$$

If G is a NIM position with heaps $a_1, a_2, \ldots, a_k$, then the **nim value** of G is defined to be

$$a_1 \oplus a_2 \oplus \cdots \oplus a_k.$$

Call G a **zero position** if its nim value is zero. Then the strategy for NIM is neatly described:

Theorem 1.1 (Bouton's Theorem). *Let G be a* NIM *position.*

- *If G is a zero position, then every move from G leads to a nonzero position.*
- *If G is not a zero position, then there exists a move from G to a zero position.*

Therefore, if G is a zero position, second player can guarantee a win by reverting each of his opponent's moves to a new zero position. Since every move reduces the total number of tokens in play, the game will eventually end and second player will have the last move.

Likewise, if G is not a zero position, then first player can guarantee a win, simply by moving to any zero position.

For example, suppose G has three heaps, of sizes 29, 21, and 11. A moment ago we noted that $29 \oplus 21 \oplus 11 = 3$. This is nonzero, so the position must be a first-player win, and it is not hard to see that the unique winning move is to take three coins from the heap of size 11, leaving $29 \oplus 21 \oplus 8 = 0$.

Proof of Bouton's Theorem. First suppose $a_1 \oplus a_2 \oplus \cdots \oplus a_k = 0$ and consider a typical move from a_1 to a_1'. Necessarily $a_1 \neq a_1'$, so

$$a_1' \oplus a_2 \oplus \cdots \oplus a_k \neq a_1 \oplus a_2 \oplus \cdots \oplus a_k = 0.$$

Conversely, suppose $x = a_1 \oplus \cdots \oplus a_k \neq 0$. Consider the *most-significant bit*, say the jth bit, of the binary representation of x. At least one of the a_i must have its jth bit equal to 1; assume without loss of generality that it is a_1. Put $a_1' = x \oplus a_1$. Necessarily $a_1' < a_1$, since its jth bit is equal to 0 and it agrees with a_1 on all higher-order bits. So there is a move available from a_1 to a_1', and

$$a_1' \oplus \cdots \oplus a_k = x \oplus a_1 \oplus \cdots \oplus a_k = x \oplus x = 0. \qquad \square$$

Outcomes and Solutions

Every NIM position G has a well-defined outcome: if G is a zero position, then second player can force a win; otherwise first player can force a win. This divides all NIM positions into two disjoint **outcome classes**.

The positions of most combinatorial games can likewise be partitioned into various outcome classes. While it's easy to prove that such outcomes *exist* (and we'll shortly do so for a wide class of games), actually computing them tends to be much harder. We can use Bouton's Theorem to determine the outcome of any NIM position; extending this sort of analysis to more complicated games is a central goal of combinatorial game theory.

Note that Bouton's Theorem applies not to any particular starting position, but to all possible NIM positions, with any number of arbitrarily large heaps. When we say that NIM has been "solved," we mean that an efficient rule exists for calculating the outcome of *any* position. For most games (including NIM) it's straightforward to translate such a rule into a perfect winning strategy.

DAWSON'S KAYLES

DAWSON'S KAYLES is played with several rows of boxes, such as those shown in Figure 1.2(a). On his turn, a player must remove exactly two *adjacent* boxes from one of the strips. For example, he might play as shown in Figure 1.2(b), splitting the strip of size 10 into two smaller strips and leaving the position (c). Eventually all remaining boxes will be contained in strips of length one (or perhaps, no boxes will be left at all), after which no further moves are available.

There are two ways to play DAWSON'S KAYLES:

- In **normal play**, the player who makes the last move *wins*. This is the same convention used in NIM.
- In **misère play**, the player who makes the last move *loses*.

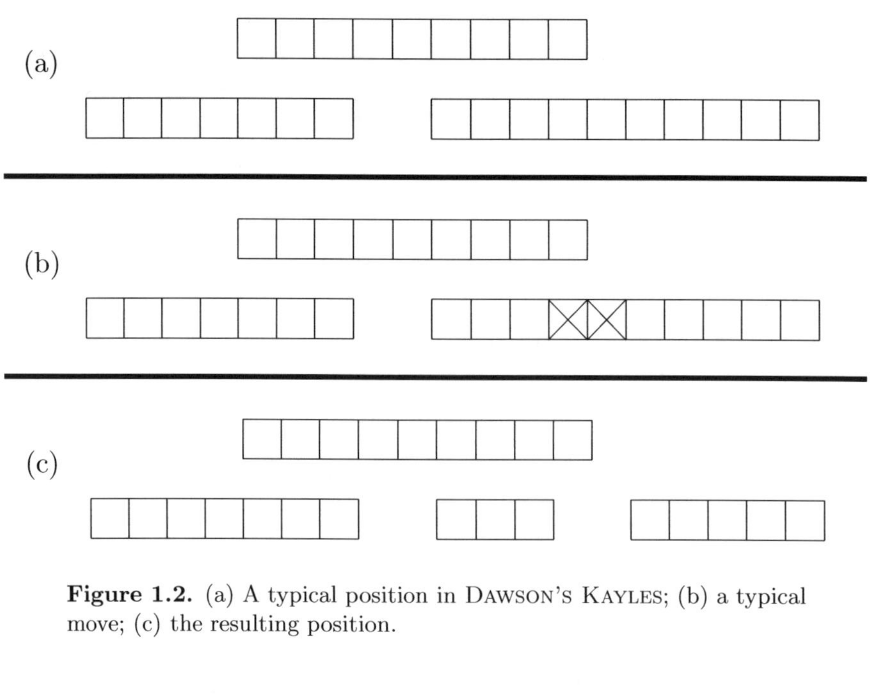

Figure 1.2. (a) A typical position in DAWSON'S KAYLES; (b) a typical move; (c) the resulting position.

A complete solution to the normal-play variant (on an arbitrary number of strips, of arbitrarily large size) was found more than fifty years ago by Richard Guy and Cedric Smith. They showed that a strip of $n + 34$ boxes behaves just like a strip of n for almost all n, the only exceptions being $n = 0, 15, 17, 32, 35, 52$. It therefore suffices to consider strips of a certain bounded size, and these can be analyzed exhaustively by generalizing the strategy for NIM. (Skip ahead to Figure IV.2.1 on page 186 for a full presentation of the Guy–Smith solution.)

DAWSON'S KAYLES, however, was first proposed as a misère game, and in misère play the situation is sharply different: no complete solution is known, despite significant effort by many authors. In fact misère DAWSON'S KAYLES is among the oldest unsolved problems in combinatorial game theory.[1]

At first glance it seems paradoxical that normal and misère play should so diverge in difficulty. After all, there is one sense in which the two forms of the game are equally hard: a given position G has exactly the same number of subpositions, regardless of the play convention. A brute-force analysis that computes the outcome of G by examining every subposition should therefore be no harder in misère play than in normal play.

The trouble is that the complexity of a brute-force analysis grows exponentially with the size of the position. Therefore, such an approach is not practical, nor does it grant much insight into the structure of the game. The key to the normal-play solution is the observation that any two strips in a DAWSON'S KAYLES position are **disjoint**—no single move can affect them both. In normal play, it is relatively easy to quantify the interactions between disjoint components. Each strip can therefore be analyzed independently, and this is what makes the Guy–Smith solution work.

Conversely, in misère play there is no simple characterization of the interaction between disjoint components. So although normal and misère play are similar from a brute-force perspective, the *combinatorics* of misère play are fundamentally more difficult. This combinatorial complexity is what distinguishes the two conventions.

HACKENBUSH

HACKENBUSH is played on an undirected graph with edges colored blue, red, or green, such as shown in Figure 1.3 on the next page. Since this book is printed in black and white, we draw soLid lines for bLue edges, paRallel lines for Red edges, and dottEd lines for grEen edges.

[1]The game was introduced in 1935 by T. R. Dawson, in a slightly different form known as DAWSON'S CHESS. The two forms of the game are isomorphic—see Exercise 1.5.

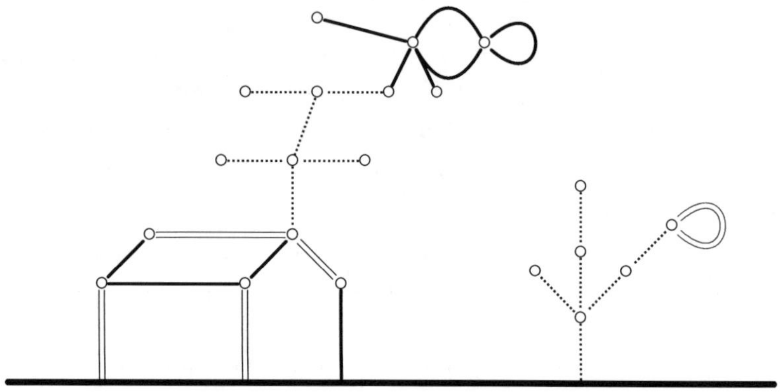

Figure 1.3. A typical HACKENBUSH position.

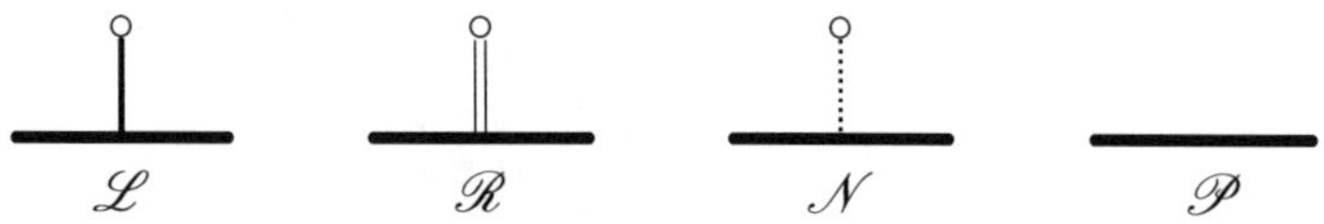

Figure 1.4. Four HACKENBUSH positions with distinct outcome classes.

The thick horizontal bar should be regarded as a single vertex of the graph, the **ground**. On her turn, Left may select any blue (solid) edge and remove it from the game; likewise Right may remove any red (parallel) edge. Green (dotted) edges may be removed by either player. As soon as an edge is removed, *all edges no longer connected to the ground are also removed.* Whoever chops the last edge wins.

There is one major difference between HACKENBUSH and the two previous examples. From a given HACKENBUSH position, the moves available to Left are distinct from those available to Right (except in the special case where every edge is green). Conversely, in both NIM and DAWSON'S KAYLES, both players have the same moves available at all times. Games such as HACKENBUSH, which distinguish between Left and Right, are called **partizan**; NIM-like games are **impartial**.

Impartial games like NIM and DAWSON'S KAYLES have just two outcome classes, but HACKENBUSH exhibits four:

- $\mathscr{N}$ first player (the *N*ext player) can force a win;
- $\mathscr{P}$ second player (the *P*revious player) can force a win;
- $\mathscr{L}$ *L*eft can force a win, no matter who moves first;
- $\mathscr{R}$ *R*ight can force a win, no matter who moves first.

$\mathscr{N}$ and $\mathscr{P}$ are familiar from NIM—they contain the nonzero and zero positions, respectively. But $\mathscr{L}$ and $\mathscr{R}$ are unique to partizan games, since they are biased in favor of one player or the other. Figure 1.4 illustrates simple HACKENBUSH positions representing each of the four possibilities.

Elements of $\mathscr{N}$ are called **$\mathscr{N}$-positions**, and likewise for $\mathscr{P}$, $\mathscr{L}$, and $\mathscr{R}$.

DOMINEERING

DOMINEERING (sometimes called CROSSCRAM) was invented by Göran Andersson around 1973. The game is usually played on an 8×8 checkerboard. Left and Right alternately place dominoes on the board, with the following restrictions:

- Each domino must cover exactly two adjacent squares of the grid. Dominoes may never overlap nor hang off the edge of the board.
- Left can only place verticaLly oriented dominoes; Right can only place hoRizontal ones.

The player who makes the last move wins. Since play is alternating, this is equivalent to placing the most dominoes, except that ties are broken in favor of second player.

DOMINEERING is a territorial game. Players will typically aim for moves like the one shown in Figure 1.5(a), which carves out space for another vertical domino and effectively locks Right out of the northeast corner.

Well-played games of DOMINEERING continue along this theme. Figure 1.5(b) shows Game 3 of the first (and so far only) World Domineering Championship match. After fourteen moves the board decomposes into six

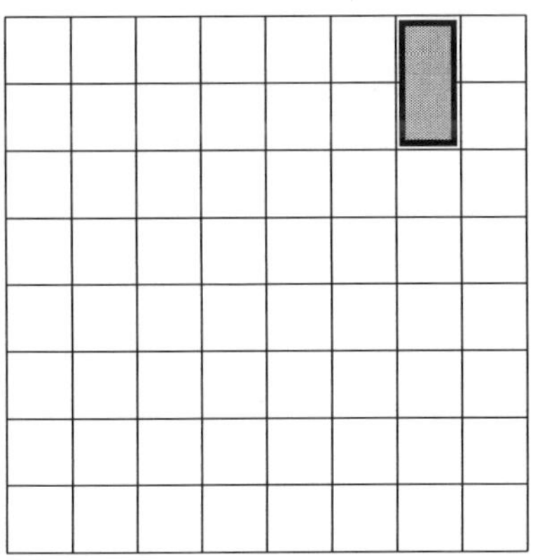

(a)

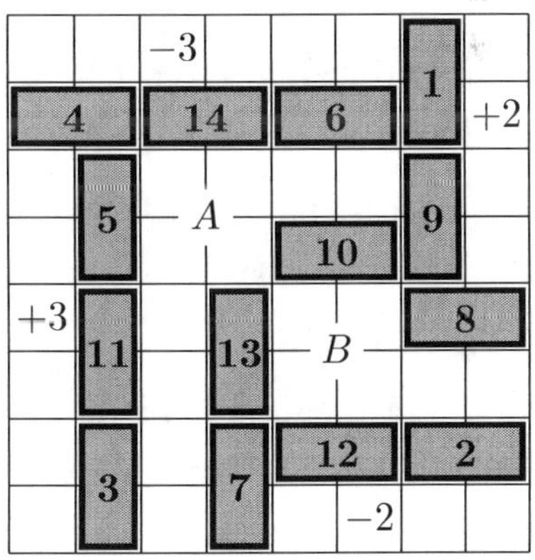

(b)

Figure 1.5. (a) A typical opening move in DOMINEERING; (b) the first fourteen moves of Wolfe–Calistrate 1994, Round 3.

independent regions. Four of them are simple $1 \times n$ or $n \times 1$ strips, each of which is clearly worth an integer number of dominoes no matter how the game progresses. The remaining two regions can't be assigned numerical values in quite the same way: the final score on region A depends upon who makes the first move *within* A, and likewise for B.

Nevertheless, it's easy to check that Left's best move on region A yields a net score of $+2$, and Right's best move there yields a net score of 0, and we can indicate this by writing $A = \{2 \mid 0\}$. By similar reasoning we find that $B = \{0 \mid -2\}$. If Left moves on A, Right can respond on B, and vice versa; so Right can guarantee a net score of 0 on the combination $A + B$. Since the other four regions also cancel each other out, we conclude that Left and Right will ultimately place equally many dominoes. Therefore Right, as second player, will win.

Games, Options, Rulesets

Throughout this text, we'll use the term **game** to refer to an *individual position* in a combinatorial game, and we'll call a system of playable rules a **ruleset** instead. For example, DOMINEERING is a ruleset, and the specific DOMINEERING position pictured in Figure 1.5(b) is a game. This convention reflects the fact that individual positions are the primitive objects of our theory. We'll keep the distinction clear by using capital Roman letters

$$G, H, \ldots$$

to denote combinatorial games and capital Greek letters

$$\Gamma, \Lambda, \ldots$$

to denote rulesets.

The following definitions are straightforward and fundamental.

Definition 1.2. Let G and H be combinatorial games. H is a **Left option** (resp. **Right option**) of G if Left (resp. Right) can move directly from G to H.

We usually denote a typical Left option of G by G^L and a typical Right option by G^R.

Definition 1.3. Let G and H be combinatorial games. H is a **subposition** of G if there exists a sequence of consecutive moves (for either player, not necessarily alternating) leading from G to H.

For example, Figure 1.5(b) is a subposition of Figure 1.5(a).

The definition of subposition includes the case where the sequence of moves is empty, so that every game G is considered a subposition of itself. We'll use the term **proper subposition** in order to exclude this case.

Definition 1.4. Let G be a combinatorial game.

(a) A **run** of G of length $k < \infty$ is a sequence of positions

$$G_0, G_1, G_2, \ldots, G_k$$

such that $G_0 = G$, and each G_{i+1} is an option of G_i. Likewise, a run of G of length ∞ is a sequence

$$G_0, G_1, G_2, \ldots.$$

(b) An **alternating run** is a run in which successive moves alternate between Left and Right, i.e., if G_{i+1} is a Left option of G_i, then G_{i+2} is a Right option of G_{i+1}, etc.

(c) An alternating run of length k is said to be a **play** of G if either $k = \infty$ or else G_k has no options for the player to move.

The Fundamental Theorem

We'll now show that a wide class of games can be partitioned into the four outcome classes $\mathscr{L}$, $\mathscr{R}$, $\mathscr{N}$, and $\mathscr{P}$. A combinatorial game G is **short** provided that:

- G has just finitely many distinct subpositions; and
- G admits no infinite run.

It's easy to check that every position in NIM, DAWSON'S KAYLES, HACKENBUSH, and DOMINEERING is short. For example, suppose G is a DOMINEERING position with n empty squares. Then G has at most 2^n distinct subpositions, and infinite runs are impossible since every move strictly decreases the number of empty spaces.

Theorem 1.5 (Fundamental Theorem of Combinatorial Game Theory). *Let G be a short combinatorial game, and assume normal play. Either Left can force a win playing first on G or else Right can force a win playing second, but not both.*

Proof. Consider a typical Left option of G, say G^L. Since G is short, G^L must have strictly fewer subpositions than G. Therefore we may assume by induction (and symmetry) that either Right can force a win playing first on G^L, or else Left can force a win playing second.

If Right can win *all* such G^L playing first, then certainly he can win G playing second regardless of Left's opening move. Conversely, if Left can win *any* such G^L playing second, then he can win G by moving to it. Exactly one of these two possibilities must hold. □

The proof of Theorem 1.5 is straightforward, but it contains several interesting features. First, the theorem has an obvious dual, in which "Left" and "Right" are interchanged. This is true of many theorems about partizan games, and we will usually state and prove just "one side" of each result.

Alert readers will also notice that while induction is central to the proof, there is no explicit mention of a base case. This is deliberate: there is an *implicit* base case in which Left has no options at all—in which case the inductive hypothesis is satisfied vacuously. Throughout this book we'll see many more inductive arguments whose base case is collapsed into the induction.

The Fundamental Theorem shows that every short game belongs to one of the four normal-play outcome classes $\mathscr{L}$, $\mathscr{R}$, $\mathscr{N}$, or $\mathscr{P}$. We denote by $o(G)$ the outcome class of G.

Disjunctive Sum

All four games—NIM, DAWSON'S KAYLES, HACKENBUSH, and DOMINEERING—share a fundamental property: a typical position can be broken down into several independent components. For example, the three nim-heaps in Figure 1.1 are disjoint, in the sense that no single move may affect more than one of them. The outcome of the full position G therefore depends only on the structure of its individual heaps, and it makes sense to write

$$G = H_3 + H_5 + H_7,$$

where H_n denotes a heap of size n.

Similarly, the DAWSON'S KAYLES position of Figure 1.2(a) consists of three disjoint strips. Moreover, after the opening move shown in Figure 1.2(b), the strip of length 10 is divided into two smaller strips. Here we may speak of a move from

$$S_7 + S_9 + S_{10}$$

to

$$S_7 + S_9 + S_3 + S_5,$$

where S_n is a strip of length n, and it is clear that any position can be written this way. In both NIM and DAWSON'S KAYLES, the problem of finding the *outcome* of an arbitrary position reduces to understanding the *structure* of individual heaps (or strips).

Likewise, in the HACKENBUSH position of Figure 1.3 the house and the flower can never interact; and in the DOMINEERING position of Figure 1.5(b) there are six disjoint components.

This modularity is the driving force of combinatorial game theory. Given a combinatorial game G, it is often impractical to undertake a brute-force analysis of G itself. Instead, we analyze the components of G individually and attempt to extract information that will tell us something useful when they are pieced back together. The precise nature of this "information" is the main topic of this book.

To obtain a cohesive theory, it's necessary to formalize the notion of "piecing together" a game's components. If G and H are any two combinatorial games, we define

$$G + H,$$

the **disjunctive sum** of G and H, as follows. Copies of G and H are placed side-by-side. On his turn, a player must move in *either* G *or* H (but not both). Likewise, in the sum

$$G_1 + G_2 + \cdots + G_k,$$

a player must move in *exactly one* of the components G_i.

It should be clear that disjunctive sum accurately models the decompositions that we've observed in all four examples. Here, rather than "decomposing" complex positions, we regard the components themselves as primitive building blocks. This approach is far more flexible. We can certainly assemble the components of a position G back up into G itself; but we can also form new sums $G + H$, where H is unrelated to G and perhaps even drawn from a different ruleset. For example,

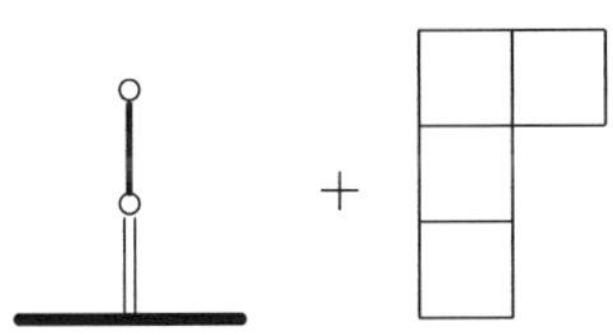

is a perfectly reasonable combinatorial game.

This flexibility has far-reaching consequences. It helps to expose shared structure across many rulesets, and it permits direct quantitative comparisons between them.

The Fundamental Equivalence

Now for the point! Let G and H be short games, and write

(†) $\quad G = H \quad \text{if} \quad o(G + X) = o(H + X)$ for all X.

Here X ranges over *all conceivable* short games—including, for example, all DOMINEERING positions, even if G or H happens to be a HACKENBUSH position. In such cases the definition of + is the same as if we were summing components drawn from the same ruleset. We'll make all of this precise in Chapter II, when we give a formal definition of short game.

The relation given by (†) is the **fundamental equivalence**, and in various forms it will drive virtually every idea that we explore in this book. It's easily seen to be an equivalence relation, and the equivalence class of G modulo = is known as the **game value** of G. More precisely, it is the **normal-play short partizan game value** of G, but normally we'll drop the cumbersome adjectives when it's clear which type of value we're talking about. The set of game values is denoted by

$$\mathbb{G}$$

and we'll spend most of Chapters II and III investigating its structure.

Notice that equality is a *defined relation* given by the fundamental equivalence. This is a standard convention in combinatorial game theory. When we wish to indicate that G and H have isomorphic structure, we'll write the notation

$$G \cong H$$

instead. Isomorphism is a stronger condition than equality: it is often the case that $G = H$ even when $G \not\cong H$. (Conversely, $G \cong H$ always implies $G = H$.)

There is a strong analogy between $\mathbb{G}$ and the set of rational numbers $\mathbb{Q}$. The statement

$$G = H$$

for distinct games G and H should be no more troubling than the assertion

$$\frac{2}{4} = \frac{3}{6}.$$

Here G and H are distinct *games* that represent the same *game value*, just as 2/4 and 3/6 are distinct *ratios* that represent the same *rational number*.

Example. We have the following HACKENBUSH identity:

This is slighly less trivial than it seems, since from the left-hand side G Left's only move (up to symmetry) is to

$G^L =$  , while from the right-hand side H she can move to either

$H^{L_1} =$ 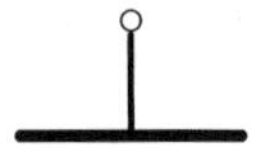 or $H^{L_2} =$.

So in fact $G \not\cong H$. But it's easy to see that in any sum

$$H + X,$$

Left's extra option H^{L_2} is no help. Any time she could win with a move to $H^{L_2} + X$, she might just as well play to $H^{L_1} + X$ instead and then ignore the extra stalk for the remainder of play.

Example. The fundamental equivalence provides a means to compare positions from different rulesets. For example, it's easy to see that

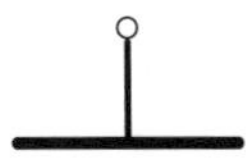 $=$  .

In fact the two positions are isomorphic, since they both have one Left option to the empty game and no Right option. There's nothing profound in this observation, of course, but you might be surprised to learn that

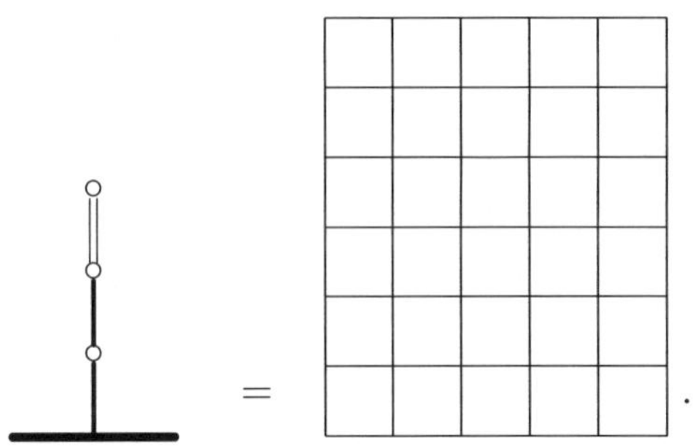

Figure 1.6. A nontrivial identity between HACKENBUSH and DOMINEERING.

Let's take a moment to appreciate what this means. The DOMINEERING grid H in Figure 1.6 is a complicated game—there are fifteen distinct opening moves (nine for Left and six for Right), even after accounting for symmetries; and we can imagine that in a sum

$$H + X$$

it might become quite difficult to work out all the possibilities needed to determine the outcome. But knowing that H is equal to the much simpler HACKENBUSH stalk G, we can instead compute the outcome of

$$G + X,$$

and the fundamental equivalence gives $o(H+X) = o(G+X)$. In Chapter II we'll develop several tools for establishing identities like the one in Figure 1.6.

Other Kinds of Values

The definition of equality

$$G = H \quad \text{if} \quad o(G+X) = o(H+X) \text{ for all } X$$

is sensitive to three primitive parameters:

- the definition of the *outcome* $o(G)$;
- the definition of the *sum* $G+H$; and
- the domain of the *for all* quantifier over X.

In arriving at $\mathbb{G}$ we took $o(G)$ to be the *normal-play outcome* of G; $G+H$ to be the *disjunctive sum* of G and H; and *for all* to range over the set of short partizan games. But there are many other natural choices for all three primitives. For example, $o(G)$ might just as well be defined as the *misère-play outcome* of G, and we'll discuss variant meanings of "sum" and "for all" in Section 4.

Astonishingly, in virtually all cases—no matter how we choose to fix meanings for these three parameters—the fundamental equivalence remains the same:

$$G = H \quad \text{if} \quad o(G+X) = o(H+X) \text{ for all } X.$$

Each time we vary the meanings of "outcome," "sum," and "for all," the fundamental equivalence yields a different theory—but it is always a coherent theory. The fundamental equivalence is the glue that holds combinatorial game theory together, and we'll turn to it for guidance again and again throughout this book.

Exercises

1.1 Determine all winning moves from the NIM position with heaps 18, 22, and 29.

1.2 Let G be a NIM position. Let n be the number of *distinct* winning moves on G. Prove that either $n = 0$ or n is odd.

1.3 Determine the strategy for misère NIM.

1.4 Let G be an impartial game. Suppose that there exists an option H of G such that every option J of H is also an option of G. Prove that G must be an $\mathscr{N}$-position.

1.5 DAWSON'S CHESS is played with two rows of pawns on a $3 \times n$ board:

The pawns move and capture like ordinary CHESS pawns, except that capture is *mandatory*. (If a choice of captures is available, then the player may select either one.) Show that DAWSON'S CHESS on a $3 \times n$ board is isomorphic to DAWSON'S KAYLES with $n + 1$ boxes, regardless of which play convention (normal or misère) is observed. (This is the original form of the game proposed by Dawson [**Daw73**].)

1.6 No position in BLUE–RED HACKENBUSH is an $\mathscr{N}$-position.

1.7 Determine the outcomes of 4×4, 5×4, and 5×5 DOMINEERING.

1.8 Let S_L and S_R be sets of positive integers. The **partizan subtraction game** on S_L and S_R is played with a single heap of n tokens. On her move, Left must remove k tokens for some $k \in S_L$; likewise, on his move Right removes k tokens for some $k \in S_R$. Denote by $o(H_n)$ the (normal-play) outcome of a heap of n tokens. Prove that the sequence $n \mapsto o(H_n)$ is periodic, i.e., there is some $p > 0$ such that $o(H_{n+p}) = o(H_n)$ for all sufficiently large n.

2. HACKENBUSH: A Detailed Example

Many of the ideas and principles of combinatorial game theory are neatly illustrated by various situations that arise in HACKENBUSH. The simplest position is the game with no options at all,

which is called 0, the **empty game**. We'll write

$$0 = \{ \mid \}$$

to indicate that there are no options for either player.

Introducing a single blue edge gives the game

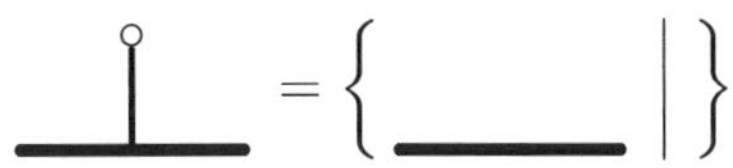

$= \{0 \mid \}$

which is an $\mathscr{L}$-position, since Left can win immediately by moving to 0, while Right has no move at all. We'll call it 1, as it behaves like one spare move for Left.

What can one say about positions like the one in Figure 2.1, in which every disjoint component is monochromatically red or blue?

Clearly Left will always play to remove just one blue edge; there is nothing to be gained by deleting more than one at a time. Likewise Right will always play to delete a single red segment. Each player is powerless to disrupt the other's plans, so the win will ultimately go to whoever started with more edges.

We can therefore replace every blue component with a positive integer number of edges, and every red component with a negative integer, and total

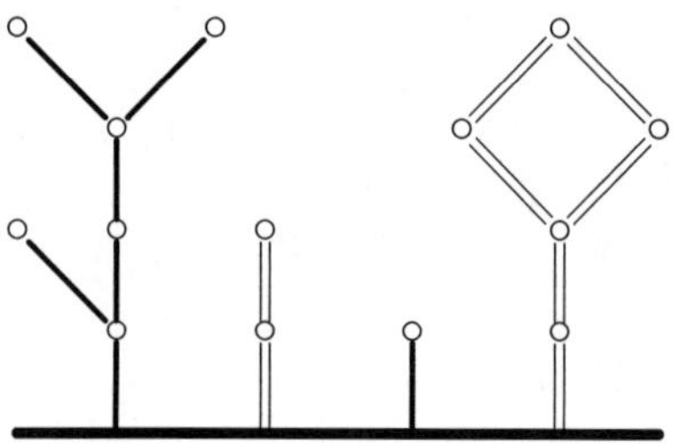

Figure 2.1. A HACKENBUSH position in which every disjoint component is monochromatic.

up the result. It's easy to see that for a position G of total value n,

$$o(G) = \begin{cases} \mathscr{L} & \text{if } n > 0; \\ \mathscr{R} & \text{if } n < 0; \\ \mathscr{P} & \text{if } n = 0. \end{cases}$$

In the third case there are an equal number of red and blue edges, and the win goes to whoever plays second.

For example, in Figure 2.1 the relevant number is

$$6 - 2 + 1 - 6 = -1,$$

so Right wins no matter who moves first.

These integer values are a sort of partizan analogue of the nim values that we saw in Section 1. Unlike nim values, they're *signed* and they add like ordinary integers. Also unlike nim values, they needn't be integers! Consider the game

$G =$ 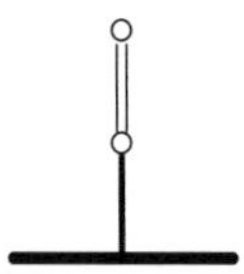 $= \{$  $\mid$ 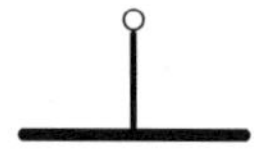 $\} = \{0 \mid 1\}$.

What is the value of G? It's weaker for Left than 1, since Right has more flexibility. In fact Right can win quite handily on $G + (-1)$:

 $+$ 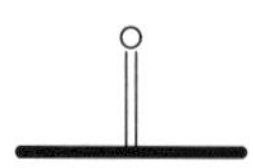

no matter who moves first, just by moving preferentially on G. We'll soon prove the remarkable identity

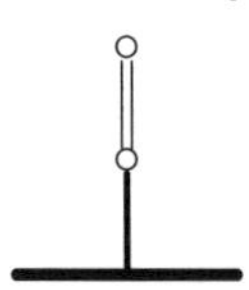 $+$ 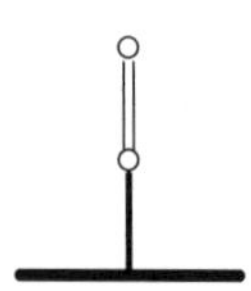 $=$ 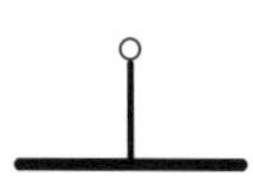

so that two copies of G behave like one copy of 1, and we may say that G has value $\frac{1}{2}$.

Zero Positions

A **zero position** is a game G with $G = 0$, that is, such that

$$o(G + X) = o(X) \quad \text{for all } X.$$

If G is a zero position, then it can be ignored in any sum where it appears. Clearly every zero position is a $\mathscr{P}$-position, since $o(G) = o(0) = \mathscr{P}$. Remarkably, the converse is also true.

Theorem 2.1. *If $o(G) = \mathscr{P}$, then G is a zero position.*

Proof. We must show that

$$o(G + X) = o(X)$$

for every short game X. By symmetry, it suffices to show that whenever Left can win X playing first (resp. second), she can also win $G + X$ playing first (resp. second).

But suppose Left can win X playing second. Then on $G + X$, she can respond to each of Right's moves with a winning move on the same component. This guarantees that Left will never run out of moves, and since $G + X$ is short, the game must eventually end. Thus Left can win $G + X$ playing second.

Likewise, suppose Left can win X playing first, say to X^L. Then on $G + X$, she can make an opening move to $G + X^L$ and thereafter follow the strategy just described. □

Now let $-G$ denote the **negative** of G, in which the roles of Left and Right are reversed. In HACKENBUSH this means interchanging blue and red edges, leaving green ones unaffected. Symbolically, we may write

$$-G = \{-G^R \mid -G^L\}$$

and we'll write $G - H$ as shorthand for $G + (-H)$. Clearly $o(G - G) = \mathscr{P}$, following the mirror image strategy, so in fact $G - G = 0$.

Theorem 2.2. *If $o(G - H) = \mathscr{P}$, then $G = H$.*

Proof. We have $G - H = 0$, so $G - H + H = H$. But $H - H = 0$, so also $G - H + H = G$. Therefore $G = H$. □

Therefore in order to prove that G and H are equal, it suffices to show that their difference is a second-player win. Let's apply this technique to HACKENBUSH.

Half a Point

To prove the identity

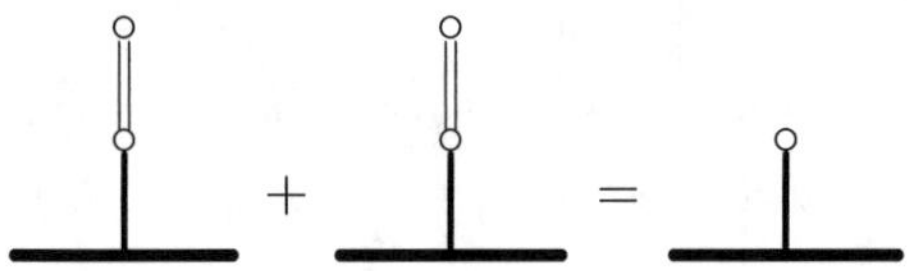

it suffices to show that the difference $\frac{1}{2} + \frac{1}{2} - 1$ is a second-player win:

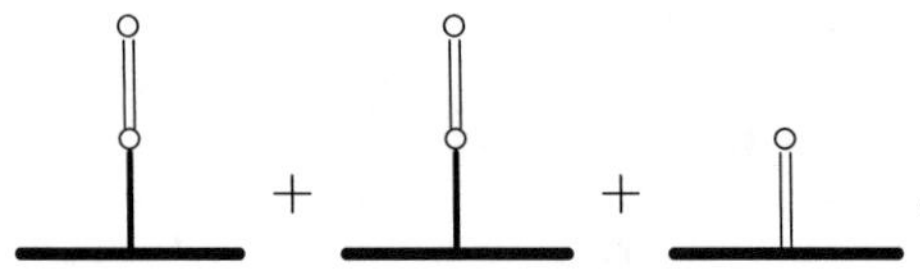

If either player moves on either copy of $\frac{1}{2}$, then her opponent can respond on the other, leaving a position of value $1 + 0 - 1 = 0$. In other words, the two highlighted moves below "cancel each other out":

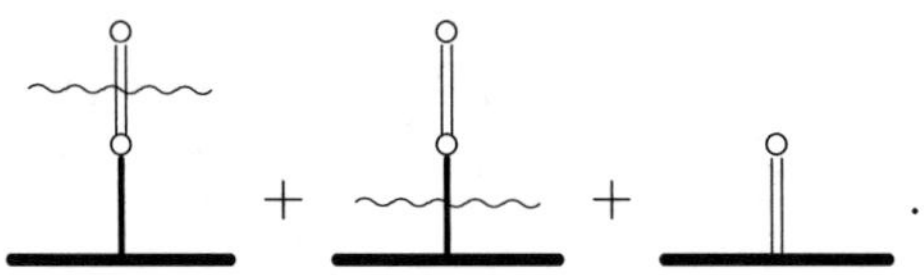

The only remaining option is Right's opening move on -1, but this clearly loses, since it leaves an $\mathscr{L}$-position.

What about the game

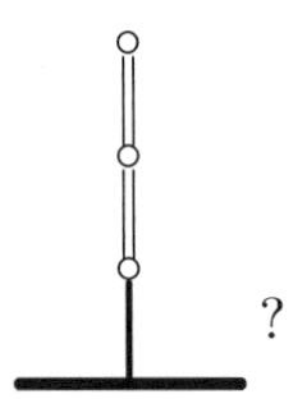

A similar argument shows that

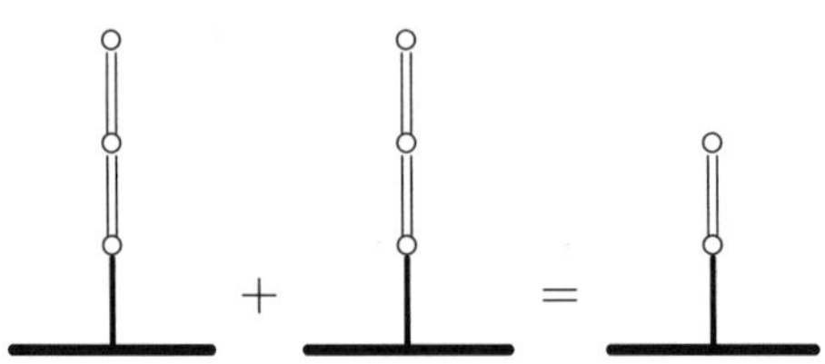

so that we might call this game $\frac{1}{4}$. Proceeding like this, we obtain games of all values $1/2^n$:

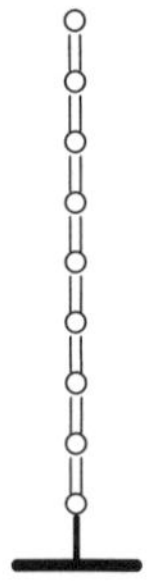 $+$ 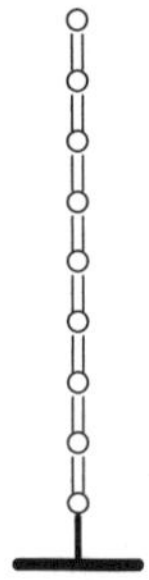 $=$ 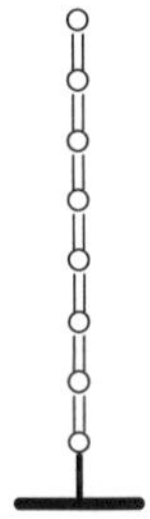, etc.,

and we can construct games of value $m/2^n$ by adding together an appropriate number of copies of $1/2^n$.

Now writing out the options of $\frac{1}{4}$, we see that

$\frac{1}{4} =$  $= \{0 \mid \frac{1}{2}, 1\}.$

However it's clear that Right should always prefer to chop the higher of his two edges first. The reasoning is the same as in the monochromatic case: if he can't eliminate any of Left's edges, then Right should strive to remove as few of his own as possible. In fact it's not hard to check directly that any time Right can win a sum of games $1 + X$ (playing first or second), then he can also win $\frac{1}{2} + X$, so that he will always be content to replace the component 1 by $\frac{1}{2}$.

So Right's move from $\frac{1}{4}$ to 1 is **dominated** by $\frac{1}{2}$, and we can write

$$\tfrac{1}{4} = \{0 \mid \tfrac{1}{2}\},$$

which is easily verified by showing that $\frac{1}{4} - \{0 \mid \frac{1}{2}\}$ is a second-player win.

A similar line of reasoning shows that

$$2^{-(n+1)} = \{0 \mid 2^{-n}\}$$

for all $n \geq 0$, since Right's remaining options $1, \frac{1}{2}, \ldots, 2^{-(n-1)}$ are all dominated.

GREEN HACKENBUSH

Recall that Either player is permitted to chop a grEen (dottEd) edge; for example, on

$G =$  $= \{0 \mid 0\}$

either player may move to 0. Consider the sum $G + 2^{-n}$:

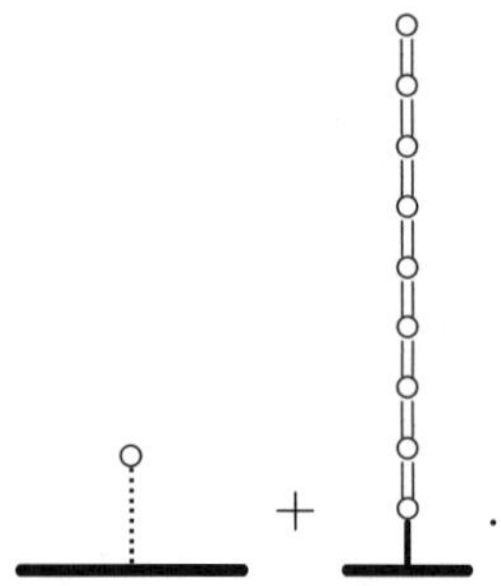

Left can win this game by following a simple strategy: chop the green edge preferentially or, if it's already gone, move to 0. In fact G is dominated by even the smallest numbers, and we might say that

$$\frac{1}{2^n} > G > -\frac{1}{2^n}$$

for all n. But $G \neq 0$, since it's a first-player win! Moreover,

is a second-player win, so that $G + G = 0$, or equivalently $G = -G$. The value of G isn't a number at all, and in fact it's **infinitesimal** with respect to numbers like 2^{-n}. It has the special name

$$*$$

(pronounced "star"), and in Chapter II we'll see that it plays a central role in the theory of combinatorial games.

Now $*$ is obviously isomorphic to a nim-heap of size 1, and if we play Green Hackenbush with long stalks, it's clear that we're just playing Nim. For example, Bouton's Theorem tells us that

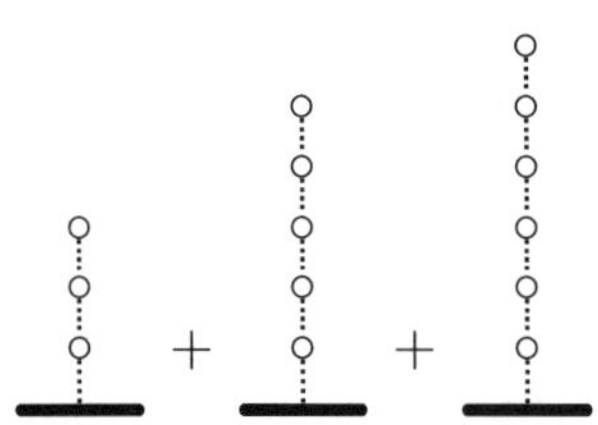

is a second-player win. We previously assigned these components **nim values** of 3, 5, and 6, respectively; but in the more general context of **game values** we need to call them something else, to distinguish them from ordinary integers. So we'll use the notation $*m$ (pronounced "star m") to denote a nim-heap of size m, and we'll write (for example)

$$*3 + *5 + *6 = 0.$$

It's clear that *any* nim-heap is dwarfed by 2^{-n}.

In BLUE–RED HACKENBUSH we could compute the value of *monochromatic* positions just by counting up the number of edges. This doesn't work in GREEN HACKENBUSH, though, and it's worth taking a moment to confirm that

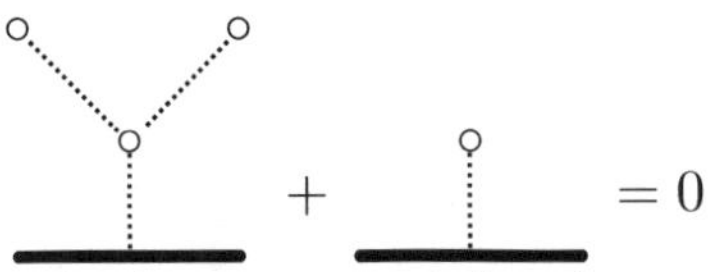

so that the value of the first component is $*1$, even though it has three edges. Fortunately there's a simple rule you can use to work out the values of arbitrary GREEN HACKENBUSH trees (see Exercise 2.1).

Tricolor HACKENBUSH

Let's consider one more example, the position

$G =$  .

G is a first-player win, but it's clearly more favorable to Left than $*$, and in fact two copies of G make an $\mathscr{L}$-position. For on

$G + G =$ 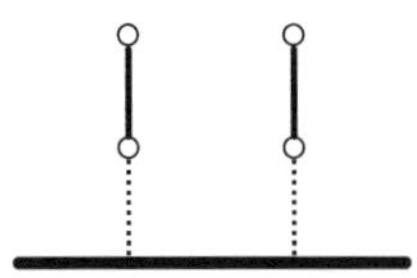

the *first* player to chop a green edge will lose, and Left can ensure this will always be Right. But $G + G$ is still infinitesimal: Right can win the sum

$G + G - 2^{-n} =$ 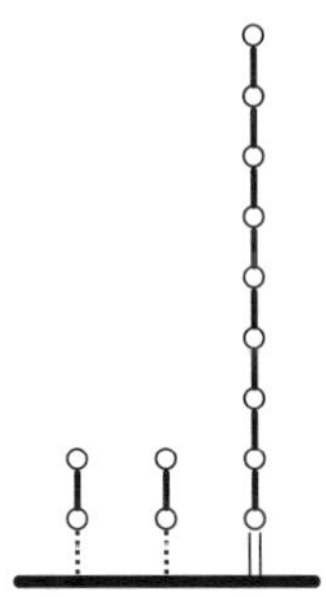

using the familiar strategy of chopping green edges preferentially.

Exercises

2.1 A HACKENBUSH position G is a **tree** if there is a unique path from the ground to each node. Let G be a tree that is completely green. Suppose that v is a vertex of G and the subtree above v consists of simple paths extending from v (i.e., every vertex above v has valence ≤ 2). Let $a_1, \ldots, a_k$ denote the lengths of these paths. Show that the subtree above v can be replaced by a single path of length $a_1 \oplus \cdots \oplus a_k$, without changing the value of G. Show furthermore that by repeatedly applying this transformation, G can be reduced to a single stalk.

3. How to Read This Book

Combinatorial Game Theory is divided into eight chapters. Chapter I is primarily expository; each of the remaining seven chapters describes a major topic in detail.

The basic theory of short partizan games is developed in Chapter II; then Chapter III applies this theory to study the abstract algebraic and combinatorial structure of $\mathbb{G}$. Short impartial games are studied in Chapters IV (normal play) and V (misère play), with brief detours into loopy impartial (Section IV.4) and misère partizan (Section V.6) games.

The remaining chapters discuss various extensions of the partizan theory: loopy games in Chapter VI; orthodox temperature theory in Chapter VII; and transfinite games, including the surreal numbers, in Chapter VIII.

These eight chapters are further subdivided into a total of thirty-nine sections. It is not necessary to read all of this material straight through; the approximate interdependence of chapters is shown in Figure 3.1.

Chapters are numbered with Roman numerals, and sections with Arabic numerals. We'll ordinarily omit the chapter number when referring to a section *within the same chapter*. For example, you are currently reading Section 3, but in later chapters it will be known as Section I.3. The same

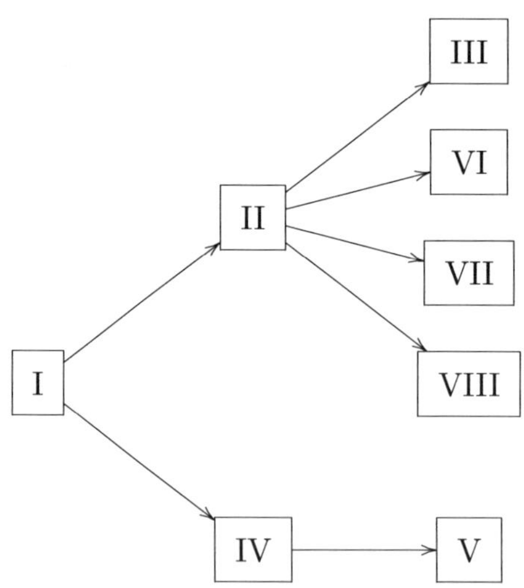

Figure 3.1. Approximate interdependence of chapters.

applies to theorems and exercises: the Fundamental Theorem (page 9) will be known as Theorem 1.5 from within Chapter I, but as Theorem I.1.5 elsewhere.

Most sections end with a list of pertinent exercises, as well as bibliographic notes summarizing results and references not covered in the body of the text. These notes are the recommended starting point for further investigation into each topic.

A full understanding of combinatorial game theory requires fluency in a wide range of mathematical subject matter—principally algebra, but also various ideas from enumerative combinatorics, set theory, and computational complexity. Some of the necessary background material is summarized in Appendix B, with an emphasis on key results used elsewhere in the text. Most results in this appendix are stated without proof and are intended primarily for review; references are included for other sources that are appropriate for more detailed study.

Notation

Figure 3.2 summarizes our naming conventions for various mathematical entitites. A particularly thorny problem is the distinction between a *game* and its *value*. Most of the time, we'll address this problem simply by suppressing the distinction—passing freely from a game to its value, or from a value to one of its representatives. This treatment requires a little extra care, but experience has shown it to be the least cumbersome approach. Therefore, we use capital Roman letters $G, H, J, \ldots$ to represent both games and game values, as needed.

Capital Roman	$G, H, J, X, Y, \ldots$	Games and values
Lowercase Roman	$x, y, z, \ldots$	Numbers
	$a, b, m, n, \ldots$	Integers
Lowercase Greek	$\alpha, \beta, \gamma, \ldots$	Ordinals
Script	$\mathscr{A}, \mathscr{B}, \mathscr{N}, \mathscr{P}, \ldots$	Sets of games
Calligraphic	$\mathcal{A}, \mathcal{B}, \ldots$	Sets of values
Capital Greek	$\Gamma, \Lambda, \ldots$	Rulesets
Boldface Roman	$\mathbf{ON}, \mathbf{PG}, \mathbf{SN}, \ldots$	Proper classes
Sans Serif	$\mathsf{P}, \mathsf{NP}, \mathsf{PSPACE}, \ldots$	Complexity classes

Figure 3.2. Default naming conventions for variables in this book.

On the other hand, we'll take care to distinguish between *sets* of games and *sets* of values, consistently using script letters $\mathscr{A}, \mathscr{B}, \ldots$ to denote sets of games, and calligraphic letters $\mathcal{A}, \mathcal{B}, \ldots$ for sets of values.

The following symbols are standard:

- $\mathbb{N}$ is the set of nonnegative integers, including 0.
- $\mathbb{N}^+$ is the set of strictly positive integers.
- $\mathbb{Z}$ is the ring of all integers.
- $\mathbb{Z}_n$ is the ring of integers modulo n (i.e., the quotient $\mathbb{Z}/n\mathbb{Z}$).
- $\mathbb{Q}$ is the field of rational numbers.
- $\mathbb{D}$ is the ring of dyadic rational numbers (those whose denominator is a power of 2).
- $\mathbb{R}$ is the field of real numbers.
- $\mathbb{R}^+$ is the set of strictly positive real numbers.

If $x \in \mathbb{R}$, then $\lfloor x \rfloor$ denotes the greatest integer $\leq x$ (the **floor** of x), and $\lceil x \rceil$ denotes the least integer $\geq x$ (the **ceiling** of x).

For an integer $n > 0$, the symbol $\mathrm{lb}(n)$ denotes the base-2 logarithm of n, rounded down—equivalently, the largest integer k such that $2^k \leq n$.

If $\mathcal{S}$ and $\mathcal{T}$ are sets, then $\mathcal{S} \subset \mathcal{T}$ *allows for* the possibility that $\mathcal{S} = \mathcal{T}$. We'll write $\mathcal{S} \subsetneq \mathcal{T}$ to mean "$\mathcal{S} \subset \mathcal{T}$ and $\mathcal{S} \neq \mathcal{T}$."

If $\mathcal{S}$ is a set, then $|\mathcal{S}|$ denotes the cardinality of $\mathcal{S}$, and $\mathrm{Pow}(\mathcal{S})$ denotes the powerset of $\mathcal{S}$ (the set of all subsets of $\mathcal{S}$).

If f is a mapping and $\mathcal{S}$ a set, then

$$f[\mathcal{S}] = \{f(x) : x \in \mathcal{S}\} \quad \text{and} \quad f^{-1}[\mathcal{S}] = \{x \in \mathrm{dom}(f) : f(x) \in \mathcal{S}\}.$$

We'll occasionally encounter sets $\mathcal{S} \in \mathrm{dom}(f)$ that also have elements in the domain of f; this is the reason for using $f[\mathcal{S}]$ instead of $f(\mathcal{S})$.

If $\mathcal{A}$ and $\mathcal{B}$ are Abelian groups, then $\mathcal{A} \times \mathcal{B}$ denotes the group of ordered pairs (a, b), with $a \in \mathcal{A}$, $b \in \mathcal{B}$, and componentwise addition. For $n \in \mathbb{N}$, we then write

$$\mathcal{A}^n = \overbrace{\mathcal{A} \times \mathcal{A} \times \ldots \times \mathcal{A}}^{n \text{ times}}.$$

We'll also write $\mathcal{A}^\omega$ for the direct *sum* of countably many copies of $\mathcal{A}$, i.e., the group of countable sequences of elements of $\mathcal{A}$ with just finitely many nonzero terms. These topics are described in more detail in Section B.1.

Standard References

The bibliography on pages 493–504 lists more than 200 references. Three of these stand out as essential companions to this volume:

- *Winning Ways for Your Mathematical Plays*, by Berlekamp, Conway, and Guy [**BCG01**], is the defining text on combinatorial games. Originally published in 1982, it is now available in a revised second edition, split into four volumes. Volume 1 is required reading. The remaining volumes are also valuable and address many topics and examples that are omitted from this book.
- Conway's *On Numbers and Games* [**Con01**] introduced the partizan theory to the world in 1976. Some of the material, particularly on the theory of surreal numbers, is still not available anywhere else.
- *Lessons in Play: An Introduction to Combinatorial Game Theory*, by Albert, Nowakowski, and Wolfe [**ANW07**], is a recent (2007) undergraduate-level textbook on combinatorial games. It is recommended as a gentle introduction and includes many examples not found elsewhere.

References to these three books will be by name, using the abbreviations *Winning Ways*, ONAG, and *Lessons in Play*. Other references will appear in square brackets: for example [**Bou01**], cross-referenced to the bibliography.

A more comprehensive combinatorial games bibliography is maintained by Aviezri Fraenkel [**Fra09**] and is recommended for references not mentioned in this book. Another valuable reference is *Games of No Chance*, a series of four volumes of papers published by Cambridge University Press [**Now96, Now02, AN09, NW13**]. The papers are freely available (search for "Games of No Chance") and are generally of very high quality.

Appendix A of this book summarizes all of the conjectures and open problems that appear throughout the text. A much longer (and ever-growing) list of unsolved problems is maintained by Guy and Nowakowski [**GN13**].

With just a few exceptions, none of the results in this book are due to the author. Attributions for specific results are generally omitted from the body of the text, in order to avoid disrupting the exposition. The historical notes at the end of each section contain an overview of the development of each topic, and an essay on the history of combinatorial game theory can be found in Appendix C.

Other Resources

Combinatorial game theory is heavily computational. Raw calculation of game values is often essential: to solve specific problems; to test or refute conjectures; and to assemble databases that can be mined for patterns. This was recognized early in the history of the subject, leading to a succession of increasingly sophisticated computer algebra systems designed by Flanigan, Allemang, Wolfe, and many others.

The most recent and comprehensive such system, *cgsuite*, is an indispensible companion to this textbook. The author strongly recommends[2] that all readers obtain a copy (it's freely available at `www.cgsuite.org`). Experimenting with *cgsuite* is one of the best ways to become fluent in the algebra of combinatorial games, and it's an invaluable tool for working through the exercises.

4. A Survey of the Landscape

Combinatorial game theory is a vast subject. Over the past forty years it has grown to encompass a wide range of games, extending far beyond the carefully selected examples from Section 1. All of those examples were **short games**, which have just finitely many subpositions and which prohibit infinite play. The combinatorial theory of short games is central to the subject and will motivate roughly half the material in this course, but its extensions to wider classes of games are equally important.

In this section we'll introduce several further examples that are *not* short games: FOX AND GEESE; GO; CHESS; and ENTREPRENEURIAL CHESS. We'll also survey various additional topics that fall outside the scope of the disjunctive theory: nondisjunctive compounds; positional games; multiplayer games; hardness results; and exhaustive and heuristic search techniques.

FOX AND GEESE

FOX AND GEESE is an old children's game played on an 8×8 checkerboard. In the starting position, shown in Figure 4.1 on the facing page, there are four **geese** (controlled by Left) arranged against a single **fox** (the white circle, controlled by Right). On his turn a player must move exactly one of the pieces under his control. The geese move like ordinary checker pieces, one space diagonally southwest or southeast. The fox moves like a checker king, one space in any of the four diagonal directions. If the geese eventually trap the fox, so that he is completely surrounded and unable to move, then Left wins. If this never happens, then Right wins.

[2]Disclosure: I'm the project administrator and primary architect of *cgsuite*.

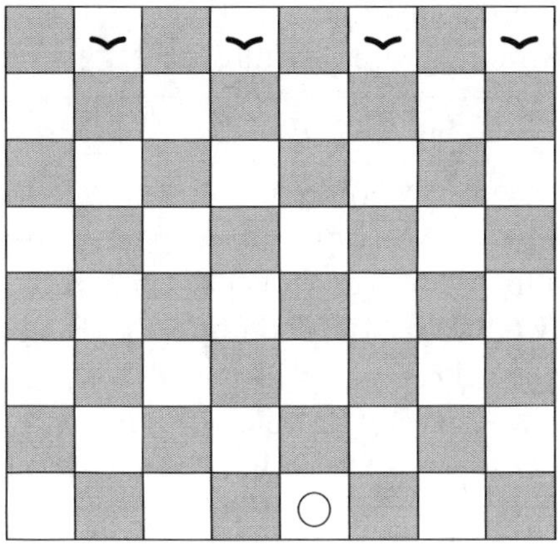

Figure 4.1. The starting position in FOX AND GEESE.

Since the geese can never retreat, the game is guaranteed to end after a finite number of moves. If the geese fail to trap the fox, then Left will eventually run out of moves, so the winning condition is really just the normal-play convention in disguise: whoever makes the last move wins. Therefore FOX AND GEESE has much in common with HACKENBUSH and DOMINEERING.

There is also a fundamental difference: FOX AND GEESE is *not* a short game, since if Right were permitted two consecutive moves, then he could return to the exact position from which he started. In the notation of Section 1, we have $G^{RR} = G$, where G is the position in Figure 4.1. So although every *play* of FOX AND GEESE is guaranteed to end (since Left can make just finitely many moves), there are infinite *runs* that involve many consecutive moves for Right.

This observation is highly relevant to the combinatorial structure of FOX AND GEESE. For consider a hypothetical sum

$$G + H,$$

in which H is a game to which *Left* can return in two consecutive moves: $H^{LL} = H$. (For example, we might take $H = -G$.) There is a four-move alternating sequence

$$G + H \to G^R + H \to G^R + H^L \to G^{RR} + H^L \to G^{RR} + H^{LL}$$

that returns to $G + H$, so in fact there exist infinite plays of $G + H$. So although *isolated* plays of G are always finite-length, its infinite runs nonetheless have combinatorial importance.

A detailed analysis of this type of disjunctive structure can sometimes yield useful insights, even for games that don't "naturally" decompose. For example, an isolated FOX AND GEESE position can never decompose in a nontrivial way, since there is just one fox! However, the combinatorial theory can still provide a compelling window into questions like:

Who is ahead, and by how much?

In Figure 4.1 the geese can force a win. It's not hard to work this out through a "brute force" analysis, without using any nontrivial mathematics. But it's much more difficult to quantify the precise extent of Left's advantage. What if the fox were permitted one pass move, to be used at any point during the game? Could the geese still win? How about two pass moves? Or $2\frac{1}{2}$ pass moves, in the sense of HACKENBUSH?

Combinatorial game theory provides the tools to answer these kinds of questions. It turns out that Left can still win if Right is given two pass moves, so that

$$G > 2.$$

But Right can win with just a slightly greater advantage: for all n,

$$G < 2 + \frac{1}{2^n}.$$

Moreover, if H is any *short* game with $2 < H < 2 + 2^{-n}$ for all n, then $G > H$. In particular, no short game has the same value as G. These types of quantitative comparisons are among the most powerful applications of combinatorial game theory, and they provide perhaps the clearest evidence that the cycles in G are an intrinsic part of its structure.

Games that admit infinite runs, such as FOX AND GEESE, are called **loopy**; those that don't (including all short games) are **loopfree**.

GO

GO is among the oldest and most widely played board games in the world. It originated in China as early as 2000 BCE, and today there are more than two thousand professional players throughout the world, and tens of millions of active amateurs.

GO has also been a motivating influence on combinatorial game theory throughout the history of the subject. The strong modularity of late-stage endgames inspired Conway to discover the theory of partizan games, which led in turn to the surreal numbers. In recent times GO continues to be a source of inspiration, spurring new developments in temperature theory and the study of capturing races.

Unfortunately GO serves as a poor introductory example, for several reasons. Most of our examples have simple, well-defined rules that are easily translated into an abstract mathematical description. By contrast, GO is highly complex, with all the messy peculiarities one might expect from a

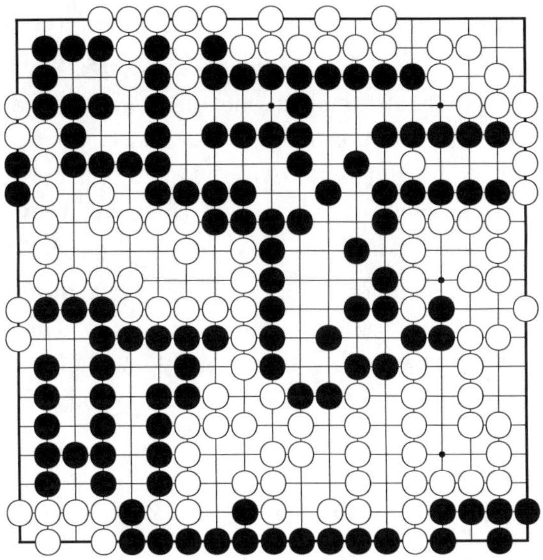

Figure 4.2. The "9-dan stumping problem" by Berlekamp and Wolfe. White to move and win.

game that's evolved over thousands of years. A full account of its rules would require several pages of discussion and would still fail to convey sufficient understanding to penetrate its mathematical structure even at a rudimentary level.

For these reasons, we'll carefully avoid saying too much about GO in this book, and the following discussion will largely be of interest to GO players. Readers who are unfamiliar with the game can safely skip this section. GO players, read on!

Figure 4.2 shows the famous "9-dan stumping problem" composed by Berlekamp and Wolfe in 1994. The problem is carefully designed so that the board decomposes into many independent regions, such as the five-point corridor in the northwest. Each such component is surrounded by living groups of stones, in such a way that play in one component cannot affect any other. (Some regions, such as the south-center of the board, are composed of several weakly interacting subcomponents. In these cases the entire region can be regarded as a single large component and can be further broken down by quantifying these weak interactions.)

The entire problem can therefore be written as the disjunctive sum of its various regional components. Each component has a game value (in the sense of Section 1), and the combinatorial theory of partizan games can be applied to find the value, and therefore the outcome, of the overall position. We'll develop the theoretical machinery for this in Chapter II; details of its specific applications to GO are published in Berlekamp and Wolfe's book, *Mathematical Go: Chilling Gets the Last Point* [**BW94**].

The Berlekamp–Wolfe analysis is one of combinatorial game theory's great success stories. Several 9-dan GO professionals (the top professional rank) have tried quite hard to solve Figure 4.2, without success. With sufficient knowledge of combinatorial game theory, however, it becomes a reasonably straightforward problem.

The 9-dan stumping problem is highly contrived and has little in common with positions that arise in real games. In most real-world situations, the Berlekamp–Wolfe analysis is relevant only in the very late endgame, and even then nontrivial applications are relatively rare. When *Mathematical Go* was first published in 1994, this was true of most applications of combinatorial game theory to GO.

Gradually, however, this situation is changing. Berlekamp has developed a rich temperature theory that applies to early-stage endgames and is relevant to a broader range of positions than the combinatorial theory of *Mathematical Go*. Over the past fifteen years, Berlekamp's theory has been investigated and extended by many other researchers and has been used to discover mistakes in games between top professional players. We'll explore the temperature theory in Chapter VII in an abstract mathematical context; references for specific applications to GO are given in the notes on page 49.

CHESS

On the surface, combinatorial game theory appears ill-suited to CHESS, since CHESS positions rarely decompose into disjunctive sums. In the fall of 1992, Elwyn Berlekamp made comments to this effect as part of an expository lecture on combinatorial games. Fortunately, Noam Elkies, who was in the audience at the time, interpreted Berlekamp's comments as a challenge, and he soon discovered several positions that are amenable to a disjunctive analysis.

Figure 4.3 on the next page illustrates a rare example from tournament play. The kings are in mutual zugzwang: the first player to make a king move will lose the game. Since the kings are effectively immobilized, the kingside and queenside pawn clusters cannot interact, so the board decomposes. Moreover, whoever makes the last pawn move will win, since her opponent will be forced to move his king. So we can regard Figure 4.3 as a normal-play combinatorial game with two disjoint components, one for each pawn cluster.

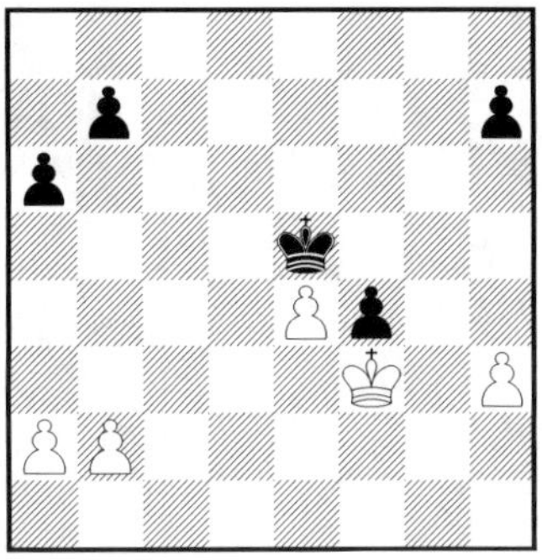

Figure 4.3. Schweda–Sika, Brno 1929.

Figure 4.3 is sufficiently simple that it can be successfully analyzed without any knowledge of combinatorial game theory. However, such a brute-force analysis provides little insight. Elkies quotes Max Euwe (a former world champion), who correctly determined that Figure 4.3 is a first-player win but had difficulty explaining why:

> Neither side appears to have any positional advantage in the normal sense .... The player with the move is able to arrange the pawn-moves to his own advantage [and win] in each case. It is difficult to say why this should be so, although the option of moving a pawn one or two squares at its first leap is a significant factor. [**EH60**]

In the language of combinatorial game theory, by contrast, one can say *exactly* why Figure 4.3 is a first-player win. We'll develop this language in Chapter II; skip ahead to Exercise II.4.9 on page 97 for the "correct" answer to Figure 4.3.

Elkies's two papers on the subject [**Elk96, Elk02**] are a nice example of how combinatorial game theory can be be applied—often in surprising contexts—to gain new insights into game situations that are otherwise baffling. The theory casts a wider net than a literal interpretation of its assumptions might suggest.

Entrepreneurial Chess

This fascinating game was invented by Elwyn Berlekamp. It is played on a quarter-infinite chessboard, with just a white king, white rook, and black king, as shown in Figure 4.4(a) on the next page. The pieces move just like their Chess counterparts, and Left (playing black) has the additional option of *cashing out*. If Left opts to cash out, then the entire position is replaced by an integer n, equal to the sum of the row and column numbers

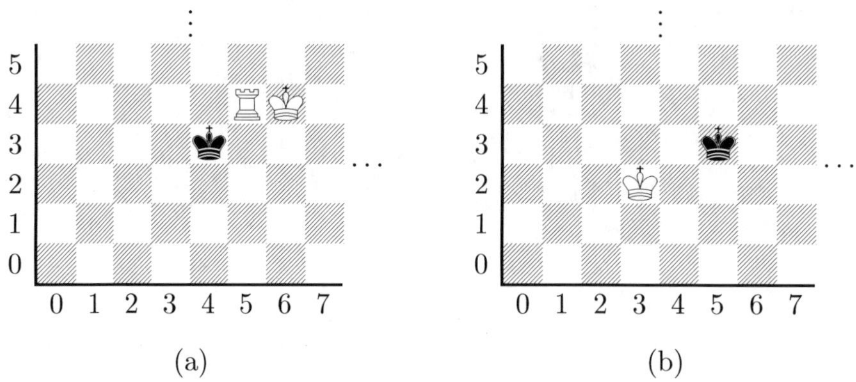

Figure 4.4. ENTREPRENEURIAL CHESS. (a) A typical position; (b) a pathological position in which the rook has been captured.

for the current location of her king. For example, in Figure 4.4(a), Left may either move her king or replace the entire position by the number 7.

ENTREPRENEURIAL CHESS is certainly loopy, like FOX AND GEESE and the other examples from this section. But there is a further difference from the classical examples from Section 1. A given ENTREPRENEURIAL CHESS position might have infinitely many subpositions; we'll call games with this property **transfinite**. There exist finite games that are loopy (think FOX AND GEESE) and transfinite games that are loopfree (we'll see one in a moment); ENTREPRENEURIAL CHESS is *neither* finite nor loopfree.

Nonetheless some positions *are* equivalent to finite values. Consider Figure 4.4(a). With best play, Left will never move her king, because there is no possibility of scoring more than 7 points (unless Right cooperates). Right can do no better than to shuttle his king around the squares adjacent to his rook. In fact it turns out that the value of Figure 4.4(a) is exactly equal to $5 + G$, where G is the FOX AND GEESE starting position from Figure 4.1. The machinery used to prove this (highly nonobvious!) fact is developed in Chapter VI.

The position K in Figure 4.4(b) is also quite interesting. In this pathological example, Right's rook has been captured, and Left is free to wander off indefinitely. It is clear that $K > n$ for any integer n: Left can systematically move her king to a square of value $n + 1$, then cash out; and Right is powerless to stop her advance.

But consider the difference of K with the *transfinite* HACKENBUSH stalk in Figure 4.5 on the facing page. The value of this HACKENBUSH position is the transfinite ordinal ω:

$$\omega = \{0, 1, 2, 3, \dots \mid \}.$$

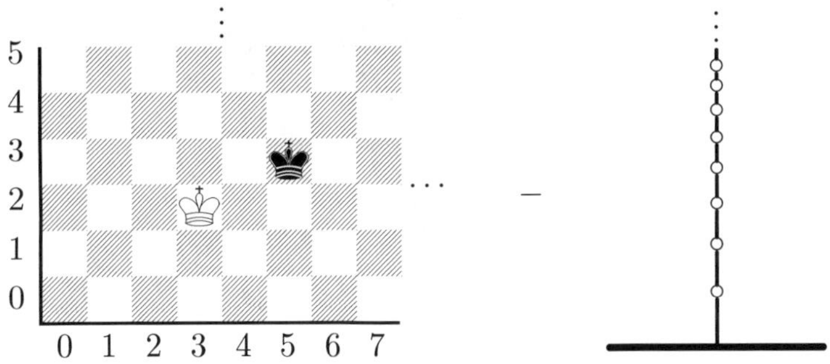

Figure 4.5. ENTREPRENEURIAL CHESS meets ω.

On $K - \omega$, Right can either move his king or replace $-\omega$ by a suitable integer $-n$. It's clear that the latter is a losing move, since it reduces the position to $K - n$, which we know Left can win. But from $K - \omega$ Left can never afford to cash out, since then the position will be $n - \omega$, from which *Right* has a move to $n - (n + 1)$! So on $K - \omega$ good players will simply shuttle their kings around aimlessly, and the outcome is a draw.

We conclude that $K - n$ is a win for Left for any finite n, but $K - \omega$ is drawn. It can be shown that no finite game has this property, so that the value of K is genuinely transfinite. Understanding precisely what *is* its value will require a lot of theory; peek ahead at Exercise VIII.1.12 on page 411 if you're curious about the answer.

Various Classes of Games

We've now seen several kinds of combinatorial games:

- *short impartial* games: NIM and DAWSON'S KAYLES;
- *short partizan* games: HACKENBUSH and DOMINEERING;
- *finite loopy* games: FOX AND GEESE, GO, and CHESS;
- *transfinite* games: ENTREPRENEURIAL CHESS and transfinite HACKENBUSH.

It turns out to be quite difficult to subsume *all* combinatorial games into a single theory. The most successful results come when we restrict ourselves to a particular subclass of games and consider the combinatorial interactions within that class.

Figure 4.6 on the following page summarizes several natural subclasses. There are three **structural constraints**, defined as follows:

Structural Constraints

Impartial Same moves for both players	**Partizan** Distinct moves for each player
Loopfree All runs are finite-length	**Loopy** Infinite runs are permitted
Finite Finitely many subpositions	**Transfinite** Infinitely many distinct subpositions

Play Convention

Normal Play Last player to move wins	**Misère Play** Last player to move loses

Figure 4.6. Various classes of combinatorial games.

Definition 4.1. Let G be a combinatorial game. G is said to be:

(a) **finite** if G has just finitely many distinct subpositions;

(b) **loopfree** if every run of G has finite length;

(c) **impartial** if Left and Right have exactly the same moves available from every subposition of G.

G is **transfinite** (resp. **loopy**, **partizan**) if it is not necessarily finite (resp. loopfree, impartial).

We say G is **short** if it is both finite and loopfree.

One can adopt any subset of these constraints (finite, loopfree, impartial) to obtain a restricted class, for a total of eight possibilities. The **play conventions**, also summarized in Figure 4.6, don't really define classes of games at all, but they are included in our table since they're a crucial ingredient in the resulting theory. Remarkably, in all cases the definition of equality is exactly the same:

$$G = H \quad \text{if} \quad o(G + X) = o(H + X) \text{ for all } X.$$

We can choose any of the eight structural classes of games for the domain of G, H, and X, and we can choose either normal or misère outcome for the meaning of $o(G)$. This gives sixteen different flavors of the fundamental equivalence, and each one yields a different theory!

In Section 1, for example, we defined $\mathbb{G}$ by choosing *short partizan* games for the domain and *normal-play* outcome for $o(G)$. $\mathbb{G}$ has a particularly rich structure that will occupy most of our attention in Chapters II through IV. Later chapters will investigate how much of this structure is retained when we vary each of these constraints (finite, loopfree, normal-play).

<table>
<tr><td></td><td>Normal-Play
Impartial</td><td>Normal-Play
Partizan</td><td>Misère-Play
Impartial</td><td>Misère-Play
Partizan</td></tr>
<tr><td>Short</td><td>Sprague–Grundy
IV.1–3</td><td>Winning Ways
II, III, VII.1–2</td><td>V.1–5</td><td>V.6</td></tr>
<tr><td>Finite
Loopy</td><td>Smith, Fraenkel
IV.4</td><td>VI, VII.3–5</td><td colspan="2" rowspan="3">Uncharted
Terrain</td></tr>
<tr><td>Transfinite
Loopfree</td><td>ONAG
VIII.4</td><td>ONAG
VIII.1–3</td></tr>
<tr><td>Transfinite
Loopy</td><td>Extends
Finite Loopy</td><td></td></tr>
</table>

Figure 4.7. A landscape of sixteen combinatorial game theories.

Figure 4.7 shows a rough roadmap of where we're headed. Short games in normal play, which are reasonably well understood, will occupy several chapters. By contrast, the theory of short partizan games in misère play was only recently isolated, and the small body of work since then has only barely scratched the surface. Finite loopy games, misère impartial games, and transfinite loopfree games fall somewhere between these extremes. You can see from Figure 4.7 that there are several classes of games (such as loopy games in misère play) about which essentially nothing is known at all!

In all these discussions, note that the *structural* classes on the right in Figure 4.6 (partizan, loopy, transfinite) are defined to include those on the left (impartial, loopfree, finite). For example, every impartial game is also *a priori* partizan. This makes the terminology far more convenient; we'll reserve the term **strictly partizan** for those rare cases when we wish to describe a game that specifically *isn't* impartial.

What's a Solution?

Much of this book is about finding *solutions* to games. Ordinarily, such a solution will take the form of a concise, closed-form algebraic description of a certain class of positions—Bouton's Theorem is the defining example. When we try to frame this sort of subjective understanding into a precise definition of *solution*, however, various subtleties arise. In Section 1 we were rather vague about our intentions:

> A **solution** to a ruleset Γ is an efficient algorithm for computing $o(G)$, for any position G of Γ.

This can be made precise using ideas from computational complexity. The rest of this discussion presupposes some background in complexity theory, which is not covered in this book; the requisite material can be found in standard references such as [**LP97, Pap93**]. However since the focus of this book is on the algebraic and combinatorial structure of games, rather than their computational complexity, it is safe to skim or skip this discussion and to stick with the above as a working (though imprecise) definition of solution.

Here's a more formal definition of solution:

A **solution** to a ruleset Γ is a polynomial-time algorithm for computing $o(G)$, for any position G of Γ.

Of course this revised definition is still not precise, since it leaves open the fundamental question: polynomial-time in terms of what? Typically, G is formalized as the tree (or graph) of its subpositions. But the size of this **game graph** isn't suitable to define input complexity, since the naïve brute-force algorithm (that examines each node once) is linear in the size of the graph. However there is usually some other natural choice—a "succinct encoding" for Γ that expresses its structure using the minimum necessary information. For example, in the case of DOMINEERING it's logical to use the number of squares on the board position; for HACKENBUSH the number of edges in the graph; and so on.

It's clear that the right notion of input complexity for Γ depends upon how we choose to represent its positions. This problem can be sidestepped by building a measure of input complexity directly into our definition of ruleset. Here, finally, is a precise definition of **ruleset** and **solution**:

Definition 4.2. A **ruleset** is a set of games Γ (the **positions** of the ruleset), together with a function $N : \Gamma \to \mathbb{N}$. The value $N(G)$ is called the **input complexity** of G (as an element of Γ).

A **solution** to $\Gamma = (\Gamma, N)$ is an algorithm for computing $o(G)$, for any $G \in \Gamma$, whose running time is polynomial in $N(G)$.

Fortunately it will rarely be necessary to specify N explicitly, since there are certain "natural" choices that are almost universally adopted:

- If Γ is played on a grid and G is an $m \times n$ position of Γ, then $N(G) = mn$.
- If Γ is played on a graph and the graph for G has v vertices and e edges, then $N(G) = v + e$.
- If Γ is played with heaps of tokens and G is the sum of heaps of sizes $a_1, a_2, \ldots, a_k$, then
$$N(G) = \mathrm{lb}(a_1) + \mathrm{lb}(a_2) + \cdots + \mathrm{lb}(a_k).$$
($N(G)$ is the number of bits of information necessary to represent the position.)

The definition of ruleset has useful implications for the algebraic theory of combinatorial games, since *for the most part* the main goals of the theory have now been precisely stated. The reason for qualifying this statement is that the consequences of Definition 4.2 are occasionally counterintuitive. Here are some of the common complications:

Positions of Bounded Size. Under Definition 4.2, any single (finite) position is *a priori* solved. For example, consider the starting position G in 8×8 CHESS. There is a trivial brute-force search algorithm that determines $o(G)$ by exhaustively examining every subposition. Since 8×8 CHESS has just finitely many possible positions, the running time of this algorithm is bounded by a constant and is therefore polynomial. The fact that this constant is enormously large, and hopelessly out of reach of present-day computational resources, is irrelevant. So although no one knows the outcome of 8×8 CHESS for certain, it's technically "solved" according to Definition 4.2.

For this reason, Definition 4.2 requires an infinite class of positions (with unbounded input complexity) in order to work. For abstract combinatorial games such as NIM and DAWSON'S KAYLES, this constraint is typically satisfied automatically. However, playable games such as CHESS are often assigned a fixed board size, usually with a specific starting position. Meaningful hardness results for such games depend on suitable generalizations to arbitrary board sizes.

In the case of CHESS, it's been shown that a suitable generalization of $n \times n$ CHESS is EXPTIME-complete and therefore admits no solution. This might strike CHESS players as a pedantic observation, but it actually says something deep about the specific case of 8×8 CHESS. It strongly suggests that there is intrinsic computational complexity embedded in the rules to CHESS that affects individual cases (such as 8×8) just as much as the asymptotic case. Searching for a clever algebraic solution to 8×8 CHESS is *probably* fruitless, because we know that such a solution *provably* cannot be extended to the $n \times n$ case.

Solutions to Certain Classes of Positions. FOX AND GEESE presents an interesting case. We can generalize 8×8 FOX AND GEESE to an $n \times 8$ board; the starting position G_n is an elongated version of Figure 4.1, still with four geese and one fox. Berlekamp proved that G_n has the exact value $1 + 2^{8-n}$ for all $n \geq 9$. The outcomes for $n \leq 8$ are easily computable by exhaustive search, so that the outcomes of all *starting* positions are known. Moreover, a complete strategy is known for enforcing a win by Left in polynomial time.

However, it remains unknown how to compute the outcome of an arbitrary position. In actual play of FOX AND GEESE, this is an academic observation: the "unresolved" positions can only arise if Left fails to play

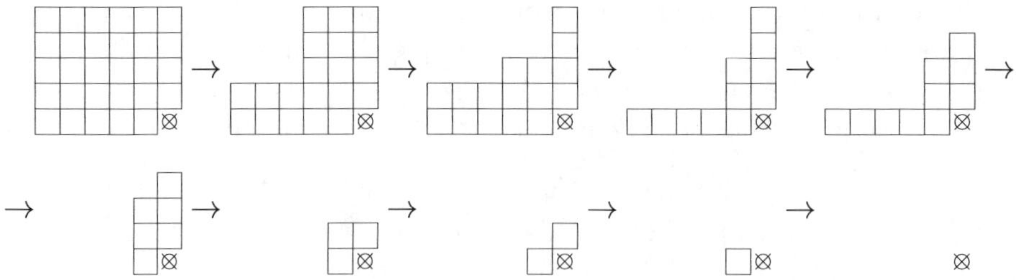

Figure 4.8. A typical game of CHOMP.

optimally or if play starts from a position other than G_n. Nonetheless, FOX AND GEESE doesn't meet our definition of "solved." This type of game is sometimes called **weakly solved**.

Another fascinating example is the impartial game CHOMP, played on an $m \times n$ grid as pictured in Figure 4.8. The players alternately remove ("chomp") a rectangle from the upper-left quadrant of the board. Denote by (i, j) the square at row i (top-to-bottom), column j (left-to-right); then a move is to select a pair (I, J) and discard all remaining squares (i, j) with $i \leq I$ and $j \leq J$. Each move must discard at least one square and *must leave at least one square remaining.* (This last rule is essential in order to prevent first player from simply removing the entire board.) The game ends when only one square remains, so that no further moves are possible, and whoever makes the last move wins.

Computing the outcome of an arbitrary CHOMP position appears to be quite difficult. However, the starting position G is easily shown to be a first-player win (except for the trivial case $m = n = 1$). The "strategy-stealing" argument goes like this: Suppose (for contradiction) that G is a second-player win. Suppose also that first player opens by removing the single square $(1, 1)$, moving to the position G'. Then second player must have a winning response, to some option G''. But G necessarily has a move directly to G'', so first player could have won by moving there initially. Contradiction!

A peculiar feature of this argument is that it gives absolutely no insight into the structure of CHOMP. We can say for certain that $m \times n$ CHOMP is a first-player win, but we can't even say what the winning moves are! Such games are sometimes called **ultra-weakly solved**. CHOMP, in particular, highlights the importance of including *all* positions in the definition of a full solution.

Enforcement of Strategies. NIMANIA is a curious game invented and studied by Fraenkel and Nešetřil [**FN85, FLN88**]. It is played with several heaps of tokens. On the kth turn of the game, exactly one token is removed from any one heap, say reducing it from size m to $m-1$; then k *new* heaps are added to the position, each of size $m-1$. (If $m=1$, this last step has no effect.)

A typical game, starting with a single heap of size 3, might proceed as follows. Here a position is described as a multiset of heaps; the notation n^a abbreviates a copies of a heap of size n, and the subscript denotes the current value of k:

$$\{3\}_1 \to \{2^2\}_2 \to \{2,1^3\}_3 \to \{2,1^2\}_4 \to \{2,1\}_5 \to \{2\}_6 \to \{1^7\}_7$$

after which the players alternately remove single-token heaps, and first player wins on the 13th move.

It's not hard to show by exhaustive analysis that starting from $\{3\}_1$, first player can force a win in 13 moves. In fact every starting position $\{n\}_1$ turns out to be a first-player win, and it's easy to describe the outcome of an arbitrary position G (with arbitrary subscript k):

- If G has the form $\{1^a\}_k$, then $o(G) = \mathscr{N}$ iff a is odd.
- If G has the form $\{2,1^a\}_k$, then $o(G) = \mathscr{N}$ iff either a or k is odd.
- Otherwise, $o(G) = \mathscr{N}$ iff k is odd.

This gives a rather trivial strategy for NIMANIA: if G has the exact form $\{2,1^a\}_k$, then play on a heap of size 1 if a and k are both odd, or 2 if either is even. Otherwise, play randomly!

This shows that NIMANIA is solved in the strongest sense of Definition 4.2. Amazingly, however—though first player has an easy win on $\{n\}_1$—second player can guarantee that the game lasts at least $A(n)$ moves, where A is an Ackermann function! Even $n=4$ is effectively intractable: first player can force a win in no fewer than 2^{44} moves.

This raises the obvious question: what is the meaning of declaring a game to be a forced win, if an inordinately large number of moves is needed to consummate that win? Some authors have addressed this problem by placing additional constraints on the definition of solution—for example, by requiring that a win be enforceable in $O(c^{N(G)})$ moves, for some c. Fortunately such constraints are rarely necessary. In particular, in all of the examples we'll see in this book (except NIMANIA), each position G will have at most $O(c^{N(G)})$ subpositions.

Games Farther Afield

We'll conclude this survey with a discussion of some related topics in combinatorial game theory. Each of them is an active and important area of research; but in order to keep the focus on the disjunctive theory (and hold this course to a manageable length) they won't be treated in detail in this book.

Nondisjunctive Compounds. Disjunctive decomposition arises naturally in games as diverse as DAWSON'S KAYLES, DOMINEERING, and GO. For this reason, disjunctive sum is a natural focal point in the study of abstract combinatorial properties of games.

Various other compounds also arise in practice, though less frequently. Aside from disjunctive sum, the **conjunctive** and **selective** compounds have historically received the most attention. In all three cases, the compound is formed by placing single copies of G and H side-by-side; but the rules for playing the compound differ:

In the **disjunctive sum** $G + H$,	move in *exactly one* of the two components.
In the **conjunctive sum** $G \wedge H$,	move in *both* components. (Play ends when *either* terminates.)
In the **selective sum** $G \vee H$,	move in *either or both* components.

These rules generalize in the obvious ways to games with three or more components; for example in

$$G_1 \wedge G_2 \wedge \cdots \wedge G_k$$

a player must make a move in *all* k of the components, and play ends as soon as *any* of them terminates.

Figure 4.9 gives precise symbolic definitions of conjunctive and selective sum, along with several additional types of compounds. All of these operations are associative, so they give rise to new variants of the fundamental equivalence and correspondingly new types of values. For example, for conjunctive sum we may write

$$G = H \quad \text{if} \quad o(G \wedge X) = o(H \wedge X) \text{ for all } X.$$

In each case, of course, $o(G)$ can mean either normal or misère outcome, and "for all" can range over any of the eight classes of games mentioned in Figure 4.6. So for each of these types of compounds we have another version of Figure 4.7, with another sixteen variants to consider!

There is evidently quite a lot of ground to cover here, and only a fraction of it has been properly explored. In this book we'll be concerned almost exclusively with the disjunctive theory; the notes on page 49 provide some references for exploring other types of compounds.

Compound	Description
Disjunctive	Move in exactly one component. $G + H \cong \{G^L + H, G + H^L \mid G^R + H, G + H^R\}$
Conjunctive	Move in all components. Play ends when any one of them terminates. $G \wedge H \cong \{G^L \wedge H^L \mid G^R \wedge H^R\}$
Selective	Move in any number of components, but at least one. $G \vee H \cong \{G^L \vee H, G \vee H^L, G^L \vee H^L \mid G^R \vee H, G \vee H^R, G^R \vee H^R\}$
Diminished Disjunctive	Move in exactly one component. Play ends immediately when any one of them terminates. $G \boxplus H \cong \begin{cases} 0 & \text{if } G \cong 0 \text{ or } H \cong 0; \\ \{G^L \boxplus H, G \boxplus H^L \mid G^R \boxplus H, G \boxplus H^R\} & \text{otherwise} \end{cases}$
Continued Conjunctive	Move in all nonterminal components. Play ends only after all components terminate. $G \triangledown H \cong \begin{cases} G + H & \text{if } G \cong 0 \text{ or } H \cong 0; \\ \{G^L \triangledown H^L \mid G^R \triangledown H^R\} & \text{otherwise} \end{cases}$
Shortened Selective	Move in any number of components. Play ends immediately when any one of them terminates. $G \vartriangle H \cong \begin{cases} 0 & \text{if } G \cong 0 \text{ or } H \cong 0; \quad \text{otherwise,} \\ \{G^L \vartriangle H, G \vartriangle H^L, G^L \vartriangle H^L \mid G^R \vartriangle H, G \vartriangle H^R, G^R \vartriangle H^R\} \end{cases}$
Ordinal	Move in G or H; any move on G annihilates H. $G : H \cong \{G^L, G : H^L \mid G^R, G : H^R\}$
Side	Move in G or H; Left's moves on H annihilate G, and Right's moves on G annihilate H. $G \diamond H \cong \{G^L \diamond H, H^L \mid G^R, G \diamond H^R\}$
Sequential	Move in G unless G has terminated; in that case move in H. $G \to H \cong \begin{cases} H & \text{if } G \cong 0; \\ \{G^L \to H \mid G^R \to H\} & \text{otherwise} \end{cases}$

Figure 4.9. Various types of compounds.

Positional combinatorial games. In a **positional game** the object is not to make the last move, but rather to fulfill some global positional objective. Tic-Tac-Toe and Hex are among the best-known examples.

The general case specifies a set X (the **board**) and a collection $\mathcal{F}$ of subsets of X (the **winning sets**). In Tic-Tac-Toe, for example, X is a nine-element set and $\mathcal{F}$ consists of all rows, columns, and diagonals of X. Each player alternately selects an element $x \in X$ and occupies it with her color; each x may only be selected once per game. The winning condition is usually defined in one of three ways:

- Under the **strong win** convention, the winner is the first player to occupy all the elements in one of the winning sets. If the board X is exhausted before either player succeeds, the game is a draw. This is how Tic-Tac-Toe is usually played.
- Under the **weak win** (or **Maker–Breaker**) convention, one player (the **Maker**) attempts to occupy a winning set, while the other (the **Breaker**) tries to prevent her. The game ends once the set X is exhausted. Maker wins just if she has occupied a winning set; otherwise Breaker wins. Draws are impossible. Hex and many other connection games fall into this category.
- Under the **reverse weak win** (or **Avoider–Enforcer**) convention, one player (the **Avoider**) strives to *avoid* occupying a winning set, while the other (the **Enforcer**) tries to ensure that she does. This is the misère version of the Maker–Breaker convention; misère Hex is a typical example.

The theory of positional games was introduced by Hales and Jewett [**HJ63**] and Erdős and Selfridge [**ES73**]. Over the past several decades it has been developed extensively, primarily through the sustained efforts of József Beck. The theory uses probabilistic methods to study the asymptotic structure of various abstract classes of games.

Surprisingly, at sufficiently *large* scales these methods can be used to deduce exact solutions! Beck cites the example of the (n, q)-Clique Game: the board X consists of all edges in the complete graph K_n on n vertices; $\mathcal{F}$ is the collection of all q-cliques of K_n. The players alternately select an edge from K_n; Maker attempts to complete a q-clique, and Breaker strives to prevent her. Remarkably, one can show that for $n = 2^{10^{10}}$, Maker has a winning strategy if and only if

$$q \leq 19{,}999{,}999{,}933.$$

Similar results are obtainable for any sufficiently large n, but for smaller values of n the theory is less conclusive. For example, if n is "only" 2^{100}, then there are values of q for which the exact outcome remains unknown.

This fascinating asymptotic theory is obviously quite different from the algebraic and inductive methods that are the focus of this book, so we'll have little more to say about it. Fortunately Beck has recently published an encyclopedic and up-to-date survey of the subject, which is the natural starting point for further investigations [**Bec08**].

Multiplayer combinatorial games. Multiplayer games introduce the prospect of **coalitions**, in which several players collude to pursue a shared goal. The classical theory of economic games elegantly allows for multi-player situations, but combinatorial settings usually require some degree of contrivance in order to exclude the possibility of collusion.

Let's consider the special case of impartial three-player games; generalizing to partizan N-player games is straightforward. Any of the impartial games discussed in this book can easily be adapted to a three-player setting. The players take turns moving on G in clockwise rotation; the game ends when the next player to move is unable to do so, in which case the entire prize goes to whoever made the last move.

The challenge is to say what is meant by $o(G)$. In some situations, one of the players can force a win no matter how the other two react, and in such cases the definition of $o(G)$ is noncontroversial. A trivial example is $G = n \cdot *$: every move is forced, and the outcome depends only on the parity of n modulo 3. But in most cases, none of the three players can force a win, because the outcome depends on how each of the other players reacts.

For example, on $*2 + *$ it is impossible for first player to win; but she can decide which of the other players wins, by moving variously to $* + *$ or $0 + *$ (say). Propp [**Pro00**] calls such positions $\mathscr{Q}$-positions and shows that the situation is quite dire: $G + *2 + *2$ is a $\mathscr{Q}$-position, for *any* game G!

Various authors have attempted to resolve this problem by making assumptions about the rationality of each player's behavior. For example:

- Li [**Li78**] supposed that if player X is unable to force a win, then she will "play for second," ensuring (if possible) that she will be the *second-to-last* player to move.
- Straffin [**Str85**] introduced a "revenge rule": if player X is faced with a kingmaker decision and chooses Y over Z, then we assume Z will favor Y over X the next time he is faced with a similar decision.
- Loeb [**Loe96**] considered probabilistic models for **stable coalitions**: minimal sets of players that can collectively guarantee a win for some player in the set.

The Li and Straffin approaches lead to elegant combinatorial theories, but they suffer from a serious defect: in practice, the resulting outcomes have little meaning, since they rely on the assumption that both *other* players will play perfectly. Loeb's approach is more resilient, but notably less "combinatorial" in nature.

Cincotti [**Cin10**] has proposed a different approach: simply leave the outcome of Propp's $\mathscr{Q}$-positions undefined, and try to identify situations under which one of the players has a true forced win. Focusing on the partizan theory, he defined multiplayer generalizations of the *recursive* characterization of $G \geq H$ and showed that certain combinations of such relations can sometimes identify the outcome.

Algorithmic combinatorial game theory. Definition 4.2 (page 36) enables one to say precisely what is meant by a "solution" to a ruleset. Consequently, one can also say that certain particular rulesets are formally **unsolvable** (or **intractable**). For example, $n \times n$ CHESS is known to be EXPTIME-hard, so that (by standard results in complexity theory) no polynomial-time solution can exist. Many similar results are known; see Figure 4.10 on the next page for a sampling.

Algorithmic combinatorial game theory studies the computational complexity of combinatorial games. The usual technique (applied to a particular ruleset Γ) is to identify a class of Γ-positions with a special structure that can be reduced to a known, computationally difficult problem. In the case of HACKENBUSH, for example, certain positions known as **redwood beds** are equivalent to bipartite graphs, and calculating the value of a redwood bed is equivalent to finding a minimum spanning tree for its associated graph—a problem that is known to be NP-complete.

The algorithmic approach naturally complements the algebraic theory. Both approaches seek to find structure in combinatorial games: in one case, to show that this structure is simple enough to admit a complete theory; and in the other, to show that it is varied enough to preclude such possibility. The edge cases emerge as the most troubling: there are games that we "expect" to admit solutions (misère DAWSON'S KAYLES) but that have a relatively deep structure; and games that we "expect" to be intractable (DOMINEERING) but have relatively little structure. Algorithmic combinatorial game theory provides a unifying framework for attacking these problems from both directions.

A novel technique known as **constraint logic** defines a common set of problems (**constraint games**) of known complexity, which are constructed in a way that simplifies reductions from combinatorial games. The constraint logic approach has led to several new hardness results (including KONANE) and simpler proofs of existing ones (AMAZONS).

AMAZONS	PSPACE-complete	Furtak et al. [**FKUB05**]; Hearn [**Hea09**]
$n \times n$ CHESS	EXPTIME-complete	Fraenkel–Lichtenstein [**FL81**]
CHOMP	?	
CLOBBER	?	
DAWSON'S KAYLES	Solved	Guy–Smith [**GS56**]
DAWSON'S KAYLES (misère)	?	
DOMINEERING	?	
DOTS AND BOXES	NP-hard	*Winning Ways* [**BCG01**]
FLOWERS	?	
$n \times 8$ FOX AND GEESE	?	
$n \times n$ GO	EXPTIME-complete	Lichtenstein–Sipser [**LS80**]; Robson [**Rob83**]
GRUNDY'S GAME	?	
HACKENBUSH	NP-hard	*Winning Ways* [**BCG01**]
KAYLES	Solved	Guy–Smith [**GS56**]
KAYLES (misère)	Solved	Conway–Sibert [**CS92**]
KONANE	PSPACE-complete	Hearn [**Hea09**]
NIM	Solved	Bouton [**Bou01**]
PHUTBALL	PSPACE-hard	Dereniowski [**Der10**]

Figure 4.10. Complexity status of various rulesets.

Surveys by Fraenkel [**Fra96b**] and Demaine and Hearn [**DH09**] provide excellent summaries of this material. The constraint logic approach is described in detail in a new book by Hearn and Demaine [**HD09**].

Exhaustive search. Some combinatorial game positions G are sufficiently simple that they are amenable to **exhaustive search**: an attempt to determine $o(G)$ by brute force, either by exhaustively visiting every subposition of G or by visiting enough of them that the value of $o(G)$ can be definitely isolated. (For example, to determine that $o(G) = \mathscr{N}$, one need only exhibit a single winning opening move for each player; once this is done, the proof is complete and the remaining opening moves needn't be considered.)

This is *in theory* possible for every finite G, by brute-force application of the Fundamental Theorem, but *in practice* its feasibility is limited by the available computing resources. The good news is that the rapid expansion in computing power over the past several decades has dramatically increased the scope of exhaustive search techniques.

The well-known game of CHECKERS is perhaps the most dramatic illustration. Long the subject of intense interest, CHECKERS was quite recently shown to be a first-player draw: starting from the standard opening position on an 8×8 board, either player can force at least a draw with best play. This finding was announced in 2007 by a team led by Jonathan Schaeffer [**Sch96, SBB$^+$07**], after a monumental 18-year search that examined more than 10^{14} subpositions.

Figure 4.11 lists several more classic games that have succumbed to exhaustive search. Some entries are partial results. For example, DOMINEERING has an indeterminate starting boardsize, and it turns out that the

CHECKERS	Draw (on 8×8 board) Schaeffer et al. [**SBB$^+$07**]
CHESS	All pawnless six-piece endgames are solved Stiller [**Sti96**]
DOMINEERING	First-player win on $n \times n$ for $6 \le n \le 10$ Breuker et. al. [**BUvdH00**]; Bullock [**Bul02**]
NINE MEN'S MORRIS	Draw (on traditional board) Gasser [**Gas96**]
TIGERS AND GOATS	Draw (on traditional board) Jin and Nievergelt [**JN09**]

Figure 4.11. A sampling of games that have fallen to exhaustive search.

computational challenge is greatest the closer the board is to being square. The largest square board that has been completely solved is 10×10.

In the case of CHESS, the full solution is not remotely within the realm of what is feasible today, but exhaustive search has proven to be enormously successful in computing the exact outcomes of various endgame positions (even resulting in several official adjustments to the "50-move rule").

Exhaustive search is an undoubtedly combinatorial pursuit, but it is tangential to the main ideas of this book, which are generally directed toward solutions for infinite sequences of positions based on an understanding of the algebraic structure of games. However, it remains an active area of research, and the interested reader is enthusiastically directed to the references in Figure 4.11.

Heuristic search. In recent years, exhaustive search has led to complete solutions for many games (such as CHECKERS) that had previously been considered infeasible. However others, such as CHESS and GO, remain far (and perhaps permanently) out of reach. Therefore, computer programs that *play* such games cannot resort to exhaustive search to find the best moves, and they typically incorporate a heuristic method for approximating the value of a given position G. Designing a good heuristic is a problem of an entirely different sort from the ones considered in this book: it relies on practical optimization and high-level understanding of the structure of specific games.

Positional heuristics are often coupled with brute-force search: for example, a program might search the game tree of G as deeply as possible in the allotted time, using its heuristic to evaluate the deepest nodes when time runs out. Likewise, an algorithm might rely on heuristics to prioritize the order of its search, analyzing the heuristically best options first.

Combinations of these sorts of techniques have proven to be enormously successful in applications to certain games, including CHESS, CHECKERS, and REVERSI. The CHESS-playing program Deep Blue, which defeated world champion Garry Kasparov in a much-publicized match in 1997, relied on highly sophisticated adaptations of such techniques.

By contrast, GO has proven to be remarkably resistant to heuristic search: although the best CHESS programs now routinely defeat reigning world champions, top GO software performs no better than a talented club player. This is due largely to GO's large branching factor and the importance of larger-scale positional strategy over local tactical calculations.

In recent years, a new suite of techniques known as *Monte Carlo* methods has revolutionized computer GO. The basic Monte Carlo algorithm evaluates a position G as follows. For each potential option G^L, the computer plays out several thousand random games all the way to the end and assigns a score to G^L based on the *probability* of a favorable outcome. The computer then selects the option that achieved the most probable win. This basic approach can be enhanced in various ways, for example, by starting random games from a deeper level of the search tree, or by introducing a heuristic bias into the random plays.

Monte Carlo algorithms are far superior to classical search algorithms at the strategic level. However they are vulnerable to tactical weakness: if the opponent has just one good reply to a particular G^L, for example, there is some risk that the randomizer will simply overlook it. Since 2006, a hybrid approach known as UCT search (Upper Confidence bounds applied to Trees) has found spectacular success and cemented the dominance of Monte Carlo methods in computer GO. In the 2011 Computer Olympiad, the top three GO programs all used a version of Monte Carlo tree search; such programs are now able to beat professional players at reasonable handicaps.

Exercises

4.1 The outcomes for NIMANIA given on page 39 are correct.

4.2 In DIVISORS a position is a finite set $\mathcal{N} \subset \mathbb{N}^+$ that is divisor-closed (if a is an element of $\mathcal{N}$, then so are all divisors of a). A move consists of selecting an integer $a \in \mathcal{N}$ and removing a from the set, together with all multiples of a. Whoever removes 1, necessarily leaving the set empty, loses.

Prove that CHOMP is isomorphic to DIVISORS played on integers of the form $2^a 3^b$.

4.3 (a) All of the operations in Figure 4.9 are associative.
(b) Which ones are commutative? For which of them is 0 an identity?

4.4 *Selective sums.* If G and H are short impartial games, then $o(G \vee H) = \mathscr{P}$ if and only if $o(G) = \mathscr{P}$ and $o(H) = \mathscr{P}$ (normal play). Conclude that there are just two normal-play short impartial values for selective sums. How about misère play?

4.5 *Conjunctive sums.* The **remoteness** $\mathscr{R}(G)$ of a short impartial game G is defined by

$$\mathscr{R}(G) = \begin{cases} 0 & \text{if } G \cong 0\text{; otherwise:} \\ 1 + \min\{\mathscr{R}(G') : \mathscr{R}(G') \text{ is even}\} & \text{if some } \mathscr{R}(G') \text{ is even;} \\ 1 + \max\{\mathscr{R}(G') : \mathscr{R}(G') \text{ is odd }\} & \text{if every } \mathscr{R}(G') \text{ is odd.} \end{cases}$$

(a) $o(G) = \mathscr{P}$ if and only if $\mathscr{R}(G)$ is even.
(b) $\mathscr{R}(G \wedge H) = \min\{\mathscr{R}(G), \mathscr{R}(H)\}$ for all G and H.
(c) Define the **misère remoteness** $\mathscr{R}^-(G)$ of G by interchanging "odd" and "even" in the definition for $\mathscr{R}(G)$, and show that it works for determining the misère outcome of conjunctive sums.

4.6 Research the references for conjunctive and selective sums in the notes section (below), and draw the analogous tables to Figure 4.7.

4.7 TIC-TAC-TOE is a win for Maker when played under the Maker–Breaker convention.

Notes

FOX AND GEESE features prominently in *Winning Ways*. The first edition incorrectly asserted that Figure 4.1 has value $1 + \mathbf{over}$; the correct answer $2 + \mathbf{over}$ was later given by Jonathan Welton and verified by Berlekamp and Siegel. Berlekamp also proved that for $n \geq 9$, the $n \times 8$ starting position has the exact value $1+2^{-(n-8)}$; this analysis was incorporated into the second edition of *Winning Ways* in 2003.

The combinatorial theory of GO endgames was introduced in Wolfe's thesis [**Wol91**], building on Berlekamp's earlier discoveries in the theory of DOMINEERING [**Ber88**]. Further applications have been explored by Kim [**Kim95, BK96**], Landman [**Lan96**], Nakamura [**BN03**], Takizawa [**Tak02**], and others.

The temperature theory of GO was introduced in a seminal article by Berlekamp [**Ber96**]; many examples were subsequently found by Berlekamp, Müller, and Spight [**BMS96**]. The theory has been further pursued by Fraser [**Fras02**], Kao [**Kao97**], and Spight [**Spi99, Spi02, Spi03**].

Nakamura [**Nak09**] has recently introduced a beautiful new theory of GO capturing races, in which liberty counts are represented by partizan game values. Nakamura's theory bears a striking resemblance to the atomic weight theory (described in Section II.7 of this book).

ENTREPRENEURIAL CHESS was invented by Berlekamp and Pearson. It was inspired by the following CHESS problem, originally due to Simon Norton and mentioned in Guy's 1991 list of unsolved problems [**Guy91c**]: from the starting position in Figure 4.12, "what is the smallest board that White can win on if Black is given a win if he walks off the North or East edges of the board?"

Berlekamp and Pearson carried out a detailed temperature analysis of ENTREPRENEURIAL CHESS, showing (for example) that Figure 4.12 has mean value 17, among many other results [**BP03**]. Pearson also used related methods to solve Norton's original problem: the answer is 8×11.

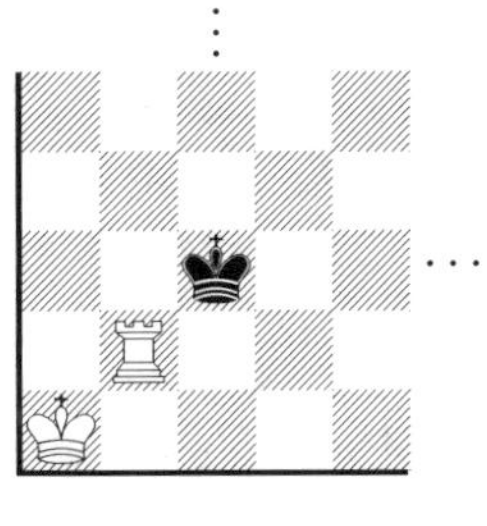

Figure 4.12. Simon Norton's CHESS problem.

CHOMP is isomorphic to the game DIVISORS, originally introduced by Frederik Schuh [**Sch52**] (see Exercise 4.2). In the form presented here, it was invented by David Gale [**Gal74**]. Transfinite generalizations, in which the board dimensions are permitted to be transfinite ordinals, have been studied by Huddleston and Shurman [**HS02**].

These examples were chosen to illustrate the breadth and variety of combinatorial games, but in truth they barely scratch the surface. Various other games are introduced and explored throughout this text, such as AMAZONS (Section II.5) and WYTHOFF (Section IV.3), but many more were left out. Here are some prominent examples in the latter category:

- DOTS AND BOXES is a well-known children's game with a surprisingly rich theory. Berlekamp's 2000 treatise [**Ber00a**] is the "bible" of DOTS AND BOXES strategy; a later advance is described by Berlekamp and Scott [**BS02**].
- The traditional Hawaiian board game KONANE meets the definition of *short combinatorial game* without any sort of contrivance. An early analysis by Ernst [**Ern95**] includes a nice account of its history; Chan and Tsai [**CT02**] later found solutions to certain sequences of $1 \times n$ positions. More recently, Santos and Silva announced the striking result that KONANE contains positions of value $*m$ for every $m \geq 0$, a rare feature for a partizan game [**SS08**].
- TOADS AND FROGS was introduced in *Winning Ways* and later studied in depth by Erickson [**Eri96**]. More recently, these efforts have been carried much further by Thanatipanonda and Zeilberger, with the help of some exciting new techniques in automated theorem proving [**TZ09, Tha11, Tha**].
- PHUTBALL ("Philosopher's Football") is an unusual board game invented by Conway and mentioned in *Winning Ways*. Demaine, Demaine, and Eppstein [**DDE02**] showed that it is NP-complete to determine whether a given player has an immediate winning move! More recently, Nowakowski, Ottaway, and Siegel [**Sie09b**] found that complicated loopy positions arise even on very small boards (such as 1×8).

Nondisjunctive compounds. Conjunctive and selective sums of short impartial games were studied by Smith [**Smi66**]. Selective sum has a trivial solution (Exercise 4.4). For the conjunctive case, Smith showed that the outcome of a sum can be computed from the **remoteness** of its components (Exercise 4.5), which is essentially a measure of how long the game will last with perfect play. Unlike the disjunctive theory, these results generalize easily to misère play.

Conway invented the diminished disjunctive, continued conjunctive, and shortened selective sums, and he showed that they can be analyzed using straightforward extensions of Smith's theories [**Con01**]. Remarkably, diminished disjunctive sum is also straightforward in misère play—so that ordinary disjunctive sum in misère play is the only really difficult theory of the bunch!

Winning Ways introduced a partizan theory for conjunctive and selective sums. In a fascinating twist, conjunctive sums can be analyzed using a trivial extension of the Steinhaus remoteness, whereas selective sums—which are trivial in the impartial case—require a deep and elegant theory all their own.

The theory of conjunctive and selective compounds can be extended to loopy games. The impartial cases were solved by Smith (they're really no harder than the loopfree analogues); the partizan cases were studied by Flanigan in the 1970s [**Fla79, Fla81, Fla83**].

Stromquist and Ullman found an elegant solution for sequential compounds of short impartial games [**SU93**]; this is explored in Exercise IV.1.6 on page 182. More recently, Fraser Stewart has investigated the partizan case [**Ste07**].

Chapter II

Short Games

Recall from Chapter I that a combinatorial game G is **short** provided that:

- G is **finite**: it has just finitely many subpositions; and
- G is **loopfree**: it admits no infinite run (equivalently, there is no sequence of moves proceeding from G that repeats a position).

These restrictions impose an inductive hierarchy that gives rise to a rich structure theory. The key results are that the values of short games form a partially ordered Abelian group $\mathbb{G}$, and each such value has a unique simplest representative. As a consequence, short games are particularly amenable to analysis and computation, and they are a central focus of study in combinatorial game theory.

The basic structure theorems are proved Sections 1 and 2. Sections 3 and 4 introduce two important classes of games, the **numbers** and **infinitesimals**. The remaining sections develop several key invariants of short games: temperature (Section 5); reduced canonical form (Section 6); and atomic weight (Section 7).

1. The Group $\mathbb{G}$

Formally, a short game G is an ordered pair $(\mathscr{G}^L, \mathscr{G}^R)$, where $\mathscr{G}^L$ and $\mathscr{G}^R$ are sets of "simpler" short games, the **Left** and **Right options** of G (we'll make this precise in a moment). We sometimes write

$$G = \{\mathscr{G}^L \mid \mathscr{G}^R\},$$

though we'll usually list the options of G explicitly:

$$G = \{G_1^L, G_2^L, \ldots, G_m^L \mid G_1^R, G_2^R, \ldots, G_n^R\}$$

or abuse notation and write simply

$$G = \{G^L \mid G^R\}$$

to mean that G^L and G^R range over the Left and Right options of G.

This is a recursive construction, and the base case is the **empty game** 0, which has no options at all:

$$0 = \{\ \mid\ \}.$$

The precise definition of short game is given through an explicit induction.

Definition 1.1. Let $\tilde{\mathbb{G}}_0 = \{0\}$, and for $n \geq 0$ put

$$\tilde{\mathbb{G}}_{n+1} = \left\{\{\mathscr{G}^L \mid \mathscr{G}^R\} : \mathscr{G}^L, \mathscr{G}^R \subset \tilde{\mathbb{G}}_n\right\}.$$

Then a **short game** is an element of

$$\tilde{\mathbb{G}} = \bigcup_{n \geq 0} \tilde{\mathbb{G}}_n.$$

Definition 1.2. Let G and H be short games. The **disjunctive sum** $G+H$ is defined recursively by

$$G + H = \{G^L + H,\ G + H^L \mid G^R + H,\ G + H^R\}.$$

Here we mean that G^L ranges over all Left options of G and H^L ranges over all Left options of H, so that the Left options of $G + H$ are given by the union

$$\{X + H : X \in \mathscr{G}^L\} \cup \{G + Y : Y \in \mathscr{H}^L\}.$$

The notation used in Definition 1.2 is almost always clearer and more succinct than set notation, and we'll use it throughout this book, without further comment.

Proposition 1.3. *Disjunctive sum is commutative and associative.*

Proof. Transparent from the definition. □

Definition 1.4. Let G be a short game. The **negative** $-G$ is defined recursively by

$$-G = \{-G^R \mid -G^L\}.$$

Proposition 1.5. $-(-G) \cong G$.

Proof. By induction on G,

$$-(-G) \cong -\{-G^R \mid -G^L\} \cong \{-(-G^L) \mid -(-G^R)\} \cong \{G^L \mid G^R\}. \quad \square$$

The following standard notational shortcuts will be used freely: $G - H$ is shorthand for $G + (-H)$; and if $n \in \mathbb{Z}$, then $n \cdot G$ denotes the sum of n copies of G. Thus

$$n \cdot G = \overbrace{G + G + \cdots + G}^{n \text{ times}} \quad \text{if } n \geq 0;$$
$$n \cdot G = \overbrace{-G - G - \cdots - G}^{|n| \text{ times}} \quad \text{if } n < 0.$$

We'll also use multiple vertical lines in order to denote "options of options" succinctly. For example:

$$\begin{aligned} \{G \parallel H \mid J\} &\quad \text{abbreviates} \quad \{G \mid \{H \mid J\}\}, \\ \{G \mid H \parallel J\} &\quad \text{abbreviates} \quad \{\{G \mid H\} \mid J\}, \\ \{G \mathbin{|||} H \mid J, K \parallel L\} &\quad \text{abbreviates} \quad \{G \mid \{\{H \mid J, K\} \mid L\}\}, \end{aligned}$$

and so on, with the number of consecutive vertical lines used to indicate precedence.

Outcomes and Values

By the Fundamental Theorem (Theorem I.1.5 on page 9), every short game belongs to one of the four outcome classes $\mathscr{N}$, $\mathscr{P}$, $\mathscr{L}$, or $\mathscr{R}$, as summarized on page 6. A formal definition can be given recursively:

Definition 1.6. We define $\mathscr{P}^L$ (games that Left can win as second player), $\mathscr{P}^R$ (likewise for Right), $\mathscr{N}^L$ (games that Left can win as first player), and $\mathscr{N}^R$ (likewise for Right) recursively by:

$$\begin{aligned} G \in \mathscr{P}^L &\quad \text{if} \quad \text{every } G^R \in \mathscr{N}^L, & G \in \mathscr{N}^L &\quad \text{if} \quad \text{some } G^L \in \mathscr{P}^L, \\ G \in \mathscr{P}^R &\quad \text{if} \quad \text{every } G^L \in \mathscr{N}^R, & G \in \mathscr{N}^R &\quad \text{if} \quad \text{some } G^R \in \mathscr{P}^R. \end{aligned}$$

Then

$$\begin{aligned} \mathscr{L} &= \mathscr{P}^L \cap \mathscr{N}^L, & \mathscr{P} &= \mathscr{P}^L \cap \mathscr{P}^R, \\ \mathscr{N} &= \mathscr{N}^L \cap \mathscr{N}^R, & \mathscr{R} &= \mathscr{P}^R \cap \mathscr{N}^R. \end{aligned}$$

The Fundamental Theorem shows that every game belongs to exactly one of these four classes. Denoting by $o(G)$ the **outcome class** (or just **outcome**) of G, we have the fundamental equivalence:

Definition 1.7. Let G and H be short games. Then we write

$$G = H \quad \text{if} \quad o(G + X) = o(H + X) \text{ for every short game } X.$$

Proposition 1.8. $=$ *is an equivalence relation.*

Proof. Equality of outcome classes is an equivalence relation. □

Definition 1.9. The **game value** of G is its equivalence class modulo $=$. The set of game values is denoted by $\mathbb{G}$.

$\mathbb{G}$ Is a Group

We'll now show that the set of game values $\mathbb{G}$ is a partially ordered Abelian group. We begin with some simple observations. Throughout this section, G, H, and J will denote arbitrary short games.

Proposition 1.10. *If $G = H$, then $G + J = H + J$.*

Proof. Since disjunctive sum is associative and $G = H$, we have

$$o\big((G+J)+X\big) = o\big(G+(J+X)\big) = o\big(H+(J+X)\big) = o\big((H+J)+X\big)$$

for all short games X. □

Proposition 1.11. $0 + G \cong G$.

Proof. 0 has no options at all, so by induction on G we have

$$0 + G \cong \{0 + G^L \mid 0 + G^R\} \cong \{G^L \mid G^R\} \cong G.$$ □

It follows that disjunctive sum induces an additive structure on $\mathbb{G}$ with identity 0. To prove that $\mathbb{G}$ is a group, we must prove that every value has an additive inverse. The following theorem is essentially the same as Theorem I.2.1 on page 17.

Theorem 1.12. *$G = 0$ if and only if $o(G) = \mathscr{P}$.*

Proof. Certainly $o(0) = \mathscr{P}$, vacuously, since 0 has no options for either player. Conversely, suppose $o(G) = \mathscr{P}$. We must show that

$$o(G + X) = o(X)$$

for every short game X. By symmetry, it suffices to show that whenever Left can win X playing first (resp. second), she can also win $G + X$ playing first (resp. second).

But suppose Left can win X playing second. Then on $G + X$, she can respond to each of Right's moves with a winning move on the same component. This guarantees that Left will never run out of moves, and since $G + X$ is short, the game must eventually end. Thus Left can win $G + X$ playing second.

Likewise, suppose Left can win X playing first, say to X^L. Then on $G + X$ she can make an opening move to $G + X^L$ and thereafter follow the strategy just described. □

Theorem 1.13. $G - G = 0$.

Proof. Either player can win on $G - G$ playing second, by following the mirror-image strategy. Therefore $o(G - G) = \mathscr{P}$, and the conclusion now follows from Theorem 1.12. $\square$

Theorem 1.14. *$\mathbb{G}$ is an Abelian group.*

Proof. Addition is commutative and associative on $\mathbb{G}$, with identity 0, and Theorem 1.13 shows that the negative of G represents its additive inverse in $\mathbb{G}$. $\square$

Corollary 1.15. *$G = H$ if and only if $o(G - H) = \mathscr{P}$.*

Proof. $G = H$ if and only if $G - H = 0$, and the result follows from Theorem 1.12. $\square$

In practice, in order to show that $G = H$ for particular games G and H, it is often easiest to prove that the difference $G - H$ is a second-player win.

Partial-Order Structure

The four outcome classes are naturally partially ordered by *favorability to Left*:

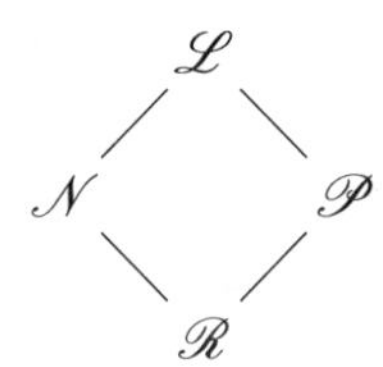

This partial order extends to $\mathbb{G}$.

Definition 1.16. Let G and H be short games. Then we write

$$G \geq H \quad \text{if} \quad o(G + X) \geq o(H + X) \text{ for every short game } X.$$

If $G \geq H$, then G is at least as favorable to Left as H: Left, playing any conceivable sum of short games, will always be content to replace the component H with G. The next propositions show that $\geq$ induces a partial order on $\mathbb{G}$ that respects its group structure.

Proposition 1.17. *$\geq$ is a partial order:*

(a) *$G \geq G$.*

(b) *If $G \geq H$ and $H \geq G$, then $G = H$.*

(c) *If $G \geq H$ and $H \geq J$, then $G \geq J$.*

Proof. Transparent from the definitions. $\square$

Proposition 1.18. *If $G \geq H$, then $G + J \geq H + J$.*

$$\begin{array}{llllll} G=0 & \Leftrightarrow & o(G)=\mathscr{P}, & G\geq 0 & \Leftrightarrow & o(G)\geq \mathscr{P},\\ G>0 & \Leftrightarrow & o(G)=\mathscr{L}, & G\leq 0 & \Leftrightarrow & o(G)\leq \mathscr{P},\\ G<0 & \Leftrightarrow & o(G)=\mathscr{R}, & G \mathrel{\|\!\!>} 0 & \Leftrightarrow & o(G)\geq \mathscr{N},\\ G \not\gtrless 0 & \Leftrightarrow & o(G)=\mathscr{N}, & G \mathrel{<\!\!\|} 0 & \Leftrightarrow & o(G)\leq \mathscr{N}. \end{array}$$

Figure 1.1. The outcome class of G is determined by its partial-order relationship to 0.

Proof. As Proposition 1.10. □

Theorem 1.19. *$G \geq H$ if and only if $o(G-H) \geq \mathscr{P}$.*

Proof. Using the fact that $\mathbb{G}$ is a partially ordered group, we have

$$G \geq H \iff G-H \geq H-H \iff G-H \geq 0. \qquad \square$$

The following recursive characterization of $\geq$ is fundamental.

Theorem 1.20. *$G \geq H$ if and only if no $G^R \leq H$ and $G \leq$ no H^L.*

Proof. By Theorem 1.19, $G \geq H$ if and only if $o(G-H) \geq \mathscr{P}$ (equivalently, Left can win $G-H$ playing second). But Right's options on $G-H$ all have the form G^R-H or $G-H^L$, so $G \geq H$ if and only if $o(G^R-H) \geq \mathscr{N}$ for every G^R, and $o(G-H^L) \geq \mathscr{N}$ for every H^L (equivalently, Left can win G^R-H and $G-H^L$ playing first). Another application of Theorem 1.19 shows that $o(G^R-H) \geq \mathscr{N}$ if and only if $G^R \not\leq H$, and likewise $o(G-H^L) \geq \mathscr{N}$ if and only if $G \not\leq H^L$. □

As a consequence of these results, the outcome class of G is determined precisely by its partial-order relationship to 0. This correspondence is summarized in Figure 1.1. The following notational conventions are used in Figure 1.1 and throughout this book.

Definition 1.21. Let G and H be short games. We write

$G \not\gtrless H$ if $G \not\geq H$ and $H \not\geq G$ (G is **confused with** H),

$G \mathrel{\|\!\!>} H$ if $G > H$ or $G \not\gtrless H$ (G is **greater than or confused with** H),

$G \mathrel{<\!\!\|} H$ if $G < H$ or $G \not\gtrless H$ (G is **less than or confused with** H).

Observe that:

$$G \mathrel{\|\!\!>} H \quad \text{if and only if} \quad G \not\leq H;$$
$$G \mathrel{<\!\!\|} H \quad \text{if and only if} \quad G \not\geq H.$$

Definition 1.22. We say that G is **positive** (resp. **negative**, **zero**, **fuzzy**) if $G > 0$ (resp. $G < 0$, $G = 0$, $G \not\gtrless 0$).

Some Simple Games

Several examples of short games are familiar from the study of HACKENBUSH in Section I.2:

$$1 = \{0 \mid \}, \qquad * = \{0 \mid 0\}, \qquad -1 = \{ \mid 0\}.$$

Clearly $o(1) = \mathscr{L}$, so that $1 > 0$, and it follows that

$$0 < 1 < 2 < 3 < \cdots$$

where $2 = 1 + 1$, as in Section I.2. Likewise

$$0 > -1 > -2 > -3 > \cdots$$

so there is a subgroup of $\mathbb{G}$ isomorphic to $\mathbb{Z}$. We'll henceforth identify $\mathbb{Z}$ with this subgroup.

What about $*$? It's certainly first-player win, so $* \gtrless 0$. But $* + *$ is a second-player win: whoever plays first from $* + *$ must make the losing move to $* + 0$. This gives the equivalence

$$* + * = 0,$$

and in particular $*$ has order 2 as an element of $\mathbb{G}$.

Of particular importance are the games

$$\uparrow = \{0 \mid *\} \qquad \text{and} \qquad \downarrow = \{* \mid 0\},$$

pronounced "up" and "down." It's easily checked that $\uparrow > 0$, since Left has a winning move to 0 and Right's only move, to $*$, is losing. We therefore have

$$0 < \uparrow < \uparrow + \uparrow < \uparrow + \uparrow + \uparrow < \cdots.$$

How does $\uparrow$ relate to $*$ and 1? The answers are somewhat surprising. $\uparrow$ is confused with $*$, but *two* (or more) copies of $\uparrow$ exceed it:

Proposition 1.23. *$\uparrow \gtrless *$, but $\Uparrow > *$.*

(Here $\Uparrow$ denotes $\uparrow + \uparrow$, pronounced "double-up.")

Proof. On $\uparrow + *$ Left can win immediately by moving to $\uparrow$, while Right has a move to $* + * = 0$. This shows that $\uparrow + * \gtrless 0$, and therefore $\uparrow \gtrless *$.

To see that $\uparrow + \uparrow \geq *$, note that Right's only moves from $\uparrow + \uparrow + *$ are to

$$\uparrow + \uparrow + 0 \quad \text{and} \quad \uparrow + * + *,$$

but these are both positive. $\square$

Moreover, $n \cdot \uparrow < 1$ for all n, and in fact this is true for a much wider class of games:

Definition 1.24. A game G is said to be **dicotic** if both players can move from every nonempty subposition of G.

For example, 0, $*$, and $\uparrow$ are dicotic, but 1 is not. Note that disjunctive sums of dicotic games are also dicotic.

Theorem 1.25. *If G is dicotic, then $G < 1$.*

Proof. We must show that Left can win the difference $1 - G$, no matter who moves first. Suppose Right moves first. He has no moves on 1, so if he has any move at all it must be to some $1 - G^L$. But G^L is dicotic, so by induction $G^L < 1$. Therefore Left has a winning move on $1 - G^L$.

Likewise, suppose Left moves first on $1 - G$. If $G \not\cong 0$, then since G is dicotic it must have a Right option G^R. By induction, $1 - G^R \geq 0$, so Left can win by moving to $1 - G^R$. Finally, if $G \cong 0$, then Left can win immediately by moving from $1 - 0$ to $0 - 0$. □

The dicotic games form an important class that will play a major role in the sequel, particularly in Section 7. The set of dicotic games is denoted by $\tilde{\mathbb{G}}^0$, and the set of *values* of dicotic games is denoted by $\mathbb{G}^0$. Since $\tilde{\mathbb{G}}^0$ is closed under disjunctive sum, it follows that $\mathbb{G}^0$ is a subgroup of $\mathbb{G}$.

Game Trees

Every short game G has an associated **game tree** that provides a useful visualization of its combinatorial structure. Each node in the tree represents a subposition H of G, with a **Left edge** to each H^L and a **Right edge** to each H^R. Some examples are shown in Figure 1.2 on the next page; in these and all succeeding diagrams, Left edges slant to the Left, and Right edges slant to the Right.

Note that a subposition H of G might appear several times in the associated game tree, since there might be several distinct runs that encounter H. In particular, every branch through the tree terminates at a different instance of 0. In the tree for $\uparrow + *$ shown in Figure 1.2, there are 12 terminal nodes, each representing a distinct instance of 0.

Birthday

The recursive structure of short games induces an important hierarchy of game values. Denote by $\mathbb{G}_n$ the set of values of games in $\tilde{\mathbb{G}}_n$. Then:

Definition 1.26. The **birthday** of a game (or value) G, denoted by $\mathrm{b}(G)$, is the least n such that $G \in \mathbb{G}_n$. We say G is **born on day** n if $\mathrm{b}(G) = n$, and it is **born by day** n if $\mathrm{b}(G) \leq n$.

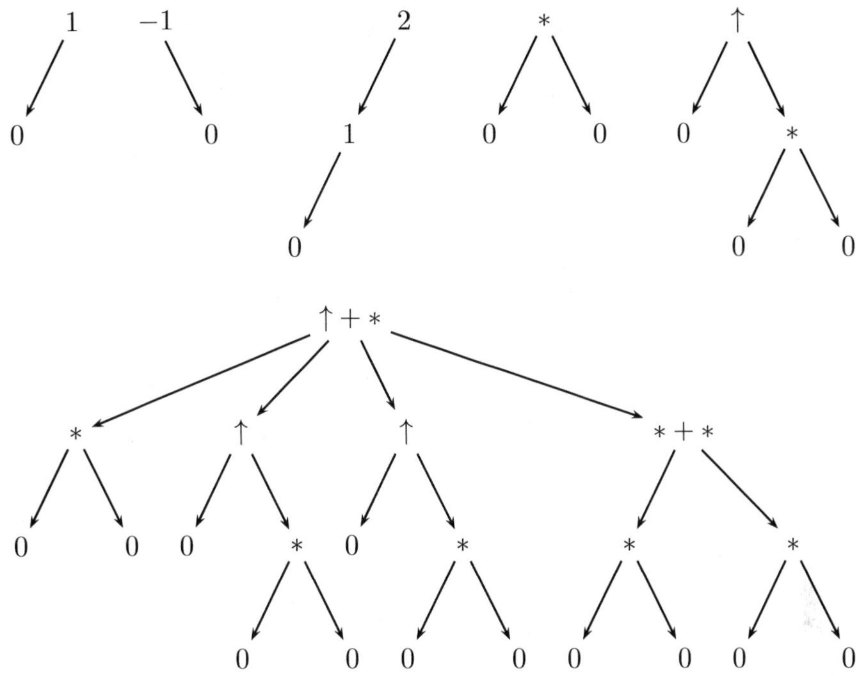

Figure 1.2. A forest of game trees.

We certainly have

$$\mathbb{G}_0 \subset \mathbb{G}_1 \subset \mathbb{G}_2 \subset \cdots$$

and

$$\mathbb{G} = \bigcup_{n \geq 0} \mathbb{G}_n.$$

Moreover, it's easily seen that each $\mathbb{G}_n$ is finite, since $|\mathbb{G}_n| \leq |\tilde{\mathbb{G}}_n|$ and

$$|\tilde{\mathbb{G}}_{n+1}| = 2^{2|\tilde{\mathbb{G}}_n|}.$$

The stratification of $\mathbb{G}$ by $\mathbb{G}_n$ will play an important role in the structure theory of Chapter III. We also write

$$\mathbb{G}_n^0 = \mathbb{G}^0 \cap \mathbb{G}_n$$

to obtain a similar stratification of $\mathbb{G}^0$.

Occasionally we'll also need the following related notions.

Definition 1.27. The **formal birthday** of a game G, denoted by $\tilde{\mathrm{b}}(G)$, is the least n such that $G \in \tilde{\mathbb{G}}_n$.

Note that $\tilde{\mathrm{b}}(G)$ is equal to the height of the game tree for G.

Definition 1.28. If $\mathcal{A} \subset \mathbb{G}$ is a set of game values, then the set of **children** of $\mathcal{A}$, denoted by $\mathrm{ch}(\mathcal{A})$, is the set of values given by

$$\mathrm{ch}(\mathcal{A}) = \Big\{\{\mathscr{G}^L \mid \mathscr{G}^R\} : \mathscr{G}^L, \mathscr{G}^R \subset \mathcal{A}\Big\}.$$

In particular, we have $\mathrm{ch}(\mathbb{G}_n) = \mathbb{G}_{n+1}$, for all $n \geq 0$.

Incentives

The **incentives** of G are a measure of how much a move on G is worth.

Definition 1.29. Let G be a short game. A **Left** (resp. **Right**) **incentive** of G is a game of the form $G^L - G$ (resp. $G - G^R$). We denote by $\Delta^L(G)$ a typical Left incentive of G and by $\Delta^R(G)$ a typical Right incentive.

The exact values of $\Delta^L(G)$ and $\Delta^R(G)$ depend on the form of G, and not just its value (see Exercise 1.9). Nonetheless, incentives are an important tool and we can prove some useful facts about them.

Theorem 1.30. *Every $G^L \lhd\!\!| \; G$ and every $G^R \;|\!\!\rhd G$.*

Proof. We know that $G^L \leq G^L$ and $G^R \geq G^R$, and the result is an immediate consequence of Theorem 1.20. (Alternatively, note that Right can win $G^L - G$ playing first, by moving to $G^L - G^L$ and thereafter following the mirror-image strategy.) □

Corollary 1.31. $\Delta \lhd\!\!| \; 0$ *for every incentive Δ (of every short game).*

Theorem 1.32 (Dominated Incentives Theorem)**.** *Let G and H be short games, and suppose that every $\Delta^L(H)$ is $\leq$ some $\Delta^L(G)$. Then Left need never move on H from $G+H$, i.e.,*

$$G + H = \{G^L + H \mid G^R + H,\ G + H^R\}.$$

Proof. Let $J = \{G^L + H \mid G^R + H,\ G + H^R\}$; we must show that $G + H - J$ is a second-player win. Every opening move from $G + H - J$ has a mirror-image response except for those of the form $G + H^L - J$. But for each H^L, there is some G^L with $G^L - G \geq H^L - H$. Therefore

$$G^L + H \geq G + H^L,$$

so Right can win from $G + H^L - J$ by moving to $G + H^L - (G^L + H)$. □

For clarity, we have stated the Dominated Incentives Theorem with just two components, but it generalizes easily to larger sums (Exercise 1.8). It is extremely useful in practice, for suppose that on the sum

$$G_1 + G_2 + \cdots + G_k,$$

every Left incentive of G_1 is $\leq$ a Left incentive of some other G_i. Then in order to find a winning move for Left on G, one can discard G_1 from consideration. Indeed, it will often be the case that the incentives of a single component G_i dominate all the others, in which case there must be a winning move on G_i (assuming that one exists at all).

Exercises

1.1 Determine all partial-order relations among the following games:

$$* \qquad \uparrow \qquad \downarrow \qquad \{1 \mid \uparrow\} \qquad \{\Uparrow \mid \downarrow\} \qquad \{1 \parallel 0 \mid -1\}$$

1.2 Confirm the following HACKENBUSH and DOMINEERING identities:

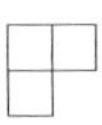 $= *$ 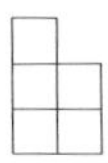 $= \{1 \mid -1\}$ 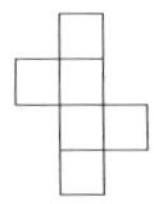 $=$ 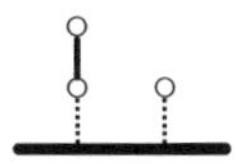 $= \uparrow$ 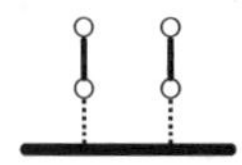 $= \Uparrow$

1.3 Show that the DOMINEERING positions

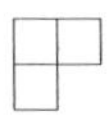 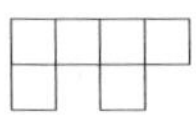 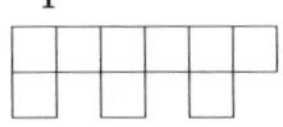 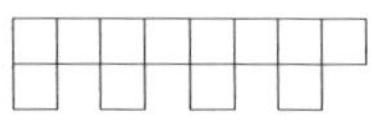 $\cdots$

are equal to

$$* \qquad 0 \qquad * \qquad 0 \qquad \cdots$$

1.4 Prove the *Gift Horse Principle*: If $H \lhd\!\!| \, G$, then $G = \{G^L, H \mid G^R\}$.

1.5 Suppose that for every subposition H of G, either H has options for both players; or else every option of H is dicotic. Prove that $G < 2$.

1.6 $\tilde{\mathrm{b}}(G) = \max\{1 + \tilde{\mathrm{b}}(G^L),\ 1 + \tilde{\mathrm{b}}(G^R)\}$, for all $G \not\cong 0$.

1.7 $\tilde{\mathrm{b}}(G + H) = \tilde{\mathrm{b}}(G) + \tilde{\mathrm{b}}(H)$, for all G and H.

1.8 (a) Every Left incentive of $G + H$ is equal to a Left incentive either of G or of H.

(b) If $G \cong G_1 + G_2 + \cdots + G_k$ and every Left incentive of G_1 is $\leq$ a Left incentive of some other G_i, then Left need never move on G_1 from G.

1.9 Give examples of short games G and H such that $G = H$, but G and H have distinct Left incentives.

Notes

The axiomatic construction of partizan games was invented by Conway, who was also the first to notice that such games form a partially ordered Abelian group. Most of the material in this section originally appeared in ONAG and has since become standard.

Theorem 1.20 is sometimes adopted as the *definition* of $\geq$. The advantages of using the fundamental equivalence instead (cf. Definition 1.16) are that it is better-motivated and also generalizes seamlessly to wider classes of games for which the recursive characterization fails.

The term *dicotic*, proposed by Michael Weimerskirch, is new. Such games are traditionally called **all-small**, because (as we'll see in Section 4) all their subpositions are infinitesimal. "Dicotic" is preferred when (as here) we wish to emphasize their structural properties. In particular, it remains appropriate in contexts where the theory of infinitesimals is inapplicable, such as the misère theory of Section V.6.

2. Canonical Form

Every short game G has many equivalent forms. For example, since $*+* = 0$, we have

$$G + * + * = G,$$

but the two forms are clearly nonisomorphic, since G is a proper subposition of $G + * + *$.

In this section we'll show that, for every short game G, there is a unique *simplest* K such that $K = G$. K is known as the **canonical form** (or **simplest form**) of G. It follows that every value is represented by exactly one game in canonical form.

It turns out that K can be obtained from G by repeated application of two fundamental simplification rules. We'll state these in a moment, after a few preliminary lemmas.

Lemma 2.1. *Let G be a short game, and suppose G' is obtained from G by removing some Left option G^L. Then $G' \le G$.*

Proof. Right can win $G' - G$ playing second, simply by following the mirror-image strategy. □

Lemma 2.2 (Replacement Lemma)**.** *Let G be a short game, and suppose G' is obtained from G by replacing some Left option G^L with a new option $G^{L'}$. If $G^{L'} \ge G^L$, then $G' \ge G$; if $G^{L'} \le G^L$, then $G' \le G$.*

Proof. Suppose $G^{L'} \ge G^L$; we must show $G' - G$ is a win for Left playing second. If Right moves to $G' - G^L$, Left wins by responding to $G^{L'} - G^L$; if Right makes any other opening move, then Left has a mirror-image response. An identical argument shows that if $G^{L'} \le G^L$, then $G' \le G$. □

Dominated and Reversible Options

The two simplification rules correspond to the following two types of "extraneous" options.

Definition 2.3. Let G be a short game.

(a) A Left option G^{L_1} is **dominated (by G^{L_2})** if $G^{L_2} \ge G^{L_1}$ for some other Left option G^{L_2}.

(b) A Right option G^{R_1} is **dominated (by G^{R_2})** if $G^{R_2} \le G^{R_1}$ for some other Right option G^{R_2}.

(c) A Left option G^{L_1} is **reversible (through** $G^{L_1R_1}$**)** if $G^{L_1R_1} \le G$ for some Right option $G^{L_1R_1}$.

(d) A Right option G^{R_1} is **reversible (through** $G^{R_1L_1}$**)** if $G^{R_1L_1} \ge G$ for some Left option $G^{R_1L_1}$.

Suppose first that G^{L_1} is dominated by G^{L_2}. Then in any situation where Left might be inclined to play from G to G^{L_1}, she would do at least as well to play to G^{L_2} instead. This suggests that dominated options do not contribute to the value of G, yielding the first simplification rule:

Theorem 2.4. *Let G be a short game, and suppose that G' is obtained from G by removing some dominated Left option G^{L_1}. Then $G' = G$.*

Proof. Certainly $G' \le G$ (by Lemma 2.1), so we must show that $G' \ge G$, i.e., that Left can win $G' - G$ playing second. Every opening move by Right has a mirror-image response on the other component, except for his move to $G' - G^{L_1}$. But G^{L_1} is dominated, so there is some $G^{L_2} \ge G^{L_1}$, and therefore Left can win by moving to $G^{L_2} - G^{L_1}$. □

The treatment of reversible moves is more subtle. If a Left option G^{L_1} is reversible through $G^{L_1R_1}$, then we can remove G^{L_1} from G, provided that we *replace it* with all positions of the form $G^{L_1R_1L}$. This operation is known as **bypassing** the reversible option G^{L_1}, and it is neatly illustrated by visualizing the game tree of G, as in Figure 2.1 on the following page.

Theorem 2.5. *Let G be a short game, and suppose that some Left option G^{L_1} is reversible through $G^{L_1R_1}$. Put*

$$G' = \{G^{L_1R_1L}, G^{L'} \mid G^R\},$$

where $G^{L_1R_1L}$ ranges over all Left options of $G^{L_1R_1}$, and $G^{L'}$ ranges over all Left options of G except *for G^{L_1}. Then $G' = G$.*

Proof. We must show that $G' - G$ is a second-player win. Every opening move has a mirror-image response on the other component, except for Left's moves to $G^{L_1R_1L} - G$ and Right's move to $G' - G^{L_1}$.

First consider $G^{L_1R_1L} - G$. By assumption $G^{L_1R_1} \le G$, so in fact

$$G^{L_1R_1L} \lhd\!\!| \; G^{L_1R_1} \le G.$$

This shows that Right must have a winning move on $G^{L_1R_1L} - G$.

Next consider $G' - G^{L_1}$. This game has $G' - G^{L_1R_1}$ as a Left option, so to complete the proof it suffices to show that $G' - G^{L_1R_1} \ge 0$.

Now Right's options from $G' - G^{L_1R_1}$ all have the form

$$G' - G^{L_1R_1L} \quad \text{or} \quad G^R - G^{L_1R_1}.$$

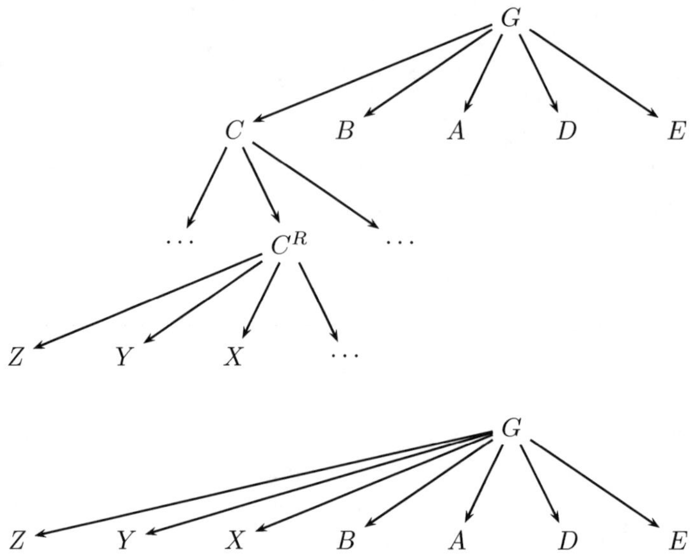

Figure 2.1. Bypassing a reversible option. If $C^R \le G$, then C can be replaced by the set of all Left options of C^R.

Each $G' - G^{L_1R_1L}$ has a mirror-image response on G', so it remains to consider $G^R - G^{L_1R_1}$. But certainly

$$G^R \mathrel{|\!\!\!>} G \ge G^{L_1R_1},$$

so Left must have a winning move on $G^R - G^{L_1R_1}$. □

Canonical Form

Definition 2.6. A short game K is in **canonical form** if no subposition of K has any dominated or reversible options.

Theorem 2.7. *For every short game G, there exists a game $K = G$ in canonical form.*

Proof. By induction, we may assume that the theorem is true for all options of G. By the Replacement Lemma, we may therefore assume that no *proper* subposition of G has any dominated or reversible options. If the same is true for G itself, then we are done.

But if G has a dominated option, then by Theorem 2.4 it can be removed without affecting the value of G, obtaining a simpler game $G' = G$. The game tree for G' has strictly fewer edges than the tree for G, so by induction, we may assume that there exists a $K = G'$ in canonical form. The same argument works if G has a reversible option, using Theorem 2.5. □

We'll now show that the canonical form given by Theorem 2.7 is unique.

Lemma 2.8. *Suppose that neither G nor H has any dominated or reversible options, and assume that $G = H$. Then for every G^L, there is an H^L such that $G^L = H^L$, and vice versa; and likewise for Right options.*

Proof. Fix G^L. Since $G - H = 0$, we have $G^L - H \lhd\!| \; 0$. So either $G^L - H^L \leq 0$ for some H^L; or else $G^{LR} - H \leq 0$ for some G^{LR}. But we cannot have $G^{LR} - H \leq 0$, since this would imply

$$G^{LR} \leq H \leq G,$$

contradicting the assumption that G has no reversible options. Therefore $G^L \leq H^L$ for some H^L. By exactly the same argument, there is some $G^{L'}$ such that $H^L \leq G^{L'}$. But now we have

$$G^L \leq H^L \leq G^{L'}.$$

Since G has no dominated options, this implies $G^L = H^L = G^{L'}$. The other cases follow by symmetry. □

Theorem 2.9 (Simplest Form Theorem)**.** *Suppose that G and H are in canonical form, and assume that $G = H$. Then $G \cong H$.*

Proof. Lemma 2.8 shows that for every G^L, there is an H^L with $G^L = H^L$. By induction, this implies that $G^L \cong H^L$. Therefore $\mathscr{G}^L \subset \mathscr{H}^L$, and likewise $\mathscr{H}^L \subset \mathscr{G}^L$, so that G and H have identical Left options. The same argument works for Right options. □

Definition 2.10. The **canonical form** (or **simplest form**) of G is the unique game $K = G$ such that no subposition of K has any dominated or reversible options.

Example. Recall that $*$ is the game $\{0 \mid 0\}$ and that $\uparrow = \{0 \mid *\}$. Let's find the canonical form of $G = \uparrow + *$. We have

$$\begin{aligned} \uparrow + * &= \{\uparrow + 0, 0 + * \mid \uparrow + 0, * + *\} \\ &= \{\uparrow, * \mid \uparrow, 0\} \end{aligned}$$

using the facts that $* + * = 0$ and $X + 0 = X$ for all X. Now we know that $\uparrow > 0$, so the right option $G^R = \uparrow$ is dominated by $G^{R'} = 0$, and we have

$$\uparrow + * = \{\uparrow, * \mid 0\}.$$

Finally, Left's option $G^L = \uparrow$ has the Right option $G^{LR} = *$. Since $\uparrow > 0$, we must have $\uparrow + * > *$, so G^L is reversible through G^{LR} and in fact

$$\begin{aligned} \uparrow + * &= \{*^L, * \mid 0\} \\ &= \{0, * \mid 0\}. \end{aligned}$$

This is canonical: there are no dominated options, since $0 \not\geq *$; and $*$ can't reverse through 0, since $\uparrow + *$ is a first-player win.

The sum $\uparrow + *$ is usually abbreviated $\uparrow *$, pronounced "up-star."

Exercises

2.1 Determine the canonical forms of $\{a \cdot \uparrow \mid b \cdot \uparrow\}$, for all $a, b \in \mathbb{Z}$.

2.2 Let G be a short game with canonical form K. Show that:
(a) For every K^L, there is a G^L such that $G^L \geq K^L$.
(b) For every Left incentive $\Delta^L(K)$, there is a $\Delta^L(G) \geq \Delta^L(K)$.
(c) For every subposition L of K, there is a subposition H of G such that $H = L$.

2.3 If G is dicotic, then so is its canonical form.

2.4 If K is in canonical form, then $\tilde{\mathrm{b}}(K) = \mathrm{b}(K)$.

2.5 $\mathrm{b}(G+H) \leq \mathrm{b}(G) + \mathrm{b}(H)$, for all games G and H. Give an example to show that the inequality might be strict.

2.6 A Left option G^L is **sensible** if $G^L \geq K^L$ for some K^L, where K is the canonical form of G. Otherwise G^L is **senseless**. Show that senseless options can be removed without changing the value of G.

2.7 A short game G is **even** if every option of G is odd; it is **odd** if $G \not\cong 0$ and every option of G is even. A value is even (resp. odd) if *any* of its representatives is even (resp. odd).
(a) 0 is even and $*$ is odd, but $\uparrow$ is neither even nor odd.
(b) If $n \in \mathbb{Z}$, then this definition agrees with the ordinary arithmetic definitions of "even" and "odd."
(c) If G is even (resp. odd), then so is its canonical form.
(d) Let $\mathcal{E} = \{G \in \mathbb{G} : G \text{ is even}\}$ and $\mathcal{O} = \{G \in \mathbb{G} : G \text{ is odd}\}$. Prove that $\mathcal{E}$ and $\mathcal{E} \cup \mathcal{O}$ are subgroups of $\mathbb{G}$, and show that $\mathcal{E}$ has index 2 in $\mathcal{E} \cup \mathcal{O}$.

2.8 (a) For each game G, there exists a game $H = G$ such that no subposition of H has more than three options.
(b) If every subposition of H has at most two options, then $H \neq \uparrow *$. (So "three" in part (a) cannot be improved to "two.")

Notes

All of the material in this section is standard. The Simplest Form Theorem and its proof were first published in ONAG.

Exercise 2.8: Michael Albert.

3. Numbers

In the study of HACKENBUSH in Section I.2, we observed that the game $G = \{0 \mid 1\}$ resembles the rational number $\frac{1}{2}$, in the sense that $G+G$ is equal to 1. Likewise $H = \{0 \mid G\}$ resembles the number $\frac{1}{4}$, and this construction can be iterated to obtain values 2^{-n} for all n.

One goal of this section will be to prove that such games do in fact behave exactly like the corresponding rational numbers. To reduce confusion, we'll temporarily enclose game values in square brackets: $[\![1]\!]$, $[\![\frac{1}{2}]\!]$, ..., to distinguish them from the ordinary rational numbers 1, $\frac{1}{2}$, ... Thus in particular we define

$$[\![\tfrac{1}{2}]\!] = \{0 \mid [\![1]\!]\}, \quad [\![\tfrac{1}{4}]\!] = \{0 \mid [\![\tfrac{1}{2}]\!]\}, \quad \ldots, \quad [\![2^{-n-1}]\!] = \{0 \mid [\![2^{-n}]\!]\}.$$

An important feature of such games is that their incentives are negative, suggesting a *disincentive* to play on them. Thus

$$0 < [\![\tfrac{1}{2}]\!] < [\![1]\!], \quad 0 < [\![\tfrac{1}{4}]\!] < [\![\tfrac{1}{2}]\!], \quad \ldots, \quad 0 < [\![2^{-n-1}]\!] < [\![2^{-n}]\!].$$

This turns out to be the basis for the most convenient *definition* of number:

Definition 3.1. A short game x is said to be a **number** if $y^L < y^R$ for every subposition y of x and every y^L and y^R.

Clearly every game of the form $[\![2^{-n}]\!]$ is a number.

Proposition 3.2. *If x is a number, then $x^L < x < x^R$ for every x^L and x^R. In particular, every incentive of x is negative.*

Proof. Certainly $x \mathrel{|\!\!\rhd} x^L$, so to show that $x > x^L$ it suffices to prove that Left can win $x - x^L$, playing second. Now on $x - x^L$ every Right option has the form

$$x^R - x^L \quad \text{or} \quad x - x^{LL}.$$

But $x^R - x^L > 0$ by definition of number, whereas $x - x^{LL}$ has a move to $x^L - x^{LL}$, which by induction is also positive. □

Proposition 3.3. *If x and y are numbers, then so is $x + y$.*

Proof. Since $x^L < x < x^R$ for every x^L and x^R, we have

$$x^L + y < x + y < x^R + y,$$

and likewise for options of y. □

This shows that numbers form a subgroup of $\mathbb{G}$. We'll prove that this subgroup is in fact isomorphic to the **dyadic rationals**:

Definition 3.4. A rational number is **dyadic** if its denominator is a power of 2 (in lowest terms). The group of dyadic rationals is denoted by $\mathbb{D}$:

$$\mathbb{D} = \left\{\frac{m}{2^n} : m \in \mathbb{Z},\ n \in \mathbb{N}\right\} = \left\{q \in \mathbb{Q} : 2^n q \in \mathbb{Z} \text{ for some } n \geq 0\right\}.$$

Proposition 3.5. *For all $n \geq 0$ we have $[\![2^{-n}]\!] > 0$ and*

$$[\![2^{-n-1}]\!] + [\![2^{-n-1}]\!] = [\![2^{-n}]\!] .$$

Proof. Let $x = [\![2^{-n}]\!]$, $y = [\![2^{-n-1}]\!]$, so that $y = \{0 \mid x\}$. We'll show that

$$y + y - x$$

is a second-player win. If either player moves on y, her opponent can respond on the other copy of y, leaving

$$0 + x - x = 0.$$

It remains to consider the options on $-x$. Right's move from $-x$ to 0 loses immediately, since $y+y > 0$. Likewise, if Left moves on $-x$, then Right can respond on y leaving the position

$$y + x - [\![2^{-n+1}]\!] .$$

But since y is a number, we have $x = y^R > y$, so

$$y + x - [\![2^{-n+1}]\!] < x + x - [\![2^{-n+1}]\!] = 0,$$

by induction. □

It follows that the map

$$2^{-n} \mapsto [\![2^{-n}]\!]$$

induces a group homomorphism $\mathbb{D} \to \mathbb{G}$. Moreover, this homomorphism respects the partial orderings on $\mathbb{D}$ and $\mathbb{G}$, so it must be injective. Therefore $\mathbb{G}$ contains a natural subgroup isomorphic to $\mathbb{D}$. From this point forward, we will identify this subgroup with $\mathbb{D}$ itself, and we'll simply write

$$\frac{m}{2^n}$$

in place of the game value

$$m \cdot [\![2^{-n}]\!] .$$

Theorem 3.6. *Let $n \geq 1$ and let m be an odd integer. Then*

$$\frac{m}{2^n} = \left\{ \frac{m-1}{2^n} \;\middle|\; \frac{m+1}{2^n} \right\}$$

in canonical form.

Proof. By symmetry we may assume $m > 0$. Then straight from the definition of disjunctive sum, we have

$$m \cdot \frac{1}{2^n} = \left\{ (m-1) \cdot \frac{1}{2^n} \;\middle|\; (m-1) \cdot \frac{1}{2^n} + \frac{1}{2^{n-1}} \right\} = \left\{ \frac{m-1}{2^n} \;\middle|\; \frac{m+1}{2^n} \right\}.$$

To see that this is canonical, suppose that $x = m/2^n$ and that x and all its subpositions are in the given form. Certainly there can be no dominated options, since there is just one option for each player. Moreover,

$$x^{LR} = \frac{m-1}{2^n} + \frac{1}{2^{n'}},$$

for some $n' < n$ (the inequality is strict since $m - 1$ is even). Therefore $x^{LR} > x$, and the same argument shows that $x^{RL} < x$, so that x has no reversible options. □

Theorem 3.7 (Archimedean Principle). *For any $G \in \mathbb{G}$, there is some integer n such that $n > G > -n$.*

Proof. Choose $n > \mathrm{b}(G)$. Then in any play of $n - G$ or $n + G$, Right can make at most $n - 1$ moves (presuming G to be in simplest form). So Left can win either game (playing first or second) by making all of her moves on n. □

The Simplicity Theorem

Fix a short game G (in any form), and suppose that $G = x$ for some $x \in \mathbb{D}$. Then certainly

(†) $\qquad G^L \lhd\!\vert\; x \lhd\!\vert\; G^R \quad$ for every G^L and G^R,

by Theorem 1.30 (page 62). We'll now prove a converse: if *any* $x \in \mathbb{D}$ satisfies the inequality (†), then G is equal to the *simplest* such x. One consequence of this theorem will be that every number (in the sense of Definition 3.1) belongs to $\mathbb{D}$.

Definition 3.8. A subset $\mathcal{I} \subset \mathbb{D}$ is said to be an **interval** if, whenever $x, y \in \mathcal{I}$ and $x \geq z \geq y$, then also $z \in \mathcal{I}$.

It is convenient to describe intervals using standard *interval notation.* For $a, b \in \mathbb{D}$, we write

$$[a,b] = \{x \in \mathbb{D} : a \leq x \leq b\}, \qquad [a,b[= \{x \in \mathbb{D} : a \leq x < b\},$$
$$]a,b] = \{x \in \mathbb{D} : a < x \leq b\}, \qquad]a,b[= \{x \in \mathbb{D} : a < x < b\}.$$

Proposition 3.9. *Let $\mathcal{I} \subset \mathbb{D}$ be a nonempty interval. Then there is a unique $x \in \mathcal{I}$ of smallest birthday.*

Proof. Suppose that $x, y \in \mathcal{I}$ have the same birthday, and $x > y$. We'll show that there is a $z \in \mathcal{I}$ of strictly smaller birthday.

Since $x > y$, Left has a winning move on $x - y$, so either $x^L \geq y$ or $x \geq y^R$, with x and y in canonical form. But we know that $x > x^L$ (since x is a number), so if $x^L \geq y$, then necessarily $x^L \in \mathcal{I}$. Likewise, if $x \geq y^R$, then necessarily $y^R \in \mathcal{I}$. Since x^L and y^R have strictly smaller birthdays than x and y, this shows that $\mathrm{b}(x)$ cannot be minimal over $\mathcal{I}$. □

Theorem 3.10 (Simplicity Theorem). *Let G be a short game and put*

$$\mathcal{I} = \{x \in \mathbb{D} : G^L \lhd\!| \ x \lhd\!| \ G^R \text{ for every } G^L \text{ and } G^R\}.$$

If $\mathcal{I} \neq \emptyset$, then G is equal to the unique $x \in \mathcal{I}$ of smallest birthday.

Proof. We will show that $G - x \geq 0$; the proof that $G - x \leq 0$ is the same. If Right moves to $G^R - x$, then since $x \in \mathcal{I}$ we have $G^R \, |\!\rhd \, x$, so Left wins *a priori*. But if Right moves to $G - x^L$, then by choice of x we have $x^L \notin \mathcal{I}$. Therefore either $x^L \leq G^L$ or $G^R \leq x^L$. But the latter would imply $G^R \leq x$, contradicting the fact that $x \in \mathcal{I}$, so necessarily $x^L \leq G^L$, whence Left's move to $G^L - x^L$ wins. □

Corollary 3.11. *If a short game x is equal to a number, then $x \in \mathbb{D}$.*

Proof. $x^L < x < x^R$ for every x^L and x^R, so this follows directly from the Simplicity Theorem. □

There is an easy rule for identifying the simplest number in an interval.

Proposition 3.12. *Let $\mathcal{I}$ be an interval and suppose x is the unique element of $\mathcal{I}$ of smallest birthday.*

- *If $\mathcal{I}$ contains an integer, then x is the unique integer in $\mathcal{I}$ of minimal magnitude.*
- *Otherwise, x is the unique element of $\mathcal{I}$ of minimal denominator (when written in lowest terms).*

The proof is left as an exercise.

Example. Let $G = \{0 \mid \frac{3}{4}\}$. Then $\mathcal{I} = \left]0, \frac{3}{4}\right[$. $\mathcal{I}$ contains no integers, so in fact $G = \frac{1}{2}$, the unique number in $\mathcal{I}$ of denominator 2.

Number Avoidance

If x is a number, then all of its incentives are negative (Proposition 3.2 on page 69), suggesting a *disincentive* to play on x. The following theorem makes that suggestion precise.

Theorem 3.13 (Number Avoidance Theorem). *Suppose that x is equal to a number and G is not. If Left (resp. Right) has a winning move on $G + x$, then she has a winning move of the form $G^L + x$ (resp. $G^R + x$).*

Proof. We may assume that x is in canonical form, since the outcomes of $G + x$ and $G^L + x$ do not depend on the form of x.

Now suppose that Left has a winning move of the form $G + x^L$. Then certainly $G + x^L \geq 0$. But G is not equal to a number, so $G \neq -x^L$, so in fact $G+x^L > 0$. Therefore Left has a winning move on $G+x^L$. By induction on x, we may assume that it has the form $G^L + x^L$, so that $G^L + x^L \geq 0$. But since x is a number in canonical form, we have $x^L < x$, so in fact

$$G^L + x > G^L + x^L \geq 0. \qquad \square$$

There is a stronger form of the Number Avoidance Theorem: if Left has a winning move on $G + H + x$, then she has a winning move of the form $(G + H)^L + x$. This is true for any choice of H, with the same assumptions on x and G as in Theorem 3.13. The reason it does not follow from Theorem 3.13 is that $G + H$ might be equal to a number, even when G and H are not. We'll prove the stronger form below (Theorem 3.22 on page 78), after developing some additional machinery.

The Number Tree

The inductive structure of $\mathbb{D}$ can be visualized as a complete binary tree, as in Figure 3.1 on the next page. Each number with birthday n has exactly two children on day $n + 1$, so there are exactly 2^n numbers born on day n. If x is a number with birthday n and if y and z are the numbers to its immediate left and right with birthday $\leq n$, then the children of x on day $n + 1$ have canonical forms $\{y \mid x\}$ and $\{x \mid z\}$.

There is a close relationship between the number tree and HACKENBUSH stalks. For each number x, there is a unique path from 0 through the tree to x. This path can be represented by a string of $+$ and $-$ symbols, indicating a left or right branch, respectively. Then the HACKENBUSH stalk with a blue edge for every $+$ and a red edge for every $-$ has value x.

Confusion Intervals

Definition 3.14. A **switch** is any game of the form $\{x \mid -x\}$, where $x \in \mathbb{D}$. A switch is usually written using the abbreviation

$$\pm x = \{x \mid -x\}.$$

If $G = \pm x$ is a switch and y is a number, then $y - G = \{y + x \mid y - x\}$ (this is easily verified by showing that the difference is a second-player win). So if $y > x$, then Left wins no matter who moves first, since both options are to positive numbers. Likewise if $y < -x$, then Right wins no matter

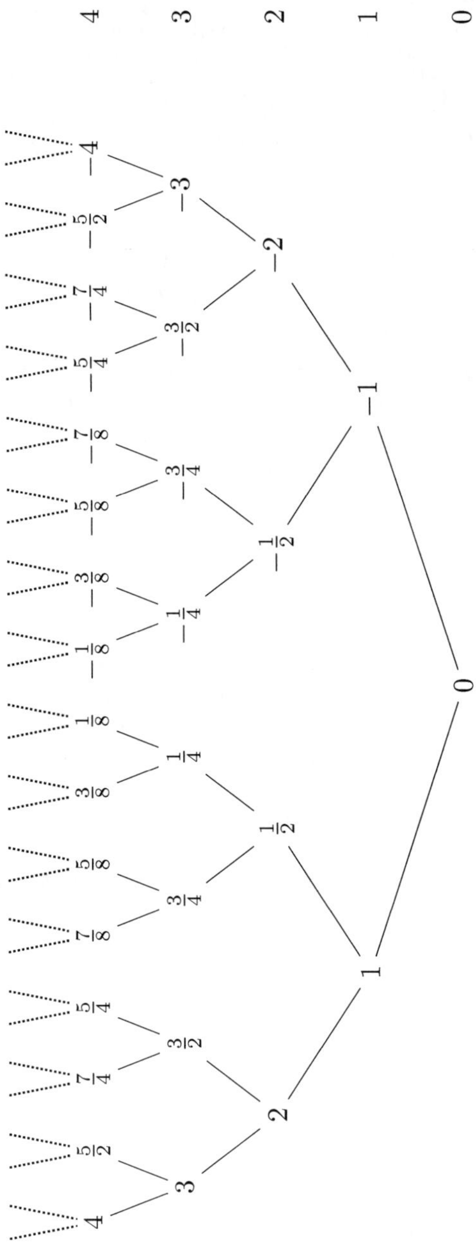

Figure 3.1. The number tree (with birthdays labeled on the right).

who moves first. But if $x \geq y \geq -x$, then $G \not\gtrless y$, since $y - G$ is a first-player win. We say that $[-x, x]$ is the **confusion interval** of G.

Definition 3.15. The **confusion interval** of G is defined by

$$\mathcal{C}(G) = \{x \in \mathbb{D} : G \not\gtrless x\}.$$

It's clear that $\mathcal{C}(G)$ is always a finite-length interval, since its complement

$$\{x \in \mathbb{D} : x \geq G\} \cup \{x \in \mathbb{D} : x \leq G\}$$

is the union of two half-bounded intervals.

Examples. The reader is invited to confirm the following examples:

- $\mathcal{C}(*) = \{0\}$, a singleton.
- $\mathcal{C}(\uparrow) = \emptyset$, the empty interval.
- $\mathcal{C}\big(\{x + * \mid -x + *\}\big) = \,]-x, x[$, an open interval.
- $\mathcal{C}\big(\{x + * \mid -x\}\big) = [-x, x[$, a half-open interval.

Stops

The endpoints of $\mathcal{C}(G)$ are fundamental invariants of G and are easily described through the following explicit recursion.

Definition 3.16. Let G be a short game. The **Left stop** $L(G)$ and the **Right stop** $R(G)$ are defined recursively by

$$L(G) = \begin{cases} G & \text{if } G \text{ is equal to a number;} \\ \max\limits_{G^L}\big(R(G^L)\big) & \text{otherwise;} \end{cases}$$

$$R(G) = \begin{cases} G & \text{if } G \text{ is equal to a number;} \\ \min\limits_{G^R}\big(L(G^R)\big) & \text{otherwise.} \end{cases}$$

Suppose Left and Right play G in isolation, with Left moving first, and agree to stop play as soon as G reaches a number x. Then Left can guarantee that $x \geq L(G)$, and Right can guarantee that $x \leq L(G)$. So assuming both play well, it will be the case that $x = L(G)$. We may think of x as the final "score" after such a play of G. Likewise, $R(G)$ is the corresponding score obtained if Right moves first.

Proposition 3.17. *Let G be a game and x a number.*

(a) $L(-G) = -R(G)$ *and* $R(-G) = -L(G)$.

(b) *If* $R(G) > x$, *then* $G > x$. *Likewise, if* $L(G) < x$, *then* $G < x$.

(c) *If* $L(G) > x$, *then* $G \mathrel{|\!\!\triangleright} x$. *Likewise, if* $R(G) < x$, *then* $G \mathrel{\triangleleft\!\!|} x$.

(d) *If* $G \geq x$, *then* $R(G) \geq x$. *Likewise, if* $G \leq x$, *then* $L(G) \geq x$.

(e) $L(G)$ *and* $R(G)$ *are the endpoints of* $\mathcal{C}(G)$.

Proof. None of the statements are affected if x is replaced by an equal game, so we may suppose x to be in canonical form.

(a) is clear from the definitions. We'll prove (b) and (c) by simultaneous induction; (d) is just the contrapositive of (c), and (e) follows immediately.

(b) Suppose that $R(G) > x$. If G is equal to a number, then $G = R(G)$ and there is nothing to prove, so assume otherwise. Then for every G^R, we have $L(G^R) \geq R(G) > x$, so by induction on (c), $G^R \mathrel{|\!\!>} x$. By the Number Avoidance Theorem, this shows that $G \geq x$. But certainly $G \neq x$, so in fact $G > x$.

(c) Suppose that $L(G) > x$. If G is equal to a number, then $G = L(G)$ and there is nothing to prove, so assume otherwise. Then for some G^L, we have $R(G^L) = L(G) > x$, so by induction on (b), $G^L > x$. Therefore $G \mathrel{|\!\!>} x$. □

Since $\mathcal{C}(G)$ is an invariant of the form of G, it follows that $L(G)$ and $R(G)$ are also invariants. Although the stops are the endpoints of $\mathcal{C}(G)$, they don't determine whether $\mathcal{C}(G)$ is open or closed; the examples on the preceding page show that all four cases are possible. Exercise 3.11 on page 81 explores a refinement of $L(G)$ and $R(G)$ that completely isolates $\mathcal{C}(G)$.

Proposition 3.18. *Let* G *be a short game.*

(a) $L(G) \geq R(G)$.

(b) $R(G^L) \leq L(G)$ *for every* G^L *and* $L(G^R) \geq R(G)$ *for every* G^R, *even if* G *is equal to a number.*

(c) $L(G + x) = L(G) + x$ *and* $R(G + x) = R(G) + x$ *for all* $x \in \mathbb{D}$.

Proof. (a) Suppose (for contradiction) that $L(G) < R(G)$. Then there exists a number x such that

$$L(G) < x < R(G),$$

so that by Proposition 3.17(b), $G < x < G$, a contradiction.

(b) is immediate from the definitions except when G is equal to a number. But in that case, every $G^L \mathrel{<\!\!|} G$ and every $G^R \mathrel{|\!\!>} G$, so the conclusion follows from Proposition 3.17(b).

(c) is trivial if G is equal to a number. Otherwise, by induction,

$$\begin{aligned} L(G + x) &= \max\left\{R(G^L + x),\ R(G + x^L)\right\} \\ &= \max\left\{R(G^L) + x,\ R(G) + x^L\right\}. \end{aligned}$$

Since G is not equal to a number, there must be a Left option G^L with $R(G^L) = L(G)$. By (a), this implies $R(G^L) \geq R(G)$, and since $x^L < x$, we then have $R(G^L) + x > R(G) + x^L$. Therefore

$$L(G+x) = \max_{G^L} \{R(G^L)\} + x = L(G) + x. \qquad \square$$

The next inequalities are fundamental.

Proposition 3.19. *Let G and H be short games.*

(a) $L(G) + L(H) \geq L(G+H) \geq L(G) + R(H)$;
$R(G) + R(H) \leq R(G+H) \leq L(G) + R(H)$.

(b) *If $G \geq H$, then $L(G) \geq L(H)$ and $R(G) \geq R(H)$.*

Proof. (a) If $G+H$ is equal to a number, say x, then $H = x - G$. By Propositions 3.17 and 3.18 we have $R(H) = x - L(G)$. Therefore

$$L(G) + R(H) = x = L(G+H),$$

and it follows from Proposition 3.18(a) that

$$L(G) + L(H) \geq L(G+H).$$

If $G+H$ is not equal to a number, then without loss of generality there is some G^L with $R(G^L + H) = L(G+H)$. Then

$$\begin{aligned} L(G+H) &= R(G^L + H) & \\ &\leq R(G^L) + L(G) & \text{(induction)} \\ &\leq L(G) + L(H) & \text{(Proposition 3.18(b))}. \end{aligned}$$

Similarly, without loss of generality we can assume that G is not equal to a number, so there exists some $G^{L'}$ (possibly identical to G^L) such that $R(G^{L'}) = L(G)$. Then

$$\begin{aligned} L(G+H) &\geq R(G^{L'} + H) & \text{(Proposition 3.18(b))} \\ &\geq R(G^{L'}) + R(H) & \text{(induction)} \\ &= L(G) + R(H). & \end{aligned}$$

The other inequalities follow by symmetry.

(b) If instead $L(G) < L(H)$, then there exists an $x \in \mathbb{D}$ such that

$$L(G) < x < L(H).$$

By Proposition 3.17(b) and (c), this implies

$$G < x \lhd\!\!| \; H,$$

contradicting the assumption $G \geq H$. $\square$

Number Avoidance, Revisited

We can now prove the strong form of the Number Avoidance Theorem (originally proved as Theorem 3.13 on page 72). The key observation is that if G is *not* equal to a number, then its incentives exceed every negative number.

Lemma 3.20. *If G is not equal to a number, then G has both a Left incentive $\Delta^L(G)$ and a Right incentive $\Delta^R(G)$ such that*

$$\Delta^L(G) \geq -x \quad \textit{and} \quad \Delta^R(G) \geq -x \quad \textit{for every number } x > 0.$$

Proof. Choose G^L such that $R(G^L) = L(G)$. Then

$$R(G^L - G) \geq R(G^L) - L(G) = 0.$$

So for every $x > 0$, we have $R(G^L - G) > -x$, and therefore $G^L - G > -x$ as well. □

Theorem 3.21 (Number Translation Theorem). *Suppose x is equal to a number and G is not. Then*

$$G + x = \{G^L + x \mid G^R + x\}.$$

Proof. By the Replacement Lemma we can assume that x is in canonical form. The incentives of x are negative numbers, so by Lemma 3.20, they are dominated by the incentives of G. The conclusion follows from the Dominated Incentives Theorem (Theorem 1.32 on page 62). □

Theorem 3.22 (Number Avoidance Theorem). *Suppose that x is equal to a number and G is not, and let H be an arbitrary game. If Left (resp. Right) has a winning move on $G + H + x$, then she has a winning move of the form $(G + H)^L + x$ (resp. $(G + H)^R + x$).*

(In particular, this is true even if $G + H$ is equal to a number.)

Proof. Suppose $G + H + x \mathrel{|\!\rhd} 0$. Now every Left incentive of G is equal to a Left incentive of $G + H$, so just as in the previous proof, the Dominated Incentives Theorem gives

$$\{(G + H)^L + x \mid (G + H)^R + x\} = G + H + x \mathrel{|\!\rhd} 0,$$

and the conclusion follows immediately. □

The Mean Value Theorem

The **mean** of G is another fundamental invariant. It is a number $m(G)$ that closely approximates the limiting behavior of many copies of G, in the sense that the difference

$$n \cdot G - n \cdot m(G)$$

is bounded by a constant independent of n. It will always be the case that

$$L(G) \geq m(G) \geq R(G) \quad \text{for all } G$$

and

$$m(G+H) = m(G) + m(H) \quad \text{for all } G \text{ and } H.$$

Theorem 3.23 (Mean Value Theorem). *For all short games G,*

$$\lim_{n\to\infty} \frac{L(n\cdot G)}{n} = \lim_{n\to\infty} \frac{R(n\cdot G)}{n},$$

and in particular both limits exist.

The Mean Value Theorem is a corollary of the following stronger assertion, which is of independent interest.

Lemma 3.24. *Let G be a short game. Then there is a number x such that*

$$L(n\cdot G) - R(n\cdot G) \leq x \quad \textit{for all } n \geq 0.$$

Proof. Put

$$x = \max_{G^L} \big(0, L(G^L - G)\big).$$

Now fix n. If $n \cdot G$ is a number, then $L(n\cdot G) = R(n\cdot G)$, and since $x \geq 0$, there is nothing to prove. Otherwise, choose G^L such that

$$L(n\cdot G) = R\big((n-1)\cdot G + G^L\big).$$

Then by Proposition 3.19(a) we have

$$\begin{aligned} L(n\cdot G) &= R\big(n\cdot G + (G^L - G)\big) \\ &\leq R(n\cdot G) + L(G^L - G) \\ &\leq R(n\cdot G) + x. \end{aligned}$$

□

Proof of Theorem 3.23. Let x be as in Lemma 3.24. Then

$$R(n\cdot G) + x \geq L(n\cdot G) \geq R(n\cdot G) \geq L(n\cdot G) - x,$$

so for all $n \geq 0$,

$$\frac{R(n\cdot G)}{n} + \frac{x}{n} \geq \frac{L(n\cdot G)}{n} \geq \frac{R(n\cdot G)}{n} \geq \frac{L(n\cdot G)}{n} - \frac{x}{n}.$$

But x is a constant independent of n, so $x/n \to 0$ as $n \to \infty$. □

Definition 3.25. The **mean** (or **mean value**) of G is given by

$$m(G) = \lim_{n\to\infty} \frac{L(n\cdot G)}{n}.$$

Theorem 3.26. *For all short games G and H,*

$$m(G+H) = m(G) + m(H).$$

Proof. For all n,

$$\begin{aligned} L(n \cdot G) + L(n \cdot H) &\geq L\big(n \cdot (G+H)\big) \\ &\geq R\big(n \cdot (G+H)\big) \\ &\geq R(n \cdot G) + R(n \cdot H). \end{aligned}$$

Therefore

$$m(G) + m(H) \geq m(G+H) \geq m(G) + m(H).$$

□

A Theorem on Incentives

We conclude with a theorem that will have important implications in the sequel.

Theorem 3.27. *If G is not equal to an integer, then G has both a Left incentive $\Delta^L(G) > -1$ and a Right incentive $\Delta^R(G) > -1$.*

Proof. This follows from Lemma 3.20 whenever G is not equal to a number, so assume that it is.

By the Simplicity Theorem there can be no integer x with

$$G \geq x \mathrel{|\!\!\rhd} G^L$$

for all G^L (otherwise G itself would be equal to an integer). By Proposition 3.17(c) there is no integer x with

$$L(G) \geq x > R(G^L)$$

for all G^L. Therefore

$$R(G^L) - L(G) > -1$$

for *some* G^L. But now Proposition 3.19 gives

$$\begin{aligned} R(G^L - G) &\geq R(G^L) + R(-G) \\ &= R(G^L) - L(G) \\ &> -1 \end{aligned}$$

whereupon $G^L - G > -1$, by Proposition 3.17(b). □

Corollary 3.28 (Integer Avoidance Theorem)**.** *Suppose that x is equal to an integer and G is not, and let H be an arbitrary game. If Left (resp. Right) has a winning move on $G + H + x$, then she has a winning move of the form $(G + H)^L + x$ (resp. $(G + H)^R + x$).*

Proof. Identical to the proof of Theorem 3.22, using Theorem 3.27 in place of Lemma 3.20. □

Exercises

3.1 Exhibit short games G and H such that

$$L(G) + L(H),\ L(G+H),\ \text{and}\ L(G) + R(H)$$

are all distinct.

3.2 The hypothesis "G is not equal to a number" is necessary in Theorems 3.13, 3.21, and 3.22.

3.3 Let $x \geq 0$ be a number. Prove that

$$\mathrm{b}(x) = \lceil x \rceil + n,$$

where $x = m/2^n$ in lowest terms.

3.4 For any two numbers $x \geq y$, there exist games G, H, and J with $\mathcal{C}(G) = [x, y]$, $\mathcal{C}(H) = [x, y[$, and $\mathcal{C}(J) =]x, y[$.

3.5 If G has no Right options, then G is equal to an integer.

3.6 Calculate the stops and mean value of each of the following games:

$$G = \{1 \mid 0\}; \qquad H = \{1 \parallel 0 \mid -100\}; \qquad J = \{1 \parallel 0 \mid -1\}.$$

3.7 Let $x_1 > x_2 > \cdots > x_n \geq 0$. Determine the Left stop, Right stop, and mean value of

$$\pm x_1 \pm x_2 \pm \cdots \pm x_n.$$

3.8 Suppose that G is not equal to a number, and assume that G is in canonical form. Show that $R(G^L) \geq L(G^R)$ for *every* G^L and G^R.

3.9 Every position in BLUE–RED HACKENBUSH is a number.

3.10 Suppose that x is equal to a number and G is not, and let H be an arbitrary game. If Left has a winning move on $G + H + x$ of the form $G + H + x^L$, then she also has a winning move of the form $G^L + H + x$. (This strengthens Theorem 3.22.)

3.11 An **adorned number** is a symbol of the form x_L or x_R, with $x \in \mathbb{D}$. The **adorned stops** of G are defined by

$$L^a(G) = \begin{cases} x_R & \text{if } G \text{ is equal to a number } x; \\ \max\limits_{G^L}\big(R^a(G^L)\big) & \text{otherwise;} \end{cases}$$

$$R^a(G) = \begin{cases} x_L & \text{if } G \text{ is equal to a number } x; \\ \min\limits_{G^R}\big(L^a(G^R)\big) & \text{otherwise.} \end{cases}$$

In taking max and min, adorned numbers are ordered lexicographically, so that $x_L > x_R > y_L > y_R$ whenever $x > y$. Prove that $\mathcal{C}(G)$ is completely determined by $L^a(G)$ and $R^a(G)$.

3.12 The following are equivalent:

(i) $G - H$ is infinitesimal.

(ii) $L(G + X) = L(H + X)$ and $R(G + X) = R(H + X)$ for every X.

3.13 Does there exist a G such that $n \cdot G \not\gtrless 0$ for all n?

3.14 If $G^L \leq G^R$ for every G^L and G^R, then G is numberish (cf. Definition 4.5 on page 84, below).

3.15 If every incentive of G is negative, then G is equal to a number.

3.16 G is **alternating** (or **consecutive move banned**) if, for every subposition H of G, there are no second options of the form H^{LL} or H^{RR}. Classify all possible values for an alternating game G.

3.17 A BLUE–RED HACKENBUSH position is a **spider** if it has just one nonground node (the **center**) incident to more than two edges. The center is connected to the ground via several paths (the **legs**). In a **redwood spider** such as Figure 3.2, each leg consists of a single blue edge touching the ground, followed by a path of red edges leading up to the center.

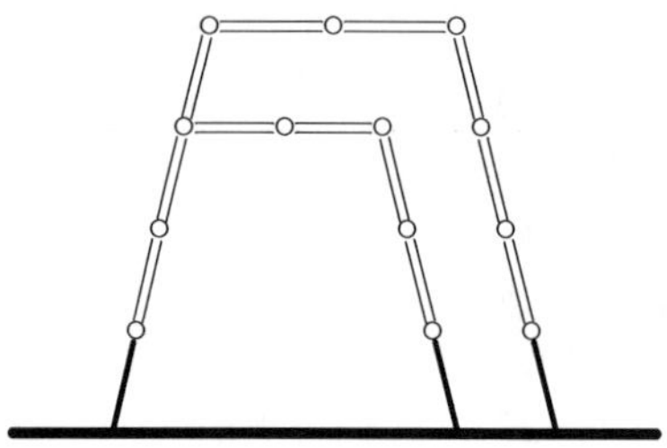

Figure 3.2. A redwood spider with three legs: $a_1 = 5$, $a_2 = 3$, $a_3 = 1$.

Let G be a redwood spider with n legs. Let
$$a_1 \geq a_2 \geq a_3 \geq \cdots \geq a_n \geq 0$$
and assume that the the ith leg of G has exactly one blue edge and $a_i + 1$ red edges, for $1 \leq i \leq n$. Show that $G = 2^{-r}$, where
$$r = \left\lfloor \frac{a_1}{2^1} + \frac{a_2}{2^2} + \cdots + \frac{a_{n-2}}{2^{n-2}} + \frac{a_{n-1} + a_n + 1}{2^{n-1}} \right\rfloor.$$

Notes

As early as the 1950s, Milnor [**Miln53**] and Hanner [**Han59**] considered sums of games whose terminal positions are assigned numerical scores. It was Conway who recognized that the explicit assignment of scores is not always necessary, because numbers themselves can be represented as particular game positions. $\mathbb{D}$ is just a fragment of a much larger number system, the **surreal numbers**, that is part of Conway's original vision; surreal numbers are described in Chapter VIII of this book.

Most of the material in this section is standard and is drawn from ONAG and *Winning Ways*. The Mean Value Theorem was proved independently in various forms by Hanner, Berlekamp, and Conway; the particularly elegant proof given here is due to Norton.

Exercise 3.16: Paul Ottaway.

Exercise 3.17: Berlekamp.

4. Infinitesimals

The theory of numbers presented in Section 3 provides a useful window into the structure of $\mathbb{G}$, but it is certainly not the whole story. For example,

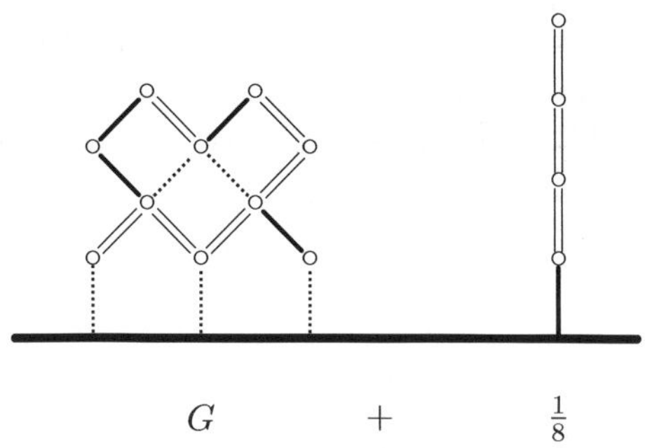

Figure 4.1. The HACKENBUSH position G is complicated, but necessarily dicotic. On the sum $G + x$, with x a positive number, Left can win by removing the ground edges of G preferentially.

in Section 1 we saw that $\uparrow < 1$ (Theorem 1.25 on page 60), and a slight refinement of that argument, which we'll shortly make precise, shows that $\uparrow < x$ for all numbers $x > 0$. In this sense, $\uparrow$ is "too small" to be measured on the scale of numbers, and in this section we'll explore a large family of **infinitesimals** with similar properties. In Section 5, we'll see how the theory of infinitesimals also contributes to an understanding of more general games.

Definition 4.1. A short game G is **infinitesimal** if $x > G > -x$ for every positive number $x \in \mathbb{D}$.

The following fundamental result shows that every dicotic game is infinitesimal, generalizing Theorem 1.25.

Theorem 4.2 (Lawnmower Theorem). *If G is dicotic, then G is infinitesimal.*

Proof. Let $x > 0$ be a positive number in canonical form. We must show that Left can win the difference $x - G$, no matter who moves first. Right's options on $x - G$ have the form $x^R - G$ or $x - G^L$. But x^R is a positive number and G^L is dicotic, so by induction all such games are positive.

Likewise, suppose Left moves first on $x - G$. If $G \not\cong 0$, then since G is dicotic, it must have a Right option G^R. By induction, Left can win by moving to $x - G^R$. If $G \cong 0$, then since $x > 0$, Left automatically has a winning move. □

Example. Let G be a HACKENBUSH position, and suppose that every edge of G that's connected to the ground is green. (An example of such a position is shown in Figure 4.1.) Then G is necessarily dicotic: every nonempty subposition of G necessarily has a green edge touching the ground, and both players have the option of removing any such edge.

This is the origin of the name *Lawnmower Theorem.* On the sum $G+x$ (with x a positive number), Left can employ an imaginary "lawnmower," systematically (and arbitrarily) cutting through the green stems that connect G to the ground. Eventually the G component will be empty, and at that point Left is guaranteed to have a move available on x. The precise structure of the component G is irrelevant.

The next proposition provides a useful characterization of infinitesimals.

Proposition 4.3. *G is infinitesimal if and only if $L(G) = R(G) = 0$.*

Proof. If $L(G) = R(G) = 0$, then Proposition 3.17(b) shows that

$$x \geq G \geq -x$$

for all $x > 0$. If $L(G) < 0$, then $L(G) < -x$ for some positive number x, so $G \leq -x$. Likewise, if $L(G) > 0$, then $L(G) > x$ for some positive number x, so by Proposition 3.17(c) we have $G \mathrel{|\!\!\triangleright} x$. The same arguments apply to $R(G)$ by symmetry. □

Definition 4.4. G and H are **infinitesimally close** if $G - H$ is infinitesimal. In this case we say that H is **G-ish**.

Definition 4.5. G is **numberish** if it is infinitesimally close to a number.

If x is a number, then G is x-ish if and only if $L(G) = R(G) = x$. This is a consequence of Proposition 4.3 and the Number Translation Theorem. In particular, G is numberish if and only if $L(G) = R(G)$.

Definition 4.6. Let $G, H > 0$. We say that G is **infinitesimal relative to H** if $n \cdot G < H$ for every integer $n \geq 0$. In this case we write $G \ll H$.

Nimbers

Every impartial game is certainly dicotic, and therefore infinitesimal. In particular, this is true for every nim-heap. For $m \geq 0$ we'll denote by $*m$ (pronounced "star m") a nim-heap of size m, so that

$$*0 = \{\,|\,\},\ *1 = \{0 \mid 0\},\ *2 = \{0, *1 \mid 0, *1\},\ *3 = \{0, *1, *2 \mid 0, *1, *2\},\ \ldots.$$

Note that $*0$ and $*1$ are just alternative names for 0 and $*$. These **nimbers** are the impartial analogues of the integers

$$0 = \{\,|\,\},\ 1 = \{0 \mid \,\},\ 2 = \{0, 1 \mid \,\},\ 3 = \{0, 1, 2 \mid \,\},\ \ldots$$

and their negations

$$0 = \{\,|\,\},\ -1 = \{\,\mid 0\},\ -2 = \{\,\mid 0, -1\},\ -3 = \{\,\mid 0, -1, -2\},\ \ldots$$

and they play a major role in the theory of infinitesimals. Note that nimbers are their own negatives: $*m \cong -(*m)$, for all m.

Bouton's Theorem (Theorem I.1.1 on page 3) generalizes easily to the partizan context:

Theorem 4.7 (Nim-Addition Rule). *For any $a, b \in \mathbb{N}$, we have*

$$*a + *b = *c, \quad \textit{where} \quad c = a \oplus b.$$

Proof. $a \oplus b \oplus c = 0$, so Bouton's Theorem implies that $*a + *b + *c$ is a $\mathscr{P}$-position. But $*c = -(*c)$, so the conclusion follows. ◻

Up and Down

Recall from Section 1 that $\uparrow$ (pronounced "up") is the game

$$\uparrow = \{0 \mid *\}$$

and $\downarrow$ (pronounced "down") is its negative

$$\downarrow = \{* \mid 0\}.$$

We also write

$$\uparrow* = \uparrow + * \quad \text{and} \quad \downarrow* = \downarrow + *.$$

In Section 2 we computed the canonical form of $\uparrow*$,

$$\uparrow* = \{0, * \mid 0\}.$$

Certainly $\uparrow > 0$, since Left can win no matter who moves first. It is also dicotic, and therefore infinitesimal. We write

$$⇑ = 2 \cdot \uparrow, \quad ⤊ = 3 \cdot \uparrow, \quad ⟰ = 4 \cdot \uparrow, \quad \ldots,$$

pronounced *double-up*, *triple-up*, etc. Likewise,

$$⇓ = 2 \cdot \downarrow, \quad ⤋ = 3 \cdot \downarrow, \quad ⟱ = 4 \cdot \downarrow, \quad \ldots.$$

We'll also write

$$\uparrow n, \qquad \uparrow n*, \qquad \uparrow n{*}m,$$

as shorthand for

$$n \cdot \uparrow, \quad n \cdot \uparrow + *, \quad n \cdot \uparrow + *m,$$

respectively.

Just as numbers provide a natural scale for measuring the size of elements of $\mathbb{G}$, the multiples of $\uparrow$ provide a scale for measuring the size of infinitesimals. The following theorem is a sort of infinitesimal analogue of the Archimedean Principle (Theorem 3.7 on page 71).

Theorem 4.8. *For any infinitesimal $G \in \mathbb{G}$, there is some integer n such that $n \cdot \uparrow > G > n \cdot \downarrow$.*

Theorem 4.8 is an immediate corollary of the following stronger theorem.

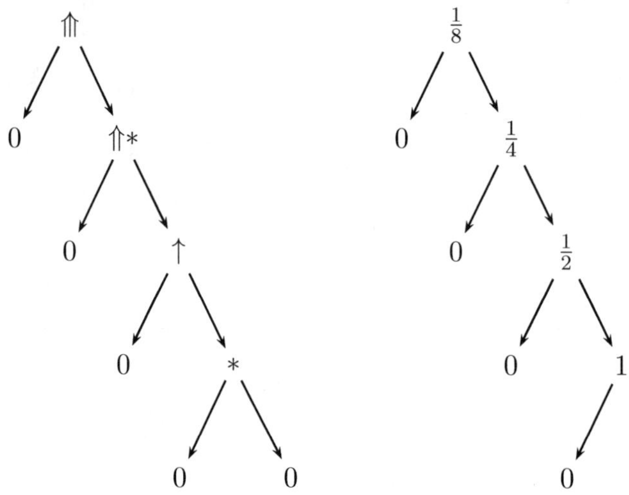

Figure 4.2. The canonical game trees of $\Uparrow$ and $\frac{1}{8}$.

Theorem 4.9. *Let G be a short game. If $R(G) \geq 0$ and $n \geq \mathrm{b}(G)+2$, then $G + {\uparrow}n > 0$ and $G + {\uparrow}n{*} > 0$.*

Proof. If $G = 0$, then the conclusion is trivial, since $\Uparrow$ and $\Uparrow{*}$ are both strictly positive. Otherwise, assume that G is in canonical form and consider $G + {\uparrow}n$. If Right moves to $G^R + {\uparrow}n$, then necessarily $L(G^R) \geq 0$. If G^R is a number, then $G^R \geq 0$, so Left can win by moving directly to G^R. Otherwise, there must be some G^{RL} with $R(G^{RL}) \geq 0$, so by induction Left can win by moving to $G^{RL} + {\uparrow}n$.

Suppose instead that Right moves to $G + {\uparrow}(n-1){*}$. If G is a number, then $G \geq 0$, so Left can win by moving directly to the positive game $G + {\uparrow}(n-1)$. Otherwise, we have $L(G) \geq R(G) \geq 0$, so there must be some G^L with $R(G^L) \geq 0$, and by induction Left can win by moving to $G^L + {\uparrow}(n-1){*}$.

This shows that Left can win $G + {\uparrow}n$ playing second; the argument for $G + {\uparrow}n{*}$ is identical. Therefore $G + {\uparrow}n \geq 0$ and $G + {\uparrow}n{*} \geq 0$, and since ${*} \not\geq 0$, both inequalities must be strict. □

There is a striking similarity between the canonical game trees of ${\uparrow}n$ and 2^{-n}. When n is odd, they are identical except for the single option ${*}^R = 0$; an example is shown in Figure 4.2. When n is even, the overall structure is still much the same (Exercise 4.1).

The Sums $\uparrow n{*}m$

Recall (from Section 1) that $\uparrow \not\gtrless *$, but $\Uparrow > *$. The following propositions describe the relationship between multiples of $\uparrow$ and higher-order nimbers.

Proposition 4.10. *If $m \geq 2$, then $\uparrow > *m$.*

Proof. Certainly $\uparrow \mathrel{|\!\!\rhd} *m$, since from $\uparrow - *m$ Left has a move to the positive game $\uparrow$. To see that $\uparrow \geq *m$, note that Right's moves from $\uparrow - *m$ are to $* - *m$ and $\uparrow - *a$ $(a < m)$. Since $m > 1$, the first of these is fuzzy; and Left can revert the second to $\uparrow$ (or to 0, if $a = 0$). □

Proposition 4.11. *If $n \geq 2$, then*

$$\uparrow n = \{0 \mid \uparrow(n-1)*\} \quad \textit{and} \quad \uparrow n* = \{0 \mid \uparrow(n-1)\}$$

in canonical form.

Proof. Put $G = \uparrow n$. From the definition of $\uparrow$, we certainly have

$$G = \{\uparrow(n-1) \mid \uparrow(n-1)*\}.$$

Left's move to $G^L = \uparrow(n-1)$ has a Right option $G^{LR} = \uparrow(n-2)*$, and since

$$G - G^{LR} = \Uparrow * > 0,$$

we see that G^L reverses through G^{LR}. By induction on n, we may assume that Left's only option from G^{LR} is 0, except in the case $n = 3$, when $G^{LR} = \uparrow *$ and Left also has a move to $*$. Therefore

$$G = \{0 \mid \uparrow(n-1)*\} \quad \text{or} \quad \{0, * \mid \uparrow(n-1)*\}.$$

But since $G > 0$, Left's option $*$ (if present at all) reverses out through 0. No further simplifications are possible, since $G^{RL} = 0 < G$.

Likewise, for $H = \uparrow n + *$, Right's incentive $\Delta^R(*) = *$ is dominated by $\Delta^R(\uparrow) = \uparrow *$, so by the Dominated Incentives Theorem we can write

$$H = \{G, \uparrow(n-1)* \mid \uparrow(n-1)\}.$$

Left's moves to G and $\uparrow(n-1)*$ reverse through $\uparrow(n-1)*$ and $\uparrow(n-2)$, repsectively. As before, Left's only options from any H^{LR} are to 0 and possibly $*$, and the rest of the argument is just as for G. □

The canonical forms of $\uparrow n{*}m$ with $m \geq 2$ are an easy generalization:

Proposition 4.12. *If $n \geq 1$ and $m \geq 2$, then*

$$\uparrow n{*}m = \{0 \mid \uparrow(n-1){*}(m \oplus 1)\}$$

in canonical form.

The proof is left as an exercise. Figure 4.3 on the following page summarizes the canonical forms of $\uparrow n{*}m$.

$$\begin{array}{rclrcl}
0 & = & \{\,\mid\,\} & * & = & \{0 \mid 0\} \\
\uparrow & = & \{0 \mid *\} & \uparrow* & = & \{0,* \mid 0\} \\
\Uparrow & = & \{0 \mid \uparrow*\} & \Uparrow* & = & \{0 \mid \uparrow\} \\
\Uparrow\!\!\!\uparrow & = & \{0 \mid \Uparrow*\} & \Uparrow\!\!\!\uparrow* & = & \{0 \mid \Uparrow\} \\
\uparrow n & = & \{0 \mid \uparrow(n{-}1)*\} & \uparrow n* & = & \{0 \mid \uparrow(n{-}1)\} \ (n \geq 2)
\end{array}$$

$$\uparrow n{*}m = \{0 \mid \uparrow(n{-}1){*}(m{\oplus}1)\} \ (m \geq 2)$$

Figure 4.3. The canonical forms of $\uparrow n{*}m$ for $n \geq 1$.

Tiny and Miny

Let's write $\uparrow$ in terms of its full game tree,

$$\uparrow = \{0 \,\|\, 0 \mid 0\},$$

and replace the rightmost 0 by something more favorable to Right, say,

$$G = \{0 \,\|\, 0 \mid -x\}$$

with x a positive number. It's still true that $G > 0$, since Left can get the last move no matter who moves first. However, G is *much* smaller than $\uparrow$, in the sense that

$$G \ll \uparrow.$$

To see this, consider the sum

$$n \cdot G + \downarrow$$

for some large value of n. Right, whenever it is his move, can play from some copy of G to $\{0 \mid -x\}$. Left cannot afford to respond in a different component, since otherwise Right will move to a position of value $-x + \epsilon$, with ϵ an infinitesimal. So Left must respond with a move from $\{0 \mid -x\}$ to 0. Right then selects another copy of G and repeats. When all copies of G have been exhausted, it will be Right's move on a position of value $\downarrow$ or $*$ (the latter if Left moved first and chose to play on $\downarrow$). In all cases Right will get the last move, so that

$$n \cdot G + \downarrow < 0.$$

Therefore G is quite small indeed, and so it is called *tiny-x*, denoted by $⧾_x$. Its negative is $⧿_x$, pronounced *miny-x*! The definition can be stated in slightly more generality:

Definition 4.13. For all $G \geq 0$, the games $⧾_G$ and $⧿_G$ are defined by

$$⧾_G = \{0 \,\|\, 0 \mid -G\} \quad \text{and} \quad ⧿_G = -(⧾_G) = \{G \mid 0 \,\|\, 0\}.$$

We can think of $⧾_x$ as an opportunity for Right to issue a threat of magnitude x. It confers a slight advantage to Left, who has the additional freedom to cancel the threat immediately. As the size of Right's threat increases, however, this advantage decreases rapidly—so rapidly, in fact, that for numbers x and y,

$$⧾_x \lll ⧾_y \quad \text{whenever} \quad x > y \geq 0.$$

The proof of this fact is left as an exercise (Exercise 4.4).

Remarkably, the sequence

$$⧾_1, ⧾_2, ⧾_3, \ldots$$

tends to 0 in a very real sense.

Theorem 4.14. *If $G > 0$, then $G > ⧾_n$ for some integer $n \geq 0$.*

Proof. Choose n such that $n + G^{RR} > 0$ for all G^{RR} (such an n necessarily exists by the Archimedean Principle, Theorem 3.7 on page 71), and consider

$$H = G + ⧿_n.$$

We will show that $H \geq 0$. Right's opening move to $G + 0$ clearly loses, by the assumption on G. But if Right moves instead to

$$G^R + ⧿_n,$$

then Left can play to

$$G^R + \{n \mid 0\}.$$

Right cannot afford to ignore this threat, since $G^{RR} + n > 0$ for all G^{RR}. So Right must revert to $G^R + 0$, which Left can win since $G^R \mathrel{|\!\triangleright} 0$. $\square$

So while the ups $n \cdot \uparrow$ are among the largest infinitesimals in the universe $\mathbb{G}$, the tinies $⧾_x$ are among the smallest. The remainder of this section explores the vast and mysterious hierarchy that lies between them.

Ordinal Sum

The **ordinal sum** is a useful operation for constructing new games, and in particular new infinitesimals. In the ordinal sum $G : H$, a player may move in either G or H (just as in the disjunctive sum), with the additional constraint that *any move on G completely annihilates the component H.* Formally:

Definition 4.15. Let G and H be short games. The **ordinal sum** $G : H$ is defined recursively by

$$G : H = \{G^L,\ G : H^L \mid G^R,\ G : H^R\}.$$

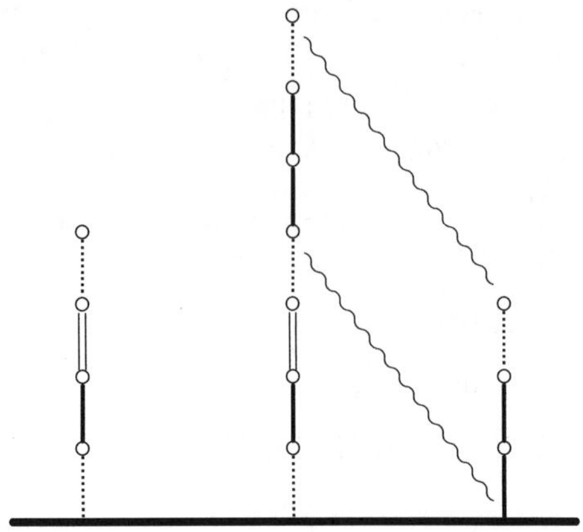

Figure 4.4. A natural visualization of ordinal sums by HACKENBUSH stalks.

Ordinal sums are perhaps most naturally visualized in terms of HACKENBUSH: if G and H are HACKENBUSH stalks, then $G : H$ is obtained by placing H atop G, as in Figure 4.4. This visualization makes the following fact obvious (cf. Theorem 4.7 on page 85).

Proposition 4.16. *For any $a, b \in \mathbb{N}$, we have*

$$*a : *b = *c, \quad \textit{where} \quad c = a + b.$$

Proof. The (Left or Right) options of $*a : *b$ are exactly

$$0,\ *,\ *2,\ \ldots,\ *(a-1),\ *a : 0,\ *a : *,\ \ldots,\ *a : *(b-1).$$

By induction on b we may rewrite these as

$$0,\ *,\ *2,\ \ldots,\ *(a-1),\ *(a+0),\ *(a+1),\ \ldots,\ *(a+b-1). \qquad \square$$

Proposition 4.17. *Let G, H, and H' be short games.*

(a) $(-G) : (-H) = -(G : H)$.

(b) $H \geq H'$ *if and only if* $G : H \geq G : H'$.

Proof. (a) is an easy induction.

(b) First suppose $H \geq H'$ and consider $(G : H) - (G : H')$. If Right moves to $(G : H^R) - (G : H')$, then $H^R \mathrel{|\!\!\rhd} H'$, so by induction

$$(G : H^R) - (G : H') \mathrel{|\!\!\rhd} 0.$$

The same argument applies if Right moves to $(G : H) - (G : (H')^L)$. Finally, if Right moves on either G or $-G$, Left has a mirror-image response on the other component.

Now suppose $(G : H) \geq (G : H')$ and consider $H - H'$. If Right moves to $H^R - H'$, then we have $(G : H^R) \mathrel{|\!\!\!>} (G : H')$, so by induction $H^R - H' \mathrel{|\!\!\!>} 0$. The same argument works if Right moves to $H - (H')^L$. □

Example. Proposition 4.17(b) shows that if $H = H'$, then $G : H = G : H'$. Conversely, $G = G'$ does *not* imply $G : H = G' : H$.

For example, $0 : H = H$ for all H, and $\{* \mid *\} = 0$, but

$$\{* \mid *\} : 1 = \{0, * \mid *\} = \uparrow.$$

This shows that the value of $G : H$ depends upon the form (and not just the value) of G.

A crucial fact about ordinal sums is that the value of $G : H$ is "close to" G, in the sense that they have the same order-relations with all games X that are not ancestors of G. Stated precisely:

Theorem 4.18 (Norton's Lemma). *Fix games G, H, and X, and suppose that no subposition of X is equal to G. Then*

$$o(G - X) = o\big((G : H) - X\big).$$

Proof. Certainly no subposition of X^L or X^R is equal to G, so we can proceed by induction on X and H. By symmetry, it suffices to prove that

$$G - X \geq 0 \iff (G : H) - X \geq 0.$$

First assume that $(G : H) - X \geq 0$. Then every $G^R - X \mathrel{|\!\!\!>} 0$ and every $(G : H) - X^L \mathrel{|\!\!\!>} 0$, since these are both Right options of $(G : H) - X$. By induction on X, the latter implies that every $G - X^L \mathrel{|\!\!\!>} 0$, so it follows that $G - X \geq 0$.

Next assume that $G - X \geq 0$. Consider the Right options of $(G : H) - X$. We know that $G^R - X \mathrel{|\!\!\!>} 0$, since it is also an option of $G - X$. For the same reason, each $G - X^L \mathrel{|\!\!\!>} 0$, so by induction on X we have $(G : H) - X^L \mathrel{|\!\!\!>} 0$. Finally, consider $(G : H^R) - X$. Since $G \neq X$, we must have $G - X > 0$, so either $G^L - X \geq 0$ or $G - X^R \geq 0$. By induction on X and H, the latter implies $(G : H^R) - X^R \geq 0$ as well. Thus in all cases, $(G : H^R) - X \mathrel{|\!\!\!>} 0$. We conclude that $(G : H) - X \geq 0$. □

Flowers

Definition 4.19. A **flower** is a game of the form

$$G = {*m} : n$$

with $m > 0$ and $n \in \mathbb{Z}$. We say G is a **blue (green, red) flower** if $n > 0$ ($n = 0$, $n < 0$). A sum of flowers is called a **flower garden**.

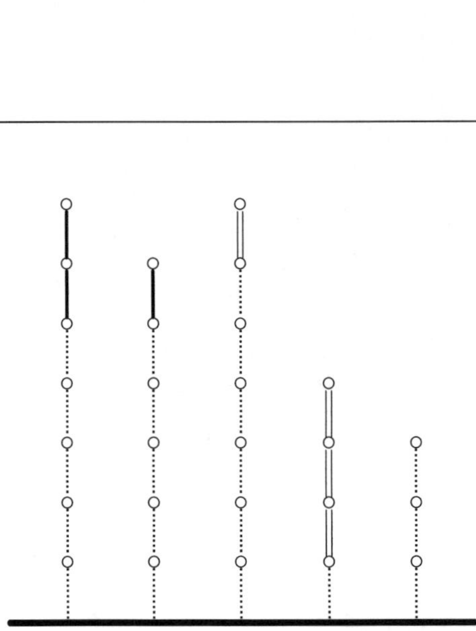

Figure 4.5. A flower garden with two blue flowers of stem length 5; red flowers of stem lengths 6 and 1; and a green flower equal to $*3$.

In HACKENBUSH every flower can be represented by a green **stem** capped by a blue or red **blossom**. A typical flower garden is shown in Figure 4.5.

Although in general flower gardens can be quite complicated, there are several special cases that are easily analyzed and that illustrate important aspects of the theory of infinitesimals. It turns out that the *number* of flowers of each color is the most relevant feature of a flower garden, and the structure of each flower is important only when the numbers of blue and red flowers differ by at most one. To be precise, define:

Definition 4.20. The **weight** $w(G)$ of a flower garden G is given by

$$w(G) = \big(\text{number of blue flowers in } G\big) - \big(\text{number of red flowers in } G\big).$$

Now if $w(G) \geq 1$ and it is Left's move, then she can always force a win, essentially by making arbitrary moves on the stems of red flowers until none remain and thereafter playing NIM on the remaining stems. The assumption $w(G) \geq 1$ guarantees that at least one blue flower remains after all red ones are eliminated, so that if the nim-sum of the remaining stems is 0, then Left can play safely on one of her blossoms. Here's the theorem:

Theorem 4.21 (The Two-Ahead Rule). *Let G be a flower garden.*

(a) *If $w(G) \geq 1$, then $G \mathrel{|\!\!\triangleright} 0$.*

(b) *If $w(G) \geq 2$, then $G > 0$.*

Proof. (a) If G contains at least one red flower, then there exists a G^L with $w(G^L) \geq 2$, obtained by making an arbitrary move on the stem of any red flower. By induction, we can assume $G^L > 0$, so that $G \mathrel{|\!\!\triangleright} 0$.

If G has no red flowers, then write

$$G = *m_1 : n_1 + \cdots + *m_k : n_k.$$

Here each $n_i \geq 0$, since there are no red flowers; and at least one $n_i > 0$, since $w(G) \geq 1$. Let

$$H = *m_1 + \cdots + *m_k.$$

Note that $G \geq H$, since G is obtained from H by introducing additional Left options. Moreover, there is some $G^L \geq H$, obtained by making an arbitrary move on any blossom. There are two cases.

Case 1: $H \mathrel{\|\!\rhd} 0$. Then we have $G \geq H \mathrel{\|\!\rhd} 0$. (Playing on G, Left can ignore the blossoms and just play NIM on the stems.)

Case 2: $H = 0$. Then $G^L \geq H \geq 0$, which also implies $G \mathrel{\|\!\rhd} 0$. (Playing on G, Left can make any move on any blossom and thereafter play NIM.)

(b) Any move by Right reduces the weight by at most one, so that $w(G^R) \geq 1$ for every G^R. By induction, each $G^R \mathrel{\|\!\rhd} 0$, so that $G \geq 0$. By (a), the inequality must be strict. □

A position in FLOWERS is a sum of games of the form $*m : 1$, $*m$, and $*m : -1$ (that is, a flower garden whose blossoms all have length 0 or 1). Despite substantial effort and several fascinating partial results (Exercises 4.18 and 4.19), FLOWERS remains unsolved.

Open Problem. Give an algorithm to determine the outcome of an arbitrary flower garden or show that the problem is NP-hard.

Uptimals

The flowers of the form $* : n$ are of particular interest. When $n = 0$ or 1, we obtain the familiar values

$$(* : 0) = \{0 \mid 0\} = *$$

and

$$(* : 1) = \{0, * \mid 0\} = \uparrow*.$$

For $n \geq 2$ we have a remarkable sequence of generalizations of $\uparrow*$ given by

$$(* : n) = \bigl\{0, \bigl(* : (n-1)\bigr) \bigm| 0\bigr\}.$$

Certainly this is an increasing sequence:

$$* < (* : 1) < (* : 2) < \cdots,$$

since on

$$G_n = \bigl(* : n\bigr) - \bigl(* : (n-1)\bigr)$$

Left has a move to 0. But the successive differences G_n must be small—much smaller than $\uparrow$, since the weight $w(G_n) = 0$ when $n \geq 2$. In fact the values G_n decrease rapidly as n increases, providing a deep window into the vast infinitesimal hierarchy between $\uparrow$ and $+_x$.

Definition 4.22. The infinitesimal $\uparrow^n$ (pronounced *up-n^{th}*) is given by

$$\uparrow^n = (* : n) - (* : (n-1)).$$

We also write

$$\uparrow^{[n]} = (* : n) - * = \uparrow + \uparrow^2 + \cdots + \uparrow^n$$

and

$$\uparrow^{[n]}* = (* : n) = \uparrow^{[n]} + *.$$

We write $\downarrow_n$ and $\downarrow_{[n]}$ for the negatives $-\uparrow^n$ and $-\uparrow^{[n]}$, respectively.

It should be noted that $\uparrow^n$ is *not* the "n^{th} power of $\uparrow$" in any meaningful sense (and in particular $\uparrow^2$ is *not* equal to $\uparrow \cdot \uparrow$ in the sense of Definition 7.6). What justifies the notation is the fact that

$$\uparrow^{n+1} \ll \uparrow^n,$$

which we'll establish momentarily.

Proposition 4.23. *We have*

$$\uparrow^{n+1} = \{0 \mid * : -n\} \quad \textit{and} \quad \uparrow^{[n+1]} = \{\uparrow^{[n]} \mid *\}$$

in canonical form.

Proof. The given forms are easily seen to be canonical, and the identities can be verified by showing that

$$G = \uparrow^{n+1} + \{* : n \mid 0\} = (* : (n+1)) + (* : -n) + \{* : n \mid 0\}$$

and

$$H = \uparrow^{[n+1]} + \{* \mid \downarrow_{[n]}\} = (* : (n+1)) + * + \{* \mid \downarrow_{[n]}\}$$

are second-player wins. □

Theorem 4.24. $\uparrow^{n+1} \ll \uparrow^n$ *for all* $n \geq 1$.

Proof. Consider the difference

$$G_k = \uparrow^n + k \cdot \downarrow_{n+1}.$$

We can write G_k as a flower garden:

$$G_k = (* : n) + (* : (-n+1)) + k \cdot (* : (n+1)) + k \cdot (* : -n).$$

If Left repeatedly removes red flowers, then by the Two-Ahead Rule Right can do no better than to remove blue ones. Eventually the position will reach

$$G_0 = (* : n) + (* : (-n+1))$$

with Left to move, which is positive. This shows that $G_k \mathrel{|\!\rhd} 0$. To see that $G_k \geq 0$, write G_k in canonical form:

$$G_k = \uparrow^n + k \cdot \{* : n \mid 0\}.$$

We know that $G_{k-1} \mathrel{|\!\!\triangleright} 0$, so Right's move on $\{* : n \mid 0\}$ loses. But if Right moves to

$$G_k^R = \big(* : (-n{+}1)\big) + k \cdot \{* : n \mid 0\},$$

then Left can revert to

$$\big(* : (-n{+}1)\big) + \big(* : n\big) + (k{-}1) \cdot \{* : n \mid 0\} = \uparrow^n + (k{-}1) \cdot \{* : n \mid 0\}$$

(or just to 0 if $k = 0$), eventually winning. □

Recall that $\uparrow > *m$ for all $m \geq 2$ (Proposition 4.10 on page 87). The following proposition shows that higher-order $\uparrow^n$ are confused with *every* nonzero nimber.

Proposition 4.25. *For all k and all $m \geq 1$, we have $k \cdot \uparrow^2 \not\gtrless *m$.*

Proof. Without loss of generality, we may assume $k \geq 0$. Write

$$G = k \cdot \uparrow^2 + *m = k \cdot (* : 2) + k \cdot (* : -1) + *m.$$

Now certainly $G \mathrel{|\!\!\triangleright} 0$ (since $G \geq *m$), so we must show that Right can win G playing first. But if Right repeatedly removes blue flowers, then by the Two-Ahead Rule, Left must respond by removing red ones. This ensures that the position $*m$ will be reached with Right to move. □

Definition 4.26. Denote by $\mathcal{U}$ the subgroup of $\mathbb{G}$ generated by

$$\{\uparrow^n : n \geq 1\}.$$

An **uptimal** is any element of $\mathcal{U}$.

By Theorem 4.24, $\mathcal{U}$ is totally ordered and free on the generators $\uparrow^n$, so every $G \in \mathcal{U}$ can be written uniquely in the form

$$G = \sum_{n \geq 1} a_n \cdot \uparrow^n \quad (a_n \in \mathbb{Z}).$$

It is often convenient to write G in **uptimal notation**:

$$G = 0.a_1 a_2 a_3 a_4 \cdots.$$

Negative a_n can be expressed using overbars, so for example

$$0.3\bar{5}4 \quad \text{is shorthand for} \quad \Uparrow - 5 \cdot \uparrow^2 + 4 \cdot \uparrow^3,$$

and combinations of ordinary numbers and $*$ can be written "before the dot," so that

$$\tfrac{3}{8}{*}.3\bar{5}4 \quad \text{is shorthand for} \quad \tfrac{3}{8} + * + 0.3\bar{5}4.$$

Since $\mathcal{U}$ is totally ordered, we can define intervals of $\mathcal{U}$ just as for $\mathbb{D}$. Then the **uptimal confusion interval** of G is the interval

$$\mathcal{UC}(G) = \{U \in \mathcal{U} : G \not\gtrless U\}.$$

Example. Let $U = 0.a_1a_2a_3\ldots$ in uptimal notation. We show that $U > *$ iff $a_n > 1$, where n is the least index with $a_n \neq 1$. By symmetry, this characterizes $\mathcal{UC}(*)$.

If such $a_n < 1$, then right away we have

$$U < \uparrow^{[n]} \mathrel{\lhd\!\!|} *.$$

Conversely, if such $a_n > 1$, then $U \geq \uparrow^{[n]} + \uparrow^n$. But on the sum

$$\uparrow^{[n]} + \uparrow^n + *$$

Right's only move of undominated incentive is to

$$\uparrow^{[n]} + (\downarrow_{[n-1]} + *) + * = \uparrow^n > 0,$$

so that $\uparrow^{[n]} + \uparrow^n > *$.

The Values $\{0 \mid +_x\}$

The previous example shows that the uptimals

$$*.1, \quad *.11, \quad *.111, \quad *.1111, \quad \ldots$$

are the largest elements of $\mathcal{U} + *$ that are confused with 0. The "next larger" values,

$$*.2, \quad *.12, \quad *.112, \quad *.1112, \quad \ldots,$$

have canonical forms

$$\{0 \mid \uparrow\}, \quad \{0 \mid \uparrow^2\}, \quad \{0 \mid \uparrow^3\}, \quad \{0 \mid \uparrow^4\}, \quad \ldots,$$

which suggests that they are "just barely" positive.

Between these two sequences lies another important hierarchy:

$$\{0 \mid +_1\}, \quad \{0 \mid +_2\}, \quad \{0 \mid +_3\}, \quad \{0 \mid +_4\}, \quad \ldots.$$

These games are close to $\uparrow*$, and their differences are smaller than any uptimal (see Exercises 4.11 and 4.12). A related sequence is given by

$$\{0 \mid +_x\}, \quad \{0 \parallel 0 \mid +_x\}, \quad \{0 \mid\mid\mid 0 \parallel 0 \mid +_x\}, \quad \ldots,$$

which is often abbreviated using the shorthand notation

$$\{0^1 \mid +_x\}, \quad \{0^2 \mid +_x\}, \quad \{0^3 \mid +_x\}, \quad \{0^4 \mid +_x\}, \quad \ldots.$$

The games in this sequence are close to

$$\uparrow*, \qquad \Uparrow, \qquad \text{⤊}*, \qquad \text{⟰}, \qquad \ldots.$$

Note that a summand of $*$ appears just in the odd-indexed terms in this sequence; this is because $\{0 \mid \Uparrow\}$ (for example) has value $\text{⤊}*$, not ⤊.

These infinitesimals play an important role in the theory of Go and some related games. They arise in situations where Right has a threat that is "several moves away"; the longer it takes for Right to make the threat, the greater the advantage to Left. The atomic weight theory in Section 7 will help to make all of these notions more precise.

Exercises

4.1 Draw the game tree of $\uparrow n$ when n is even, and compare with Figure 4.2 on page 86. Do the same for $\uparrow^{[n]}$ and $\uparrow^{[n]}*$.

4.2 Prove Proposition 4.12.

4.3 Determine the uptimal confusion intervals of $\{\uparrow \mid \downarrow *\}$ and $\{\uparrow \mid *, \uparrow\}$.

4.4 (a) Show that if x and y are numbers and $x > y > 0$, then $+_x \ll +_y$. Is the same true if x and y are replaced by arbitrary games?
(b) State necessary and sufficient conditions on $x, y, z \in \mathbb{D}$ such that
$$\{x \mid y\} + +_z = \{x + +_z \mid y + +_z\}.$$
(c) Let G and H be games with $G > H$ and $R(H) > 0$. State necessary and sufficient conditions on G and H such that $+_G \ll +_H$. Does $+_G < +_H$ necessarily imply $+_G \ll +_H$?

4.5 $\{* \mid *\} : n = \uparrow^{[n]}$ for all n.

4.6 If $G > 0$, then for some n we have $G > +_n$ and $G \mathrel{|\!\rhd} \{0 \mid +_n\}$.

4.7 True or false:
(a) If $G > 0$ and $G \ll 1$, then G is infinitesimal.
(b) If $G > 0$ is dicotic and $R(H) > 0$, then $G > +_H$.
(c) If $G > 0$ is dicotic, then $G > \uparrow^n$ for some n.
(d) If $G > \uparrow^{[n]}$ for all n, then necessarily $G > *$.

4.8 Determine the canonical forms of $k \cdot \uparrow^{[n]}$, for all $k, n \geq 1$.

4.9 (a) Show that the queenside of Figure I.4.3 (page 31) has value $\uparrow$ and the kingside has value $\Downarrow *$. Conclude that the position is a first-player win.
(b) Compute the values of each component of Figure 4.6, and determine its outcome class.

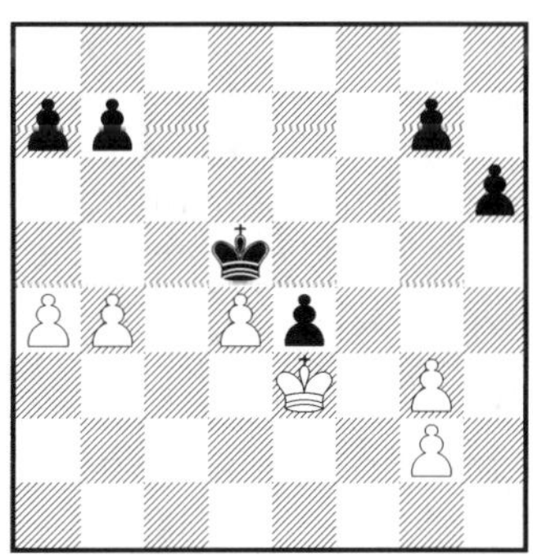

Figure 4.6. Popov–Dankov, Albena 1978.

4.10 Find all G such that $+_{+_{+_G}} = +_G$.

4.11 Let $G = \{0^k \mid \text{+}_x\}$, with $x > 0$ and $k \geq 1$. Show that for all n,
$$0 < G - k \cdot (\uparrow^{[n]}*) \ll \uparrow^n.$$

4.12 Let $y > x > 0$ and fix $k \geq 0$. Show that for every positive uptimal U,
$$0 < \{0^k \mid \text{+}_x\} - \{0^k \mid \text{+}_y\} < U.$$

4.13 Let G be a short game. Suppose that $H^L \leq H^R$ for every subposition H of G and every H^L and H^R. Determine all possible values for G.

4.14 Let Γ be the partizan subtraction game with $S_L = \{1\}$, $S_R = \{2, 3\}$ (cf. Exercise I.1.8 on page 15). Determine the value of a heap of n tokens, for all $n \geq 0$.

4.15 True or false: There exist distinct games $G_1, \ldots, G_k$, with each $R(G_i) > 0$, such that 0 can be expressed as a nontrivial linear combination
$$0 = a_1 \cdot \text{+}_{G_1} + a_2 \cdot \text{+}_{G_2} + \cdots + a_k \cdot \text{+}_{G_k} \quad \text{(each } a_i \in \mathbb{Z}\text{, each } a_i \neq 0\text{)}.$$

4.16 Express $(\uparrow m)^n$, $(\uparrow m)^{[n]}$, $(\uparrow^m)^n$, and $(\uparrow^m)^{[n]}$ as uptimals, for all $m, n \geq 1$.

4.17 (a) Prove the *Dicotic Translation Theorem*: If $L(G) > R(G)$ and H is dicotic, then $G + H = \{G^L + H \mid G^R + H\}$.
(b) Is the same true if H is an infinitesimal, but not necessarily dicotic?

4.18 Let G be a FLOWERS position with weight $w(G) = 0$. Show that:
(a) There exists an integer $m_0 \geq 0$ such that
$$G + *m \mathrel{|\!\rhd} 0 \quad \Leftrightarrow \quad m \geq m_0, \quad \text{for all } m \geq 0.$$
(b) If Right has a flower of stem length 1 and $G \mathrel{|\!\rhd} 0$, then chopping the flower of stem length 1 must be a winning move for Left.
(c) If Right has the (strictly) longest flower, then necessarily $G \mathrel{|\!\rhd} 0$.
(d) If Right has a flower of stem length $m = 2^k$ or $2^k + 1$, all of Left's flowers are shorter than m, and there are no green flowers, then $G > 0$. Is this necessarily true for any other values of m?

4.19 Let G_m be the FLOWERS position $(*(m+1) : 1) - (*m : 1)$. Draw the partial order of $\{G_m : 1 \leq m \leq 15\}$. For which m and n is $G_m \ll G_n$?

4.20 Let $\mathcal{S} \subset \mathbb{N}$ be a nonempty finite set. The **superstar** $\uparrow_{\langle \mathcal{S} \rangle}$ is defined by
$$\uparrow_{\langle \mathcal{S} \rangle} = \{0, *, *2, \ldots, *m \mid *a\}_{a \in \mathcal{S}}$$
where $m = \operatorname{mex}(\mathcal{S})$ and a ranges over the elements of $\mathcal{S}$. (Here $\operatorname{mex}(\mathcal{S})$ denotes the **min**imal **ex**cluded value of $\mathcal{S}$, as in Definition IV.1.1 on page 180.) Then a position in SUPERNIM is a sum of superstars and their negatives.
(a) $\uparrow_{\langle \mathcal{S} \rangle} \not\gtrless *a$ for $a \in \mathcal{S}$, and $\uparrow_{\langle \mathcal{S} \rangle} > *a$ for $a \notin \mathcal{S}$.
(b) Every FLOWERS position occurs in SUPERNIM, and the Two-Ahead Rule generalizes (with positive and negative superstars taking the place of blue and red flowers).
(c) If $\mathcal{T} \subset \mathbb{N}$ and there is some integer m such that $\mathcal{T} = \mathcal{S} \oplus m$, then for the *least* such m we have $\uparrow_{\langle \mathcal{T} \rangle} = \uparrow_{\langle \mathcal{S} \rangle} + *m$.

4.21 A **flowering tree** is a HACKENBUSH tree (cf. Exercise I.2.1 on page 21) that is completely green, except for a single blue or red leaf (the **blossom**). The path from the ground to the blossom is the **trunk**. Any move on the trunk yields a nimber, and the associated set of nim values is the **trunk set**.

Let G be a flowering tree with a blue blossom and trunk set $\mathcal{S}$. Prove that $G \not\gtrless *a$ for each $a \in \mathcal{S}$, and $G > *a$ for each $a \notin \mathcal{S}$. Prove furthermore that if no option of G has trunk set $\mathcal{S}$, then in fact $G = \uparrow_{\langle \mathcal{S} \rangle}$ (cf. Exercise 4.20).

4.22 What is the outcome of Figure I.1.3 on page 6? (Use Exercise 4.21.)

4.23 *Generalized uptimals.* Let $G = \{0 \mid G^R\}$ be a positive infinitesimal (so that each $G^R \mathrel{|\!\!\!>} 0$). The infinitesimals G^n and $G^{[n]}$ are given by

$$G^{[0]} = 0; \quad G^{[n+1]} = \{G^{[n]} \mid G^R\}; \quad G^{n+1} = \{0 \mid G^R - G^{[n]}\}.$$

(a) $G^{n+1} = G^{[n+1]} - G^{[n]}$ for all $n \geq 0$.
(b) $0 < G^{n+1} \ll G^n$ for all n.
(c) If $x > y > 0$, then $+_x \ll +_y^n$ for all n.
(d) What is the canonical form of $G^R + G^{[n]}$ (expressed recursively in n)?
(e) True or false: If $G \geq H$, then $G^{[n]} \geq H^{[n]}$ and $G^n \geq H^n$.

4.24 *Fractional uptimals.* For all positive $x \in \mathbb{D}$, define

$$\uparrow^{[x]} = (* : x) - * \quad \text{and} \quad \uparrow^{x+1} = \{0 \mid \downarrow_{[x]} *\}.$$

(a) If $x > y \geq 1$, then $\uparrow^x \ll \uparrow^y$.
(b) If $x > 1$ and $m \geq 1$, then $\uparrow^x \not\gtrless *m$.
(c) $\uparrow^{x+1} = \uparrow^{[n]} - \uparrow^{[x]}$, where $n = \lfloor x \rfloor + 1$.
(d) Part (c) shows that $\uparrow^{x+1} \neq \uparrow^{[x+1]} - \uparrow^{[x]}$, except when x is an integer. Investigate the differences $D_x = \uparrow^{[x+1]} - \uparrow^{[x]}$. How does D_x compare with each $\uparrow^y$?
(e) Show that $\uparrow^{[x]} = \uparrow : (x - 1)$ for $x \geq 1$. If $0 \leq x < 1$, show that $\uparrow^{[x]}$ is also of the form $\uparrow : y$ for some y. Determine y as a function of x.
(f) Let $\mathcal{U}^+$ be the group generated by $\{\uparrow^x : x \geq 1\}$. Show that $\mathcal{U}^+$ is totally ordered and free on the generators $\uparrow^x$.

4.25 A position in TOPPLING DOMINOES consists of a row of dominoes, each colored bLue or Red, represented as a string of L's and R's. On her turn, Left may "topple" any bLue domino; this removes the toppled domino, along with all dominoes to the east or to the west (Left's choice). For example, from LRLRR Left could move to either LR (toppling the middle domino eastward), RR (westward), RLRR (toppling the end domino westward), or the empty position (eastward). Right's moves are similar, toppling an R domino in either direction.

(a) $\text{L}^n = n$; $(\text{LR})^n = *n$ $(n \geq 0)$; and $\text{LR}^{n+2}\text{L} = +_n$ $(n \geq 0)$.
(b) Every $x \in \mathbb{D}$ is represented by a TOPPLING DOMINOES position. (For $x > 0$, define $G_x = G_y\text{LR}G_z$, where $x = \{y \mid z\}$ in canonical form.)
(c) Let X be a position whose value is a number $x > 0$. Prove that X contains no two consecutive R's.
(d) Fix a position X, and let Y be obtained by replacing every instance of R in X with RL. For example, if $X = \text{LRR}$, then $Y = \text{LRLRL}$. Prove that if X has value G, then LY has value $1 : G$.
(e) Prove that the representation in part (b) is unique. (Assume $x > 0$, show that $x = 1 : y$ for a unique number y, and proceed by induction, using (c) and (d).)
(f) Show furthermore that this position is a palindrome.

Notes

The games $\Uparrow n$, $*m$, $\uparrow^n$, and $\uparrow^{[n]}$ are fundamental to the theory of infinitesimals and are treated extensively in *Winning Ways* and ONAG. Although all sources agree on the notation for $*m$ and $\uparrow^n$, the meanings of $\Uparrow n$ and $\uparrow^{[n]}$ have been inconsistent in the literature. The game we call $\uparrow^{[n]}$ was denoted by $\Uparrow n$ in ONAG; elsewhere in the literature it has variously been written as $\uparrow[n]$ or $\uparrow^{\rightarrow n}$. Our notation reflects the modern convention, which is used by *cgsuite*.

Uptimal notation was introduced by Conway and Ryba in the 1980s. Very recently, Neil McKay has isolated the canonical forms of arbitrary uptimals. Specifically, McKay showed that if $a_n > 0$, then

$$0.a_1a_2\cdots a_n = \left\{*.a_1^+a_2^+\cdots a_j^+,\ 0.a_1a_2\cdots a_k \;\middle|\; *.a_1^-a_2^-\cdots a_j^-\right\},$$
$$*.a_1a_2\cdots a_n = \left\{0.a_1^+a_2^+\cdots a_j^+,\ *.a_1a_2\cdots a_k \;\middle|\; 0.a_1^-a_2^-\cdots a_j^-\right\},$$

where j is the largest $i \le n$ with $a_i < 0$ (or 0 if none exists) and

$$k = \begin{cases} n-1 & \text{if } a_n = 1; \\ \text{the largest } i \le n \text{ with } a_i \le 0 \text{ (or 0 if none exists)} & \text{otherwise.} \end{cases}$$

(Here we write a^+ and a^- as shorthand for $a+1$ and $a-1$, respectively.) These forms are canonical, except in certain cases where one of the two Left options reverses out. The full details are found in [**McK11**].

McKay also notes that uptimal notation can be extended by changing the "base." Let G be a positive infinitesimal of the form $\left\{0 \;\middle|\; G^R\right\}$. Then

$$\text{we write}\quad 0.a_1a_2\cdots a_n \ \ (\text{base } G) \quad \text{to mean} \quad \sum_{i=1}^{n} a_i \cdot G^i,$$

with G^i as defined in Exercise 4.23. McKay has offered the following pretty generalization of his results on uptimals.

Conjecture. Let G be a positive infinitesimal. Then writing all uptimals in base G,

$$0.a_1a_2\cdots a_n = \left\{-G^R + 0.a_1^+a_2^+\cdots a_j^+,\ 0.a_1a_2\cdots a_k \;\middle|\; G^R + 0.a_1^-a_2^-\cdots a_j^-\right\}.$$

Games of the form $\{0^k \mid +_x\}$ play an important role in the theory of Go [**BW94**]. A sweeping desirability theorem, due to Berlekamp, settles the strategies for sums of such games:

Theorem (Desirability Theorem). *Let G be a sum of games of the form $\{0^k \mid -x\}$ or $\{x \mid 0^k\}$ $(x > 0,\ k \ge 1)$, no two of which sum to 0. If $G \mathrel{\|\!\!\triangleright} 0$, then Left has a winning move on the component of highest* **desirability**, *where the components are ranked with decreasing desirability as follows:*

$$-_y,\ \ \{y \mid 0\},\ \ -_x,\ \ \{x \mid 0\},\ \ \{-_x \mid 0^k\},\ \ \{-_x \mid 0^{k+1}\},\ \ \{-_y \mid 0\},\ \ \textit{positive terms}$$

for $y > x > 0$.

See [**BW94**, pp. 68–69] for an elegant visual presentation of this theorem. David Moews has generalized the Desirability Theorem to include games of the form $\{0^k \mid -x\}^{[n]}$ (cf. Exercise 4.23); details and proofs are found in [**Moe96a**].

Exercise 4.9: (a) Noam Elkies; (b) Fabian Mäser.

Exercise 4.10: Conway.

Exercises 4.18, 4.19, 4.21, and 4.22: Berlekamp.

Exercise 4.23: David Moulton.

Exercise 4.24: Neil McKay.

Exercise 4.25: Fink, Nowakowski, Siegel, and Wolfe.

5. Temperature

The mean value (Definition 3.25 on page 79) is one answer to the question:

Who is ahead, and by how much?

A related question has great practical relevance:

How much is a move worth?

In a sum such as

$$\{8 \mid -8\} + \{2 \mid -2\}$$

one might say that a move on ± 8 is worth eight points, while a move on ± 2 is worth only two. The component ± 8 is said to be *hotter* than ± 2, because it is more urgent to play there.

It turns out that every game G has a numerical **temperature** $t(G)$ that quantifies the urgency of a move on G. Skilled play on a sum will often, but not always, concentrate on the hottest component. And just as

$$m(G+H) = m(G) + m(H),$$

it will always be the case that

$$t(G+H) \leq \max\big\{t(G), t(H)\big\}.$$

In this section, we'll state the definition of temperature along with some related notions. Later, in Chapter VII, we'll pursue a more detailed study of temperature theory and its implications.

Cooling

There are several equivalent ways to measure the temperature of G. The classical method is to **cool** G by reducing the value of a move.

Consider an example such as $G = \pm 8$. If we impose a penalty of two points to move on G, both players will still be eager to play there, since the payoff (eight points) exceeds the penalty. As the size of the penalty increases, play on G becomes increasingly less attractive. If the penalty exceeds 8, then neither player will care to move on G at all, since playing there now incurs a net loss.

A penalty of exactly 8 is borderline: numerically speaking, there is neither a net gain nor a net loss to playing on G, and G behaves just like the infinitesimal $*$. It turns out that every game has a unique "freezing point" at which it cools to an infinitesimal. This observation motivates the following definitions.

Definition 5.1. Let $t \geq -1$. We define G **cooled by** t, denoted by G_t, as follows. If G is equal to an integer n, then simply $G_t = n$. Otherwise, put

$$\tilde{G}_t = \{G^L_t - t \mid G^R_t + t\}.$$

Then $G_t = \tilde{G}_t$, *unless* there is some $t' < t$ such that $\tilde{G}_{t'}$ is infinitesimally close to a number x. In that case, fix the *smallest* such t' and put

$$G_t = x.$$

Definition 5.2. The **temperature** of G, denoted by $t(G)$, is the smallest $t \geq -1$ such that G_t is infinitesimally close to a number.

Our definition of G_t (and hence also of $t(G)$) is not really proper, since there is no guarantee that such a "smallest" t' necessarily exists. We'll soon prove that it does, though, and we'll take care not to *assume* that G_t is well-defined until after we've finished the proof, except in specific examples where its existence is clear.

Intuitively, one may think of G_t as a variant of G in which each player must pay a penalty of t in order to move. Sometimes this penalty is so large that good players will not wish to play on G at all. In such cases we must be careful to replace G_t by an equitable number. (If $t < 0$, then the "penalty" is actually a reward, but the theory is identical.)

Example. Let $G = \pm 8$. We have $G_t = \pm(8 - t)$ whenever $0 \leq t \leq 8$. For example, $G_4 = \pm 4$, $G_6 = \pm 2$, and $G_8 = *$. In particular, G_8 is infinitesimally close to 0, so $t(G) = 8$ and $G_t = 0$ for all $t > 8$.

Example. Let $G = \{2 \mid \pm 1\}$. If $0 \leq t \leq 1$, then $G_t = \{2 - t \parallel 1 \mid -1 + 2t\}$. In particular, we have $G_1 = \{1 \parallel 1 \mid 1\} = 1 + \uparrow$, which is infinitesimally close to 1. Therefore $t(G) = 1$ and $G_t = 1$ for all $t > 1$.

Example. Let $x = m/2^n$ in lowest terms, with $m \neq 0$, and assume that x is in canonical form. We will show that $t(x) = -1/2^n$. When $n = 0$, this is by definition. Otherwise, $x^L = (m-1)/2^n$ and $x^R = (m+1)/2^n$ (cf. Theorem 3.6 on page 70). By induction, $t(x^L) < -1/2^n$ and $t(x^R) < -1/2^n$. Thus if $t = -1/2^n$, then $x^L_t = x^L$ and $x^R_t = x^R$, and we have

$$\{x^L - t \mid x^R + t\} = \{x \mid x\} = x + *,$$

which is infinitesimally close to the number x.

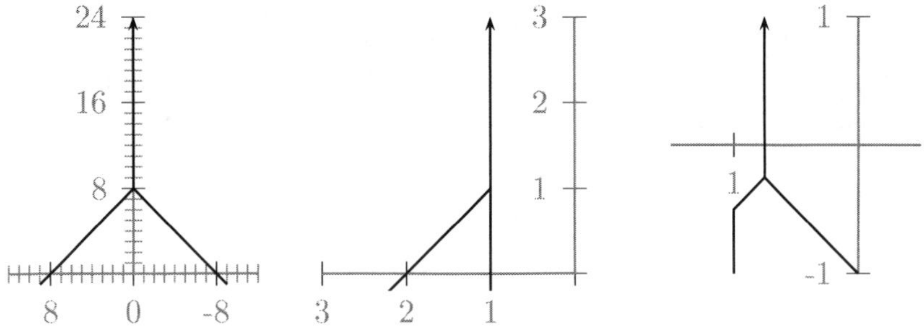

(a) ± 8 (b) $\{2 \mid \pm 1\}$ (c) $\frac{3}{4}$

Figure 5.1. A few simple thermographs.

Thermographs

Consider the Left and Right stops of G_t, as functions of t.

Definition 5.3. The **Left** and **Right scores** of G, denoted by $L_t(G)$ and $R_t(G)$, are defined by

$$L_t(G) = L(G_t) \quad \text{and} \quad R_t(G) = R(G_t),$$

regarded as functions of t. The **thermograph** of G is the ordered pair $(L_t(G), R_t(G))$ (again, regarded as a pair of functions of t).

We still haven't proven that G_t is well-defined, so for the time being we'll discuss $L_t(G)$ and $R_t(G)$ in the context of specific examples.

Example. If $G = \pm 8$, then $G_t = \pm(8 - t)$ whenever $t \le 8$, and $G_t = 0$ for all $t > 8$. Therefore

$$L_t(G) = \begin{cases} 1 - t & \text{if } t \le 8; \\ 0 & \text{otherwise;} \end{cases} \qquad R_t(G) = \begin{cases} -8 + t & \text{if } t \le 1; \\ 0 & \text{otherwise.} \end{cases}$$

Figure 5.1 illustrates the thermographs of our three prior examples. Such diagrams use the following plotting conventions:

- $L_t(G)$ and $R_t(G)$ are plotted together on the same axes. It must always be the case that $L_t(G) \ge R_t(G)$, for all t (Proposition 3.18(a) on page 76), so it will always be clear which trajectory is which.
- The independent variable t is plotted along the vertical axis, and the dependent variables $L_t(G)$ and $R_t(G)$ are plotted along the horizontal axis. This inverts the usual (Cartesian) convention but conforms to the intuition of temperatures "rising" as t increases.
- The numbers on the horizontal axis are laid out positive-to-negative. This also inverts the Cartesian convention but ensures that positive numbers (favorable to *Left*) appear on the *left* side of the diagram.

- Figure 5.1(a) and (b) only show values for $t \geq 0$, even though the thermograph is defined for all $t \geq -1$. This is because the structure of a thermograph for $t < 0$ is completely determined by its behavior "just below" 0 (Exercise 5.9). In Figure 5.1(c), we've chosen to extend the diagram down to $t = -1$ in order to illustrate the behavior of numbers (which have negative temperature).

In each of these examples, the thermograph is topped by an infinite vertical mast, because G_t is a number for all sufficiently large t. This number is called the **mast value** of G, denoted by G_∞. It turns out that every game has such a mast value and that it's always equal to $m(G)$:

$$G_\infty = m(G).$$

Also observe that when G_t is not a number, we have

$$(G_t)^L = G_t^L - t \quad \text{and} \quad (G_t)^R = G_t^R + t,$$

so that

$$L_t(G) = \max_{G^L} \big(R_t(G^L) - t\big) \quad \text{and} \quad R_t(G) = \min_{G^R} \big(L_t(G^R) + t\big).$$

This suggests a recursive relationship between the thermographs of G and its options. For an example, see Figure 5.2. The Right score of G^L is "tilted" by subtracting t, yielding the **scaffold** (light gray trajectory) in the leftmost diagram of Figure 5.2. Separately, the Left score of G^R is tilted by adding t. The two scaffolds are then plotted together on the same axes (middle diagram) to obtain the thermograph of G. The scaffolds' point of intersection becomes the base of the vertical mast.

To prove that G_t exists, we'll introduce a recursive calculus for defining auxiliary functions $\lambda_t(G)$ and $\rho_t(G)$ (the **walls** of G), essentially generalizing the example from Figure 5.2. We'll then show (inductively) that the walls

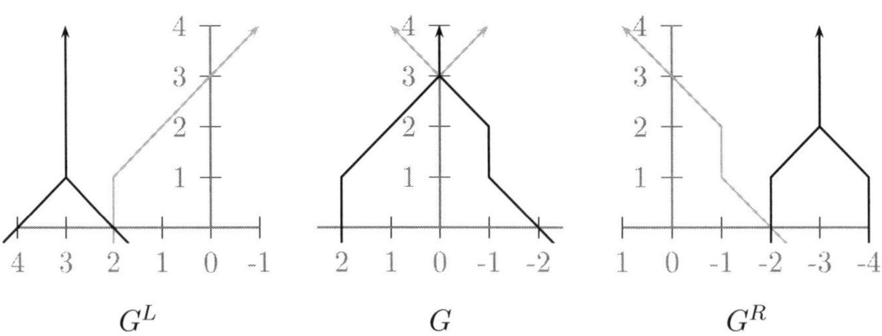

Figure 5.2. The thermograph of $G = \{4 \mid 2 \;|||\; 0 \mid -2 \;||\; -4 \mid -6\}$ constructed from its options, $G^L = \{4 \mid 2\}$ and $G^R = \{0 \mid -2 \;||\; -4 \mid -6\}$.

are precisely equal to the scores $L_t(G)$ and $R_t(G)$, which implies that the latter are well-defined. Since $\lambda_t(G)$ and $\rho_t(G)$ are easy to compute, we also obtain an algorithm for calculating $m(G)$ and $t(G)$.

Rational Trajectories

The elements of the thermographic calculus are rational-valued functions over the interval $[-1, \infty[$. Examples include the Left and Right walls of the thermographs from Figure 5.1. It is convenient to frame the discussion in a more general setting, allowing arbitrary intervals for the domain and values over all of $\mathbb{Q}$ (rather than just $\mathbb{D}$).

For the purposes of this discussion, a **closed interval** will be a subset of $\mathbb{Q}$ of the form $[x, y]$ or $[x, \infty[$, for some $x, y \in \mathbb{Q}$. Note that we are specifically including intervals that are half-open and unbounded above: $[x, \infty[$ is called a "closed interval" even though ∞ is not properly a member of the set. For convenience of notation and discussion, we will also write $[x, \infty]$ as a synonym for $[x, \infty[$.

Definition 5.4. A **line segment** on a closed interval $\mathcal{I} \subset \mathbb{Q}$ is a function $f : \mathcal{I} \to \mathbb{Q}$ given by

$$f(t) = at + b$$

for some $a, b \in \mathbb{Q}$.

Definition 5.5. Let $\mathcal{I} \subset \mathbb{Q}$ be a closed interval. A function $f : \mathcal{I} \to \mathbb{Q}$ is a **trajectory** if $\mathcal{I}$ is the union of finitely many adjacent closed intervals,

$$\mathcal{I} = [t_1, t_2] \cup [t_2, t_3] \cup \cdots \cup [t_{k-1}, t_k]$$

(possibly with $t_k = \infty$), such that f is a line segment on each $[t_i, t_{i+1}]$.

If $t_k = \infty$, then the **mast** of f is the segment on $[t_{k-1}, \infty]$. If the mast has slope 0, then its constant value is called the **mast value** of f.

A value $t \in \mathcal{I}$ is a **junction point** of f if f changes slope at t.

Note that every trajectory is continuous: individual line segments are certainly continuous, and the definition of trajectory requires that successive segments coincide at their endpoints. The trajectories of primary interest are walls of thermographs, for which the domain $\mathcal{I} = [-1, \infty]$.

Theorem 5.6 (Intermediate Value Theorem). *Let $f : \mathcal{I} \to \mathbb{Q}$ be a decreasing trajectory, and suppose that $f(t_1) \geq 0 \geq f(t_2)$ for some $t_1 < t_2$. Then there is a* smallest *value of $t \in [t_1, t_2]$ such that $f(t) = 0$.*

Proof. Since f is composed of just finitely many line segments, we can choose t_1 and t_2 so that

$$f(t) = at + b \quad \text{for } t \in [t_1, t_2],$$

for some $a, b \in \mathbb{Q}$. If $a = 0$, then $f(t_1) = f(t_2) = 0$, so $t = t_1$ suffices. Otherwise $a < 0$, and we put

$$t = t_1 - \frac{f(t_1)}{a}.$$

Then

$$at + b = at_1 - f(t_1) + b = 0$$

and so if $t' < t$, then $at' + b > 0$. Since $f(t_2) \leq 0$, this implies $t \in [t_1, t_2]$, and hence $f(t) = at + b = 0$. □

Proposition 5.7. *If $f, g : \mathcal{I} \to \mathbb{Q}$ are trajectories, then so are $-f$, $f + g$, $\max\{f, g\}$, and $\min\{f, g\}$.*

Proof. It suffices to assume that f and g are line segments, since the general case can be obtained by restricting to each subinterval over which f and g are both line segments and then amalgamating the results.

So assume $f(t) = at + b$ and $g(t) = ct + d$ for $t \in \mathcal{I}$. Then certainly

$$(-f)(t) = -at - b \qquad \text{and} \qquad (f + g)(t) = (a + c)t + (b + d).$$

In particular, $f - g$ is a trajectory. If $f - g$ is strictly positive, then $\max\{f, g\} = f$. If $f - g$ is strictly negative, then $\max\{f, g\} = g$. Suppose instead that there are t_1 and t_2 with

$$(f - g)(t_1) \geq 0 \geq (f - g)(t_2).$$

Without loss of generality we can assume $t_1 < t_2$ (interchanging f and g if necessary). By the Intermediate Value Theorem, there is a $t_0 \in [t_1, t_2]$ with $f(t_0) = g(t_0)$. Then

$$\max\{f, g\}(t) = \begin{cases} at + b & \text{if } t \in [t_1, t_0]; \\ ct + d & \text{if } t \in [t_0, t_2], \end{cases}$$

and the proof for $\min\{f, g\}$ is the same. □

The Thermographic Calculus

We can now describe the thermographic calculus and prove that it's correct.

Definition 5.8. Let G be a game. We define trajectories $\lambda_t(G)$ and $\rho_t(G)$ (the **walls** of G), for $t \geq -1$, as follows. If G is equal to an integer n, then $\lambda_t(G) = \rho_t(G) = n$ for all t. Otherwise, first define the **scaffolds** $\tilde{\lambda}_t(G)$ and $\tilde{\rho}_t(G)$ by

$$\tilde{\lambda}_t(G) = \max_{G^L} \left(\rho_t(G^L) - t\right) \quad \text{and} \quad \tilde{\rho}_t(G) = \min_{G^R} \left(\lambda_t(G^R) + t\right).$$

Then $\lambda_t(G) = \tilde{\lambda}_t(G)$ and $\rho_t(G) = \tilde{\rho}_t(G)$, *unless* there is some $t' < t$ such that $\tilde{\lambda}_{t'}(G) = \tilde{\rho}_{t'}(G)$. In that case, let $x = \tilde{\lambda}_{t'}(G)$ for the *smallest* such t', and put

$$\lambda_t(G) = \rho_t(G) = x.$$

Note the similarity to Definition 5.1. However, $\lambda_t(G)$ and $\rho_t(G)$ are automatically well-defined, since the "smallest" t' in Definition 5.8 is guaranteed to exist by the Intermediate Value Theorem.

Theorem 5.9. *For all G:*

(a) *Every segment of $\lambda_t(G)$ has slope 0 or -1.*

(b) *Every segment of $\rho_t(G)$ has slope 0 or $+1$.*

(c) *$\lambda_t(G)$ and $\rho_t(G)$ have masts of slope 0, with the same mast value.*

Proof. All of these assertions are trivial when G is equal to an integer. Otherwise, by induction we may assume that every $\lambda_t(G^R)$ satisfies (a) and every $\rho_t(G^L)$ satisfies (b). Since $\tilde{\lambda}_t(G)$ is derived from $\max(\rho_t(G^L))$ by translation by $-t$, each of its segments must have slope 0 or -1, and likewise every segment of $\tilde{\rho}_t(G)$ has slope 0 or $+1$. Moreover, $\lambda_t(G)$ is obtained from $\tilde{\lambda}_t(G)$ by introducing a new segment of slope 0, so this proves (a), and likewise (b).

Now consider some G^R. By induction on (c), every $\rho_t(G^L)$ has a mast of slope 0, so $\tilde{\lambda}_t(G)$ has a mast of slope -1. Likewise, $\tilde{\rho}_t(G)$ has a mast of slope $+1$. Therefore $\tilde{\lambda}_{t_0}(G) \le \tilde{\rho}_{t_0}(G)$ for some t_0, and Definition 5.8 ensures that $\lambda_t(G)$ is constant for $t > t_0$ (and likewise for $\rho_t(G)$). $\square$

Theorem 5.10. *For all G:*

(a) *G_t is well-defined, for all $t \in \mathbb{D}$ with $t \ge -1$.*

(b) *$L_t(G) = \lambda_t(G)$ and $R_t(G) = \rho_t(G)$, for all $t \in \mathbb{D}$ with $t \ge -1$.*

(c) *There is some t such that G_t is a number.*

Proof. This is trivial when G is equal to an integer. Otherwise, by induction on G we may assume that all G^L_t and G^R_t are well-defined, with

$$R_t(G^L) = \rho_t(G^L) \quad \text{and} \quad L_t(G^R) = \lambda_t(G^R).$$

Therefore

$$\tilde{\lambda}_t(G) = \max_{G^L}\left(R_t(G^L) - t\right) \quad \text{and} \quad \tilde{\rho}_t(G) = \min_{G^R}\left(L_t(G^R) + t\right).$$

Now

$$\tilde{G}_t = \left\{G^L_t - t \;\middle|\; G^R_t + t\right\},$$

so $\tilde{G}_t$ is numberish *if and only if* $\tilde{\lambda}_t(G) \le \tilde{\rho}_t(G)$. Now consider the trajectory f given by

$$f(t) = \tilde{\lambda}_t(G) - \tilde{\rho}_t(G).$$

The mast of f has slope -2, so there must be some t with $f(t) \le 0$. By the Intermediate Value Theorem, there is necessarily a smallest such t, say t_0. Moreover, $t_0 \in \mathbb{D}$ since the slope of every segment of f is an integer of magnitude ≤ 2. Therefore t_0 is the smallest temperature with $\tilde{G}_{t_0}$ infinitesimally close to a number, say x.

It follows that G_t is well-defined for all t, with $G_t = \tilde{G}_t$ for $t \le t_0$ and $G_t = x$ for $t > t_0$. This proves (a) and (c).

Now when $t \le t_0$, we have

$$\lambda_t(G) = \tilde{\lambda}_t(G) = L(\tilde{G}_t) = L(G_t)$$

and for $t > t_0$, we have $\lambda_t(G) = x = L(G_t)$. So in all cases $\lambda_t(G) = L_t(G)$, and likewise $\rho_t(G) = R_t(G)$. This proves (b) and the theorem. □

In light of Theorem 5.10, we'll drop the notation $\lambda_t(G)$ and $\rho_t(G)$ in favor of $L_t(G)$ and $R_t(G)$. We'll also write

$$\tilde{L}_t(G) = \max_{G^L} \left(R(G^L) - t\right) \quad \text{and} \quad \tilde{R}_t(G) = \min_{G^R} \left(L(G^R) + t\right)$$

so that by Theorem 5.10, we have $\tilde{L}_t(G) = \tilde{\lambda}_t(G)$ and $\tilde{R}_t(G) = \tilde{\rho}_t(G)$.

Some basic properties of thermographs emerge directly from the thermographic calculus:

Theorem 5.11. *For all G:*

(a) *$L_t(G)$ and $R_t(G)$ are piecewise linear in t. Every segment of $L_t(G)$ has slope 0 or -1, and every segment of $R_t(G)$ has slope 0 or $+1$.*

(b) *$L_t(G)$ is decreasing in t, and $R_t(G)$ is increasing in t.*

(c) *$L_t(G) \ge R_t(G)$ for all t, with equality if and only if $t \ge t(G)$.*

(d) *$L_t(G) \ge L(G) - t$ and $R_t(G) \le R(G) + t$, for all t.*

Proof. These are immediate consequences of Theorems 5.9 and 5.10. □

Another important consequence of the preceding analysis:

Proposition 5.12. *For all G and $t \ge -1$:*

(a) *If $\tilde{L}_t(G) \ge \tilde{R}_t(G)$, then $L_t(G) = \tilde{L}_t(G)$ and $R_t(G) = \tilde{R}_t(G)$.*

(b) *If $\tilde{L}_t(G) \le \tilde{R}_t(G)$, then $L_t(G) = R_t(G) = G_\infty$.*

Proof. Theorem 5.10(b) shows that $L_t(G) = \lambda_t(G)$ and $R_t(G) = \rho_t(G)$, so this is essentially a restatement of Definition 5.8. □

Later we'll see that $G_\infty = m(G)$, so that G_∞ can be replaced by $m(G)$ in Proposition 5.12(b).

Properties of Cooling

We can now prove several key results. For all G, H, t, and u:

- Cooling is a homomorphism: $(G+H)_t = G_t + H_t$.
- Cooling respects inequalities: $G \geq H$ implies $G_t \geq H_t$.
- Successive cooling operations commute: $(G_t)_u = G_{t+u}$.
- Mean equals mast: $G_\infty = m(G)$.
- Temperatures are submaximal: $t(G+H) \leq \max\{t(G), t(H)\}$.

We'll need one lemma before proceeding.

Lemma 5.13. *If $G+H$ is equal to a number but G is not, then there is some G^L with*

$$R(G^L + H) = G + H.$$

Proof. Choose G^L with $R(G^L) = L(G)$. Then

$$\begin{aligned} R(G^L + H) &= R(G^L - G + G + H) \\ &\geq R(G^L) + R(-G) + G + H \\ &= L(G) - L(G) + G + H. \end{aligned}$$

Since $L(G+H) \geq R(G^L + H)$ (cf. Proposition 3.18(b) on page 76) and $L(G+H) = G+H$, we must have equality. □

Theorem 5.14. *For all games G and H and all $t \geq -1$, we have*

$$(G+H)_t = G_t + H_t.$$

Proof. The proof is by induction on G and H. If both are equal to integers, the conclusion is obvious, so assume at least one is not. There are two cases.

Case 1: $t \leq \max\{t(G), t(H)\}$. We first show that

$$G_t + H_t = (G+H)_t^\sim.$$

(Here and throughout the sequel, we'll write $G_t^\sim$ as a synonym for $\tilde{G}_t$, for clarity.) We can assume without loss of generality that $t \leq t(G)$, so that $G_t = \tilde{G}_t$. If also $t \leq t(H)$, then $H_t = \tilde{H}_t$, and by induction

$$\begin{aligned} G_t + H_t &= \{G_t^L + H_t - t, G_t + H_t^L - t \mid G_t^R + H_t + t, G_t + H_t^R + t\} \\ &= \{(G^L + H)_t - t, (G + H^L)_t - t \mid (G^R + H)_t + t, (G + H^R)_t + t\} \\ &= (G+H)_t^\sim. \end{aligned}$$

Suppose instead that $t > t(H)$. Then H_t is necessarily a number. Since G_t is not equal to a number, there must be some G^L such that $L(G_t) = R(G_t^L) - t$, so for every H^L we have

$$\begin{aligned} R(G_t^L + H_t) &= R(G_t^L) + H_t && (H_t \text{ is a number}) \\ &= L(G_t) + t + H_t \\ &= L(G_t + H_t) + t && (H_t \text{ is a number}) \\ &> R(G_t + H_t^L). \end{aligned}$$

Therefore

$$(G+H)_t^{\sim} = \{G_t^L + H_t - t \mid G_t^R + H_t + t\}$$

since the options of the form $G_t + H_t^L - t$ and $G_t + H_t^R + t$ are dominated. Now since H_t is a number, the Number Translation Theorem gives

$$(G+H)_t^{\sim} = \{G_t^L - t \mid G_t^R + t\} + H_t = G_t + H_t.$$

If $G_t + H_t$ is not equal to a number, then we are done. Otherwise, Lemma 5.13 guarantees a G^L and G^R with $R(G_t^L + H_t - t) \geq L(G_t^R + H_t + t)$. Thus necessarily $(G+H)_t^{\sim} = (G+H)_t$.

Case 2: $t > \max\{t(G), t(H)\}$. Put $t' = \max\{t(G), t(H)\}$. Then $G_{t'}$ and $H_{t'}$ are infinitesimally close to numbers, say x and y, respectively. Furthermore, by Case 1, we have

$$G_{t'} + H_{t'} = (G+H)_{t'}.$$

Therefore $(G+H)_{t'}$ is infinitesimally close to $x+y$, and it follows that $G_t = x$, $H_t = y$, and $(G+H)_t = x+y$. □

Theorem 5.15. *For all games G and H and all $t \geq -1$:*

(a) *If $G \geq H$, then $G_t \geq H_t$.*

(b) *If $G \,|\!\!\triangleright\, H$, then $G_t + t \,|\!\!\triangleright\, H_t$, provided that $t > -1$.*

Proof. By Theorem 5.14, it suffices to assume that $H = 0$. We prove (a) and (b) together, by simultaneous induction on G.

(a) Suppose that $G \geq 0$. Since $R(G) \geq 0$, we have $R_t(G) \geq 0$ for all t, so if G_t is a number, then we are done. Otherwise, $G_t = \{G_t^L - t \mid G_t^R + t\}$. Since each $G^R \,|\!\!\triangleright\, 0$, we have $G_t^R + t \,|\!\!\triangleright\, 0$ by induction on (b), and therefore $G_t \geq 0$.

(b) Suppose that $G \,|\!\!\triangleright\, 0$. If G is equal to an integer, then $G \geq 1$, so that $G_t + t = G + t > 0$.

So assume that G is not equal to an integer. Then some $G^L \geq 0$, so by induction on (a) we have $G_t^L \geq 0$. If $t \leq t(G)$, then $G_t^L - t$ is an option of G_t, and we are done. Otherwise, put $t' = t(G)$. Then by the preceding argument we have $G^L_{t'} + t' \mathrel{|\!\!\triangleright} 0$. But $t' < t$ and $G^L_{t'}$ is infinitesimally close to G^L_t, so necessarily $G^L_t + t \mathrel{|\!\!\triangleright} 0$. □

Theorem 5.16. *For all games G and H and all $t \geq -1$ and $u \geq 0$, we have*

$$(G_t)_u = G_{t+u}.$$

Proof. If G_t is equal to a number, then $(G_t)_u = G_t = G_{t+u}$.

Otherwise, $(G_t)_u$ (resp. G_{t+u}) depends only on the values of $(G_t)^\sim_u$ (resp. $\tilde{G}_{t+u}$) for $u \geq 0$, and moreover

$$\begin{aligned}
(G_t)^\sim_u &= \{G^L_t - t \mid G^R_t + t\}_u \\
&= \{(G^L_t - t)_u - u \mid (G^R_t + t)_u + u\} \\
&= \{G^L_{t+u} - (t+u) \mid G^R_{t+u} + (t+u)\} \\
&= \tilde{G}_{t+u}.
\end{aligned}$$

□

Theorem 5.17. $G_t = m(G)$ *for all* $t > t(G)$.

Proof. $G_{t(G)}$ is infinitesimally close to a number x, and hence $G_t = x$ for all $t > t(G)$. It therefore suffices to prove that $x = m(G)$.

Fix $n > 0$ and $t > t(G)$; by Theorem 5.14 we have

$$(n \cdot G)_t = n \cdot G_t = nx.$$

But $L_t(n \cdot G) \geq L(n \cdot G) - t$ and $R_t(n \cdot G) \leq R(n \cdot G) + t$, so

$$L(n \cdot G) - t \leq nx \leq R(n \cdot G) + t$$

and therefore

$$\frac{L(n \cdot G)}{n} - \frac{t}{n} \leq x \leq \frac{R(n \cdot G)}{n} + \frac{t}{n}.$$

Since $t/n \to 0$ as $n \to \infty$, this shows that $x = m(G)$. □

Theorem 5.18. *For all G and H,*

$$t(G + H) \leq \max\{t(G), t(H)\}.$$

Moreover, if $t(G) \neq t(H)$, then necessarily $t(G + H) = \max\{t(G), t(H)\}$.

Proof. Let $t = \max\{t(G), t(H)\}$. Then G_t and H_t are numberish. Since $(G + H)_t = G_t + H_t$, this shows that $(G + H)_t$ is also numberish, so that $t(G + H) \leq t$.

Moreover, suppose that $t(G) < t(H)$. Then for any u with $t(G) \leq u < t$, we have that G_u is numberish but H_u is not. Therefore $(G + H)_u$ is *not* numberish, and so $t(G+H) > u$. This shows that $t(G+H)$ is exactly equal to t, and the same argument works if $t(G) > t(H)$. □

Cold	$\Leftrightarrow$	$t(G) < 0$	$\Leftrightarrow$	G is a number	$\Rightarrow$	$L(G) = R(G)$
Tepid	$\Leftrightarrow$	$t(G) = 0$	$\Leftrightarrow$	G is numberish, but not a number	$\Rightarrow$	$L(G) = R(G)$
Hot	$\Leftrightarrow$	$t(G) > 0$	$\Leftrightarrow$	G is not numberish	$\Leftrightarrow$	$L(G) > R(G)$

Figure 5.3. Hot, Cold, or Tepid?

Hot, Cold, and Tepid

It is useful to divide $\mathbb{G}$ into three **temperature classes**, based on the sign of $t(G)$.

Definition 5.19. We say that G is **cold** if $t(G) < 0$; **tepid** if $t(G) = 0$; or **hot** if $t(G) > 0$.

The games in each temperature class have quite different properties. The following proposition provides familiar characterizations of each.

Proposition 5.20. *For all G:*

(a) *G is cold if and only if G is equal to a number.*

(b) *G is tepid if and only if G is infinitesimally close to a number (but not equal to a number).*

(c) *G is hot if and only if $L(G) > R(G)$.*

Proof. (a) If $G = m/2^n$, then $t(G) = -1/2^n < 0$, as observed on page 102. Conversely, if G is not equal to a number, then since $G = G_0$ it follows that G_t is not infinitesimally close to a number for any $t < 0$. Therefore $t(G) \geq 0$.

(b) and (c) are immediate from (a) and the definition of G_0. □

Figure 5.3 summarizes the properties of each temperature class.

Heating

Just as G can be cooled by imposing a penalty, it can be heated up by giving the players an extra incentive to move.

Definition 5.21. For all games G and T, we define G **heated by** T by

$$\int^T G = \begin{cases} G & \text{if } G \text{ is equal to a number;} \\ \left\{ T + \int^T G^L \;\middle|\; -T + \int^T G^R \right\} & \text{otherwise.} \end{cases}$$

When a game H is cooled by some temperature t, some of its subpositions might freeze to numbers. When this happens, the combinatorial structure of the frozen subpositions is lost, and only their mean values are preserved. Therefore for any particular choice of G, there are numerous distinct values of H for which $H_t = G$. Here's a simple example:

Example. Let $G = \{1 \mid -1\}$. Then $H_2 = J_2 = G$, where

$$H = \{3 \mid -3\} \quad \text{and} \quad J = \{3, \{4 \mid 2\} \parallel -3\}.$$

However it is clear that $H \neq J$, since on $J - H$ Left can win by moving to $\{4 \mid 2\} - H$.

As this example demonstrates, there are many "reasonable" choices for $\int^t G$. The one we've adopted for the definition is the simplest such choice, requiring the least amount of additional information. But the presence of the "others" means that heating is *not* a homomorphism and it's *not* independent of form.

Example. $G = H$ need not imply $\int^T G = \int^T H$. Let $G = \{1 \mid 0\}$ and $H = \{1, * \mid 0\}$. Then clearly $G = H$, since the extra option $*$ is dominated, but

$$\int^2 G = \{3 \mid -2\} \quad \text{whereas} \quad \int^2 H = \{3, \{4 \mid 0\} \parallel -2\}.$$

However it *is* true that the temperature of $\int^t G - \int^t H$ must be less than t (Exercise 5.12) in a sense, all the uncertainty about the value of $\int^t G$ resides in the temperature region through which G is heated.

Although heating is defined for all games T, the most common usage is the case where T is a positive number. In this case heating acts on temperatures much as expected. The following proposition summarizes some of its natural properties.

Proposition 5.22. *Let G be a game and let $t, u \geq 0$ be numbers.*

(a) $\int^t \int^u G = \int^{t+u} G$.

(b) *If $t \leq u$, then* $\left(\int^u G\right)_t = \int^{u-t} G$.

(c) *If $t \geq u$, then* $\left(\int^u G\right)_t = G_{t-u}$.

(d) *If G is not equal to a number, then* $t(\int^t G) = t(G) + t$.

The proof is left as an exercise.

Temperature Can Be Misleading!

In a sum of games, it is *not* necessarily true that the best move is on the hottest component. For example, consider the sum

$$G = \int^{100} *3 + \int^{99} *5 + \int^{98} *8.$$

Left, playing first on G, should prefer to move on the component of temperature 98, to

$$G^L = \int^{100} *3 + \int^{99} *5 + \left(98 + \int^{98} *6\right).$$

From the study of NIM, we can see that Left will get the last move on G, collecting at least an extra 98 points. By declining to move on the hottest component, she's given up only two points. Left might be forced to do this several times in order to win the NIM battle; but the game will certainly last fewer than 98/2 moves, no matter how it's played.

Contrast with the following position:

$$H = \int^{100} *3 + \int^{99} +*5 + \int^{1} *8.$$

Now the best move is on the component of temperature 99, to

$$H^L = \int^{100} *3 + \left(99 + \int^{99} *3\right) + \int^{1} *8.$$

On H it's most critical to get the last move at temperature 99, so the correct strategy is to play NIM just on the two hottest components.

We'll return to this example in Chapter VII.

Overheating

G can also be **overheated** by extending the heating operation all the way down to temperature -1.

Definition 5.23. For all games G and T, we define G **overheated by** T by

$$\oint^T G = \left\{T + \oint^T G^L \,\middle|\, -T + \oint^T G^R\right\}$$

(even when G is a number).

Overheating does not invert cooling, but it enjoys many useful properties that ordinary heating lacks.

Theorem 5.24. *Fix a value of T. Then:*

(a) *If $G \geq H$, then $\oint^T G \geq \oint^T H$.*

(b) $\oint^T G + \oint^T H = \oint^T (G + H)$.

(c) *If* $0 \le T \lhd\!\!| \; 1$, *then* $\oint^T n = n$ *for all integers* n.

Proof. For (a), Left can follow the same strategy on $\oint^T G - \oint^T H$ that she uses on $G - H$, canceling out copies of T and $-T$ in pairs. (b) is a fairly trivial induction. For (c), assuming $n > 0$ we have

$$\oint^T n = \{T + (n-1) \mid \}$$

by induction. Since $n - 1 \le T + (n-1) \lhd\!\!| \; n$, this is equal to n by the Simplicity Theorem. □

Berlekamp has further generalized the overheating operator, as follows.

Definition 5.25. For all games G, S, and T, we define G **overheated from** S **to** T by

$$\oint_S^T G = \begin{cases} n \cdot S & \text{if } G \text{ is equal to an integer } n; \\ \left\{ T + \oint_S^T G^L \,\middle|\, -T + \oint_S^T G^R \right\} & \text{otherwise.} \end{cases}$$

Proposition 5.26. *If* s *is an integer and* $s - 1 \le T \lhd\!\!| \; s$, *then*

$$\oint^T G = \oint_s^T G.$$

Proof. By induction on G. It suffices to assume that G is equal to an integer, since otherwise the two definitions are identical. We can also assume G to be in canonical form, since the value of $\oint^T G$ is independent of the form of G, and the value of $\oint_s^T G$ depends only on the fact that G is equal to an integer. For $G = 0$ the result is trivial, so assume instead that $G > 0$. Then

$$\begin{aligned} \oint^T G &= \left\{ T + \oint^T (G-1) \,\middle|\, \right\} \\ &= \left\{ T + \oint_s^T (G-1) \,\middle|\, \right\} && \text{(by induction)} \\ &= \left\{ T + (G-1) \cdot s \,\middle|\, \right\}. \end{aligned}$$

But $G \cdot s - 1 \le T + (G-1) \cdot s \lhd\!\!| \; G \cdot s$, by the assumption on T, so by the Simplicity Theorem

$$\{T + (G-1) \cdot s \mid \} = G \cdot s.$$

□

Thermal Dissociation

Definition 5.27. A **thermal dissociation** for a short game G is an expression of the form

$$G = m(G) + \int^{t_1} \epsilon_1 + \int^{t_2} \epsilon_2 + \cdots + \int^{t_n} \epsilon_n,$$

where $t_1 > t_2 > \cdots > t_n \geq 0$ are numbers and $\epsilon_1, \epsilon_2, \ldots, \epsilon_n$ are nonzero infinitesimals in canonical form. The numbers $t_1, t_2, \ldots, t_n$ are the **critical temperatures** of G, and the infinitesimal ϵ_i is the **particle released by** G at temperature t_i. We refer to ϵ_1, ϵ_2, ... as the **first particle**, **second particle**, ... of G.

We will show that every short game G admits a unique thermal dissociation, so that it is safe to talk about *the* thermal dissociation, critical temperatures, and particles of G. First an example:

Example. Let $G = \{4 \mid 2 \mid\mid\mid 0 \mid -2 \mid\mid -4 \mid -6\}$ (cf. Figure 5.2 on page 104). Then by direct calculation,

$$G_t = \begin{cases} 0 & \text{if } t > 3; \\ \{3 - t \mid -3 + t\} & \text{if } 2 < t \leq 3; \\ \{3 - t \mid\mid -1 \mid -5 + 2t\} & \text{if } 1 < t \leq 2; \\ \{4 - 2t \mid 2 \mid\mid\mid -t \mid -2 + t \mid\mid -4 + t \mid -6 + 3t\} & \text{if } t \leq 1. \end{cases}$$

In particular, $m(G) = 0$ and $t(G) = 3$. Now $G_3 = *$. To compute the next lower critical temperature, we let

$$H = G - \int^3 * = G \pm 3$$

and (using *cgsuite*, say) calculate $t(H) = 2$, and

$$H_2 = \{\{2 \mid 0\}, \{2 \mid *\} \mid\mid \{0 \mid -2*\}, \{* \mid -2*\}\}.$$

Finally, we let

$$J = H - \int^2 \{\{2 \mid 0\}, \{2 \mid *\} \mid\mid \{0 \mid -2*\}, \{* \mid -2*\}\}$$

and (again using *cgsuite*) find that $J = \pm 1 = \int^1 *$. Therefore

$$G = \int^3 * + \int^2 \{\{2 \mid 0\}, \{2 \mid *\} \mid\mid \{0 \mid -2*\}, \{* \mid -2*\}\} + \int^1 *$$

is the thermal dissociation of G.

The proof that thermal dissociations exist relies on the following bit of machinery.

Definition 5.28. Let $x \in \mathbb{D}$. A short game G is **x-based** if $m(H)$ is an integer multiple of x, for every subposition H of G.

If G and H are x-based, then so is $G+H$, since mean values are additive. The following lemmas introduce key properties of x-based games that help establish the existence of thermal dissociations.

Lemma 5.29. *Suppose that G is x-based. Then $t(G)$ is also an integer multiple of x.*

Proof. We'll show that every junction point of the thermograph of G has coordinates that are integer multiples of x.

By induction, we can assume that this is true for all G^L and G^R. In particular, the trajectories $R_t(G^L)$ have junction points whose coordinates are integer multiples of x; therefore so do the trajectories $R_t(G^L) - t$. Now recall that

$$\tilde{L}_t(G) = \max_{G^L} R_t(G^L) - t.$$

Therefore every junction point P of $\tilde{L}_t(G)$ is either a junction point of some $R_t(G^L) - t$ or a point of intersection of two trajectories $R_t(G^{L_1}) - t$ and $R_t(G^{L_2}) - t$. But in the latter case, P is the intersection of two lines, which necessarily have slopes 0 and -1 (since these are the only possible slopes for trajectories of the form $R_t(G^L) - t$) and intercepts that are integer multiples of x. The coordinates of any such P must also be integer multiples of x.

Now every junction point of $L_t(G)$ is also a junction point of $\tilde{L}_t(G)$, except for the point $(t(G), m(G))$. But since G is x-based, we know that $m(G)$ must be an integer multiple of x; hence $t(G)$ must be as well. It follows that the junction points of $L_t(G)$ have coordinates that are integer multiples of x, and the argument for $R_t(G)$ is the same. □

Lemma 5.30. *If G is x-based, then so is its canonical form. Moreover, if t is an integer multiple of x, then G_t and $\int^t G$ are also x-based.*

Proof. Removing a dominated option from a subposition H of G, or bypassing a reversible one, does not affect the value of $m(H)$. This shows that the canonical form of G is x-based.

Now if G_t is equal to a number, then it must be equal to $m(G)$, which is an integer multiple of x. Otherwise, every option of G_t has the form $G^L_t - t$ or $G^R_t + t$. By induction, G^L_t and G^R_t are x-based, and they certainly remain so after translation by an integer multiple of x. Since $m(G_t) = m(G)$, this shows that G_t is x-based. A similar argument works for $\int^t G$. □

Theorem 5.31. *Every short game G admits a unique thermal dissociation.*

Proof. Fix any $x \in \mathbb{D}$; we'll first prove that every x-based game G admits a thermal dissociation, by induction on $t(G)/x$. This suffices to prove existence, since every game must certainly be x-based for some value of x (for example, one could take x to be the greatest common divisor of $m(H)$, as H ranges over the subpositions of G).

Put $t = t(G)$. Then G_t is infinitesimal and $G_t \neq 0$. Let ϵ be the canonical form of G_t, and put

$$J = G - \int^t \epsilon.$$

Then by linearity of cooling, we have $J_t = G_t - \epsilon = 0$, so that $t(J) < t$.

Now since G is x-based, Lemma 5.29 implies that t must be an integer multiple of x. Moreover, $\int^t \epsilon$ is obtained from G by successively cooling by t, passing to the canonical form, and then reheating by t; therefore Lemma 5.30 implies that $\int^t \epsilon$ is also x-based. It follows that J is x-based.

Therefore, we can assume by induction on $t(G)/x$ that J has a thermal dissociation, say,

$$J = m(J) + \int^{t_2} \epsilon_2 + \cdots + \int^{t_n} \epsilon_n,$$

with $t_2 > \cdots > t_n$. Since $t > t(J)$ and $m(G) = m(J)$, it follows that

$$G = m(G) + \int^{t_1} \epsilon_1 + \int^{t_2} \epsilon_2 + \cdots + \int^{t_n} \epsilon_n$$

is a thermal dissociation for G, with $t_1 = t$ and $\epsilon_1 = \epsilon$.

For uniqueness, suppose also that

$$G = m(G) + \int^{u_1} \zeta_1 + \int^{u_2} \zeta_2 + \cdots + \int^{u_n} \zeta_n$$

with $u_1 > u_2 > \cdots > u_n \geq 0$ and each ζ_i a nonzero infinitesimal in canonical form. Then $G_{u_1} = m(G) + \zeta_1$, so necessarily $u_1 = t(G) = t_1$ and $\zeta_1 = \epsilon_1$. Since ϵ_1 and ζ_1 are in canonical form, we have $\zeta_1 \cong \epsilon_1$, so that

$$\int^{t_1} \epsilon_1 = \int^{u_1} \zeta_1.$$

Uniqueness follows by induction on n. □

Example: AMAZONS

AMAZONS is a combinatorial game played on a rectangular grid. In the starting position on a 10×10 board, shown in Figure 5.4(a), we see four **black Amazons** (controlled by Left) facing four **white Amazons** (controlled by

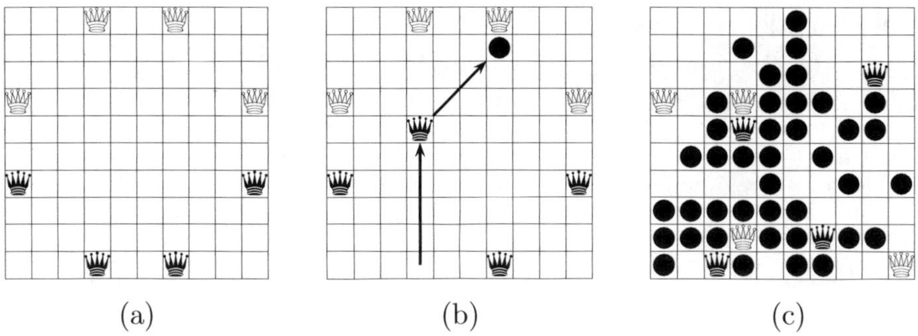

Figure 5.4. (a) The starting position of 10×10 AMAZONS; (b) a typical opening move for Left; (c) an endgame position.

Right). A move for Left consists of the following two actions, which must be taken in order:

(i) First, move any black Amazon as a CHESS queen: any number of squares in any direction (orthogonally or diagonally). It must pass entirely over empty squares and land on an empty square, known as the **landing square**. Capture is not permitted.

(ii) Then *fire an arrow*: place a black stone on any square that is a queen's move away from the landing square. The target square is permanently **burned**: no Amazon may land on it or cross over it, nor any arrow fired through it, for the remainder of the game.

The game ends when a player is unable to move (because all of her Amazons are surrounded on all eight sides by burned squares or other Amazons), and whoever makes the last move wins.

AMAZONS is fundamentally a territorial game: a competent player will strive to constrain the movement of her opponent's Amazons, carving out space for her own use. A typical opening move is shown in Figure 5.4(b). As the board fills up with burned squares—such as in Figure 5.4(c)—it decomposes into independent subregions and can be written as a disjunctive sum.

AMAZONS was invented in 1988 by Walter Zamkauskas and has gradually gained currency as an important example of a classical combinatorial game. Each move strictly increases the number of burned squares on the board, so that AMAZONS is a short game; and it fits well into the disjunctive theory, due to the endgame modularity.

The full 10×10 game is nearly intractable, but simpler positions are a source of useful insights into the temperature theory. Consider the following position on a $2 \times n$ board:

$G =$ 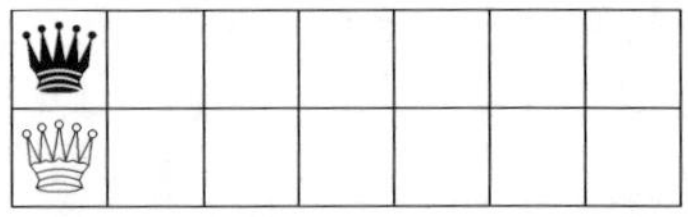 .

Left has an "obvious" opening move, blockading Right:

$G^{L_1} =$ 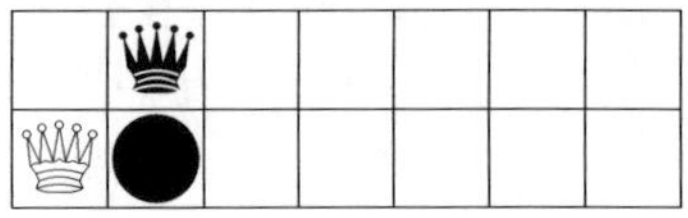 .

From this position Left has ten available moves and Right only one, so its value is 9 (this can easily be confirmed by showing that $G^{L_1} - 9$ is a second-player win). But there is another option that, somewhat surprisingly, is canonically relevant:

$G^{L_2} =$  .

The reason G^{L_2} is canonical is that if Left is permitted a consecutive move, she can move to a position of value 10, trapping Right tightly in the corner:

$G^{L_2L} =$ 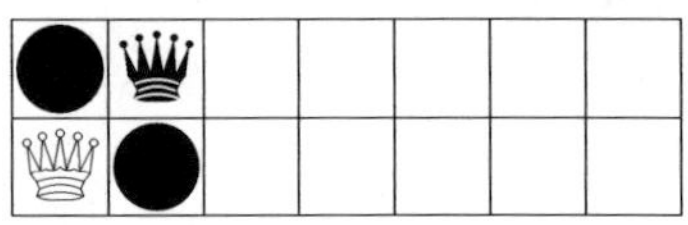 .

Of course G^{L_2} is a risky move, in the sense that Right can now threaten to escape in any of a variety of ways, say

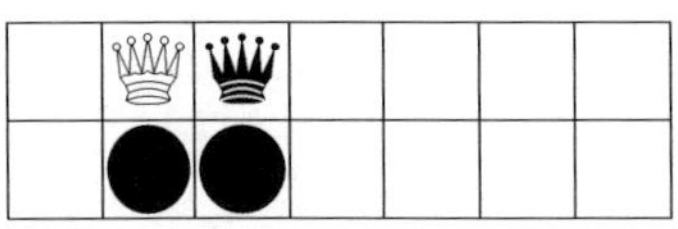 or  .

The first of these has value 6; the second is another "risky" move that creates more canonical confusion. From this point on the difficulties multiply; according to *cgsuite*, the full canonical form of G has more than 200 distinct stopping positions.

However, it's not hard to show that all of these extra canonical options are thermographically irrelevant. Consider G^{L_2}. Certainly $\tilde{L}_t(G^{L_2}) = 10 - t$, since 10 is the maximum possible stop for Left (using up all the available territory). But $\tilde{R}_t(G^{L_2}) \leq 6 + t$, since G^{L_2} has a Right option equal to 6. Since the thermographic intersection of these trajectories has a mast at 8, this gives an upper bound $m(G^{L_2}) \leq 8$, and in particular $R_t(G^{L_2}) \leq 8$ for all t.

Since $G^{L_1} = 9$, we certainly have $R_t(G^{L_1}) = 9$ for all t, so this shows that G^{L_2} cannot contribute to the thermograph of G. The same argument works for any other move by Left: for any position $G^L \neq G^{L_1}$, Left's maximum upside is 10, but Right can revert to a position whose value is ≤ 6.

We conclude that the thermograph of G is entirely determined by G^{L_1} and the symmetric Right option G^{R_1}. Therefore:

$$G \quad \text{has the same thermograph as} \quad \pm 9.$$

The argument generalizes easily to show that the analogous position on a $2 \times n$ board has the same thermograph as $\pm(2n-5)$, for all $n \geq 3$. This is a nice illustration of the power of the temperature theory: the canonical forms of such positions explode in complexity as n increases, but a thermographic analysis provides simple and clear insights.

Exercises

5.1 Let $G = \{51 \pm 50 \parallel \{0, \pm 50 \mid -200\}\}$. Describe the thermograph of $n \cdot G$ for $n = 1, 2, \ldots, 51$.

5.2 Let $\mu = \frac{1}{2}(L(G) + R(G))$ and $\tau = \frac{1}{2}(L(G) - R(G))$.
(a) If G is not equal to a number, then $t(G) \geq \tau$ and $|m(G) - \mu| \leq t(G) - \tau$.
(b) If $t(G) = \tau$, then $m(G) = \mu$; if $m(G) = L(G)$, then $t(G) \geq 2\tau$.

5.3 (a) Let $G = \{4 \mid 1\}$. Calculate $\tilde{G}_t$ for all t. For which t does $\tilde{G}_t = G_t$?
(b) For all G, there is a number $\delta > 0$ such that
$$\tilde{G}_t = m(G) \quad \text{whenever} \quad t(G) < t \leq t(G) + \delta.$$
(c) Does $G = H$ necessarily imply $\tilde{G}_t = \tilde{H}_t$?

5.4 If G is not equal to an integer, then there exist G^L and G^R such that
$$m(G^L) \geq m(G) + t(G) \quad \text{and} \quad m(G^R) \leq m(G) - t(G).$$

5.5 If G is not equal to an integer and $t = t(G)$, then G_t is not equal to a number.

5.6 For any two numbers $x \geq y \geq 0$, there exist games G and H such that $t(G) = t(H) = x$ and $t(G + H) = y$.

5.7 (a) Suppose that $G = \{A \mid B\}$ and $t(A), t(B) < \frac{1}{2}(m(A) - m(B))$. Then
$$t(G) = \frac{m(A) - m(B)}{2} \quad \text{and} \quad m(G) = \frac{m(A) + m(B)}{2}.$$
(b) If the hypothesis is weakened to allow $t(A) = \frac{1}{2}(m(A) - m(B))$, do either (or both) of the conclusions still hold?
(c) What if we allow both $t(A) = t(B) = \frac{1}{2}(m(A) - m(B))$?

5.8 Suppose that the only numbers that occur as subpositions of G are integers.
(a) Show that if $m(G) = a/2^b$ in lowest terms, then $t(G) \geq 1 - 1/2^b$.
(b) Determine all possible values for $t(G)$.

5.9 Let a be the slope of $L_t(G)$ just below $t = 0$. Prove that the values of $L_t(G)$ for $t \leq 0$ are completely determined by a and $L(G)$.

5.10 $L_t(G) \leq \lceil L(G) \rceil$ and $R_t(G) \geq \lfloor R(G) \rfloor$ for all $t \geq -1$ (cf. Theorem 5.11(d)).

5.11 Every noninteger G has at least one Left incentive Δ such that $t(\Delta) \geq t(G)$.

5.12 (a) If $t \geq 0$ and G is not equal to a number, then $t(\int^t G) = t + t(G)$.
(b) If $G = H$, then $t(\int^t G - \int^t H) < t$.
(c) If $G = H$ and G and H are dicotic, then $\int^t G = \int^t H$.

5.13 Let $s, t \geq 0$. Then s is a critical temperature of G if and only if $s + t$ is a critical temperature of $\int^t G$. Moreover, $\int^t G$ has no critical temperatures on the interval $[0, t[$.

5.14 Suppose that $t(G) = t(H) = t$, and let ϵ and ζ be the first particles of G and H, respectively. Show that $t(G + H) = t$ if and only if $\epsilon + \zeta \neq 0$.

5.15 Is the operator $\oint_{1*}^{1}$ a homomorphism? Is it invariant of form? How about $\oint_{1}^{1}$?

5.16 (a) Does the value of $\int^T G$ depend on the form of T?
(b) Does the value of $\oint_S^T G$ depend on the form of S and/or T?

5.17 True or false: If x and $t \geq 0$ are numbers, $H = x + G$, and G and H are in canonical form, then $\int^t H = x + \int^t G$.

5.18 (a) Show that $\oint_0^0 G$ is dicotic for all G.
(b) If x is a number, show that $\oint_0^0 x$ is an uptimal, and express it in uptimal notation. (Reduce to the case $0 < x < \frac{1}{2}$, and consider the binary expansion of x.)
(c) YELLOW–BROWN HACKENBUSH is a restricted form of ordinary BLUE–RED HACKENBUSH. Left may remove any yeLLow edge, provided that her move disconnects *at least one* brown edge and *no other* yellow edges. Right may remove any bRown edge, with the converse restrictions. (In particular, neither player can move from any *monochromatic* component.) Determine the complete solution for YELLOW–BROWN HACKENBUSH stalks.

5.19 (a) If $L_t(G)$ or $R_t(G)$ changes slope at temperature t, then t is necessarily a critical temperature of G.
(b) Give an example of a game G with the same thermograph as ± 2, such that $t = 1$ is a critical temperature for G.

5.20 True or false:
(a) If $t = t(H)$ for some subposition H of the canonical form of G, then t is necessarily a critical temperature of G.
(b) If $G = H$, then G and H have the same critical temperatures.

Notes

In the 1950s, Milnor [**Miln53**] and Hanner [**Han59**] studied two-player games in which the terminal positions are assigned explicit scores, and proved an early version of the Mean Value Theorem. Later, once Conway recognized that his axiomatic system of combinatorial games intrinsically incorporates a number system (see Notes to Section 3 on page 82), the *Winning Ways* authors were quick to develop an appropriate temperature theory in the new context.

Most of the material in this section is classical and appeared (in various forms) in ONAG and *Winning Ways*. Generalized overheating operators (Definition 5.25 on page 115) were introduced later by Berlekamp, in conjunction with the study of DOMINEERING [**Ber88**]; the notation $\oint$ used here is new and is intended to distinguish from ordinary heating when both operators are used in the same context.

The combinatorial theory of AMAZONS has been studied extensively by Berlekamp [**Ber00b**] and Snatzke [**Sna02, Sna03, Sna04**]. Computational aspects have been investigated by Furtak et al. [**FKUB05**], Hearn [**Hea06, Hea09**], and Müller and Tegos [**MT02**].

The temperature theory has also proved to be successful in the study of DOMINEERING. In the early 1980s Berlekamp's student Robin Pemantle calculated many explicit values of $2 \times n$ and $3 \times n$ DOMINEERING positions. Berlekamp then observed that these values can be concisely described in terms of overheated infinitesimals [**Ber88**]. This led to the introduction of a simplified form of $2 \times n$ DOMINEERING known as BLOCKBUSTING, which has a striking "chilling-invariance" property: if G and H are BLOCKBUSTING positions and $G_1 = H_1$, then in fact G is infinitesimally close to H. Moreover, G can be recovered from G_1 by overheating, up to an infinitesimal:

$$G \text{ is infinitesimally close to } \oint_1^{1*} G_1, \text{ for every BLOCKBUSTING position } G.$$

Berlekamp and Wolfe [**BW94**] subsequently discovered that the vast majority of GO positions are similarly chilling-invariant, leading to the spectacularly successful combinatorial theory of GO endgames discussed in Section I.4. For example, the "9-dan stumping problem" (Figure I.4.2 on page 29) is carefully constructed so that each component G has temperature $t(G) \leq 1$. Every chilled value G_1 is therefore numberish, and their interactions can be studied using the theory of infinitesimals from Section 4. Due to the chilling-invariance properties of GO, the outcome $o(G_1)$ of the cooled position G_1 is sufficient to determine the stops $L(G)$ and $R(G)$ of the original position G.

YELLOW–BROWN HACKENBUSH was invented by Berlekamp [**Ber09**].

Exercises 5.1 and 5.18: Berlekamp.

6. Reduced Canonical Form

The **partizan subtraction game** $\Gamma = \text{SUBTRACTION}(1,3 \mid 2,3)$ is played with several heaps of tokens, as with NIM. On her turn, Left must remove exactly 1 or 3 tokens from any one heap; Right must remove exactly 2 or 3. Let H_n be a Γ-heap of size n. The canonical forms of the H_n become frighteningly complicated even for small values of n; for example, using *cgsuite* we can calculate

$$H_6 = \Big\{1 \parallel 1, \{1 \mid 0\} \,\Big|\!\Big|\!\Big|\, 0, \{1 \mid 0\} \parallel 0 \,\Big|\!\Big|\!\Big|\!\Big|\, 0, \{1 \mid 0\} \parallel 0\Big\}.$$

A more careful inspection, however, reveals a striking regularity. We find that

$$(\dagger) \qquad H_n \text{ is } \begin{cases} 0\text{-ish} & \text{if } n \equiv 0 \pmod 3; \\ 1\text{-ish} & \text{if } n \equiv 1 \pmod 3; \\ \{1 \mid 0\}\text{-ish} & \text{if } n \equiv 2 \pmod 3. \end{cases}$$

This characterization doesn't determine $o(H_n + X)$ for arbitrary X, of course—for that, we need to know the full canonical forms of H_n. But it *does* determine $L(H_n + X)$ and $R(H_n + X)$ for all X, since stops are invariant up to an infinitesimal. This, in turn, is good enough to compute $o(H_n + X)$ except in the particular case where at least one stop happens to be 0.

It turns out that for every game G, there is a unique *simplest* game that is infinitesimally close to G. This game $\mathrm{rcf}(G)$ is the **reduced canonical form** of G, and it can be computed in much the same way that we calculate ordinary canonical forms.

Recall that the full canonical form carries exactly the information needed to determine $o(G+X)$ for all X. Likewise, the reduced canonical form carries exactly the information needed to determine $L(G + X)$ and $R(G + X)$ for all X. We can view $\mathrm{rcf}(G)$ as *preserving* information about the final score of $G + X$, while *discarding* details about the number of moves it took to get there.

We'll prove all these assertions in this section. When we're done, it will be easy to establish (†) and similar results.

Infinitesimal Equivalence

Recall (from Section 4) that G and H are **infinitesimally close** if $G - H$ is an infinitesimal, that is, if

$$x \geq G - H \geq -x \quad \text{for every positive number } x.$$

When G and H are infinitesimally close, we'll write

$$G \equiv H \pmod{\mathrm{Inf}},$$

and we can define a corresponding inequality as follows.

Definition 6.1. We write

$$G \geqq H \pmod{\mathrm{Inf}}$$

to mean

$$G - H \geq -x \quad \text{for every positive number } x.$$

Throughout this section we'll abbreviate by writing simply "$G \equiv H$" in place of "$G \equiv H \pmod{\mathrm{Inf}}$" and "$G \geqq H$" in place of "$G \geqq H \pmod{\mathrm{Inf}}$." Observe the additional horizontal bar under the $=$ and $\geq$ symbols: $G \equiv H$ and $G \geqq H$ both imply "mod Inf," whereas $G = H$ and $G \geq H$ denote ordinary (in-)equality as in Section 1.

Proposition 6.2. *If $G \geqq H$ and $H \geqq J$, then $G \geqq J$.*

Proof. For all $x > 0$, we have $G - H \geq -x/2$ and $H - J \geq -x/2$, so that $G - J \geq -x$. □

Proposition 6.3. *If $G \geqq H$, then $G + J \geqq H + J$.*

Proof. Immediate, since $G - H = (G + J) - (H + J)$. □

Lemma 6.4. *The following are equivalent.*

(i) $G \geqq 0$.

(ii) $R(G) \geq 0$.

(iii) $G \geq \epsilon$ *for some infinitesimal* ϵ.

Proof. (i) ⇒ (ii): We have $G \geq -x$ for every $x > 0$, so that

$$R(G) \geq R(-x) = -x.$$

Since $R(G)$ is a number, this necessarily implies $R(G) \geq 0$.

(ii) ⇒ (iii): This follows from Theorem 4.9 on page 86, taking $\epsilon = n \cdot \downarrow$ for sufficiently large n.

(iii) ⇒ (i): Since ϵ is infinitesimal, we have $\epsilon \geq -x$ for every $x > 0$, and the conclusion follows. □

Lemma 6.5 (Inf-Replacement Lemma; cf. Lemma 2.2 on page 64). *Assume that G is not equal to a number, and suppose G' is obtained from G by replacing some Left option G^L with a new option $G^{L'}$. If $G^{L'} \geqq G^L$, then $G' \geqq G$; if $G^{L'} \leqq G^L$, then $G' \leqq G$.*

Proof. Suppose $G^{L'} \geqq G^L$; we must show that $G' - G + x$ is a win for Left playing second, for any positive number x. Since x is a number and $-G$ is not, we can disregard Right's opening moves on x, by the Number Avoidance Theorem. All of Right's moves on G' or $-G$ have mirror-image responses on the other component, except for Right's move to $G' - G^L + x$. But this move has a response to $G^{L'} - G^L + x$, which is positive since $G^{L'} \geqq G^L$. All remaining moves on G' or $-G$ have mirror-image responses on the other component.

An identical argument shows that if $G^{L'} \leqq G^L$, then $G' \leqq G$. □

Inf-Dominated and Inf-Reversible Options

We'll now show that for every G, there is a unique simplest game K with $K \equiv G$. The proof is similar to the development of ordinary canonical forms in Section 2, with some additional complications.

Definition 6.6 (cf. Definition 2.3 on page 64). Let G be a short game.

(a) A Left option G^{L_1} is Inf-**dominated** (**by** G^{L_2}) if $G^{L_2} \geqq G^{L_1}$ for some other Left option G^{L_2}.

(b) A Right option G^{R_1} is Inf-**dominated** (**by** G^{R_2}) if $G^{R_2} \leqq G^{R_1}$ for some other Right option G^{R_2}.

(c) A Left option G^{L_1} is Inf-**reversible** (**through** $G^{L_1R_1}$) if $G^{L_1R_1} \leqq G$ for some Right option $G^{L_1R_1}$.

(d) A Right option G^{R_1} is Inf-**reversible** (**through** $G^{R_1L_1}$) if $G^{R_1L_1} \geqq G$ for some Left option $G^{R_1L_1}$.

Theorem 6.7 (cf. Theorem 2.4 on page 65). *Assume that G is not equal to a number, and suppose G' is obtained from G by removing some* Inf*-dominated Left option G^{L_1}. Then $G' \equiv G$.*

Proof. Certainly $G' \leqq G$ (in fact $G' \leq G$, by Lemma 2.1 on page 64), so we must show that $G' \geqq G$, i.e., that Left can win $G' - G + x$ moving second. Since x is equal to a number and $-G$ is not, we can disregard Right's opening moves on x (by the Number Avoidance Theorem). Now every opening move by Right on G' or $-G$ has a mirror-image response on the other component, except for his move to $G' - G^{L_1} + x$. But G^{L_1} is Inf-dominated, so there is some $G^{L_2} \geqq G^{L_1}$, and therefore Left can win by moving to $G^{L_2} - G^{L_1} + x$. □

The generalization to reversible moves requires more care. It turns out that Inf-reversible options of G can be safely bypassed only in the case where G is hot; the weaker assumption that G is not equal to a number is insufficient. The essence of the argument is captured by the following lemma, which uses a slightly different hypothesis.

Lemma 6.8 (cf. Theorem 2.5 on page 65). *Let G be a short game, and suppose that some Left option G^{L_1} is* Inf*-reversible through $G^{L_1R_1}$. Let*

$$G' = \{G^{L_1R_1L}, G^{L'} \mid G^R\},$$

where $G^{L_1R_1L}$ ranges over all Left options of $G^{L_1R_1}$ and $G^{L'}$ ranges over all Left options of G except *for G^{L_1}. If G' is not equal to a number, then $G' \equiv G$.*

Proof. We must show that $G - G' \geq -x$ and $G' - G \geq -x$ for every positive number x, i.e., that Left can win both $G - G' + x$ and $G' - G + x$ moving second. Since G' is not equal to a number (and therefore neither is $-G'$), we may disregard Right's opening moves on x in both cases. All of his other opening moves have mirror-image responses, except for those of the form

$$G - G^{L_1R_1L} + x \quad \text{or} \quad G' - G^{L_1} + x.$$

First consider $G - G^{L_1R_1L} + x$. By assumption, $G^{L_1R_1} \leqq G$, so in fact

$$G^{L_1R_1L} \lhd\!\!| \; G^{L_1R_1} \leq G + x.$$

This shows that Left must have a winning move on $G - G^{L_1R_1L} + x$.

Next consider $G' - G^{L_1} + x$. This game has $G' - G^{L_1R_1} + x$ as a Left option, so to complete the proof it suffices to show that $G' - G^{L_1R_1} + x \geq 0$.

Now since x is equal to a number and G' is not, we may assume that Right's next move is not on x. All of his other options from $G' - G^{L_1R_1} + x$ have the form

$$G' - G^{L_1R_1L} + x \quad \text{or} \quad G^R - G^{L_1R_1} + x.$$

Each $G' - G^{L_1R_1L} + x$ has a mirror-image response on G', so it remains to consider $G^R - G^{L_1R_1} + x$. But since $-G^{L_1R_1} \geqq -G$, we have

$$G^R - G^{L_1R_1} + x \geq G^R - G \; |\!\!\rhd \; 0,$$

so Left must have a winning move on $G^R - G^{L_1R_1} + x$. □

Theorem 6.9 (cf. Theorem 2.5 on page 65). *Assume that G is hot, and suppose that some Left option G^{L_1} is* Inf*-reversible through $G^{L_1R_1}$. Let G' be as in Lemma 6.8. Then $G' \equiv G$.*

Proof. It suffices to show that G' is not equal to a number, for then it will fulfill the hypothesis of Lemma 6.8. Now since G is hot, we know that $L(G) > R(G)$. We'll show that $L(G') \geq L(G)$ and $R(G') \leq R(G)$, establishing that G' is also hot.

First choose some G^R with $L(G^R) = R(G)$. Then since G^R is also a Right option of G', we have $L(G^R) \geq R(G')$ (using, say, Proposition 3.18(b) on page 76). This establishes $R(G') \leq R(G)$.

For the inequality $L(G') \geq L(G)$, there are two cases.

Case 1: $R(G^{L_1}) = L(G)$. Then $L(G^{L_1R_1}) \geq L(G)$. Furthermore, since $G^{L_1R_1} \leqq G$, we have $R(G^{L_1R_1}) \leq R(G)$. Thus

$$L(G^{L_1R_1}) \geq L(G) > R(G) \geq R(G^{L_1R_1}),$$

so $G^{L_1R_1}$ is not equal to a number. So there is some $G^{L_1R_1L}$ with

$$R(G^{L_1R_1L}) = L(G^{L_1R_1}) \geq L(G).$$

Since $G^{L_1R_1L}$ is a Left option of G', this implies $L(G') \geq L(G)$.

Case 2: $R(G^{L_1}) \neq L(G)$. Then we must have $R(G^{L'}) = L(G)$ for some other $G^{L'}$, but $G^{L'}$ is also a Left option of G'.

In both cases, $L(G') \geq L(G)$. □

Reduced Canonical Form

Definition 6.10. A short game K is in **reduced canonical form** if, for every subposition H of K, either:

(i) H is a number in canonical form; or

(ii) H is hot and contains no Inf-dominated or Inf-reversible options.

Theorem 6.11. *For every short game G, there exists a game $K \equiv G$ in reduced canonical form.*

Proof. If G is infinitesimally close to a number x, then K is simply the canonical form of x. So suppose that G is hot. By induction, we may assume that the theorem is true for all options of G. By the Inf-Replacement Lemma, we may therefore assume that no *proper* subposition of G has any Inf-dominated or Inf-reversible options. If the same is true for G itself, then we are done.

But if G has an Inf-dominated option, then by Theorem 6.7 it can be removed to obtain a simpler game $G' \equiv G$. The game tree for G' has strictly fewer edges than the tree for G, so by induction, we may assume that there exists a $K \equiv G'$ in reduced canonical form. The same argument works if G has an Inf-reversible option, using Theorem 6.9. □

The following lemma is the analogue of Lemma 2.8 on page 67.

Lemma 6.12. *Assume that G and H are hot. Suppose that neither G nor H has any* Inf*-dominated or* Inf*-reversible options, and assume $G \equiv H$. Then for every G^L, there is an H^L such that $G^L \equiv H^L$, and vice versa; and likewise for Right options.*

Proof. Fix G^L. By Lemma 6.4, we have $G - H + \epsilon \leq 0$ for some infinitesimal ϵ, so in particular $G^L - H + \epsilon \lhd\!\!| \; 0$. Since $-H$ is hot and ϵ is tepid, we can disregard Right's responses on ϵ, so either $G^L - H^L + \epsilon \leq 0$ for some H^L, or else $G^{LR} - H + \epsilon \leq 0$ for some G^{LR}. But the latter would imply

$$G^{LR} \leqq H \leqq G,$$

contradicting the assumption that G has no Inf-reversible options. Therefore $G^L \leqq H^L$ for some H^L. By exactly the same argument, there is some $G^{L'}$ such that $H^L \leq G^{L'}$. But now we have

$$G^L \leqq H^L \leqq G^{L'}.$$

Since G has no Inf-dominated options, this implies $G^L \equiv H^L \equiv G^{L'}$. The remaining cases follow by symmetry. □

Theorem 6.13 (Inf-Simplest Form Theorem)**.** *Suppose that G and H are in reduced canonical form, and assume that $G \equiv H$. Then $G \cong H$.*

Proof. If either G or H is infinitesimally close to a number x, then both must be, so that $G \cong H \cong x$ (assuming x to be in canonical form).

Otherwise, Lemma 6.12 shows that for every G^L, there is an H^L with $G^L \equiv H^L$. By induction, this implies that $G^L \cong H^L$. Therefore $\mathscr{G}^L \subset \mathscr{H}^L$, and likewise $\mathscr{H}^L \subset \mathscr{G}^L$, so that G and H have identical Left options. The same argument works for Right options. □

Definition 6.14. The **reduced canonical form** (or Inf-**simplest form**) of G is the unique game $K \equiv G$ such that K is in reduced canonical form.

We write $\mathrm{rcf}(G)$ for the reduced canonical form of G. If $G = \mathrm{rcf}(G)$ (not necessarily $G \cong \mathrm{rcf}(G)$), then we say G is **reduced**.

The Analysis of Subtraction$(1, 3 \mid 2, 3)$

We're now able to prove the assertions about Subtraction$(1, 3 \mid 2, 3)$ from the beginning of this section (page 123). Recall the main claim:

$$H_n \text{ is } \begin{cases} 0\text{-ish} & \text{if } n \equiv 0 \pmod 3; \\ 1\text{-ish} & \text{if } n \equiv 1 \pmod 3; \\ \{1 \mid 0\}\text{-ish} & \text{if } n \equiv 2 \pmod 3. \end{cases}$$

This can be verified directly for $n = 0, 1, 2$, so assume $n \geq 3$. Presuming that H_a has the desired form for all $a < n$ and using the Inf-Replacement Lemma, we have

$$\begin{aligned} H_n &= \{H_{n-1}, H_{n-3} \mid H_{n-2}, H_{n-3}\} \\ &\equiv \begin{cases} \{\{1 \mid 0\}, 0 \parallel 1, 0\} & \text{if } n \equiv 0 \pmod 3; \\ \{0, 1 \parallel \{1 \mid 0\}, 1\} & \text{if } n \equiv 1 \pmod 3; \\ \{1, \{1 \mid 0\} \parallel 0, \{1 \mid 0\}\} & \text{if } n \equiv 2 \pmod 3. \end{cases} \end{aligned}$$

The expressions for $n \equiv 0$ and $n \equiv 1$ are numberish and so can be replaced by the corresponding numbers, 0 and 1. For $n \equiv 2$, we observe that

$$1 \geqq \{1 \mid 0\} \geqq 0 \pmod{\mathrm{Inf}}$$

so that by eliminating Inf-dominated options,

$$\{1, \{1 \mid 0\} \parallel 0, \{1 \mid 0\}\} \equiv \{1 \mid 0\}$$

as needed.

Temper

If Left moves first on $G + X$ and both players play well, then eventually the game will reach a position of value $L(G + X)$. If $L(G + X) \neq 0$, then the outcome depends only on its sign. However if $L(G + X) = 0$, then the outcome depends on which player made the last move—or, put another way, on the parity of the total number of moves.

In particular, let $\mathcal{A} \subset \mathbb{G}$ be the Inf-equivalence class of G:

$$\mathcal{A} = \{H \in \mathbb{G} : H \equiv G\}.$$

Then each $H \in \mathcal{A}$ satisfies $L(H + X) = L(G + X)$, for all X. But if this number is 0, then it is possible that $o(H + X) \neq o(G + X)$, because the optimal plays on $H + X$ and $G + X$ (leading to the stopping position 0) might have opposite parity.

The reduction $\mathrm{rcf}(G)$ explicitly discards such parity information. It is determined uniquely (as an element of $\mathcal{A}$) by a *minimality* condition: each stopping position is reached after the fewest possible number of moves. If $G \equiv H$ and G and H are both reduced, then $L(G + X) = L(H + X)$ implies $o(G+X) = o(H+X)$, essentially because this minimality constraint predetermines the parity of each play.

There is another natural way to isolate a unique element of $\mathcal{A}$, which is to impose a *parity* condition: each stopping position is reached after a sequence of moves of fixed parity, even or odd. Such games are called **even-tempered** and **odd-tempered**, respectively. If $G \equiv H$ and G and H are both even-tempered (or both odd-tempered), then $L(G + X) = L(H + X)$ implies $o(G + X) = o(H + X)$, this time because of the explicit parity constraint.

Definition 6.15. Let G be a short game.

(a) G is **even-tempered** if G is equal to a number or every option of G is odd-tempered.

(b) G is **odd-tempered** if G is not equal to a number and every option of G is even-tempered.

Examples. Any switch $\{x \mid y\}$, with $x \geq y$, is odd-tempered. (This is equal to $x*$ in the case $x = y$.) Likewise, if $x > y$, then $\{x* \mid y*\}$ is even-tempered. $\{x* \mid y\}$ is neither even-tempered nor odd-tempered.

We've seen that each Inf-equivalence class of $\mathbb{G}$ contains (the value of) exactly one reduced game; we'll now prove that every such class also contains exactly one even-tempered game and exactly one odd-tempered game. First we show that temper behaves as expected in sums.

Theorem 6.16. *Let G and H be short games.*

(a) *If G and H are both even-tempered or both odd-tempered, then $G+H$ is equal to an even-tempered game.*

(b) *If G is even-tempered and H is odd-tempered (or vice versa), then $G+H$ is equal to an odd-tempered game.*

Proof. If G and H are both numbers, then so is $G+H$, and the conclusion follows. If G is a number and H is not, then by the Number Avoidance Theorem

$$G+H = \{G+H^L \mid G+H^R\},$$

and by induction $G+H^L$ and $G+H^R$ have the same temper as H^L and H^R. Finally, if neither G nor H is a number, then

$$G+H = \{G^L+H, G+H^L \mid G^R+H, G+H^R\}.$$

By induction and assumption on G and H, all options have the same temper. Furthermore, they are odd-tempered if G and H have the same temper and even-tempered otherwise. □

We proceed with the explicit construction of an even-tempered game infinitesimally close to G. The idea is to inspect each stopping position x of G, replacing it by $x*$ if it was reached after an odd number of moves.

Definition 6.17. We define G **reduced by** $*$, denoted by G_*, recursively:

$$G_* = \begin{cases} G & \text{if } G \text{ is equal to a number;} \\ \{G_*^L + * \mid G_*^R + *\} & \text{otherwise} \end{cases}$$

Note that reduction is exactly the same operation as heating by $*$ (Definition 5.21 on page 112). The terminology and notation used here are more natural in the context of reductions.

Proposition 6.18. *G_* is even-tempered and $G_* \equiv G$, for every game G.*

Proof. The proposition is trivial if G is a number. Otherwise, we have

$$G_* = \{G_*^L + * \mid G_*^R + *\}.$$

By induction, every G_*^L and G_*^R is even-tempered, so by Theorem 6.16, every $G_*^L + *$ and $G_*^R + *$ is odd-tempered. Therefore G_* is even-tempered.

Also, $G_*^L \equiv G^L$; hence $G_*^L + * \equiv G^L$. Likewise, $G_*^R + * \equiv G^R$, so from the Inf-Replacement Lemma we have $G_* \equiv G$. □

Uniqueness:

Theorem 6.19. *Suppose that G and H are both even-tempered and $G \equiv H$. Then $G = H$.*

Proof. By Theorem 6.16, it suffices to show that if $G \equiv 0$, then $G = 0$. We'll prove that $G \geq 0$; it follows by symmetry that $G \leq 0$.

Since $G \geqq 0$, we have $R(G) \geq 0$. Now if G is a number, then we are done. Otherwise, we must show that every $G^R \mathrel{|\!\!>} 0$. But $L(G^R) \geq 0$, and since G^R is odd-tempered, it cannot be a number. So there is some G^{RL} with $R(G^{RL}) \geq 0$. Since G^{RL} is even-tempered, we conclude by induction that $G^{RL} \geq 0$. □

Corollary 6.20. *If* $G \equiv H$, *then* $G_* = H_*$.

Proof. Applying Proposition 6.18 to G and H, we have

$$G_* \equiv G \equiv H \equiv H_*.$$

But G_* and H_* are both even-tempered, so by Theorem 6.19 they must be equal. □

Corollary 6.20 shows that G_* depends only on the value of G, and not its form. A nice consequence is that the mapping $G \mapsto G_*$ is a homomorphism $\mathbb{G} \to \mathbb{G}$. This is *not* the case for the mapping $G \mapsto \mathrm{rcf}(G)$ (Exercise 6.2).

Theorem 6.21. $(G + H)_* = G_* + H_*$.

Proof. By Proposition 6.18, we have that

$$(G + H)_* \equiv G + H \equiv G_* + H_*.$$

Furthermore, $(G+H)_*$ is even-tempered, and by Theorem 6.16 so is G_*+H_*. Therefore $(G + H)_* = G_* + H_*$. □

The Group $\mathbb{G}/\mathrm{Inf}$

Since Inf is a subgroup of $\mathbb{G}$, we may consider the quotient

$$\mathbb{G}/\mathrm{Inf},$$

the group of **reduced game values**. We can see that

$$\mathbb{G} \cong (\mathbb{G}/\mathrm{Inf}) \oplus \mathrm{Inf},$$

with the isomorphism given by

$$G \mapsto \big([G], G - G_*\big),$$

where $[G]$ is the equivalence class of G in $\mathbb{G}/\mathrm{Inf}$.

We also have the usual hereditary hierarchy:

$$\mathbb{G}_n/\mathrm{Inf} = \{[G] : G \in \mathbb{G}_n\}.$$

We'll study this group again in Chapter III.

Transitive Games

The transitive games are a special class of positions that provide a nice application for reduced canonical forms.

Definition 6.22. G is **transitive** if every G^{LL} is also a Left option of G itself and every G^{RR} is also a Right option of G itself.

G is **hereditarily transitive** if every subposition of G is transitive.

Lemma 6.23. *Let G be hereditarily transitive, and suppose G is not equal to a number. Then $L(G)$ is equal to a Left option of G and $R(G)$ is equal to a Right option of G.*

Proof. Let $x = L(G)$ and fix G^L with $R(G^L) = x$. If G^L is equal to a number, then in fact $G^L = x$ and there is nothing more to prove. Otherwise, let $y = L(G^L)$. Then $y \geq x$ (since y is the Left stop of G^L and x is the Right stop). Moreover, by induction y is a Left option of G^L and hence, since G is transitive, of G as well. But this shows that $y \leq x$ (since $x = L(G)$), and therefore $y = x$.

The proof for $R(G)$ is identical. □

Lemma 6.24. *If G is transitive, then $L(G^L) \leq L(G)$ for every G^L.*

Proof. If G^L is a number, then certainly $G^L \leq L(G)$. Otherwise there is some G^{LL} with $R(G^{LL}) = L(G^L)$. But G^{LL} must also be a Left option of G (since G is transitive), so that $R(G^{LL}) \leq L(G)$. □

Here's the main theorem:

Theorem 6.25 (Nowakowski–Ottaway). *Let G be hereditarily transitive. Then* $\mathrm{rcf}(G)$ *is either a number or a game of the form $\{a \mid b\}$ (for numbers a and b).*

Proof. Let $a = L(G)$ and $b = R(G)$. If $a = b$, then G is infinitesimally close to a number, so $\mathrm{rcf}(G)$ is a number. Otherwise a is a Left option of G, and by Lemma 6.24 each other Left option is Inf-dominated by a. The same argument applies to b as a Right option of G, so that $\mathrm{rcf}(G) = \{a \mid b\}$. □

Theorem 6.26 (Nowakowski–Ottaway). *Let G be hereditarily transitive. Then*

$$\Downarrow\!* \leq G - \mathrm{rcf}(G) \leq \Uparrow\!*.$$

Proof. First suppose $\mathrm{rcf}(G)$ is a number, say a. Then $G = \left\{a, G^L \,\middle|\, a, G^R\right\}$. Moreover, each G^R has Left stop $a' \geq a$ and therefore has a' as a Left option. So on the sum

$$H = G - \mathrm{rcf}(G) + \Uparrow\!*$$

Left can win playing second: either by reverting a move G^R to some $a' \geq a$; or by responding on G if Right moves from $\Uparrow*$ to $\uparrow$. (By the Number Avoidance Theorem, we needn't consider moves on $\mathrm{rcf}(G)$.)

Next suppose $\mathrm{rcf}(G) = \{a \mid b\}$, so that $G = \{a, G^L \mid b, G^R\}$. Now on the sum H, Right's opening move on $\mathrm{rcf}(G)$ has an immediate counter on G; likewise Right's opening move from G to b.

If instead Right moves to some other G^R, then Left can revert to some $a' \geq a$, as before; while if Right opens by moving on $\Uparrow*$, Left can respond by moving immediately from G to a. In either case, after Left's move the position has the form

$$a' - \{a \mid b\} + \Uparrow* \quad (a' \geq a),$$

from which Right's best move is to $a' - a + \Uparrow* \geq 0$.

This shows that $G - \mathrm{rcf}(G) \leq \Uparrow*$; the other inequality is by symmetry. □

The bounds in Theorem 6.26 can be strengthened (see Exercise 6.15).

Exercises

6.1 (a) The hypothesis "G is not a number" is necessary in Lemma 6.5.
(b) The hypothesis "G is hot" is necessary in Theorem 6.9 and cannot be weakened to "G is not a number."

6.2 $G \mapsto \mathrm{rcf}(G)$ is *not* a group homomorphism of $\mathbb{G}$.

6.3 The **$*$-projection** of G, denoted by $p(G)$, is defined by

$$p(G) = \begin{cases} x & \text{if } G = x \text{ or } x* \text{ for some number } x; \\ \{p(G^L) \mid p(G^R)\} & \text{otherwise.} \end{cases}$$

Prove that $p(G_*) = \mathrm{rcf}(G)$. This gives an alternative construction for $\mathrm{rcf}(G)$.

6.4 Prove that $G_{0+} = \mathrm{rcf}(G)$ for all G. Conclude that G_t is reduced for all but finitely many t. This gives another alternative construction for $\mathrm{rcf}(G)$.

6.5 Let G be a game with canonical form K. Prove that if G is even-tempered, then so is K.

6.6 For which numbers s and t $(0 \leq s \leq t)$ is $\int^s \uparrow + \int^t \uparrow$ reduced?

6.7 Assume that G is not equal to a number, and suppose G' is obtained from G by replacing every Left option G^L with a new option $G^{L'} \geqq G^L$, and every Right option G^R with a $G^{R'} \geqq G^R$. Show that $G' \geqq G$. (Why does this *not* follow from inductive application of Lemma 6.5?)

6.8 Assume that K is in reduced canonical form and K is not a number. Let Δ be an incentive of K. Prove that $\Delta \not\leqq 0$.

6.9 Suppose that $G^{L_1R_1} \leqq G$, and assume that G is not equal to a number (but is not necessarily hot). Prove that we can bypass G^{L_1} through $G^{L_1R_1}$, as in

Theorem 6.9, *provided that* $G^{L_1R_1}$ is not equal to a number. (This might be false when $G^{L_1R_1}$ is a number; cf. Exercise 6.1(b).)

6.10 Let G be a game with reduced canonical form K (cf. Exercise 2.2 on page 68).
(a) Show that for every K^L, there is a G^L such that $G^L \geqq K^L$.
(b) Show that for every subposition T of K, there is a subposition H of G such that $H \equiv T$.

6.11 A Left option G^L is Inf-**sensible** if $G^L \geqq K^L$ for some K^L, where K is the reduced canonical form of G. Otherwise G^L is Inf-**senseless**. Show that removing Inf-senseless options changes the value of G by at most an infinitesimal (cf. Exercise 2.6 on page 68).

6.12 Show that an Inf-sensible option need not be sensible. Is the converse true?

6.13 Let $\Gamma = \text{SUBTRACTION}(1, n \mid n{-}1, n)$, for fixed $n \geq 3$. Determine the reduced canonical forms of all Γ-heaps.

6.14 If G is dicotic and hereditarily transitive, then $G = * : H$ for some hereditarily transitive game H.

6.15 If G is hereditarily transitive, then $-U < G - \text{rcf}(G) < U$ for any uptimal U of the form

$$U = \uparrow^{[n]} + \uparrow^{n} + * \quad (n \geq 1).$$

(This is proved for $n = 1$ as Theorem 6.26 on page 133.)

6.16 Let $\mathcal{E} = \{G \in \mathbb{G} : G \text{ is even-tempered}\}$ and $\mathcal{O} = \{G \in \mathbb{G} : G \text{ is odd-tempered}\}$. Prove that $\mathcal{E}$ and $\mathcal{E} \cup \mathcal{O}$ are subgroups of $\mathbb{G}$, and show that $\mathcal{E}$ has index 2 in $\mathcal{E} \cup \mathcal{O}$.

6.17 Let $\varepsilon \mathrel{|\!\triangleright} 0$ be an infinitesimal, and define

$$G_\varepsilon = \begin{cases} G & \text{if } G \text{ is a number;} \\ \{G^L_\varepsilon - \varepsilon \mid G^R_\varepsilon + \varepsilon\} & \text{otherwise.} \end{cases}$$

Note that this generalizes our definition of G_*. Prove that:
(a) $G_\varepsilon \equiv G$, and if $G \equiv 0$, then $G_\varepsilon = 0$.
(b) If $G = H$, then $G_\varepsilon = H_\varepsilon$.
(c) $(G{+}H)_\varepsilon = G_\varepsilon{+}H_\varepsilon$. (Hint: First show that $L(G_\varepsilon) = L(G)$ and $R(G_\varepsilon) = R(G)$. Use this to prove the special case when H is a number. Then prove the full assertion by induction.)
(d) If G_ε is not a number, then G_ε has at least one Left incentive and one Right incentive that are $\geq -\varepsilon$.
(e) The *Dicotic Avoidance Theorem*: Let $\varepsilon \mathrel{|\!\triangleright} 0$ be dicotic, and assume that G_ε is not a number. Then G_ε has at least one Left incentive and one Right incentive that are $\geq$ every incentive (Left or Right) of ε.

Notes

Reduced canonical form was introduced in a 1996 paper by Dan Calistrate [**Cal96, Cal98**], who gave the construction in the form of Exercise 6.3. Inf-dominated and Inf-reversible moves first appeared earlier, in a paper by David Moews [**Moe91**], but in a totally different context (the group structure of $\mathbb{G}_3$; cf. Section III.3 of this book) that didn't involve the Inf-Simplest Form Theorem.

Finally in 2009, J.P. Grossman and Aaron Siegel gave a proof of the Inf-Simplest Form Theorem using the methods outlined here [**GS09**]. Temper was introduced in the same paper.

Transitive games were introduced by Nowakowski and Ottaway [**NO11, Ott09**] and have also been studied by Angela Siegel [**Sie11**] (no relation to the author); see also the notes to Section III.2 on page 166. They are often known as **option-closed** games elsewhere in the literature.

Exercise 6.3: Dan Calistrate.

Exercises 6.14 and 6.15: Neil McKay.

Exercise 6.17: J.P. Grossman.

7. Atomic Weight

If G is an infinitesimal, then

$$L(G) = R(G) = m(G) = t(G) = \mathrm{rcf}(G) = 0.$$

This is quite convenient when one's goal is to suppress infinitesimal differences. In order to study the structure of infinitesimals, however, a different suite of techniques is required.

The analysis of flower gardens is a simple example of such a technique. In Section 4 we saw that in a sum of flowers, the *number of flowers* of each color is more important than their composition, the latter being relevant only when the number of flowers of each color differs by at most one. In particular, the **weight** of a garden G is given by

$$w(G) = (\text{number of blue flowers}) - (\text{number of red flowers})$$

with the following implications:

$$\text{if } w(G) \geq 1, \text{ then } G \mathrel{|\!\!>} 0; \quad \text{and} \quad \text{if } w(G) \geq 2, \text{ then } G > 0.$$

Therefore $w(G)$ is sufficient to determine $o(G)$ except when $|w(G)| \leq 1$.

In this section, we'll see that $w(G)$ is a special case of a much more general invariant, the **atomic weight** $\mathrm{aw}(G)$ of an infinitesimal G. As we showed in Theorem 4.8 (page 85), every infinitesimal G satisfies

$$n \cdot \downarrow < G < n \cdot \uparrow$$

for some sufficiently large n, so the sequence

$$\uparrow,\ \Uparrow,\ \Uparrow\!\!\!\uparrow,\ \ldots$$

establishes a scale that can be used to calibrate the size of any infinitesimal. The atomic weight of G describes precisely where G should be placed on this scale. We'll see that:

- $\mathrm{aw}(n \cdot \uparrow) = n$;
- $\mathrm{aw}(G) = w(G)$ for flower gardens G;

- $\mathrm{aw}(G+H) = \mathrm{aw}(G)+\mathrm{aw}(H)$, whenever $\mathrm{aw}(G)$ and $\mathrm{aw}(H)$ are defined.

But although the atomic weight of every flower garden is an integer, this is *not* the case for more general infinitesimals. In fact the mapping $G \mapsto \mathrm{aw}(G)$ is a surjection onto $\mathbb{G}$, so that *any* game is equal to the atomic weight of some infinitesimal. For example:

- $\mathrm{aw}\big(\{\Uparrow \mid \downarrow\}\big) = \frac{1}{2}$;
- $\mathrm{aw}\big(\{\Uparrow \mid \Downarrow\}\big) = *$;
- $\mathrm{aw}\big(\{\Uparrow \,\|\, 0 \mid \downarrow 6\}\big) = +_2$.

Remote Stars

If G is a flower garden, then $o(G)$ is determined by $w(G)$, except when $|w(G)| \leq 1$. However if G contains a sufficiently large green flower, then in fact $o(G)$ is completely determined even when $w(G)$ is small. In such cases, either player can achieve her optimal outcome by playing randomly on her opponent's flowers; the "remote" green flower ensures that a winning move will automatically be available once all of the red and blue flowers are gone. The following proposition makes this argument precise; see Figure 7.1 for the picture.

Proposition 7.1. *If G is a flower garden, then*

$$o(G + *m) = \begin{cases} \mathscr{L} & \text{if } w(G) > 0; \\ \mathscr{N} & \text{if } w(G) = 0; \\ \mathscr{R} & \text{if } w(G) < 0, \end{cases}$$

for all m such that $*m$ is not equal to a subposition of G.

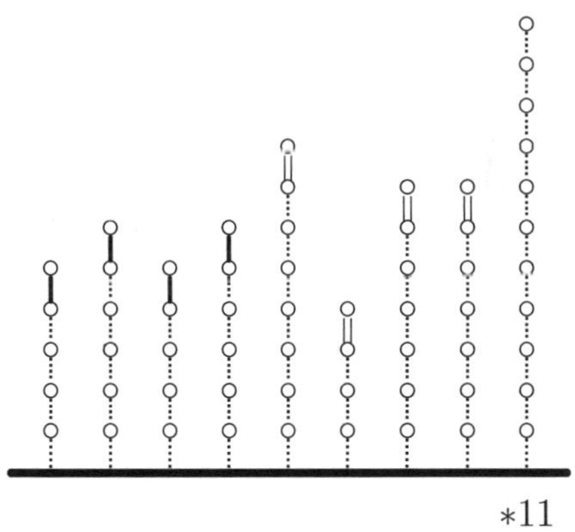

Figure 7.1. This flower garden of weight 0 is necessarily an $\mathscr{N}$-position. The exact structure of the red and blue flowers is immaterial, because the green flower of value $*11$ is sufficiently remote.

Proof. Suppose $w(G) = 0$. If G contains only green flowers, then $G = *a$ for some a. The assumption on m guarantees that $a \neq m$, so $o(*a + *m) = \mathscr{N}$.

Otherwise, G must have at least one red flower, so Left has a move to some $G^L + *m$ with $w(G^L) > 0$. By induction, $G^L + *m > 0$, for any such G^L. The same argument shows that some $G^R + *m < 0$, so that $o(G + *m) = \mathscr{N}$.

Suppose instead that $w(G) > 0$. Then on $G + *m$ Right's options all have the form $G + *a$ for some $a < m$, or $G^R + *m$ for some G^R. But $G + *a \mathrel{|\!\!\triangleright} 0$ by the Two-Ahead Rule, and $G^R + *m \mathrel{|\!\!\triangleright} 0$ by induction on G (since necessarily $w(G^R) \geq 0$). □

It turns out that Proposition 7.1 generalizes quite nicely to arbitrary G: the atomic weight $\text{aw}(G)$ is sufficient to determine $o(G + *m)$, for all m not occurring as a subposition of G. The presence of a sufficiently remote $*m$ has a simplifying effect, eliminating details about the fine structure of G from consideration and enabling the players to focus just on the weights of each component. This is the central idea of the atomic weight theory, and it will essentially motivate the *definition* of atomic weight.

Definition 7.2. A nimber $*m$ is **remote for** G if it is not equal to any subposition of G.

Proposition 7.3. *Suppose $*m_0$ is remote for G. Then for all $m \geq m_0$,*

$$o(G + *m) = o(G + *m_0).$$

Proof. This is a direct consequence of Norton's Lemma (Theorem 4.18 on page 91), since we can write

$$*m = *m_0 : *(m - m_0),$$

and $*m_0$ is not a subposition of G. □

Proposition 7.3 shows that if $*m$ is remote for G, then the exact value of m is immaterial—all that matters is that it's sufficiently large. It is therefore convenient to adopt the symbol $☆$ as shorthand for a sufficiently remote nimber. We refer to $☆$ as a **remote star** (or sometimes **far star**). Technically speaking, $☆$ does not represent a specific game, but rather an arbitrary $*m$ where m is sufficiently large to carry out the argument at hand.

Definition 7.4. For all G and H:

$$G \sim H \quad \text{if} \quad o(G + X + ☆) = o(H + X + ☆) \text{ for all } X.$$

$$G \gtrsim H \quad \text{if} \quad o(G + X + ☆) \geq o(H + X + ☆) \text{ for all } X.$$

Here the expression

$$o(G + X + ☆) = o(H + X + ☆)$$

is shorthand for

$$o(G + X + *m) = o(H + X + *m) \quad \text{for all sufficiently large } m,$$

with the understanding that the meaning of "sufficiently large" depends on the choice of X.

Although $G \sim H$ is defined for all games G and H, it will primarily be of interest when G and H are infinitesimals. (In fact the general case reduces to the infinitesimal case; see Exercise 7.10 on page 149.) It should be recognized as another variation on the fundamental equivalence.

Clearly $G \geq H$ implies $G \gtrsim H$, and $G = H$ implies $G \sim H$, so the relation $\sim$ partitions $\mathbb{G}$ into equivalence classes. Each equivalence class contains at most one representative of the form $n \cdot \uparrow$, since if $m > n$, then

$$m \cdot \uparrow - n \cdot \uparrow + ☆ > 0, \quad \text{whereas} \quad n \cdot \uparrow - n \cdot \uparrow + ☆ \not\gtrsim 0.$$

Moreover, if G is a flower garden and $n = w(G)$, then $G \sim n \cdot \uparrow$. We'll prove this shortly. In the meantime, note that if X is also a flower garden, then the identity $o(G + X + ☆) = o(n \cdot \uparrow + X + ☆)$ follows immediately from Proposition 7.1, since $w(G + X) = w(n \cdot \uparrow + X)$.

Various notational shortcuts will help to simplify many arguments involving remote stars. For example, we can write

$$☆ + *m = ☆$$

for all m, since if $*2^k$ is not equal to a subposition of G (say), then neither is $*(2^k \oplus m)$. Likewise, we can write

$$☆ + ☆ = ☆,$$

signifying that if $*m$ is remote for G and if $*n$ is remote for $G + *m$, then $*(m \oplus n)$ is remote for G.

The identity $☆ + * = ☆$, together with Proposition 4.12 (page 87), implies

$$\uparrow☆ = \{0 \mid ☆\}, \quad \Uparrow☆ = \{0 \mid \uparrow☆\}, \quad \ldots, \quad \uparrow n☆ = \{0 \mid \uparrow(n-1)☆\}.$$

All of these shortcuts will be used freely in arguments that involve a sufficiently remote nimber $*m$ and that are otherwise not sensitive to the exact value of m.

Theorem 7.5. *$G \gtrsim 0$ if and only if $G > \downarrow☆$.*

Proof. If $G \gtrsim 0$, then $o(G + \uparrow☆) \geq o(\uparrow☆)$. Since $\uparrow☆ > 0$ (Proposition 4.10 on page 87), this implies $G + \uparrow☆ > 0$, and hence $G > \downarrow☆$.

Conversely, suppose that $G > \downarrow☆$. Then since $\uparrow☆ = \{0 \mid ☆\}$, we have in particular that $G \mathrel{|\!\rhd} ☆$, and $G^R \mathrel{|\!\rhd} \downarrow☆$ for all G^R.

Now to establish $G \gtrsim 0$, we must show that $o(G + X + ☆) \geq o(X + ☆)$, for all X. For fixed X, it suffices to prove that

$$X + ☆ \mathrel{|\!\!>} 0 \quad \text{implies} \quad G + X + ☆ \mathrel{|\!\!>} 0 \quad \text{and}$$
$$G + X + ☆ \mathrel{<\!\!|} 0 \quad \text{implies} \quad X + ☆ \mathrel{<\!\!|} 0.$$

Suppose first that $G + X + ☆ \mathrel{<\!\!|} 0$. There are three cases.

Case 1: $G + X^R + ☆ \leq 0$ for some X^R. Then by induction on X, we may assume $X^R + ☆ \leq 0$, so that $X + ☆ \mathrel{<\!\!|} 0$.

Case 2: $G + X + *m \leq 0$ for some $*m$. Then

$$\begin{aligned} X + ☆ &\mathrel{<\!\!|} X + {\downarrow}☆ && (\text{since } {\downarrow}☆ = \{☆ \mid 0\}) \\ &= X + {\downarrow}☆ + *m && (\text{since } ☆ + *m = ☆) \\ &< G + X + *m && (\text{since } G > {\downarrow}☆) \\ &\leq 0. \end{aligned}$$

Case 3: $G^R + X + ☆ \leq 0$ for some G^R. Then since $G^R \mathrel{|\!\!>} {\downarrow}☆$, we have either $G^R \geq 0$ or else $G^{RL} > {\downarrow}☆$ for some G^{RL}. If $G^R \geq 0$, then $X + ☆ \mathrel{<\!\!|} 0$ follows immediately. If instead $G^{RL} > {\downarrow}☆$, then since

$$G^{RL} + X + ☆ \mathrel{<\!\!|} 0,$$

we may assume that $X + ☆ \mathrel{<\!\!|} 0$ by induction on G.

The proof that $X + ☆ \mathrel{|\!\!>} 0$ implies $G + X + ☆ \mathrel{|\!\!>} 0$ is similar; here Case 2 uses the identity $G \mathrel{|\!\!>} ☆$, and there is no Case 3. □

Theorem 7.5 yields a constructive test for $G \sim H$, since

$$G \sim H \iff G - H \sim 0 \iff {\uparrow}☆ > G - H > {\downarrow}☆.$$

Atomic Weight

As a first application of Theorem 7.5, we can now establish that $G \sim H$ whenever G and H are flower gardens of equal weight. For then

$${\uparrow}☆ > G - H > {\downarrow}☆,$$

since $G - H + {\uparrow}☆$ and $H - G + {\uparrow}☆$ are flower gardens of weight 1 that contain remote stars. In particular, this shows that

$$G \sim w(G) \cdot {\uparrow}$$

for every flower garden G.

A similar relationship holds between an arbitrary game G and its atomic weight $\mathrm{aw}(G)$. Now $\mathrm{aw}(G)$ is not necessarily an integer (as we remarked at the beginning of this section), so the following generalization of $n \cdot \uparrow$ is a key ingredient of the theory.

Definition 7.6. For all G, the **Norton product** of G by $\uparrow$ is given by

$$G \cdot \uparrow = \begin{cases} n \cdot \uparrow & \text{if } G \text{ is equal to an integer } n; \\ \{G^L \cdot \uparrow + \Uparrow* \mid G^R \cdot \uparrow + \Downarrow*\} & \text{otherwise.} \end{cases}$$

We will show that for any G, there is at most one H such that

$$G \sim H \cdot \uparrow.$$

When such an H exists, it is defined to be the atomic weight of G. Now there *are* certain games for which no such atomic weight exists—we'll see some examples shortly. Remarkably, however, every *dicotic* game has an atomic weight, and there is a simple calculus that can be used to determine it. This is the central result of the atomic weight theory, and it simplifies many calculations involving dicotic games.

We begin by describing the fundamental properties of Norton product.

Proposition 7.7. *For all G and H:*

(a) $(-G) \cdot \uparrow = -(G \cdot \uparrow)$.

(b) $(G + H) \cdot \uparrow = G \cdot \uparrow + H \cdot \uparrow$.

(c) $G \geq H$ *if and only if* $G \cdot \uparrow \geq H \cdot \uparrow$.

Proof. (a) is clear from the definition. (b) and (c) are consequences of (a) together with the following assertion:

(†) If $G + H + J \geq 0$, then $G \cdot \uparrow + H \cdot \uparrow + J \cdot \uparrow \geq 0$.

We will prove (†) simultaneously with the following similar statement:

(‡) If $G + H + J \mathrel{|\!\!\rhd} 0$, then $G \cdot \uparrow + H \cdot \uparrow + J \cdot \uparrow + \Downarrow* \mathrel{|\!\!\rhd} 0$.

The proof is by induction on the *noninteger* components of $G + H + J$. At each stage of the induction, we first prove (‡) and then use this to establish (†).

We first show (‡). If all three of G, H, and J are equal to integers, the conclusion is clear (since $\downarrow* \mathrel{|\!\!\rhd} 0$), so assume without loss of generality that G is not equal to an integer. Now $G + H + J \mathrel{|\!\!\rhd} 0$, so by the Integer Avoidance Theorem (Corollary 3.28 on page 80), we may assume that some $G^L + H + J \geq 0$. By induction on (†), this implies

$$G^L \cdot \uparrow + H \cdot \uparrow + J \cdot \uparrow \geq 0,$$

and since $(G \cdot \uparrow)^L + \Downarrow* = G^L \cdot \uparrow$, this suffices.

We now prove (†). By symmetry it suffices to show that

$$(G \cdot \uparrow)^R + H \cdot \uparrow + J \cdot \uparrow \mathrel{|\!\!\rhd} 0.$$

If G is not equal to an integer, then in fact $(G \cdot \uparrow)^R = G^R \cdot \uparrow + \Downarrow *$. Since necessarily $G^R + H + J \mathrel{|\!\!>} 0$, the desired conclusion follows by induction on (‡).

Now suppose G is equal to an integer. Then $(G \cdot \uparrow)^R$ is either $(G-1) \cdot \uparrow + *$ or $(G+1) \cdot \uparrow$, depending on whether G is positive or negative. In either case we have

$$(G \cdot \uparrow)^R \geq (G+1) \cdot \uparrow + \Downarrow *.$$

But $G + H + J \geq 0$, so $(G+1) + H + J > 0$. Since the *noninteger* components of $(G+1) + H + J$ are equally simple as those of $G + H + J$, it is safe to assume (‡) for $(G+1) + H + J$. Therefore

$$(G \cdot \uparrow)^R + H \cdot \uparrow + J \cdot \uparrow \geq (G+1) \cdot \uparrow + H \cdot \uparrow + J \cdot \uparrow + \Downarrow * \mathrel{|\!\!>} 0. \qquad \square$$

Lemma 7.8. *If $G \cdot \uparrow \gtrsim 0$, then $G \geq 0$.*

Proof. Suppose instead that $G \mathrel{<\!\!|} 0$; we will show that $G \cdot \uparrow \not\gtrsim 0$. By Theorem 7.5, it suffices to show that $G \cdot \uparrow + \uparrow\!\star \mathrel{<\!\!|} 0$. If G is equal to an integer, then necessarily $G < 0$, so $G + 1 \leq 0$ and we have

$$G \cdot \uparrow + \uparrow\!\star = (G+1) \cdot \uparrow + \star \leq \star \not\gtrsim 0.$$

Suppose G is not equal to an integer. Then since $G \mathrel{<\!\!|} 0$, Right has a winning move, say to $G^R \leq 0$. By Proposition 7.7(c), $G^R \cdot \uparrow \leq 0$, so

$$(G^R \cdot \uparrow + \Downarrow *) + \uparrow\!\star = G^R \cdot \uparrow + \downarrow\!\star \leq \downarrow\!\star < 0.$$

Therefore Right has a winning move from $G \cdot \uparrow + \uparrow\!\star$. $\square$

Corollary 7.9. *For any G, there is at most one H such that $G \sim H \cdot \uparrow$ (up to equality of game values).*

Proof. If $G \sim H \cdot \uparrow \sim J \cdot \uparrow$, then $(H - J) \cdot \uparrow \sim 0$, so by Lemma 7.8 $H - J = 0$. $\square$

We are now finally positioned to give the definition of atomic weight.

Definition 7.10. G is said to be **atomic** if $G \sim H \cdot \uparrow$ for some H.

Definition 7.11. If G is atomic, then the **atomic weight** of G, denoted $\mathrm{aw}(G)$, is the unique value $H \in \mathbb{G}$ such that $G \sim H \cdot \uparrow$.

Examples. Consider $\multimap_1 = \{1 \mid 0 \| 0\}$. It's easy to see that

$$\downarrow\!\star \leq \multimap_1 \leq \uparrow\!\star$$

since on $\multimap_1 + \uparrow\!\star$ Right's moves are to $\uparrow\!\star$ and $\multimap_1 + \star$, which are both $\mathrel{|\!\!>} 0$; while on $\multimap_1 + \downarrow\!\star$ Left's moves are to $\{1 \mid 0\} + \downarrow\!\star$ and $\multimap_1$, both $\mathrel{<\!\!|} 0$. This shows that $\multimap_1 \sim 0$, so that $\mathrm{aw}(\multimap_1) = 0$.

Likewise, consider $G = \{1 \mid \uparrow \| 0\}$. We'll show that $G \sim \uparrow$, so that $\mathrm{aw}(G) = 1$. For this we must check that

$$G + \downarrow + \downarrow\star \le 0 \quad \text{and} \quad G + \downarrow + \uparrow\star \ge 0.$$

But on $G + \Downarrow\star$ Left's moves are to $G + \downarrow\star$ and $\{1 \mid \uparrow\} + \Downarrow\star$, and Right can revert either of these to $\downarrow\star$, which is negative. Likewise, on $G + \star$ Right's moves are to $\star$, which is fuzzy, and $G + *m$. But Left can move from $G + *m$ to $\{1 \mid \uparrow\} + *m$, forcing Right's reply to $\uparrow *m \mathrel{|\!\!>} 0$.

Example. Let $G = \{1 \mid \Uparrow \| 0\}$. Then G is certainly an infinitesimal, and from the previous example we might be tempted to conclude that $G \sim \Uparrow$. But in fact G is not even atomic. To see this, we must show that no H satisfies both

$$G + H \cdot \uparrow + \downarrow\star \le 0 \quad \text{and} \quad G + H \cdot \uparrow + \uparrow\star \ge 0.$$

So fix H and suppose $G + H \cdot \uparrow + \downarrow\star \le 0$. Then if Left moves from G to $\{1 \mid \Uparrow\}$, Right must have a winning response. Now $H \cdot \uparrow + \downarrow\star$ is dicotic, so by the Lawnmower Theorem, Right's winning move must be from $\{1 \mid \Uparrow\}$ to $\Uparrow$. This shows that

$$\Uparrow + H \cdot \uparrow + \downarrow\star \le 0.$$

But $\Uparrow + \downarrow\star = \uparrow\star$, so Right can win on

$$G + H \cdot \uparrow + \uparrow\star$$

by moving from G to 0. Therefore $G + H \cdot \uparrow + \uparrow\star \mathrel{<\!\!|} 0$, as needed.

Left's move from G to $\{1 \mid \Uparrow\}$ behaves like an "infinitely large threat": Right is compelled to respond before he may consider moving in a dicotic component. It is the presence of this threat, not commensurate with any dicotic game, that causes G to fail to be atomic. However, it is crucial that Left's payoff for completing the threat be at least $\Uparrow$; comparing this with the previous example shows how subtle this type of argument can be.

The following proposition summarizes the fundamental properties of atomic weight.

Proposition 7.12. *Let G and H be atomic. Then:*

(a) $\mathrm{aw}(-G) = -\mathrm{aw}(G)$.

(b) $\mathrm{aw}(G + H) = \mathrm{aw}(G) + \mathrm{aw}(H)$.

(c) $G \gtrsim H$ *if and only if* $\mathrm{aw}(G) \ge \mathrm{aw}(H)$.

Proof. (a) and (b) are restatements of Proposition 7.7.

For (c), first assume $G \gtrsim H$. Then $\mathrm{aw}(G) \cdot \uparrow \gtrsim \mathrm{aw}(H) \cdot \uparrow$, and by Lemma 7.8 this implies $\mathrm{aw}(G) \ge \mathrm{aw}(H)$.

Conversely, suppose $\mathrm{aw}(G) \geq \mathrm{aw}(H)$. Then from Proposition 7.7(c), we have $\mathrm{aw}(G) \cdot \uparrow \geq \mathrm{aw}(H) \cdot \uparrow$, whereupon $G \gtrsim H$. □

If $G \geq 0$, then certainly $\mathrm{aw}(G) \geq 0$. The converse is clearly false: for example, $\mathrm{aw}(\uparrow *) = \mathrm{aw}(\uparrow) + \mathrm{aw}(*) = 1$, but $\uparrow * \not\geq 0$. The following theorem is a sweeping generalization of the Two-Ahead Rule (Theorem 4.21 on page 92) and Proposition 7.1, essentially extending the theory of flower gardens to arbitrary atomic games.

Theorem 7.13. *Let G be atomic. Then:*

(a) *If $\mathrm{aw}(G) \mathrel{|\!\!\triangleright} 0$, then $G \mathrel{|\!\!\triangleright} 0$.*

(b) *If $\mathrm{aw}(G) \geq 2$, then $G > 0$.*

(c) $\mathrm{aw}(G) \geq 1$ *if and only if* $G > \text{☆}$.

Proof. (a) is trivial, since if $G \leq 0$, then necessarily $G \lesssim 0$. For (b), suppose that $\mathrm{aw}(G) \geq 2$. Then $G \gtrsim \Uparrow$, so

$$G > \Uparrow + \downarrow\text{☆} = \uparrow\text{☆} > 0.$$

For (c) we have, by Theorem 7.5 and Proposition 7.12,

$$\mathrm{aw}(G) \geq 1 \text{ iff } G \gtrsim \uparrow \text{ iff } G + \downarrow \gtrsim 0 \text{ iff } G + \downarrow > \downarrow\text{☆} \text{ iff } G > \text{☆}.$$ □

The hypotheses in Theorem 7.13(a) and (b) cannot be weakened (see Exercise 7.3).

The Atomic Weight Calculus

Every dicotic game G is necessary atomic. To prove this, we'll introduce a recursive calculus for defining an auxiliary function $\upsilon(G)$ (denoted by the Greek letter *upsilon*) and then show inductively that $\upsilon(G) = \mathrm{aw}(G)$. This strategy resembles our approach to the thermographic calculus in Section 5.

Definition 7.14. Let G be dicotic. We define $\upsilon(G)$ and $\tilde{\upsilon}(G)$ recursively as follows. Put

$$\tilde{\upsilon}(G) = \{\upsilon(G^L) - 2 \mid \upsilon(G^R) + 2\}.$$

Then $\upsilon(G) = \tilde{\upsilon}(G)$ *unless* $\tilde{\upsilon}(G)$ is equal to an integer. In that case, let

$$\mathcal{I} = \{n \in \mathbb{Z} : \upsilon(G^L) - 2 \mathrel{\triangleleft\!\!|} n \mathrel{\triangleleft\!\!|} \upsilon(G^R) + 2 \text{ for all } G^L \text{ and } G^R\}$$

and define

$$\upsilon(G) = \begin{cases} \text{the smallest element of } \mathcal{I} & \text{if } G < \text{☆}; \\ \text{the largest element of } \mathcal{I} & \text{if } G > \text{☆}; \\ 0 & \text{if } G \not\gtrless \text{☆}. \end{cases}$$

Theorem 7.15. *Let G be dicotic. Then G is atomic and $\upsilon(G) = \mathrm{aw}(G)$.*

The following lemma is used in the proof of Theorem 7.15:

Lemma 7.16. *If G is not equal to an integer, then $G \cdot \uparrow$ has both a Left incentive and a Right incentive that are $\geq \uparrow*$.*

Proof. By Theorem 3.27 (page 80), there is some G^L with $G^L - G \geq -1$. Therefore $G^L \cdot \uparrow - G \cdot \uparrow \geq \downarrow$, so that

$$\begin{aligned}(G \cdot \uparrow)^L - G \cdot \uparrow &= G^L \cdot \uparrow + \Uparrow* - G \cdot \uparrow \\ &\geq \Uparrow* + \downarrow \\ &= \uparrow*.\end{aligned}$$

The same argument works for Right incentives. □

Proof of Theorem 7.15. By symmetry, it suffices to show $\upsilon(G) \cdot \uparrow \gtrsim G$. By Theorem 7.5, we must show that Left can win playing second on

$$X = \upsilon(G) \cdot \uparrow - G + \uparrow☆.$$

First suppose $\upsilon(G)$ is not equal to an integer, so that $\upsilon(G) = \tilde{\upsilon}(G)$. Now

$$\Delta^R(\uparrow☆) = \uparrow*,$$

so by Lemma 7.16 the Right incentives of $\uparrow☆$ are dominated by those of $\upsilon(G) \cdot \uparrow$. By the Dominated Incentives Theorem (Theorem 1.32 on page 62), we can disregard Right's moves from X on the component $\uparrow☆$. But

$$\upsilon(G) \cdot \uparrow = \{\upsilon(G^L) \cdot \uparrow + * \mid \upsilon(G^R) \cdot \uparrow + *\},$$

so by induction (and the fact that $☆ + * = ☆$) Right's moves from X on $\upsilon(G) \cdot \uparrow$ and $-G$ have mirror-image responses on the other component.

Now assume $\upsilon(G)$ is equal to an integer. Then $X = \uparrow n☆ - G$, where $n = \upsilon(G) + 1$. We may assume that $\uparrow n☆$ is in canonical form (since the assertion $X \geq 0$ does not depend on the form of X), so by Proposition 4.12 (page 87),

$$\uparrow n☆ = \begin{cases} \{\uparrow(n{+}1)☆ \mid 0\} & \text{if } n < 0; \\ ☆ & \text{if } n = 0; \\ \{0 \mid \uparrow(n{-}1)☆\} & \text{if } n > 0. \end{cases}$$

There are four cases.

Case 1: Right moves to $\uparrow n☆ - G^L$. Then by induction, $G^L \sim \upsilon(G^L) \cdot \uparrow$, so we may replace $-G^L + ☆$ by $-\upsilon(G^L) \cdot \uparrow + ☆$. Now $\upsilon(G^L) - 2 \lhd\!\!\mid \upsilon(G)$. If $\upsilon(G^L)$ is also an integer, then $\upsilon(G^L) - 1 \leq \upsilon(G)$, so $\upsilon(G^L) \leq n$ and hence

$$\uparrow n☆ - \upsilon(G^L) \cdot \uparrow \geq ☆ \mid\!\!\rhd 0.$$

If $v(G^L)$ is not an integer, then by the Integer Avoidance Theorem there is some $v(G^L)^R \leq n+1$ and Left can move to

$$\uparrow n\star - v(G^L)^R \cdot \uparrow + \Uparrow\!* \geq \big(n - (n+1) + 2\big) \cdot \uparrow + \star = \uparrow\star > 0.$$

Case 2: $v(G) < 0$ and Right moves on $\uparrow n\star$. Since $v(G)$ is a negative integer, we have $n \leq 0$. Thus Right's move leaves a position of the form $*m - G$ (where in fact $m = 0$ except possibly when $n = 0$). But since $v(G) < 0$, we know that $G < \star$. Since $\star \not\gtrless *m$, this implies $*m - G \mathrel{|\!\rhd} 0$.

Case 3: $v(G) = 0$ and Right moves on $\uparrow n\star$. Then $n = 1$, so Right's move leaves the position $\star - G$. But since $v(G) = 0$, we have $G \not\gtrless \star$, so that $\star - G \mathrel{|\!\rhd} 0$.

Case 4: $v(G) > 0$ and Right moves on $\uparrow n\star$. Then Right's move leaves the position $X^R = v(G) \cdot \uparrow + \star - G$. But in this case, $G > \star$, so $v(G)$ is the largest element of $\mathcal{I}$ (with $\mathcal{I}$ as in Definition 7.14). Therefore $v(G) + 1 \notin \mathcal{I}$. By definition of $\mathcal{I}$, there is some G^R with $v(G^R) + 2 \leq v(G) + 1$. Therefore Left can move from X^R to

$$\begin{aligned} X^{RL} &= v(G) \cdot \uparrow + \star - G^R \\ &\geq \big(v(G^R) + 1\big) \cdot \uparrow + \star - G^R \\ &= v(G^R) \cdot \uparrow - G^R + \uparrow\star. \end{aligned}$$

By induction, $X^{RL} \geq 0$. □

Clobber

In Clobber every position is an infinitesimal! The game is played on an $m \times n$ grid such as shown in Figure 7.2. On her turn, Left must move a black stone onto an orthogonally adjacent white stone. The white stone is *clobbered* and removed from the game. An isolated black stone, or a black stone that only touches other black stones, is unable to move. Right's moves are the same with colors interchanged.

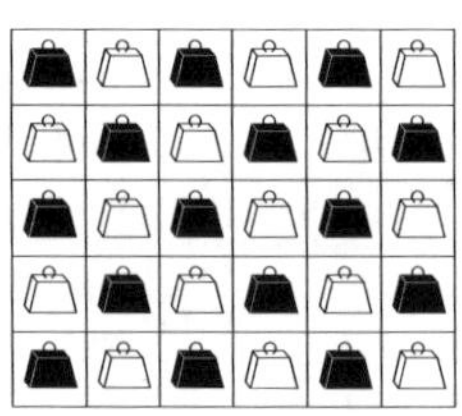

Figure 7.2. Starting position on 5×6 Clobber.

Every position in CLOBBER is dicotic, since whenever Left can clobber Right, then Right can just as easily do the same to Left. Familiar values arise naturally; for example:

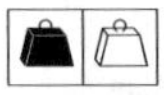 $= \{0 \mid 0\} = *$,

 $= \{0 \mid *\} = \uparrow$,

 $= \{0 \mid \uparrow\} = \Uparrow*$,

 $= \{0 \mid \Uparrow*\} = \Uparrow\!\!\uparrow$,

and so on. But other sequences of positions exhibit great combinatorial complexity. For example, *cgsuite* cheerfully tells us that

 $= \pm(*, \uparrow)$,

 $= 0$,

 $= \pm\left(\uparrow^{[2]}, \{\uparrow*, \Uparrow \mid 0, \uparrow*, \pm(0, \uparrow*)\}\right)$,

 $= \pm\left(\uparrow, \uparrow^{[2]}*, \{0 \mid \uparrow, \pm(*, \uparrow)\}, \{\Uparrow* \mid \downarrow, \pm(*, \uparrow)\}\right)$,

and it only gets worse from there. (Here $(X)^n$ denotes the juxtaposition of n copies of X, so that  $^3 =$ , etc.) In such cases the atomic weight calculus can be a powerful tool.

As an example, we'll use the atomic weight calculus to show that every position  ^{n} in the previous sequence has atomic weight 0. We'll show simultaneously that every position in the related sequence  ^{n} has atomic weight 1, except for  $= 0$ and  $= *$. First we give a preliminary lemma.

Lemma 7.17. *Let G be dicotic and let n be an integer, and suppose that $\mathrm{aw}(G^L) \le n+1$ for all G^L and $\mathrm{aw}(G^R) \ge n-1$ for all G^R. Suppose furthermore that there exist G^{L_0} and G^{R_0} that attain both bounds. Then $\mathrm{aw}(G) = n$.*

Proof. The assumptions are such that $\mathrm{aw}(G^{L_0}) \ge \mathrm{aw}(G^L)$ for all G^L and $\mathrm{aw}(G^{R_0}) \le \mathrm{aw}(G^R)$ for all G^R. Thus the atomic weights of G^{L_0} and G^{R_0} dominate all others, and

$$\tilde{v}(G) = \left\{\mathrm{aw}(G^{L_0}) - 2 \;\middle|\; \mathrm{aw}(G^{R_0}) + 2\right\} = \{n-1 \mid n+1\} = n.$$

Therefore $\mathrm{aw}(G)$ is an integer, and since n is the *only* integer such that $n - 1 \lhd\!\!| \; n \lhd\!\!| \; n + 1$, we necessarily have $\mathrm{aw}(G) = n$. □

Consider $G =$  ^{n}. For $n = 1$ we can check directly that $G = *$, so assume $n > 1$. Then every G^L has the form

$X_{ab} =$  $^a +$   ^{b} or $Y_{ab} =$  ^{a}  $+$   ^{b}.

Now if $b = 0$, then $\mathrm{aw}(X_{ab}) = 0$. If $b > 0$, then X_{ab} has the Right option

$$X^R_{ab} = (\bullet\circ)^a + \bullet\circ + (\bullet\circ)^{b-1},$$

and by induction $\mathrm{aw}(X^R_{ab}) = 0$. This shows that $\mathrm{aw}(X_{ab}) \le 1$.

Likewise, if $a = 0$, then $\mathrm{aw}(Y_{ab}) = 0$ or 1, depending on whether $b \le 1$ or $b > 1$. If $a = 1$, then $\bullet\circ\bullet\bullet = {\downarrow}{*}$, so $\mathrm{aw}(Y_{ab}) = -1$ or 0. If $a = 2$, then we have $Y^R_{ab} = \bullet\circ\circ + \circ(\bullet\circ)^b$; and since $\bullet\circ\circ = {\downarrow}$, $\mathrm{aw}(Y^R_{ab}) = -1$ or 0. Finally, if $a > 2$, then Y_{ab} has the Right option

$$Y^R_{ab} = (\bullet\circ)^{a-1}\bullet + \circ\bullet + \circ(\bullet\circ)^b$$

with $\mathrm{aw}(Y^R_{ab}) = -1$ or 0. So in all cases, $\mathrm{aw}(Y_{ab}) \le 1$.

We have shown that $\mathrm{aw}(G^L) \le 1$ for every G^L. Furthermore, since $\bullet\bullet\circ = {\uparrow}$, we have $\mathrm{aw}(X_{n-2,b}) = 1$, so this bound is attained. By symmetry every $\mathrm{aw}(G^R) \ge -1$, and this bound is also attained. By Lemma 7.17, we must have $\mathrm{aw}(G) = 0$.

Now consider $H = \circ(\bullet\circ)^n$. For $n = 2$ it is easily checked that $H = {\uparrow}^{[2]}$, so assume $n > 2$. Then every H^L has the form

$$Z_{ab} = \circ(\bullet\circ)^a + \bullet(\bullet\circ)^b.$$

If $b = 0$, then $\mathrm{aw}(Z_{ab}) \le 1$, while if $b > 0$, we have

$$Z^R_{ab} = \circ(\bullet\circ)^a + \bullet\circ + (\bullet\circ)^{b-1}$$

with $\mathrm{aw}(Z^R_{ab}) \le 1$. Therefore $\mathrm{aw}(Z_{ab}) \le 2$, and since $n > 2$, this bound is clearly met when $b = 1$.

Finally, every H^R has the form

$$W_{ab} = \circ(\bullet\circ)^a\circ + (\bullet\circ)^b.$$

When $a = 0$, we have $\mathrm{aw}(W_{ab}) = 0$. When $a = 1$, we have $\circ\bullet\circ\circ = {\uparrow}{*}$ and hence $\mathrm{aw}(W_{ab}) = 1$. If $a = 2$, then $W^L_{ab} = \circ\bullet\circ\circ + (\bullet\circ)^b$ has atomic weight 1. Finally, if $a > 2$, then there is the Left option

$$W^L_{ab} = \circ(\bullet\circ)^{a-1} + \bullet\circ + (\bullet\circ)^b$$

with $\mathrm{aw}(W^L_{ab}) = 1$. In all cases, $\mathrm{aw}(W_{ab}) \ge 0$, and the bound is attained when $a = 0$.

We have shown that $\mathrm{aw}(H^L) \le 2$ for all H^L, that $\mathrm{aw}(H^R) \ge 0$ for all H^R, and that there exist options that attain both bounds. By Lemma 7.17, we have $\mathrm{aw}(H) = 1$.

Exercises

7.1 Determine the atomic weights of ${\uparrow}^n$, ${\uparrow}^{[n]}$, $+_x$, and $\{0^n \mid +_x\}$, for all $x > 0$ and $n \ge 1$.

7.2 If $\mathrm{aw}(G) = 0$, then $G \ll {\Uparrow}$.

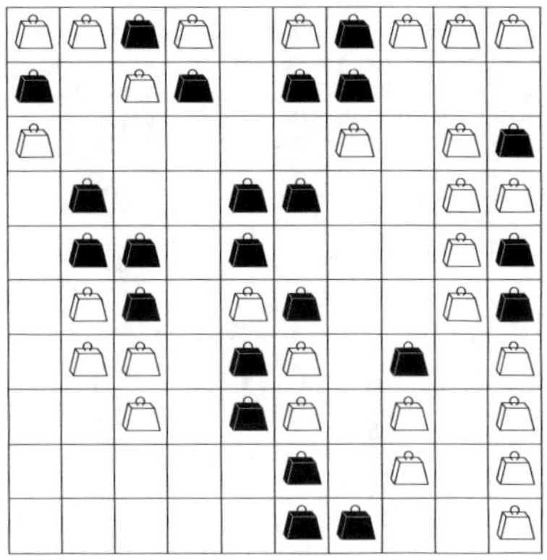

Figure 7.3. A tricky CLOBBER problem (Duffy and Kolpin, 2002).

7.3 (a) If $H \le 0$, then there exists a $G \le 0$ with $\mathrm{aw}(G) = H$.
(b) If $H \lhd\!| \; 2$, then there exists a $G \lhd\!| \; 0$ with $\mathrm{aw}(G) = H$.

7.4 If $R(\mathrm{aw}(G)) > 0$, then $n \cdot G > 0$ for some n.

7.5 True or false:
(a) If $G > 0$ and $G > *$, then $\mathrm{aw}(G) \ne 0$.
(b) If $\mathrm{aw}(G) = 0$, then $G \lhd\!| \uparrow^x$ for some number $x > 1$ (cf. Exercise 4.24).
(c) If $\mathrm{aw}(G) > 0$, then $n \cdot G \ge 0$ for some n.

7.6 Exhibit a sequence of games $G_0 < G_1 < G_2 < \cdots$ of atomic weight 0 such that whenever $\mathrm{aw}(H) = 0$, we have $H < G_n$ for some n.

7.7 Show that G is atomic if and only if there exist dicotic games H and J such that $H \gtrsim G \gtrsim J$ and $\mathrm{aw}(H) = \mathrm{aw}(J)$.

7.8 If $G^L \le G^R$ for every G^L and G^R, then $G - m(G)$ is atomic with integer atomic weight. (This strengthens Exercise 3.14 on page 81.)

7.9 The value of every HACKENBUSH position has the form $x + G$, where x is a number and G is dicotic with integer atomic weight.

7.10 If $G \sim H$, then $G \equiv H \pmod{\mathrm{Inf}}$. Conclude that the general theory of $\sim$ reduces to the case where G and H are infinitesimals.

7.11 If G is hereditarily transitive, then $G - \mathrm{rcf}(G)$ is atomic with atomic weight 1, 0, or -1.

7.12 The **bynumbers** B_n are given recursively by $B_n = \{0, B_a + B_{n-a} \mid *\}_{1 \le a < n}$. (In particular $B_1 = \{0 \mid *\} = \uparrow$.) Show that $\mathrm{aw}(B_n) = 1$ for all $n \ge 1$.

7.13 Analyze the CLOBBER sequences ()^n and ()^n.

7.14 Determine the outcome class of the CLOBBER position in Figure 7.3. (Hint: The western component has value $*2$. Compute the atomic weights of the other components and invoke Theorem 7.13, presuming $*2$ to be remote.)

7.15 Let $U > 0$. The **generalized Norton product** of G by U is given by

$$G{\cdot}U = \begin{cases} n \cdot U & \text{if } G \text{ is equal to an integer } n; \\ \{G^L \cdot U + U + \Delta \mid G^R \cdot U - U - \Delta\} & \text{otherwise,} \end{cases}$$

where Δ ranges over all Left and Right incentives of U.

(a) Prove Proposition 7.7 with $\uparrow$ replaced by arbitrary $U > 0$.

(b) Show that if $U \lhd\!\!| \; 0$, then Proposition 7.7(b) and (c) both fail.

(c) Show that if U is infinitesimal (resp. dicotic), then so is $G \cdot U$.

(d) True or false: If $U > 0$ and $H \cdot U > 0$, then $(G \cdot H) \cdot U = G \cdot (H \cdot U)$.

(e) True or false: $U = V$ implies $G \cdot U = G \cdot V$.

7.16 Let G be dicotic. The **positive galvanized atomic weight** $\text{aw}^+(G)$ is defined as follows. Put

$$\tilde{v}^+(G) = \{\text{aw}^+(G^L) - 2 \mid \text{aw}^+(G^R) + 2\}.$$

Then $\text{aw}^+(G) = \tilde{v}^+(G)$ *unless* $\tilde{v}^+(G)$ is an integer. In that case, $\text{aw}^+(G)$ is the *largest* integer n such that

$$\text{aw}^+(G^L) - 2 \lhd\!\!| \; n \lhd\!\!| \; \text{aw}^+(G^R) + 2$$

for all G^L and G^R. The **negative galvanized atomic weight** $\text{aw}^-(G)$ is defined the same way, but we choose the *smallest* integer n.

(a) Show that $G = H$ does *not* imply $\text{aw}^+(G) = \text{aw}^+(H)$.

(b) In the **galvanized sum** $G \circledast H$, the winner is determined based on the *component* in which the last move is made. Left wins if the last move is made on G, and Right wins if it is made on H, *regardless of who actually makes the last move.* Prove that the outcome of

$$(G_1 + G_2 + \cdots + G_m) \circledast (H_1 + H_2 + \cdots + H_n)$$

is exactly the same as the (normal-play) outcome of

$$\text{aw}^+(G_1) + \cdots + \text{aw}^+(G_m) + \text{aw}^-(H_1) + \cdots + \text{aw}^-(H_n) + ☆.$$

(c) Show that if G is any impartial game, then $G \circledast G$ is a first-player win.

Notes

The atomic weight theory was invented by Simon Norton in an extraordinary display of mathematical imagination. The theory was sketched in ONAG and first appeared completed in *Winning Ways.* Its curious nature and peculiar relationship to the nimbers remain deeply mysterious.

The atomic weight theory obviously has little to say about the vast hierarchy of infinitesimals $G \ll \uparrow$. Conway wrote in ONAG:

> Perhaps we can hope for an extended version of this theory which would enable us to measure still smaller games in terms of $\uparrow^2$, $\uparrow^3$, and so on. But the curious and complicated nature of the atomic weight algorithm suggests that any such theory will be very difficult to find. What do we expect to play the role of the remote stars, which enter so mysteriously and essentially into our theory? [**Con01**]

Scant progress has been made in understanding this problem, so we leave it as a question.

Question. Can the atomic weight theory be generalized to higher-order uptimals?

Extensions of the atomic weight calculus are similarly underexplored. Atomic weight is defined for all atomic games, but the calculus applies only to dicotic ones. The following problem seems natural (and feasible):

Open Problem. Extend the atomic weight calculus to a procedure for computing $\mathrm{aw}(G)$ for arbitrary atomic G.

A more subtle question concerns the structure of games that are *not* atomic. We've seen that aw is a surjective homomorphism

$$\mathrm{aw} : \mathcal{A} \to \mathbb{G}$$

where $\mathcal{A} = \{G \in \mathbb{G} : G \text{ is atomic}\}$. The atomic weight calculus implies $\mathbb{G}^0 \subset \mathcal{A}$, but the example on page 143 shows that its domain cannot be extended to all of $\mathbb{G}$. However, aw can be viewed as the *restriction* of a more general homomorphism

$$\mathrm{av} : \mathbb{G} \to \mathbb{G}^+$$

into a suitable extension $\mathbb{G}^+$ of $\mathbb{G}$. This extension is the group of **atomic values**, which we now describe.

Let $\mathcal{Z} = \{G \in \mathbb{G} : G \sim 0\}$ be the group of game values with atomic weight 0. (Note that $\mathcal{Z}$ is just the kernel of the atomic weight homomorphism.) Let

$$\mathrm{av} : \mathbb{G} \to \mathbb{G}/\mathcal{Z}$$

be the quotient map sending each $G \in \mathbb{G}$ to its equivalence class modulo $\mathcal{Z}$. We say that $\mathrm{av}(G)$ is the **atomic value** of G. The atomic weight calculus shows that $\mathrm{av}[\mathcal{A}] \cong \mathbb{G}$; this isomorphism identifies the atomic value of each $G \in \mathcal{A}$ with its corresponding atomic weight. Put another way: given a value $x \in \mathbb{G}/\mathcal{Z}$, the isomorphism selects the unique $H \in \mathbb{G}$ such that $H \cdot \uparrow \in x$.

Then the extension $\mathbb{G}^+$ is simply the quotient $\mathbb{G}/\mathcal{Z}$, with the subgroup $\mathrm{av}[\mathcal{A}]$ identified with $\mathbb{G}$ via this natural isomorphism. $\mathbb{G}^+$ contains atomic values for games like

$$\{1 \mid \Uparrow \| 0\},$$

which isn't equivalent to any dicotic game.

Question. What can be said about the structure of $\mathbb{G}^+$? What is its isomorphism type?

Question. Are there natural representatives of arbitrary elements of $\mathbb{G}^+$ that help clarify its structure, in the same way atomic weight clarifies the structure of $\mathbb{G}^0/\mathcal{Z}$?

CLOBBER was invented in 2002 by Albert, Grossman, Nowakowski, and Wolfe [**AGNW05**]. The problem shown in Figure 7.3 was composed by Adam Duffy and Garrett Kolpin; it was the winning entry in a 2002 Clobber Problem Composition contest held in Edmonton, Alberta.

Exercise 7.16(c): Fred Galvin.

Chapter III

The Structure of $\mathbb{G}$

The study of $\mathbb{G}$ in Chapter II introduced powerful tools for analyzing specific game values, including key invariants such as mean, temperature, reduced canonical form, and atomic weight. These invariants expose a high degree of regularity in the structure of $\mathbb{G}$ itself. In this chapter, the focus will be on understanding the abstract structure of $\mathbb{G}$ as a problem of independent interest, rather than as a tool for studying specific examples.

There are two principal aspects to the structure theory of $\mathbb{G}$. One is combinatorial: the hereditary structure of games built up from their subpositions. The other is algebraic: the abstract group structure of $\mathbb{G}$. The interplay between the algebraic and combinatorial structure of $\mathbb{G}$ remains poorly understood, and a fuller understanding of their interaction is one of the great challenges of combinatorial game theory.

Sections 1 and 2 focus on the hereditary structure of $\mathbb{G}$, in particular the sets $\mathbb{G}_n$ and their partial-order structure. We also undertake a less detailed study of certain related hierarchies, including $\mathbb{G}_n^0$ and $\mathbb{G}_n/\text{Inf}$. Section 3 is concerned primarily with the group structure of $\mathbb{G}$ itself, with a slight digression into the group structure generated by $\mathbb{G}_2$.

The lattice structure of $\mathbb{G}_n$ and its subclasses has been a particularly active area of research in recent years; some of the most exciting advances, and topics for further study, are summarized in the notes for Section 2 on page 166.

1. Hereditary Structure

Recall that the **birthday** $\text{b}(G)$ of a game G is the smallest integer $n \geq 0$ such that $G \in \mathbb{G}_n$. We'll say that G is **born on day** n if $\text{b}(G) = n$, and

<table>
<tr><td></td><td colspan="7">Right Options</td></tr>
<tr><td rowspan="7">Left Options</td><td></td><td>-1</td><td>$0,*$</td><td>0</td><td>$*$</td><td>1</td><td>—</td></tr>
<tr><td>1</td><td>± 1</td><td>$\{1 \mid 0,*\}$</td><td>$\{1 \mid 0\}$</td><td>$\{1 \mid *\}$</td><td>$1*$</td><td>2</td></tr>
<tr><td>$0,*$</td><td>$\{0,* \mid -1\}$</td><td>$*2$</td><td>$\uparrow*$</td><td rowspan="2">$\uparrow$</td><td rowspan="2">$\frac{1}{2}$</td><td rowspan="2">1</td></tr>
<tr><td>0</td><td>$\{0 \mid -1\}$</td><td>$\downarrow*$</td><td>$*$</td></tr>
<tr><td>$*$</td><td>$\{* \mid -1\}$</td><td colspan="2">$\downarrow$</td><td colspan="3" rowspan="3">0</td></tr>
<tr><td>-1</td><td>$-1*$</td><td colspan="2">$-\frac{1}{2}$</td></tr>
<tr><td>—</td><td>-2</td><td colspan="2">-1</td></tr>
</table>

Figure 1.1. The 22 games born by day 2.

born by day n if $\mathrm{b}(G) \leq n$. A basic goal of the structure theory is to count the number of *unequal* games born by day n—equivalently, the number of distinct game values born by day n; the number of canonical forms born by day n; or simply the cardinality of $\mathbb{G}_n$.

For $n = 0$ and 1 this is a simple matter.

Proposition 1.1. $\mathbb{G}_0 = \{0\}$ *and* $\mathbb{G}_1 = \{0, 1, -1, *\}$.

Proof. $\mathbb{G}_0 = \{0\}$ by definition. If $\mathrm{b}(G) = 1$, then 0 is the only option of G, so the only possibilities are $\{0 \mid \}$, $\{ \mid 0\}$, and $\{0 \mid 0\}$, which are defined to be 1, -1, and $*$, respectively. □

Now suppose that $G \in \mathbb{G}_n$, and write $G = \left\{\mathscr{G}^L \mid \mathscr{G}^R\right\}$ in canonical form. By identifying each element of $\mathbb{G}_{n-1}$ with its canonical form, we can regard $\mathscr{G}^L$ and $\mathscr{G}^R$ as subsets of $\mathbb{G}_{n-1}$. Since G is in canonical form, no two elements of $\mathscr{G}^L$ can be pairwise comparable (otherwise one of them would dominate the other). So $\mathscr{G}^L$ must be an antichain of $\mathbb{G}_{n-1}$, and likewise $\mathscr{G}^R$. Using this idea, we can prove:

Theorem 1.2. *There are* 22 *games born by day* 2.

Proof. By inspection, there are six antichains of $\mathbb{G}_1$: four one-element antichains; the empty antichain $\emptyset$; and $\{0, *\}$. Every element of $\mathbb{G}_2$ is represented by a pair of antichains from $\mathbb{G}_1$, so there are 36 such pairs to consider. To isolate the exact size of $\mathbb{G}_2$, we must determine which pairs of antichains $G = \left\{\mathscr{G}^L \mid \mathscr{G}^R\right\}$ are canonical—that is, contain no reversible option.

Suppose that G^L is reversible. Then some $G^{LR} \leq G$. Since $\mathrm{b}(G) \leq 2$, we necessarily have $\mathrm{b}(G^{LR}) = 0$, whence $G^{LR} = 0$. Therefore $G^L = *$ or -1. Since $G^L \in \mathscr{G}^L$ and $\mathscr{G}^L$ is an antichain of $\mathbb{G}_1$, this implies that $\mathscr{G}^L = \{*\}$, $\{0, *\}$, or $\{-1\}$. But we also have $G \geq G^{LR} = 0$, so $0 \notin \mathscr{G}^R$ and $-1 \notin \mathscr{G}^R$. Therefore $\mathscr{G}^R = \emptyset$, $\{*\}$, or $\{1\}$.

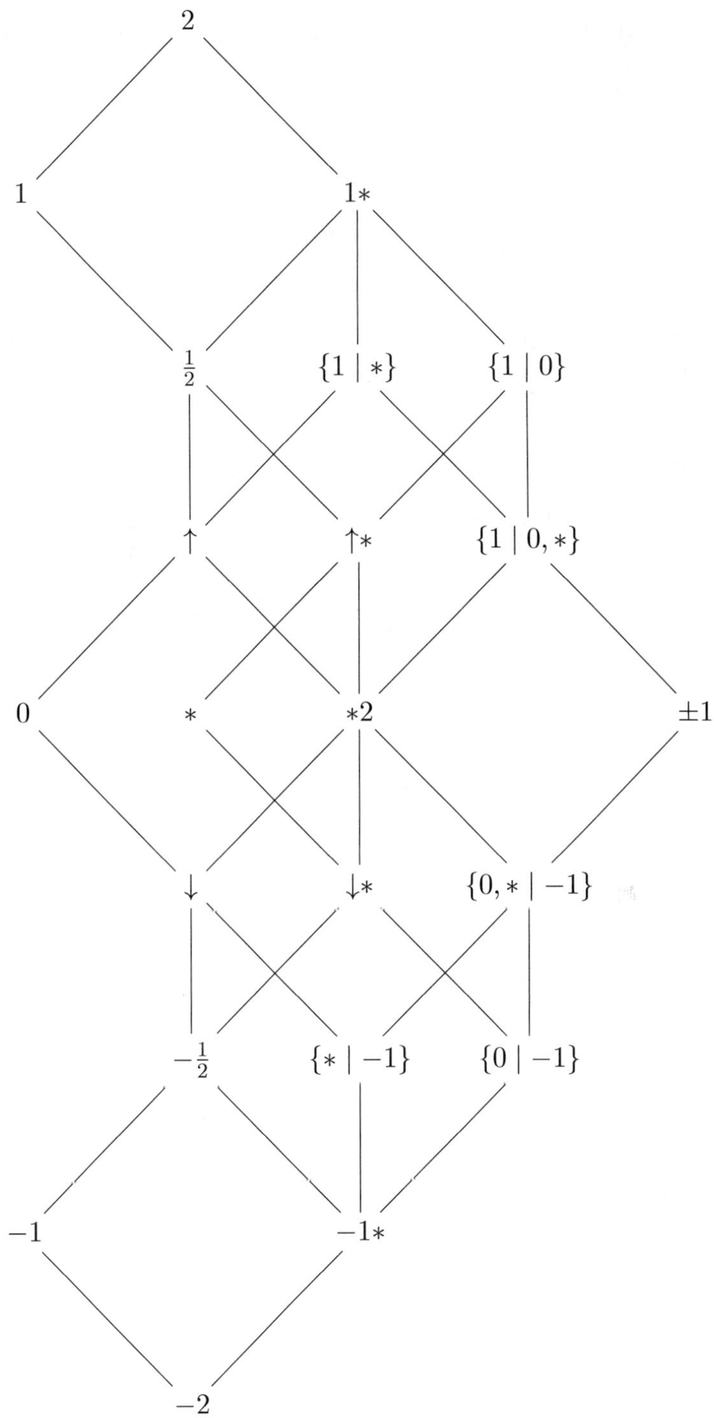

Figure 1.2. The partial-order structure of $\mathbb{G}_2$.

This yields nine pairs with a reversible Left option. By symmetry, there is a reversible G^R iff $\mathscr{G}^R = \{*\}$, $\{0,*\}$, or $\{1\}$ and $\mathscr{G}^L = \emptyset$, $\{*\}$, or $\{-1\}$. Thus there are nine pairs with a reversible Right option.

Finally, note that G has a reversible option for *both* players precisely when ($\mathscr{G}^L = \{*\}$ or $\{-1\}$) and ($\mathscr{G}^R = \{*\}$ or $\{1\}$). This gives four possibilities, which we have double-counted, so

$$|\mathbb{G}_2| = 36 - 9 - 9 + 4 = 22. \qquad \square$$

Figure 1.1 on page 154 tabulates the 22 elements of $\mathbb{G}_2$. Their partial-order structure is shown in Figure 1.2; we'll investigate this structure in detail in Section 2.

Extrema

The extreme values of the posets $\mathbb{G}_n$ grant some insight into their structure. The following theorem is a refinement of the Archimedean Principle (Theorem II.3.7 on page 71).

Theorem 1.3. *If $G \in \mathbb{G}_n$, then $n \geq G \geq -n$.*

Proof. Suppose $G \in \mathbb{G}_{n+1}$ and consider $(n+1) - G$. All of Right's options have the form $(n+1) - G^L$, from which Left can respond to $n - G^L$. But $G^L \in \mathbb{G}_n$ (taking G to be in canonical form), so by induction $n - G^L \geq 0$. $\square$

Likewise, we can show that 2^{-n} is the smallest positive number born by day $n+1$.

Theorem 1.4. *If $x \in \mathbb{G}_{n+1}$ is a positive number, then $x \geq 2^{-n}$.*

Proof. For $n = 0$ this is clear. Otherwise, fix a positive number $x \in \mathbb{G}_{n+1}$; we may assume that x is in canonical form. Consider $x - 2^{-n}$, and suppose Right moves to some $x^R - 2^{-n}$. Then $x^R \in \mathbb{G}_n$ and $x^R > x \geq 0$, so by induction $x^R \geq 2^{-(n-1)}$. Therefore $x^R - 2^{-n} > 0$ and Left has a winning move.

But Right's only other move from $x - 2^{-n}$ is to $x - 0$, which is assumed to be positive. $\square$

Along a similar theme is the following refinement of Theorem II.4.8.

Theorem 1.5. *If $G \in \mathbb{G}_{n+1}$ is infinitesimal, then either $G \leq {\uparrow}n$ or else $G \leq {\uparrow}n{*}$. Likewise, either $G \geq {\downarrow}n$ or else $G \geq {\downarrow}n{*}$.*

Proof. We'll prove that if $G \in \mathbb{G}_{n+1}$ and $R(G) \geq 0$, then either $G \geq {\downarrow}n$ or $G \geq {\downarrow}n{*}$. The theorem clearly follows from this claim (and its dual).

If $b(G) \leq 2$, then the claim is true by inspection of Figure 1.2. So suppose that $b(G) \geq 3$; then in particular $n \geq 2$.

If G is equal to a number, then $G = R(G) \geq 0$, and the conclusion is trivial. So suppose that G is not equal to a number, and assume it is given in canonical form. Then there exists a G^L such that $R(G^L) = L(G)$. Now $L(G) \geq R(G) \geq 0$, so by induction, either $G^L \geq \downarrow(n-1)$ or $G^L \geq \downarrow(n-1)*$.

First suppose that $G^L \geq \downarrow(n-1)*$; we will show that $G \geq \downarrow n$. On the difference game $G + \uparrow n$, Right's options have the form

$$G + \uparrow(n-1)* \quad \text{or} \quad G^R + \uparrow n.$$

From $G + \uparrow(n-1)*$ Left can revert to $G^L + \uparrow(n-1)*$, which we know is ≥ 0. So consider $G^R + \uparrow n$. Here $L(G^R) \geq R(G) \geq 0$. If G^R is equal to a number, then $G^R \geq 0$, and we are done. Otherwise, there is some G^{RL} with $R(G^{RL}) = L(G^R)$. By induction, either $G^{RL} \geq \downarrow(n-2)$ or $G^{RL} \geq \downarrow(n-2)*$; but either of these implies $G^{RL} \geq \downarrow n$. Therefore Left can revert $G^R + \uparrow n$ to $G^{RL} + \uparrow n$.

The same argument shows that if $G^L \geq \downarrow(n-1)$, then $G \geq \downarrow n*$. □

Exercises 1.1 and 1.2 on page 159 give some further results of this type.

$\mathbb{G}_n^0$ and $\mathbb{G}_n/\text{Inf}$

The dicotic and reduced games born by day n also exhibit interesting structure. Since $\uparrow n$ and $\uparrow n*$ are both maximal in $\mathbb{G}_{n+1}^0$ (Theorem 1.5), it follows that $\mathbb{G}_n^0$ has no largest element for $n \geq 1$. This contrasts with $\mathbb{G}_n$, which has n as its largest element. The partial-order structure of $\mathbb{G}_2^0$ is shown in Figure 1.3 on the next page.

Recall (from Section II.6) that each element of $\mathbb{G}_n/\text{Inf}$ is represented by a game in reduced canonical form with birthday $\leq n$. Now since $*$ is not in reduced canonical form, the only *reduced* games born by day 1 are 0, 1, and -1. If G is a reduced game born by day 2, then either G is a number or $L(G) > R(G)$. This gives

$$\begin{array}{ll} \text{the numbers} & -2,\ -1,\ -\tfrac{1}{2},\ 0,\ \tfrac{1}{2},\ 1,\ 2, \quad \text{and} \\ \text{the switches} & \{1 \mid 0\},\ \{0 \mid -1\},\ \text{and}\ \pm 1 \end{array}$$

as the only ten elements of $\mathbb{G}_2/\text{Inf}$. The elements of $\mathbb{G}_3/\text{Inf}$ can be computed similarly and are summarized in Figure 1.4 on the following page.

Computer searches have determined the cardinality of $\mathbb{G}_4^0$ and $\mathbb{G}_4/\text{Inf}$; these are summarized in Figure 1.5, along with a few other similar observations.

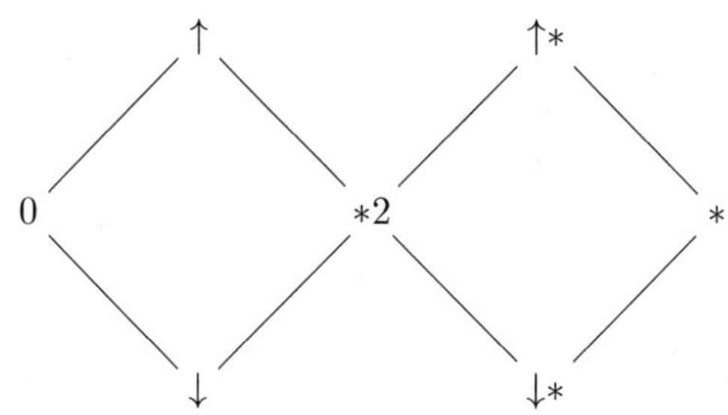

Figure 1.3. The partial-order structure of $\mathbb{G}_2^0$.

$t<0$	$3\quad 2\quad \frac{3}{2}\quad 1\quad \frac{3}{4}\quad \frac{1}{2}\quad \frac{1}{4}\quad 0$
$t=\frac{1}{4}$	$\{1 \mid \frac{1}{2}\}\quad \{\frac{1}{2} \mid 0\}$
$t=\frac{1}{2}$	$\{2 \mid 1\}\quad \{1 \mid 0\}\quad \{\frac{1}{2} \Vert 0 \mid -1\}$ $\{\frac{1}{2}, \{1 \mid 0\} \Vert -\frac{1}{2}\}\quad \pm\frac{1}{2}\quad \pm\left(\frac{1}{2}, \{1 \mid 0\}\right)$
$t=\frac{3}{4}$	$\{2 \Vert \frac{1}{2}, \{1 \mid 0\}\}\quad \{2 \mid \frac{1}{2}\}\quad \{2 \Vert 1 \mid 0\}\quad \{1 \mid -\frac{1}{2}\}$ $\{1 \Vert 0 \mid -1\}\quad \{1 \Vert \{0 \mid -1\}, -\frac{1}{2}\}$
$t=1$	$\{2 \mid \frac{1}{2}, \pm 1\}\quad \{2 \mid 0, \pm 1\}\quad \{2 \mid 0\}\quad \{2 \mid \pm 1\}\quad \pm 1$
$t=\frac{5}{4}$	$\{2 \Vert 0 \mid -1\}\quad \{2 \mid -\frac{1}{2}\}\quad \{2 \Vert \{0 \mid -1\}, -\frac{1}{2}\}\quad \{2 \Vert -\frac{1}{2}, \pm 1\}$
$t=\frac{3}{2}$	$\{2 \mid -1\}$
$t=2$	± 2

Figure 1.4. All the elements of $\mathbb{G}_3/\mathrm{Inf}$, grouped by temperature.

Birthday	1	2	3	4
Games	4	22	1,474	
Infinitesimals	2	7	292	
Dicotic Games	2	7	67	534,483
Reduced Games	3	10	61	15,845
Hereditarily Transitive Games	4	18	176	11,363

Figure 1.5. The number of games of various types born by day n.

Exercises

1.1 If $G \in \mathbb{G}_{n+2}$ and $G > 0$, then $G \geq +_n$.

1.2 For $n \geq 2$, there is a $T \in \mathbb{G}_n^0$ such that if $G \in \mathbb{G}_n^0$ and $G > 0$, then $G \geq T$. (This is the dicotic analogue of Exercise 1.1.)

1.3 True or false: If $G \in \mathbb{G}_{n+1}$ is infinitesimal, then either $\Uparrow n \geq G \geq \Downarrow n$ or else $\Uparrow n* \geq G \geq \Downarrow n*$. (Compare to Theorem 1.5.)

1.4 If there are k hereditarily transitive games born by day n, then there are $k+1$ dicotic hereditarily transitive games born by day $n+1$. (Use Exercise II.6.14 on page 135.)

1.5 Enumerate the 67 dicotic games born by day 3, and arrange them into a table like Figure 1.1.

1.6 Draw the partial-order structure of $\mathbb{G}_2/\text{Inf}$.

1.7 Fix an integer $n \geq 0$, and for each $G, H \in \mathbb{G}$ define

$$G \equiv_n H \quad \text{if} \quad o(G + X) = o(H + X) \quad \text{for each } X \in \mathbb{G}_n.$$

Prove that there is a one-to-one correspondence between elements of $\mathbb{G}_{n+1}$ and equivalence classes modulo $\equiv_n$.

Notes

The enumeration of $\mathbb{G}_2$ is a classical result. Figure 1.2 was given in the form shown here by Calistrate, Paulhus, and Wolfe [**CPW02**], improving upon earlier efforts by Guy [**Guy91d**].

The number of games born by day 3 was first isolated by Dean Hickerson and Robert Li in 1974, by a direct hand calculation. The count of day 4 games remains well (and perhaps permanently) out of reach.

The counts of dicotic and reduced games in Figure 1.5 were obtained using *cgsuite*; the hereditarily transitive statistics were first calculated by Angela Siegel [**Sie11**] (day ≤ 3) and Neil McKay (day 4).

Exercise 1.4: Neil McKay.

Exercise 1.7: Geoff Cruttwell.

2. Lattice Structure

Figure 1.2 (page 155) suggests that $\mathbb{G}_2$ has a rich poset structure. We'll now prove that *every* poset $\mathbb{G}_n$ has the structure of a distributive lattice. Throughout the following discussion, n will be a fixed positive integer.

Definition 2.1. For $G \in \mathbb{G}_n$ define

$$\lceil G \rceil = \{X \in \mathbb{G}_{n-1} : X \mathrel{|\!\!>} G\},$$
$$\lfloor G \rfloor = \{X \in \mathbb{G}_{n-1} : X \mathrel{<\!\!|} G\}.$$

Definition 2.2. For $G, H \in \mathbb{G}_n$ define

$$G \vee H = \{G^L, H^L \mid \lceil G \rceil \cap \lceil H \rceil\},$$
$$G \wedge H = \{\lfloor G \rfloor \cap \lfloor H \rfloor \mid G^R, H^R\}.$$

Note that these definitions depend on n; the value of n should always be clear from the context. The following result (and its dual with $\vee$ replaced by $\wedge$) show that $\mathbb{G}_n$ is a lattice, with join and meet given by $\vee$ and $\wedge$.

Theorem 2.3. *For every $G, H, J \in \mathbb{G}_n$:*

(a) *$G \vee H \geq G$ and $G \vee H \geq H$.*

(b) *If $J \geq G$ and $J \geq H$, then $J \geq G \vee H$.*

Proof. (a) To see that $G \vee H \geq G$, it suffices to show that Left, playing second, can win the difference game

$$(G \vee H) - G.$$

If Right moves to some $(G \vee H) - G^L$, Left can simply revert to $G^L - G^L$. Conversely, if Right moves to $X - G$, where X is a Right option of $G \vee H$, then by definition of $G \vee H$ we have $X \in \lceil G \rceil$. Therefore $X \mathrel{|\!\!\!>} G$, and it follows that Left has a winning move on $X - G$. The proof that $G \vee H \geq H$ is identical.

(b) Consider $J - (G \vee H)$. Right's moves to $J - G^L$ are losing: since $G \mathrel{|\!\!\!>} G^L$ and $J \geq G$, we know that $J - G^L \mathrel{|\!\!\!>} 0$. An identical argument suffices for Right's moves to $J - H^L$. Suppose instead that Right moves to

$$J^R - (G \vee H).$$

Then since $J \geq G$, we know that $J^R \mathrel{|\!\!\!>} G$. Moreover, since $J \in \mathbb{G}_n$, we have $J^R \in \mathbb{G}_{n-1}$, and hence $J^R \in \lceil G \rceil$. Likewise, since $J \geq H$, we have $J^R \in \lceil H \rceil$. Therefore J^R is a Right option of $G \vee H$, and Left can win on $J^R - (G \vee H)$ by making the mirror-image move to $J^R - J^R$. □

We next show that $\mathbb{G}_n$ is distributive. First we prove a lemma:

Lemma 2.4. *$\lfloor G \vee H \rfloor = \lfloor G \rfloor \cup \lfloor H \rfloor$ and $\lceil G \wedge H \rceil = \lceil G \rceil \cup \lceil H \rceil$.*

Proof. If $X \in \lfloor G \rfloor$, then

$$X \mathrel{<\!\!\!|} G \leq G \vee H,$$

so we have immediately that $X \in \lfloor G \vee H \rfloor$. This shows that $\lfloor G \rfloor \subset \lfloor G \vee H \rfloor$, and by an identical argument $\lfloor H \rfloor \subset \lfloor G \vee H \rfloor$.

Conversely, suppose $X \notin \lfloor G \rfloor \cup \lfloor H \rfloor$. Then $X \geq G$ and $X \geq H$. By minimality of $G \vee H$, we have $X \geq G \vee H$, whence $X \notin \lfloor G \vee H \rfloor$.

The proof for $G \wedge H$ is the same. □

Theorem 2.5. *The lattice* $\mathbb{G}_n$ *is distributive.*

Proof. Let $A = G \wedge (H \vee J)$ and $B = (G \wedge H) \vee (G \wedge J)$ for $G, H, J \in \mathbb{G}_n$; we must show that $A = B$. Now writing $\mathscr{G}^R$ for the set of Right options of G and applying Lemma 2.4, we have

$$\begin{aligned} A &= \Big\{ \lfloor G \rfloor \cap \lfloor H \vee J \rfloor \;\Big|\; \mathscr{G}^R \cup (\lceil H \rceil \cap \lceil J \rceil) \Big\}, \\ B &= \Big\{ (\lfloor G \rfloor \cap \lfloor H \rfloor) \cup (\lfloor G \rfloor \cap \lfloor J \rfloor) \;\Big|\; \lceil G \wedge H \rceil \cap \lceil G \wedge J \rceil \Big\} \\ &= \Big\{ \lfloor G \rfloor \cap \lfloor H \vee J \rfloor \;\Big|\; \lceil G \rceil \cup (\lceil H \rceil \cap \lceil J \rceil) \Big\}. \end{aligned}$$

Since $\mathscr{G}^R \subset \lceil G \rceil$, this immediately implies $A \geq B$. To complete the proof, we must show that Left can win $B - A$ playing second.

Now all of Right's moves on $B - A$ have mirror-image responses, except for those of the form $B^R - A$ with $B^R \in \lceil G \rceil$. But in that case, $B^R \mathrel{|\!\!\rhd} G$. Since $G \geq A$ (by definition of A), this implies $B^R \mathrel{|\!\!\rhd} A$, so Left has a winning move *a priori*. □

Join-Irreducible Elements of $\mathbb{G}_n$

Recall that $G \in \mathbb{G}_n$ is **join-irreducible** if there are no $H < G$ and $J < G$ with $G = H \vee J$ (see Definition B.2.5 on page 464). Put

$$\mathscr{J}_n = \{G \in \mathbb{G}_n : G \text{ is join-irreducible}\}.$$

Then $\mathbb{G}_n$ is isomorphic to the lattice of ideals of the poset $\mathscr{J}_n$ (Theorem B.2.7 on page 465). In particular, the lattice structure of $\mathbb{G}_n$ is completely determined by $\mathscr{J}_n$.

Remarkably, there is a simple characterization of the poset $\mathscr{J}_{n+1}$ in terms of the previous lattice $\mathbb{G}_n$.

Theorem 2.6. *For all* n,

$$\mathscr{J}_{n+1} = \mathbb{G}_n \cup \Big\{ \{G \mid -n\} : G \in \mathbb{G}_n \Big\}.$$

Proof. First suppose $G \in \mathbb{G}_n$. To see that G is join-irreducible in $\mathbb{G}_{n+1}$, fix $X, Y \in \mathbb{G}_{n+1}$ with $X, Y < G$. Then $G \in \lceil X \rceil \cap \lceil Y \rceil$, so G is a Right option of $X \vee Y$. Therefore $X \vee Y \neq G$.

Likewise, to see that $\{G \mid -n\}$ is join-irreducible, fix $X, Y \in \mathbb{G}_{n+1}$ with $X, Y < \{G \mid -n\}$. By definition of join, we have $X \vee Y \leq \{G \mid -n\}$, so it remains to show that $X \vee Y \mathrel{\lhd\!\!|} \{G \mid -n\}$. In particular, it suffices to prove that $X \vee Y \leq G$. This is equivalent to showing that $X \leq G$ and $Y \leq G$, and by symmetry we need only consider $X \leq G$.

But $\{G \mid -n\} - X > 0$ (by assumption), so

$$\text{either} \quad G - X \geq 0 \quad \text{or else} \quad \{G \mid -n\} - X^R \geq 0 \text{ for some } X^R.$$

But the latter is impossible: Theorem 1.3 implies that $-X^R \leq n$, so that $-n - X^R \leq 0$, and hence $\{G \mid -n\} - X^R \lhd\!\!| \; 0$. Therefore $G - X \geq 0$.

To complete the proof, we must show that these are the only join-irreducibles in $\mathbb{G}_{n+1}$. So suppose $G \neq H$ and $G \neq \{H \mid -n\}$ for any $H \in \mathbb{G}_n$, and put

$$J = \bigvee \{X \in \mathbb{G}_{n+1} : X < G\}.$$

We claim that $G = J$. Clearly $J \leq G$; thus it suffices to prove that Left can win $J - G$ playing second.

Suppose Right moves to $J^R - G$. Now $J^R \in \mathbb{G}_n$, so by assumption on G we have $J^R \neq G$. Furthermore, by definition of J, we have $J \geq X$, and hence $J^R \; |\!\!\rhd X$, for all $X < G$. In particular, it cannot be the case that $J^R < G$. Thus $J^R \; |\!\!\rhd G$, so Left has a winning move on $J^R - G$.

Finally, suppose instead that Right moves to $J - G^L$. Since $G \; |\!\!\rhd G^L$, we know that

$$G \geq \{G^L \mid -n\},$$

and therefore (by assumption on G)

$$G > \{G^L \mid -n\}.$$

By definition of J, this implies

$$J \geq \{G^L \mid -n\}$$

and in particular $J \; |\!\!\rhd G^L$. □

Theorem 2.6 reveals an abstract recursive structure in the lattices $\mathbb{G}_n$ that is independent of any game-theoretic interpretation: each successive lattice $\mathbb{G}_{n+1}$ can be constructed from the previous one by defining the poset $\mathcal{J}_{n+1}$ and passing to the lattice of its order-ideals. This idea is further explored in Exercise 2.10 on page 166.

Symmetries of $\mathbb{G}_n$

Every lattice $\mathbb{G}_n$ has an obvious order-reversing symmetry, given by

$$G \mapsto -G.$$

Less obvious is that $\mathbb{G}_n$ admits an order-*preserving* symmetry too. This symmetry is best understood by inspecting $\mathbb{G}_2^0$, which is included as a subposet of $\mathbb{G}_2$ (see Figure 1.3 on page 158).

The poset $\mathbb{G}_2^0$ admits a symmetry that fixes $*2$ and interchanges every other G with $G + *$. As it turns out, this symmetry extends to all of $\mathbb{G}_2$ and from there to all $\mathbb{G}_n$.

Definition 2.7. Let G be a short game. The **companion** of G, denoted by G^c, is given by

$$G^c = \begin{cases} \{0, (G^L)^c \mid 0, (G^R)^c\} & \text{if } G = 0; \\ \{0, (G^L)^c \mid (G^R)^c\} & \text{if } G > 0; \\ \{ (G^L)^c \mid 0, (G^R)^c\} & \text{if } G < 0; \\ \{ (G^L)^c \mid (G^R)^c\} & \text{if } G \not\gtrless 0. \end{cases}$$

G is said to be **lonely** if $G = G^c$.

In particular, note that $0^c = \{0 \mid 0\} = *$ and $*^c = \{* \mid *\} = 0$.

Proposition 2.8. *$G \geq H$ if and only if $G^c \geq H^c$.*

Proof. Suppose that $G \geq H$. To show that $G^c \geq H^c$, it suffices (by symmetry) to show that $G^c \mathrel{|\!\!>} \text{each } (H^c)^L$. Now for each H^L we have $G \mathrel{|\!\!>} H^L$, so that by induction $G^c \mathrel{|\!\!>} (H^L)^c$. The only remaining case occurs when H^c has an extra Left option to 0. But in that case, necessarily $H \geq 0$, so $G \geq 0$. Therefore G^c also has an extra Left option to 0, and we have $G^c \mathrel{|\!\!>} 0$ automatically.

Conversely, if $G \mathrel{<\!\!|} H$, then by symmetry we can assume $G \leq$ some H^L, so that by induction $G^c \leq (H^L)^c$. Since $(H^L)^c$ is always a Left option of H^c, this implies $G^c \mathrel{<\!\!|} H^c$. ☐

It follows that $G^c = H^c$ whenever $G = H$, so that the value of G^c depends only on the value of G (and not on its form).

Proposition 2.9. *$(G^c)^c = G$, for all G.*

Proof. If $G = 0$, then $G^c = *$, so $(G^c)^c = \{* \mid *\} = 0$.

If $G > 0$, then $G^c \mathrel{|\!\!>} 0$ since G^c has 0 as a Left option. So by induction

$$(G^c)^c = \text{either} \quad \{*, G^L \mid G^R\} \quad \text{or} \quad \{0, *, G^L \mid G^R\}.$$

But since $G > 0$, there is some $G^L \geq 0$, so the Left option 0 is dominated (if present). Moreover, $(G^c)^c > 0$ by Proposition 2.8, so the Left option $*$ reverses out through 0. So in all cases $(G^c)^c = G$, and an identical argument handles the $G < 0$ case.

Finally, if $G \not\gtrless 0$, then

$$(G^c)^c = \text{either} \quad \{G^L \mid G^R\} \quad \text{or} \quad \{0, G^L \mid G^R\} \quad \text{or} \quad \{G^L \mid 0, G^R\}.$$

But since $G \not\gtrless 0$, there is a $G^L \geq 0$ as well as a $G^R \leq 0$. So the option 0, if present for either player, is dominated. ☐

Lemma 2.10. *If $G \neq 0$ and $G \neq *$, then* $\mathrm{b}(G^c) = \mathrm{b}(G)$.

Proof. We first show that $\mathrm{b}(G^c) \leq \mathrm{b}(G)$ for all such G. If $\mathrm{b}(G) = 1$, then necessarily $G = 1$ or -1, which are easily seen to be lonely. So suppose that $\mathrm{b}(G) \geq 2$.

Then every option of G^c is either 0 or of the form $(G^L)^c$ or $(G^R)^c$. Certainly $\mathrm{b}(0) < \mathrm{b}(G)$. Now if $(G^L)^c = 0$ or $(G^L)^c = *$ (for a fixed G^L), then $\mathrm{b}\big((G^L)^c\big) \leq 1 < \mathrm{b}(G)$. Otherwise, $G^L \neq 0$ and $G^L \neq *$, so by induction

$$\mathrm{b}\big((G^L)^c\big) \leq \mathrm{b}(G^L) < \mathrm{b}(G).$$

The same argument applies to Right options, so that every option of G^c has birthday strictly less than $\mathrm{b}(G)$. This shows that $\mathrm{b}(G^c) \leq \mathrm{b}(G)$.

But certainly $G^c \neq 0$ and $G^c \neq *$, so that

$$\mathrm{b}\big((G^c)^c\big) \leq \mathrm{b}(G^c) \leq \mathrm{b}(G),$$

and equality follows since $(G^c)^c = G$. □

Theorem 2.11. *Each lattice $\mathbb{G}_n$ $(n \geq 1)$ admits a nontrivial order-preserving automorphism, given by $G \mapsto G^c$.*

Proof. $G \mapsto G^c$ is order-preserving by Proposition 2.8, nontrivial since it interchanges 0 and $*$, and closed over $\mathbb{G}_n$ by Lemma 2.10. □

We'll now show that, in fact, the companion automorphism $G \mapsto G^c$ is the *unique* nontrivial order-preserving symmetry of $\mathbb{G}_n$, for all $n \geq 1$.

Lemma 2.12. *Every integer $n \neq 0$ is lonely.*

Proof. Every $n > 0$ is the largest element of $\mathbb{G}_n$ and hence must be fixed by the companion automorphism. Negative integers follow by symmetry. □

Theorem 2.13. *$G \mapsto G^c$ is the unique nontrivial automorphism of $\mathbb{G}_n$.*

Proof. The theorem is true for $\mathbb{G}_1$ by inspection. The proof proceeds by induction on n.

Let $\sigma : \mathbb{G}_{n+1} \to \mathbb{G}_{n+1}$ be any automorphism and assume that $\sigma(G) \neq G$ for some G. We will show that $\sigma(G) = G^c$ for all G.

Now every automorphism of $\mathbb{G}_{n+1}$ is uniquely determined by its action on $\mathcal{J}_{n+1}$, so we can assume that $\sigma(J) \neq J$ for some $J \in \mathcal{J}_{n+1}$ and show that $\sigma(J) = J^c$ for all such J.

Let $\mathcal{M}_{n+1} = \{-J : J \in \mathcal{J}_{n+1}\}$, the set of meet-irreducibles of $\mathbb{G}_{n+1}$, and put

$$\mathcal{A} = \mathcal{J}_{n+1} \cap \mathcal{M}_{n+1}.$$

Since σ is an automorphism, it must fix $\mathcal{A}$. By Theorem 2.6 we have

$$\mathcal{J}_{n+1} = \mathbb{G}_n \cup \Big\{\{G \mid -n\} : G \in \mathbb{G}_n\Big\}$$

and by symmetry

$$\mathcal{M}_{n+1} = \mathbb{G}_n \cup \Big\{\ \{n \mid G\}\ : G \in \mathbb{G}_n\Big\}.$$

It follows that $\mathcal{A} = \mathbb{G}_n \cup \{\pm n\}$. Moreover, every $G \in \mathbb{G}_n$ is comparable with n and $-n$, while $\pm n$ is not. Therefore $\pm n$ is the unique element of $\mathcal{A}$ incomparable with all other elements of $\mathcal{A}$, so it must be fixed by σ.

It follows that σ fixes $\mathbb{G}_n$, and therefore $\mathcal{J}_{n+1} \setminus \mathbb{G}_n$ as well. By induction, either σ is the identity on $\mathbb{G}_n$ or else $\sigma(J) = J^c$ for all $J \in \mathbb{G}_n$. The same must be true of $\mathcal{J}_{n+1} \setminus \mathbb{G}_n$ as well, since

$$\mathbb{G}_n \cong \mathcal{J}_{n+1} \setminus \mathbb{G}_n$$

via the isomorphism $J \mapsto \{J \mid -n\}$, which is companion-preserving by Lemma 2.12.

Now σ cannot be the identity on both parts of $\mathcal{J}_{n+1}$, by hypothesis; and if it's the identity on neither part, then we are done. To complete the proof, we must show that it can't be the identity on one and the companion automorphism on the other.

But if $\sigma(0) = *$ while $\sigma(\{0 \mid -n\}) = \{0 \mid -n\}$, then we have a contradiction, since

$$\{0 \mid -n\} \le * \quad \text{but} \quad \{0 \mid -n\} \not\le 0.$$

A similar contradiction occurs when $\sigma(0) = 0$ while $\sigma(\{0 \mid -n\}) = \{* \mid -n\}$. ∎

Exercises

2.1 Show that $\mathbb{G}$ is not a lattice (under the usual partial order).

2.2 Determine the companions of $\uparrow n$, $\uparrow^n$, $\uparrow^{[n]}$, $*m$, and $+_x$.

2.3 Determine all lonely elements of $\mathbb{G}_2$.

2.4 True or false: $(G+H)^c = G^c + H^c$ for all G and H.

2.5 G^c is infinitesimally close to G, for all G.

2.6 (a) If G is dicotic, then so is G^c, and $\mathrm{aw}(G) = \mathrm{aw}(G^c)$.
(b) Is the same true with "dicotic" replaced by "atomic"?

2.7 If $L(G) \neq 0$ and $R(G) \neq 0$, then G is lonely.

2.8 If G is an uptimal, then $G^c = G + *$. How does this extend to generalized uptimals (cf. Exercise II.4.23 on page 99)?

2.9 Every finite poset $\mathcal{P}$ is order-isomorphic to a subposet of $\mathbb{G}$.

2.10 Define lattices $\mathcal{L}_n$ as follows. $\mathcal{L}_0 = \{0\}$. For $n \geq 1$, let $\mathcal{K}_{n+1}$ be the set
$$\mathcal{K}_{n+1} = \mathcal{L}_n \times \{0, 1\},$$
with the following partial order: $(x, 0) \geq (y, 0) \Leftrightarrow x \geq y \Leftrightarrow (x, 1) \geq (y, 1)$; $(x, 0) \geq (y, 1) \Leftrightarrow x \not\leq y$; and $(x, 1) \not\geq (y, 0)$ for all x and y. Then let $\mathcal{L}_{n+1}$ be the lattice of order-ideals of $\mathcal{K}_{n+1}$. Prove that $\mathcal{L}_n \cong \mathbb{G}_n$ for all n (as posets).

2.11 Let $\mathcal{A} \subset \mathbb{G}$ be hereditarily closed (cf. Definition 3.1 on page 168). Prove that $\mathcal{L} = \mathrm{ch}(\mathcal{A})$ is a distributive lattice. (This generalizes Theorem 2.5.) Show also that if $\mathcal{A}$ is closed under companionship, then $G \mapsto G^c$ is an automorphism of $\mathcal{L}$.

2.12 *Lattices of dicotic games.*
   (a) For $n \geq 1$, show that $\mathbb{G}_n^0$ is not a lattice.
   (b) Let $\mathcal{L}_n^0 = \mathbb{G}_n^0 \cup \{\triangledown, \vartriangle\}$, where $\triangledown$ and $\vartriangle$ are new symbols defined so that $\triangledown \geq G \geq \vartriangle$ for all $G \in \mathbb{G}_n^0$. Prove that $\mathcal{L}_n^0$ is a distributive lattice.
   (c) Characterize the join-irreducibles of $\mathcal{L}_{n+1}^0$ in terms of $\mathcal{L}_n^0$.
   (d) Prove that $\mathcal{L}_n^0$ admits exactly one nontrivial automorphism.

2.13 *Lattices of reduced game values.* Consider the poset for $\mathbb{G}_n/\mathrm{Inf}$ with order given by $\geqq$ (cf. Section II.6).
   (a) Prove that $\mathbb{G}_n/\mathrm{Inf}$ is a lattice.
   (b) Show that $\mathbb{G}_n/\mathrm{Inf}$ is not distributive.
   (c) Let $\mathcal{A} = \{\mathrm{rcf}(G) : G \in \mathbb{G}_n$ and G is not numberish$\}$. Prove that $\mathcal{A}$ is a distributive sublattice of $\mathbb{G}_n/\mathrm{Inf}$.

Notes

The distributive lattice structure of $\mathbb{G}_n$ was first described by Calistrate, Paulhus, and Wolfe in a 2002 paper [**CPW02**]. This breakthrough was followed by a flurry of activity. The structure of join-irreducibles was described in [**FHW05**]. Shortly afterward, Fraser and Wolfe used these results to obtain bounds on the number of games born by day n. In [**FW04**] they showed that, writing g_n for the cardinality of $\mathbb{G}_n$,
$$2^{g_n/2g_{n-1}} \leq g_{n+1} \leq g_n + 2^{g_n} + 2$$
and in particular gave a (fairly weak) bound of $g_4 \geq 3 \cdot 10^{12}$. They also proved that
$$g_n = 2^{g_{n-1}^{\alpha(n)}}$$
where $0.51 < \alpha(n) < 1$ and $\alpha(n) \to 1$ as $n \to \infty$.

Open Problem. Improve upon the Fraser–Wolfe bounds for the cardinality of $\mathbb{G}_n$ (and in particular $\mathbb{G}_4$).

Several other natural classes of games are also known to have a lattice structure. $\mathbb{G}_n^0$ has many of the same properties as $\mathbb{G}_n$: although $\mathbb{G}_n^0$ is not a lattice (for example, $\uparrow(n-1)$ and $\uparrow(n-1)*$ have no join), a distributive lattice structure can be recovered by adjoining "synthetic" top and bottom elements [**Sie05**]. This is explored in Exercise 2.12.

More recently, Albert and Nowakowski found a sweeping generalization of results of this type [**AN11**]. If $\mathcal{A}$ is any hereditarily closed set of values, then the set of children $\mathrm{ch}(\mathcal{A})$ is a distributive lattice. This generalizes Theorem 2.5, since in particular $\mathbb{G}_{n+1} = \mathrm{ch}(\mathbb{G}_n)$.

Open Problem. Investigate symmetries of the generalized lattices $\mathrm{ch}(\mathcal{A})$.

The reduced game values $\mathbb{G}_n/\mathrm{Inf}$ have a fascinating structure too. $\mathbb{G}_n/\mathrm{Inf}$ is a nondistributive lattice. However, when the chain of numbers is removed from $\mathbb{G}_n/\mathrm{Inf}$, the remainder of the lattice is distributive [**Siea**]. See Exercise 2.13.

Angela Siegel has studied the lattices of hereditarily transitive games $\mathcal{HT}_n$. Her doctoral thesis [**Sie11**] announced the striking result that $\mathcal{HT}_n$ is **planar**: every element has at most two upper covers and two lower covers, so that the Hasse diagram of $\mathcal{HT}_n$ can be drawn on a plane with no overlapping connections. A related problem is the following:

Open Problem. Describe the hereditary structure of $\mathcal{HT}_n$ (via join-irreducibles).

The hereditarily transitive games appear to have a deeper structure that is not fully understood. The number of *antichains* of $\mathcal{HT}_n$ follows the pattern

$$2,\ 6,\ 62,\ 1048574,\ \ldots,$$

in which each element has the form $2^k - 2$. We have the following conjecture:

Conjecture. $\mathcal{HT}_n$ has exactly $2^{a_n} - 2$ antichains, where a_n is the number of antichains of the complete Boolean algebra on n variables.

In particular, we conjecture that $\mathcal{HT}_4$ has exactly $2^{168} - 2$ antichains.

Exercise 2.11: Albert and Nowakowski.

3. Group Structure

In this section we'll investigate the group structure of $\mathbb{G}$. Recall the following notation (discussed in more detail in Section B.1):

- For a subset $\mathcal{S} \subset \mathbb{G}$, the symbol $\langle \mathcal{S} \rangle$ denotes the subgroup of $\mathbb{G}$ generated by $\mathcal{S}$.
- If $\mathcal{A}$ is a group, then $\mathcal{A}^\omega$ denotes the direct sum of countably many copies of $\mathcal{A}$. (See Section B.1 for further discussion.)

The main result of this section characterizes the isomorphism type of $\mathbb{G}$:

$$\mathbb{G} \cong \mathbb{D}^\omega \times \left(\frac{\mathbb{D}}{\mathbb{Z}}\right)^\omega.$$

In particular, $\mathbb{G}$ is 2-divisible and contains no elements of odd order.

As a warm-up exercise, we'll also characterize the subgroup $\langle \mathbb{G}_2 \rangle$:

$$\langle \mathbb{G}_2 \rangle \cong \mathbb{Z}^3 \times \mathbb{Z}_2^3 \times \mathbb{Z}_4.$$

The following basic definitions will be useful.

Definition 3.1. Let $\mathcal{A} \subset \mathbb{G}$. We say that $\mathcal{A}$ is **hereditarily closed** if for every $G \in \mathcal{A}$ and every option H of the canonical form of G, we also have $H \in \mathcal{A}$.

Definition 3.2. A subset $\mathcal{A} \subset \mathbb{G}$ is **closed** if it is both hereditarily closed and a subgroup of $\mathbb{G}$.

Definition 3.3. The **closure** of $\mathcal{S} \subset \mathbb{G}$, denoted $\mathrm{cl}(\mathcal{S})$, is the smallest closed set containing $\mathcal{S}$.

Group Structure of $\langle \mathbb{G}_2 \rangle$

Since each $\mathbb{G}_n$ is finite, the group $\langle \mathbb{G}_n \rangle$ is a finitely generated Abelian group, so by the structure theorem (Theorem B.1.6 on page 461) it can be written as a finite direct sum of cyclic groups. Moreover, we'll show below that $\mathbb{G}$ has no elements of odd order, so that all the cyclic components of $\langle \mathbb{G}_n \rangle$ have the form $\mathbb{Z}$ or $\mathbb{Z}_{2^k}$ for some k.

The exact isomorphism type of $\langle \mathbb{G}_n \rangle$ is known only for $n \le 2$. The $n = 0$ case is trivial; for $n = 1$, we have $\langle \mathbb{G}_1 \rangle = \langle 1, * \rangle$, so that

$$\langle \mathbb{G}_1 \rangle \cong \mathbb{Z} \times \mathbb{Z}_2.$$

The case $n = 2$ is given by the following theorem.

Theorem 3.4. $\langle \mathbb{G}_2 \rangle \cong \mathbb{Z}^3 \times \mathbb{Z}_2^3 \times \mathbb{Z}_4$, *with basis*

$$\mathcal{B} = \left\{ \tfrac{1}{2}, \uparrow, \zeta, \pm 1, \pm \tfrac{1}{2}, *2, \epsilon \right\}.$$

Here ϵ and ζ are infinitesimals defined by

$$\epsilon = \{1 \mid 0\} - \{1 \mid *\},$$
$$\zeta = \{1 \mid 0\} - \{1 \mid 0, *\}.$$

Proof. First we show the inclusion $\mathcal{B} \subset \langle \mathbb{G}_2 \rangle$. We have $\frac{1}{2}, \uparrow, \pm 1, *2 \in \mathbb{G}_2$. That $\epsilon, \zeta \in \langle \mathbb{G}_2 \rangle$ follows from their definitions. This leaves only $\pm\frac{1}{2}$, for which we observe $\pm\frac{1}{2} = \{1 \mid 0\} - \frac{1}{2}$.

Next note that $\frac{1}{2} > 0$ and $\uparrow > 0$, trivially. We also have $\zeta > 0$, since Left has a winning move on ζ to $\{1 \mid 0\} - *$. This shows that $\frac{1}{2}$, $\uparrow$, and ζ each have order ∞ in $\langle \mathbb{G}_2 \rangle$.

It is clear that ± 1, $\pm\frac{1}{2}$, and $*2$ each have order 2. To see that ϵ has order 4, observe that

$$2 \cdot \epsilon = 2 \cdot \{1 \mid 0\} - 2 \cdot \{1 \mid *\} = 1 - 1* = *.$$

The inclusion $\mathbb{G}_2 \subset \langle \mathcal{B} \rangle$ is easy: $* = 2 \cdot \epsilon$, as above; $\{1 \mid 0\} = \frac{1}{2} \pm \frac{1}{2}$; $\{1 \mid *\}$ and $\{1 \mid 0, *\}$ can then be recovered from ϵ and ζ; and all other inclusions are trivial.

This shows that $\mathcal{B}$ induces a surjective map $\mathbb{Z}^3 \times \mathbb{Z}_2^3 \times \mathbb{Z}_4 \to \langle \mathbb{G}_2 \rangle$. To complete the proof we must show that $\mathcal{B}$ is a basis.

Suppose that

$$G = a \cdot \tfrac{1}{2} + b \cdot \uparrow + c \cdot \zeta + d \cdot \pm 1 + e \cdot \pm\tfrac{1}{2} + f \cdot *2 + g \cdot \epsilon,$$

where $a, b, \ldots, g$ are integers. Assume that $G = 0$; we must show that $a = b = c = 0$, that $d \equiv e \equiv f \equiv 0 \pmod 2$, and that $g \equiv 0 \pmod 4$.

First of all, $m(B) = 0$ for each $B \in \mathcal{B}$, except for $B = \frac{1}{2}$. Since $m(G) = 0$, this implies $m(a \cdot \frac{1}{2}) = 0$, so necessarily $a = 0$.

Likewise, $t(B) \leq 0$ for each $B \in \mathcal{B}$, except for $t(\pm 1) = 1$ and $t(\pm\frac{1}{2}) = \frac{1}{2}$. If d were odd, we would have $t(d \cdot \pm 1) = 1$. Since $t(B) < 1$ for all $B \in \mathcal{B} \setminus \{\pm 1\}$, this would imply $t(G) = 1$ (Theorem II.5.18 on page 111), contradicting the assumption $G = 0$. Thus $d \equiv 0 \pmod 2$. An identical argument then shows that $e \equiv 0 \pmod 2$.

This leaves just the infinitesimal part of G. Now it is easily checked that

$$0 < \zeta < \uparrow*2,$$

so that ζ has atomic weight 0 (Theorem II.7.5 on page 139). It follows that if $b \neq 0$, then $b \cdot \uparrow + c \cdot \zeta$ has infinite order and hence cannot be equal to $-(f \cdot *2 + g \cdot \epsilon)$. Thus $b = 0$; and this implies that $c \cdot \zeta$ cannot have infinite order, so $c = 0$.

This leaves $G = f \cdot *2 + g \cdot \epsilon$. But it is easily checked that $*2 \neq \epsilon$, 2ϵ, or 3ϵ, so that $f \equiv 0 \pmod 2$ and $g \equiv 0 \pmod 4$. □

Divisibility of $\mathbb{G}$

We now investigate the full group structure of $\mathbb{G}$. The first step is to show that $\mathbb{G}$ is 2-divisible but has no elements of odd order (and in particular is not m-divisible for any odd m).

The proof that G is 2-divisible is by an explicit construction. For each $G \in \mathbb{G}$, define

$$h(G) = \{n + G \mid -n\},$$

where $n = 3 \cdot \mathrm{b}(G)$. Then:

Lemma 3.5. *$h(G) + h(G) = G$ for every $G \in \mathbb{G}$.*

Proof. We'll show that

$$h(G) + h(G) - G$$

is a second-player win, presuming G to be in canonical form. If either player moves first on $h(G)$, her opponent can respond on the other copy of $h(G)$, leaving the position $(n+G)-n-G=0$. Suppose instead that Right opens with a move to

$$h(G)+h(G)-G^L.$$

Left may then respond to $n+G+h(G)-G^L$. Right's play on the other copy of $h(G)$ now leaves a position equal to $G-G^L \mathrel{|\!\!\!>} 0$. But if Right makes any other move, Left may counter by playing on the other copy of $h(G)$, leaving a game of the form

$$2n+X$$

for some subposition X of $G+G+G$. By choice of n, we have $2n+X \geq 0$.

If Left opens by moving to $h(G)+h(G)-G^R$, then Right wins by a similar strategy. □

Corollary 3.6. *$\mathbb{G}$ is 2-divisible.*

Proof. Immediate from Lemma 3.5. □

Next we show that $\mathbb{G}$ contains no elements of odd order. For a finite-order $G \in \mathbb{G}$ of birthday n, the key idea is to replace $2^n \cdot G$ by successive equivalent forms. These forms will in fact be linear combinations of subpositions of G. We'll then show that this process must eventually terminate and can only do so at 0, so that $2^n \cdot G = 0$.

The following lexicographic order will be used to show that this process necessarily terminates.

Definition 3.7. For $k \geq 1$, we define an ordering $<_{\text{lex}}$ on $\mathbb{Z}^k$ as follows. Let $\mathbf{a}=(a_1,a_2,\ldots,a_k) \in \mathbb{Z}^k$ and $\mathbf{b}=(b_1,b_2,\ldots,b_k) \in \mathbb{Z}^k$; then

$$\mathbf{a} <_{\text{lex}} \mathbf{b} \quad \text{if} \quad |a_i| < |b_i| \text{ for the largest } i \text{ such that } a_i \neq b_i.$$

Notice that $<_{\text{lex}}$ is not a total ordering: if $a_i = -b_i$ for the "largest such i" in Definition 3.7, then $\mathbf{a}$ and $\mathbf{b}$ are incomparable.

Lemma 3.8. *$<_{\text{lex}}$ is well-founded (i.e., admits no infinite descending chain).*

Proof. The mapping

$$(a_1,a_2,\ldots,a_k) \mapsto (|a_1|,|a_2|,\ldots,|a_k|)$$

translates $<_{\text{lex}}$ onto the standard lexicographic order on $\mathbb{N}^k$, in a way that preserves strict descending chains. □

The next few lemmas show that the incentives of games with finite order have a special structure.

Definition 3.9. We say G is in **quasi-canonical form** if we can write

$$G \cong G_1+G_2+\cdots+G_k$$

with each G_i in canonical form.

Lemma 3.10. *Suppose G has finite order and is in quasi-canonical form. Then for every Left incentive $\Delta = \Delta^L(G)$, there is a Right incentive with $\Delta^R(G) \geq \Delta$.*

Proof. Write $\Delta = G^L - G$, and fix m so that $m \cdot G = 0$. Then in particular,

$$(m-1) \cdot G + G^L \lhd\!\!| \; 0.$$

There are two cases.

Case 1: $(m-2) \cdot G + G^L + G^R \leq 0$ for some G^R. Then since $0 = m \cdot G$, we have $G^L + G^R \leq 2 \cdot G$, so that $G^L - G \leq G - G^R$.

Case 2: $(m-1) \cdot G + G^{LR} \leq 0$. Then $G^{LR} \leq G$. Moreover, since G is quasi-canonical, we can write

$$G \cong G_1 + G_2 + \cdots + G_k$$

with each G_i in canonical form. Since no G_i has any reversible options, we have without loss of generality

$$G^{LR} \cong G_1^L + G_2^R + G_3 + \cdots + G_k.$$

Thus $G_1^L + G_2^R \leq G_1 + G_2$, so that $G_1^L - G_1 \leq G_2 - G_2^R$. But clearly, $G_1^L - G_1 = \Delta$ and $G_2 - G_2^R$ is equal to a Right incentive of G. □

Replacing G by $-G$ in Lemma 3.10, we also see that for every Right incentive $\Delta = \Delta^R(G)$, there is a Left incentive $\Delta^L(G) \geq \Delta$.

Lemma 3.11. *Suppose G has finite order and is in quasi-canonical form. Then some Δ is equal to both a Left incentive and a Right incentive of G.*

Proof. Choose Δ to be maximal among the set of all Left incentives of G. By successive applications of Lemma 3.10, there exists a Right incentive $\Delta^R(G) \geq \Delta$ and a Left incentive $\Delta^L(G) \geq \Delta^R(G)$. Then

$$\Delta^L(G) \geq \Delta^R(G) \geq \Delta,$$

so that by maximality of Δ all three values are equal. □

Theorem 3.12. *If G has finite order, then $2^n \cdot G = 0$, where $n = \mathrm{b}(G)$. In particular, the order of G is a power of 2.*

Proof. Since the order of G is invariant of form, we may assume that G is in canonical form. Enumerate the nonzero subpositions of G

$$G_1, G_2, \ldots, G_k$$

in such a way that if $i \le j$, then $\mathrm{b}(G_i) \le \mathrm{b}(G_j)$. (In particular, $G_k \cong G$.) Let $\mathcal{Z} \subset \mathbb{Z}^k$ be the set of all k-tuples of integers $(a_1, \dots, a_k)$ such that:

(i) each a_i is an integer multiple of $2^{\mathrm{b}(G_i)}$; and

(ii) $a_1G_1 + a_2G_2 + \cdots + a_kG_k = 2^nG$.

$\mathcal{Z}$ is clearly nonempty: $G_k = G$, so $(0, 0, \dots, 0, 2^n) \in \mathcal{Z}$. We will show that if $\mathbf{a} \in \mathcal{Z}$ and $\mathbf{a} \neq \mathbf{0}$, then there is some $\mathbf{b} \in \mathcal{Z}$ with $\mathbf{b} <_{\text{lex}} \mathbf{a}$. Since the ordering on $\mathcal{Z}$ is well-founded and $\mathcal{Z}$ is nonempty, it follows by induction that $\mathbf{0} \in \mathcal{Z}$.

So fix $\mathbf{a} \in \mathcal{Z}$ with $\mathbf{a} \neq \mathbf{0}$, and put $X \cong a_1G_1 + \cdots + a_kG_k$. Then X is in quasi-canonical form, so by Lemma 3.11, there are incentives $\Delta^L(X)$ and $\Delta^R(X)$ such that $\Delta^L(X) = \Delta^R(X)$. Write

$$\Delta^L(X) = G_i^L - G_i \quad \text{and} \quad \Delta^R(X) = G_j - G_j^R.$$

Now G_i^L and G_j^R are both subpositions of G, say G_l and G_r, respectively. Since $\Delta^L(X) = \Delta^R(X)$, we then have

$$G_i + G_j = G_l + G_r.$$

There are three cases.

Case 1: $j = i$. Then $2 \cdot G_i = G_l + G_r$, so we can define $\mathbf{b} \in \mathbb{Z}^k$ by replacing copies of $2 \cdot G_i$ with $G_l + G_r$ in the expression for X. That is:

$$b_i = 0, \qquad b_l = a_l + a_i/2,$$
$$b_r = a_r + a_i/2,$$

and $b_t = a_t$ for all other t. Then since $\mathbf{a} \in \mathcal{Z}$ and $2 \cdot G_i = G_l + G_r$, it follows that $\mathbf{b} \in \mathcal{Z}$. Moreover a_i must be even, since $\mathrm{b}(G_i) \ge 1$, so that $\mathbf{b} <_{\text{lex}} \mathbf{a}$.

Case 2: $j < i$. Then we can define $\mathbf{b} \in \mathbb{Z}^k$ by replacing copies of G_i with $-G_j + G_l + G_r$ in the expression for X. That is:

$$b_i = 0, \qquad b_l = a_l + a_i,$$
$$b_j = a_j - a_i, \qquad b_r = a_r + a_i,$$

and $b_t = a_t$ for all other t. As in Case 1 it is clear that $\mathbf{b} \in \mathcal{Z}$ and $\mathbf{b} <_{\text{lex}} \mathbf{a}$.

Case 3: $j > i$. Identical to Case 2. □

Group Structure of $\mathbb{G}$

We now characterize the full group structure of $\mathbb{G}$. The proof requires a generalization of Theorem 3.12. One can view Theorem 3.12 as asserting that 0 has no m-divisors in $\mathbb{G}$, for any odd m. The extension that follows, Theorem 3.15, characterizes the m-divisors of an arbitrary $H \in \mathbb{G}$.

The following lemma generalizes Lemma 3.10. Note that it reduces to Lemma 3.10 in the case $\mathcal{A} = \{0\}$ (since no incentive can be equal to 0).

Lemma 3.13. *Let $\mathcal{A} \subset \mathbb{G}$ be closed. Suppose that $m \cdot G \in \mathcal{A}$ for some $m \geq 1$, and assume that G is in quasi-canonical form. Then for every maximal Left incentive $\Delta = \Delta^L(G)$, either:*

(i) *there is a Right incentive $\Delta^R(G) \geq \Delta$; or else*

(ii) *$\Delta \in \mathcal{A}$.*

Proof. Write $\Delta = G^L - G$. Let K be the canonical form of $m \cdot G$. Then $m \cdot G - K = 0$, so we have

$$(m-1) \cdot G + G^L - K \lhd\!\!\mid 0.$$

Considering Right's winning response, we have three possibilities.

Case 1: $(m-2) \cdot G + G^L + G^R - K \leq 0$ for some G^R. Then since $K = m \cdot G$, we have $G^L + G^R \leq 2 \cdot G$, so that $G^L - G \leq G - G^R$.

Case 2: $(m-1) \cdot G + G^{LR} - K \leq 0$. Then $G^{LR} \leq G$. Moreover, since G is quasi-canonical, we can write

$$G \cong G_1 + G_2 + \cdots + G_k$$

with each G_i in canonical form. Since no G_i has any reversible options, we have without loss of generality

$$G^{LR} \cong G_1^L + G_2^R + G_3 + \cdots + G_k.$$

Thus $G_1^L + G_2^R \leq G_1 + G_2$, so that $G_1^L - G_1 \leq G_2 - G_2^R$. But clearly, $G_1^L - G_1 = \Delta$ and $G_2 - G_2^R$ is equal to a Right incentive of G.

Case 3: $(m-1) \cdot G + G^L - K^L \leq 0$ for some K^L. Then

$$G^L - G \leq K^L - K. \tag{$\dagger$}$$

Moreover, since $m \cdot G - K = 0$, we have $m \cdot G - K^L \mid\!\!\rhd 0$. It cannot be the case that $m \cdot G - K^{LR} \geq 0$ (since K is in canonical form), so necessarily

$$(m-1) \cdot G + G^{L'} - K^L \geq 0$$

for some $G^{L'}$, and hence $G^{L'} - G \geq K^L - K$. Combining this with ($\dagger$), we have

$$G^{L'} - G \geq K^L - K \geq \Delta,$$

so maximality of Δ implies $K^L - K = \Delta$. But since $K \in \mathcal{A}$ and $\mathcal{A}$ is closed, we have $K^L - K \in \mathcal{A}$. □

Lemma 3.14. *Let $\mathcal{A} \subset \mathbb{G}$ be closed. Suppose that $m \cdot G \in \mathcal{A}$ for some $m \geq 1$, and assume G is in quasi-canonical form. Then either:*

(i) *some Δ is equal to both a Left incentive and a Right incentive of G; or*

(ii) *some incentive of G is an element of $\mathcal{A}$.*

Proof. Fix a maximal Left incentive Δ. By Lemma 3.13, either $\Delta \in \mathcal{A}$ or there is some $\Delta^R(G) \geq \Delta$. If the latter, choose Δ' to be maximal among all Right incentives of G that are $\geq \Delta$. Then either $\Delta' \in \mathcal{A}$ or there is some $\Delta^{L'}(G) \geq \Delta'$. But the latter implies

$$\Delta^{L'}(G) \geq \Delta' \geq \Delta,$$

so that by maximality of Δ all three values are equal. □

Theorem 3.15. *Let $\mathcal{A} \subset \mathbb{G}$ be closed. Suppose that $m \cdot G \in \mathcal{A}$ for some $m \geq 1$. Then $2^n \cdot G \in \mathcal{A}$, where $n = \mathrm{b}(G)$.*

Proof. Since the assumption $m \cdot G \in \mathcal{A}$ is invariant of the form of G, we may assume that G is in canonical form. Enumerate the nonzero subpositions of G

$$G_1, G_2, \ldots, G_k$$

in such a way that if $i \leq j$, then $\mathrm{b}(G_i) \leq \mathrm{b}(G_j)$. (In particular, $G_k \cong G$.) Let $\mathcal{Z} \subset \mathbb{Z}^k$ be the set of all k-tuples of integers $(a_1, \ldots, a_k)$ such that:

(i) each a_i is an integer multiple of $2^{\mathrm{b}(G_i)}$; and

(ii) $a_1G_1 + a_2G_2 + \cdots + a_kG_k = 2^nG + H$ for some $H \in \mathcal{A}$.

$\mathcal{Z}$ is clearly nonempty: $G_k = G$ and $0 \in \mathcal{A}$, so $(0, 0, \ldots, 0, 2^n) \in \mathcal{Z}$. We will show that if $\mathbf{a} \in \mathcal{Z}$ and $\mathbf{a} \neq \mathbf{0}$, then there is some $\mathbf{b} \in \mathcal{Z}$ with $\mathbf{b} <_{\mathrm{lex}} \mathbf{a}$. Since the ordering on $\mathcal{Z}$ is well-founded and $\mathcal{Z}$ is nonempty, it follows by induction that $\mathbf{0} \in \mathcal{Z}$.

So fix $\mathbf{a} \in \mathcal{Z}$ with $\mathbf{a} \neq \mathbf{0}$, and put $X \cong a_1G_1 + \cdots + a_kG_k$. Then X is in quasi-canonical form. Moreover,

$$m \cdot X = m \cdot (2^nG + H) \in \mathcal{A},$$

so that X satisfies the hypotheses of Lemma 3.14. Therefore either:

(i) there are incentives $\Delta^L(X)$ and $\Delta^R(X)$ with $\Delta^L(X) = \Delta^R(X)$; or

(ii) some incentive Δ of X is an element of $\mathcal{A}$.

If (i) holds, then the construction of $\mathbf{b}$ is identical to the corresponding construction in the proof of Theorem 3.12.

Suppose instead that (ii) holds. Assume that Δ is a Left incentive (the proof is identical if it is a Right incentive), so that $\Delta = G_i^L - G_i$ for some i.

Now G_i^L is a subposition of G, say G_l. We define $\mathbf{b} \in \mathbb{Z}^k$ by replacing copies of G_i with G_l. That is:

$$b_i = 0, \qquad\qquad b_l = a_l + a_i,$$

and $b_t = a_t$ for all other t. Certainly $\mathbf{b} <_{\text{lex}} \mathbf{a}$. To see that $\mathbf{b} \in \mathcal{Z}$, note that $X = 2^n G + H$, so

$$X - a_i G_i + a_i G_l = 2^n G + H + a_i \Delta.$$

Since $\Delta \in \mathcal{A}$, we have $H + a_i \Delta \in \mathcal{A}$ as well. □

The following theorem is a key consequence of Theorem 3.15.

Theorem 3.16. *Let $\mathcal{A} \subset \mathbb{G}$ be closed. If $m \cdot G \in \mathcal{A}$ for some odd integer m, then $G \in \mathcal{A}$.*

Proof. Let $n = \mathrm{b}(G)$. Then m and 2^n are coprime, so by standard results of number theory (Euler's Theorem), $(2^n)^k \equiv 1 \pmod{m}$ for some k. So there are integers a and b such that

$$1 = am + b2^n,$$

and in particular

$$G = a \cdot (m \cdot G) + b \cdot (2^n \cdot G).$$

Theorem 3.15 implies $2^n \cdot G \in \mathcal{A}$, so that $G \in \mathcal{A}$ as well. □

One more tool is required before proving the main structure theorem:

Lemma 3.17. *Let G be a short game with canonical form K. For every subposition L of K, there is some subposition H of G with $H = L$.*

Proof. Suppose that G' is obtained from G by eliminating dominated options or bypassing reversible options from some subposition H of G. Then for every subposition L of G', either L is itself a subposition of G or else $H = L$. The lemma follows by inductive application of this argument. □

Lemma 3.18. *Let $\mathcal{A} \subset \mathbb{G}$ and fix $G \in \mathcal{A}$. Suppose that $\mathcal{A}$ contains every subposition of G, for* some *form of G. Then $\mathcal{A}$ contains every subposition of the* canonical *form of G.*

Proof. First suppose that G' is obtained from G by eliminating dominated options or bypassing reversible options from some subposition H of G. Then for every subposition L of G', either L is itself a subposition of G, or else $H = L$.

Now let K be the canonical form of G and let L be any subposition of K. Inductive application of the previous argument shows that $H = L$, for some subposition H of G. Therefore $L \in \mathcal{A}$. □

Theorem 3.19.

$$\mathbb{G} \cong \mathbb{D}^\omega \times \left(\frac{\mathbb{D}}{\mathbb{Z}}\right)^\omega.$$

Proof. We will construct a sequence

$$\mathcal{A}_0 \subset \mathcal{A}_1 \subset \mathcal{A}_2 \subset \cdots$$

of closed subsets of $\mathbb{G}$ such that $\mathbb{G} = \bigcup_n \mathcal{A}_n$, and for all n

$$\mathcal{A}_n \cong \mathbb{D}^p \times \left(\frac{\mathbb{D}}{\mathbb{Z}}\right)^q \quad \text{for some } p, q \in \mathbb{N}.$$

The construction is by induction on n. Put $\mathcal{A}_0 = \mathbb{D}$. Given $\mathcal{A}_n$, fix an element $X_n \in \mathbb{G} \setminus \mathcal{A}_n$ of minimal birthday, so that $\mathcal{A}_n$ contains all proper subpositions of X_n (with X_n expressed in canonical form).

Now fix an element $Y \in \mathcal{A}_n + X_n$ of minimal order. Put

$$\mathcal{B}_n = \langle h^k(Y) : k \geq 0 \rangle$$

and

$$\mathcal{A}_{n+1} = \mathcal{A}_n + \mathcal{B}_n.$$

(Here $h(Y)$ was defined on page 169, $h^{k+1}(Y) = h(h^k(Y))$, and $\mathcal{B}_n$ is the group generated by $h^k(Y)$ for $k \geq 0$.) Note that $X_n \in \mathcal{A}_n + Y$, so in particular $X_n \in \mathcal{A}_{n+1}$.

Now Lemma 3.5 implies that either $\mathcal{B}_n \cong \mathbb{D}$ (if Y has infinite order) or $\mathcal{B}_n \cong \mathbb{D}/\mathbb{Z}$ (if Y has finite order). To complete the induction, we will prove that $\mathcal{A}_{n+1}$ is closed and $\mathcal{A}_{n+1} \cong \mathcal{A}_n \times \mathcal{B}_n$.

Claim 1: $\mathcal{A}_{n+1}$ is closed. First of all, every option of X_n is an element of $\mathcal{A}_n$, by minimality of $\mathrm{b}(X_n)$. It follows that for any $A \in \mathcal{A}_n$, every option of $A + X_n$ is an element of $\mathcal{A}_{n+1}$. By Lemma 3.18, every option of Y is an element of $\mathcal{A}_{n+1}$.

Now for each k, the options of $h^{k+1}(Y)$ have the form $r + h^k(Y)$ or $-r$, for some $r \in \mathbb{N}$. Since $\mathbb{D} \subset \mathcal{A}_n$, each of these options is an element of $\mathcal{A}_{n+1}$.

Finally, consider a typical element $G \in \mathcal{A}_{n+1}$. We can write $G = A + B$, with $A \in \mathcal{A}_n$ and $B \in \mathcal{B}_n$. We have shown that every option of $A + B$ is an element of $\mathcal{A}_{n+1}$. Therefore by Lemma 3.18, the same is true for G in canonical form.

Claim 2: $\mathcal{A}_{n+1} \cong \mathcal{A}_n \times \mathcal{B}_n$. It suffices to show that $\mathcal{A}_n \cap \mathcal{B}_n = \{0\}$, since then the homomorphism $(A, B) \mapsto A + B$ from $\mathcal{A}_n \times \mathcal{B}_n \to \mathcal{A}_{n+1}$ is injective.

So suppose $G \in \mathcal{A}_n \cap \mathcal{B}_n$. Since $G \in \mathcal{B}_n$, we have $m \cdot h^k(Y) = G$ for some $k, m \geq 0$, and in particular $m \cdot h^k(Y) \in \mathcal{A}_n$.

Write $m = 2^a b$ with b odd. By Theorem 3.16, $2^a \cdot h^k(Y) \in \mathcal{A}_n$. Since $Y \notin \mathcal{A}_n$ and since $2^k \cdot h^k(Y) = Y$, it follows that $a > k$. Therefore

$$2^a \cdot h^k(Y) = 2^{a-k} \cdot Y.$$

Now since $\mathcal{A}_n$ is 2-divisible, there must be some $A \in \mathcal{A}_n$ such that

$$2^{a-k} \cdot Y = 2^{a-k} \cdot A.$$

In particular,

$$2^{a-k} \cdot (Y - A) = 0.$$

But $Y - A \in \mathcal{A}_n + X_n$, so the minimality assumption on the order of Y gives $2^{a-k} \cdot Y = 0$, and therefore $G = 0$.

This proves that every $\mathcal{A}_n \cong \mathbb{D}^p \times (\mathbb{D}/\mathbb{Z})^q$ for some $p, q \in \mathbb{N}$. Moreover, the birthdays of the X_n must be unbounded (since there are just finitely many games of bounded birthday), so the minimality assumptions on the X_n imply $\mathbb{G} = \bigcup_n \mathcal{A}_n$. Therefore $\mathbb{G} \cong \mathbb{D}^p \times (\mathbb{D}/\mathbb{Z})^q$ for some $p, q \in \mathbb{N} \cup \{\omega\}$.

To complete the proof, we must show that neither p nor q is finite. But if q were finite, then $\mathbb{G}$ would have just 2^q unequal elements of order 2, and the infinite list

$$*, *2, *3, *4, \ldots$$

shows this to be false. Likewise, the games

$$\uparrow, \uparrow^2, \uparrow^3, \uparrow^4, \ldots$$

represent an infinite list of linearly independent torsion-free elements of $\mathbb{G}$ (cf. Theorem II.4.24 on page 94), so p cannot be finite. □

Exercises

3.1 In the definition of $h(G)$ on page 169, we put $n = 3 \cdot \mathrm{b}(G)$. Show that this can be improved to $n = \left\lceil \frac{3}{2} \cdot \mathrm{b}(G) \right\rceil$. Give an example to show that in general, this is the smallest choice of n for which Lemma 3.5 can still be proved.

3.2 Determine the group structure of $\langle \mathbb{G}_2^0 \rangle$.

3.3 Determine the partial-order structure of $\langle \mathbb{G}_2 \rangle$. (Which elements are ≥ 0?)

3.4 There is no short game G with $3 \cdot G = \uparrow$.

3.5 (a) If G has finite order, then for every n, either $n \cdot G = 0$ or $n \cdot G \not\gtrless 0$.

(b) If $G \mathrel{|\!\!>} 0$ has infinite order, then $\{n \in \mathbb{N} : n \cdot G \geq 0\}$ is isomorphic to a submonoid of $\mathbb{N}$ (as a partially ordered monoid).

(c) Let G have infinite order and put $\mathcal{A} = \{n \cdot G : n \in \mathbb{Z}\}$. Prove that the isomorphism type of $\mathcal{A}$ is determined by the partially ordered monoids $\{n \cdot G : n \geq 0\}$ and $\{n \cdot G : n \leq 0\}$.

Notes

The isomorphism type of $\langle \mathbb{G}_2 \rangle$ was described in a 1991 paper by David Moews [**Moe91**]. In the same paper, Moews also characterized the isomorphism type of $\langle \mathbb{G}_3 \rangle/\mathrm{Inf}$, the group of reduced values generated by day 3:

$$\langle \mathbb{G}_3 \rangle/\mathrm{Inf} \cong \mathbb{Z}^7 \times \mathbb{Z}_2^8 \times \mathbb{Z}_4.$$

The full structure of $\langle \mathbb{G}_3 \rangle$ remains unknown.

Open Problem. Characterize the structure of $\langle \mathbb{G}_3 \rangle$ and $\langle \mathbb{G}_3^0 \rangle$ as Abelian groups.

Theorem 3.12 was proved by Simon Norton; our treatment of this result (and its generalization, Theorem 3.15) essentially follows ONAG. The full group structure of $\mathbb{G}$ was isolated by Moews in 2002; our proof of Theorem 3.19 follows [**Moe02**].

A related problem is to characterize the structure of $\mathbb{G}$ as a partially ordered Abelian group. Moews has made partial progress by characterizing the *cyclic* partially ordered subgroups of $\mathbb{G}$. In particular, for every submonoid $\mathcal{S}$ of $(\mathbb{N}, +, \geq)$, there is a $G \in \mathbb{G}$ such that

$$\{n \cdot G : n \geq 0\} \cong \mathcal{S}$$

as a partially ordered monoid. Exercise 3.5 establishes a converse, so this characterizes the cyclic submonoids, and hence the cyclic subgroups, of $\mathbb{G}$. See [**Moe02**] for further discussion.

The full problem remains wide open:

Open Problem. Characterize the structure of $\mathbb{G}$ as a partially ordered Abelian group.

In an interesting twist, the corresponding problem has been solved for **PG**, the transfinite equivalent of $\mathbb{G}$ (see Notes to Section VIII.1 on page 411).

Exercise 3.5: David Moews.

Chapter IV

Impartial Games

A game G is impartial if Left and Right have exactly the same moves available from every subposition of G. Short impartial games have a particularly simple structure, and a complete structure theory was given in the 1930s with the *Sprague–Grundy Theorem*: every short impartial game is equal to $*m$ for some m. Each such game G is therefore associated with a unique natural number, its **nim value**, and knowing the nim value of G gives a complete view of its (normal-play) combinatorial properties.

Nonetheless, isolating the *patterns* of nim values for specific rulesets Γ is often quite challenging, and the study of impartial games remains a subject of intense and continuing interest. Their close relationship to the integers gives rise to fascinating connections to combinatorial number theory and, in some cases, a rich geometric structure.

Section 1 of this chapter is an introduction to nim values, including a proof of the Sprague–Grundy Theorem. Two major classes of impartial games, the **heap games** and **Wythoff-type games**, are discussed in Sections 2 and 3, respectively. Section 4 introduces a generalization of the theory to loopy impartial games. Finally, in Section 5 we show that nim values admit a natural field structure.

1. Nim Values

Recall from Section II.4 that the **nimbers** are the single-heap Nim positions:

$$*m = \{0, *, *2, \ldots, *(m-1) \mid 0, *, *2, \ldots, *(m-1)\}.$$

In the context of impartial games, some of the notation used in Chapter II is cumbersome or misleading. If G is impartial, then there is no point in referring individually to the "Left options" and "Right options" of G; instead we refer simply to the **options** of G and denote a typical option by G'. Likewise, we write simply

$$G = \{G'_1, \ldots, G'_k\}$$

in order to list the options of G; for example,

$$*m = \{0, *, *2, \ldots, *(m-1)\}.$$

We will identify each impartial game G with the set of its options. For example, we'll freely write $G' \in G$ to mean that G' is an option of G.

The Sprague–Grundy Theorem

In Theorem II.4.7 on page 85, we generalized Bouton's Theorem to prove that any sum of nim-heaps is equal to a single heap, with the relationship given by the *nim-addition rule*

$$*a + *b = *c, \quad \text{where} \quad c = a \oplus b.$$

We will now prove the Sprague–Grundy Theorem: *every* short impartial game is equal to a nim-heap. That is, for every such game G there is a (unique) integer $m \geq 0$ with

$$G = *m.$$

The integer m is known as the **nim value** (or sometimes the **$\mathscr{G}$-value** or **Grundy value**) of G, and we write

$$\mathscr{G}(G) = m.$$

Definition 1.1. Let $\mathcal{S} \subsetneqq \mathbb{N}$ be a proper subset of the natural numbers. We define the **mex** of $\mathcal{S}$ (**m**inimal **ex**cluded value) by

$$\text{mex}(\mathcal{S}) = \text{the least integer } m \geq 0 \text{ such that } m \notin \mathcal{S}.$$

Theorem 1.2 (Mex Rule). *Let $a_1, a_2, \ldots, a_k \in \mathbb{N}$, and suppose that*

$$G = \{*a_1, *a_2, \ldots, *a_k\}.$$

Then $G = *m$, where $m = \text{mex}\{a_1, a_2, \ldots, a_k\}$.

Proof. It suffices to show that $G + *m$ is a $\mathscr{P}$-position. First, for all i we have $a_i \neq m$, so necessarily $*a_i + *m$ is an $\mathscr{N}$-position. Conversely, suppose first player moves to $G + *m'$ for some $m' < m$. Then by minimality of m, we have $m' = a_i$ for some i, so second player wins by reverting to $*a_i + *m'$. □

Theorem 1.3 (Sprague–Grundy Theorem). *Let G be any short impartial game. Then $G = *m$ for some m.*

Proof. By induction, we may assume that every option G_i' is equal to some $*a_i$. By the Replacement Lemma (Lemma II.2.2 on page 64) we have

$$G = \{G_1', \ldots, G_k'\} = \{*a_1, \ldots, *a_k\}$$

and the result follows from the mex rule. □

The Sprague–Grundy Theorem is a powerful result with far-reaching consequences. We have the following recurrence for $\mathscr{G}(G)$:

$$\mathscr{G}(G) = \operatorname*{mex}_{G' \in G} \mathscr{G}(G').$$

The **excludents** of G are those integers of the form $\mathscr{G}(G')$, for $G' \in G$. In particular, each G is the mex of its excludents. This rule provides a simple recursive computation of $\mathscr{G}(G)$.

The Sprague–Grundy Theorem also characterizes the group structure of impartial games. Consider the impartial analogue of $\mathbb{G}$,

$$\mathbb{G}^{\mathrm{I}} = \big\{G \in \mathbb{G} : \text{(some form of) } G \text{ is impartial}\big\}.$$

The nim values $\mathscr{G}(G)$ induce a bijection

$$\mathscr{G} : \mathbb{G}^{\mathrm{I}} \to \mathbb{N}.$$

According to the nim-addition rule, $\mathscr{G}$ is a group isomorphism from $(\mathbb{G}^{\mathrm{I}}, +)$ to $(\mathbb{N}, \oplus)$. Moreover, it is easily seen that

$$(\mathbb{N}, \oplus) \cong \mathbb{Z}_2^{\omega},$$

the direct sum of countably many copies of $\mathbb{Z}_2$: each copy of $\mathbb{Z}_2$ corresponds to one digit in the binary representation of $\mathbb{N}$. We therefore have a complete description of the group structure of $\mathbb{G}^{\mathrm{I}}$.

The following characterizations of nim-sum are often useful.

Proposition 1.4. *For all $a, b \geq 0$, we have*

$$a \oplus b = \operatorname*{mex}_{\substack{a' < a \\ b' < b}} \big(a' \oplus b,\ a \oplus b'\big).$$

Proof. $*a + *b$ has nim value $a \oplus b$, while its options have nim values $a' \oplus b$ and $a \oplus b'$. □

Definition 1.5. A **2-power** is an integer of the form 2^k $(k \geq 0)$.

Theorem 1.6. *$\oplus$ is the unique operation such that:*

(i) *$(\mathbb{N}, \oplus)$ is an Abelian group;*

(ii) *$a \oplus a = 0$ for every 2-power a; and*

(iii) *$a \oplus b = a + b$ for distinct 2-powers a and b.*

Proof. $\oplus$ is already known to have these properties; uniqueness is left as an exercise. □

Exercises

1.1 Let G be a short partizan game, and suppose that every option of G is equal to a nim-heap, so that

$$G = \{*a_1, *a_2, \ldots, *a_k \mid *b_1, *b_2, \ldots, *b_l\}.$$

Prove that if $m = \operatorname{mex}\{a_1, a_2, \ldots, a_k\} = \operatorname{mex}\{b_1, b_2, \ldots, b_l\}$, then $G = *m$.

1.2 Prove uniqueness in Theorem 1.6.

1.3 In NIM$_k$ a move is to remove any number of tokens from each of j heaps, for some $j \le k$. It is not necessary to remove the same number of tokens from each heap. (So NIM$_1$ is just NIM.) Determine the $\mathscr{P}$-positions of NIM$_k$.

1.4 MARK is played with heaps of tokens. A player may either remove one token from a heap or reduce the heap size by half (rounding *down* to the new heap size). Thus for $n > 0$ we have $H_n = \{H_{n-1}, H_{\lfloor n/2 \rfloor}\}$. Determine the nim value of H_n for all n.

1.5 For all integers $a, b \ge 0$, define

$$a \circledast b = \operatorname{mex}\left(\begin{array}{c} \{a' \circledast b \ : a' < a\} \cup \{a \circledast b' : b' < b\} \\ \cup \ \{a' \circledast b' : a' < a,\ b' < b,\ \text{and } a' \circledast b = a \circledast b'\} \end{array}\right).$$

Prove that $(\mathbb{N}, \circledast)$ is an Abelian group isomorphic to $\mathbb{Z}_3^\omega$, with each copy of $\mathbb{Z}_3$ corresponding to one digit in the ternary representation of $\mathbb{N}$.

1.6 The **sequential sum** $G \to H$ was described in Section I.4 (cf. Figure I.4.9 on page 41). Assume throughout this exercise that G, H, X, and Y are short impartial games. Prove that:

(a) $\mathscr{G}(*m \to *n) = n$ if $n > m$; or $n - 1$ otherwise.

(b) If $\mathscr{G}(Y) = m$, then $\mathscr{G}(G \to Y) = \mathscr{G}(G \to *m)$. Conversely, $\mathscr{G}(X) = m$ need *not* imply that $\mathscr{G}(X \to G) = \mathscr{G}(*m \to G)$.

Now, for each impartial G we define the **signature** $\chi_G : \mathbb{N} \to \mathbb{N}$ by

$$\chi_G(m) = \mathscr{G}(G \to *m).$$

(c) If $\chi_G = \chi_H$, then $\mathscr{G}(X \to G \to Y) = \mathscr{G}(X \to H \to Y)$ for all X and Y.

(d) $\chi_G(m) = m$ if $G \cong 0$, and $\operatorname{mex}\{\chi_{G'}(m) : G' \in G\}$ otherwise.

(e) $\chi_{G \to H} = \chi_G \circ \chi_H$.

1.7 *Turning games.* Let $\mathcal{F}$ be a collection of nonempty finite subsets of $\mathbb{N}^+$. TURNING($\mathcal{F}$) is played with a row of n coins, labelled $C_1, \ldots, C_n$ from left to right, each of which may be arranged either heads-up or tails-up. Each position can be represented by a vector $(c_1, \ldots, c_n) \in \mathbb{Z}_2^n$, with $c_i = 1$ if and only if C_i is heads-up. A legal move is to select any $\mathcal{A} \in \mathcal{F}$ and turn over coins C_i for all $i \in \mathcal{A}$, provided that the *rightmost* coin is turned heads-to-tails. We write TURNING(d) as shorthand for the special case where $\mathcal{F}$ consists of all nonempty sets of cardinality $< d$. Prove that:

(a) For any $\mathcal{F}$, every position in TURNING($\mathcal{F}$) can be expressed as a disjunctive sum of single-heads positions.

(b) Every position in TURNING(2) has nim value 0 or 1.

(c) TURNING(3) reduces to NIM, with $c_i = 1$ corresponding to the presence of a heap of size i.

(d) The *Mock Turtle Theorem.* Fix an even integer $d \geq 2$. Then the following are equivalent: (i) $(c_1, \ldots, c_n)$ is a $\mathscr{P}$-position for TURNING(d); and (ii) $c_1 \oplus \cdots \oplus c_n = 0$ and $(c_2, \ldots, c_n)$ is a $\mathscr{P}$-position for TURNING$(d-1)$.

1.8 *Lexicodes.* For $\mathbf{a}, \mathbf{b} \in \mathbb{Z}_2^n$, the **Hamming distance** $\delta(\mathbf{a}, \mathbf{b})$ is equal to the number of indices at which $\mathbf{a}$ and $\mathbf{b}$ differ:

$$\delta(\mathbf{a}, \mathbf{b}) = \big|\{i \leq n : a_i \neq b_i\}\big|.$$

The **binary lexicographic code** (or **lexicode**) of distance d and length n is the subset $\mathcal{L} \subset \mathbb{Z}_2^n$, defined as follows by $<_{\text{lex}}$-induction: $\mathbf{a} \in \mathcal{L}$ if and only if $\delta(\mathbf{a}, \mathbf{b}) \geq d$, for all $\mathbf{b} <_{\text{lex}} \mathbf{a}$ with $\mathbf{b} \in \mathcal{L}$. Elements of $\mathcal{L}$ are known as **codewords**. (Note that each codeword is the lexicographically least element of $\mathbb{Z}_2^n$ that is "sufficiently distant" from all prior codewords.)

(a) Show that $\mathcal{L}$ coincides with the set of $\mathscr{P}$-positions of TURNING(d) with n coins. Conclude that $\mathcal{L}$ is necessarily a subgroup of $\mathbb{Z}_2^n$.

The **dimension** of $\mathcal{L}$ is the unique k such that $\mathcal{L} \cong \mathbb{Z}_2^k$. Show that:

(b) The binary lexicode with $d = 3$ and $n = 2^r - 1$ has dimension $2^r - r - 1$. Write out a list of all 16 codewords for $n = 7$. (This is the **Hamming code**, a perfect n-bit one-error-correcting code.)

(c) $d = 7$ and $n = 23$ has dimension 12. (This is the celebrated **Golay code**, a perfect three-error-correcting code.)

(d) If $d \geq 2$ is even, then the dimension for distance d, length n is the same as for distance $d-1$, length $n-1$. (Use Exercise 1.7(d). The cases $d = 4$, $n = 2^r$ and $d = 8$, $n = 24$ are the **extended Hamming** and **extended Golay codes**.)

1.9 The *Lexicode Theorem.* Let $s \geq 2$. The **base-s lexicode** of distance d and length n is the subset $\mathcal{L} \subset \mathbb{Z}_s^n$, defined just as in Exercise 1.8 (with "2" replaced by "s"). Prove that if s is a 2-power, then $\mathcal{L}$ is necessarily closed under componentwise nim-sum. (Consider the game TURNING$_s(d)$, played with s-sided "coins" and otherwise analogous to TURNING(d).)

Notes

The Sprague–Grundy Theorem was first proved in the 1930s, independently by Roland Sprague and Patrick Grundy, and came to prominence with the work of Richard Guy a decade later. Its history is discussed in detail in Appendix C.

Nim values have surprising connections to the theory of error-correcting codes. Some of these connections are explored in Exercises 1.8 and 1.9, and in Exercise 5.9 on page 221. For further results, see [**CS86, Con90, Ple91, Fra96b, FR03**].

Exercise 1.3: E. H. Moore.

Exercise 1.4: Aviezri Fraenkel.

Exercise 1.5: Simon Norton; it has been generalized to base n by François Laubie [**Lau99**].

Exercise 1.6: Stromquist and Ullman.

Exercise 1.7: Hendrik Lenstra.

Exercises 1.8 and 1.9: Conway and Sloane.

2. Heap Games

Many impartial games are naturally represented using heaps of tokens. In such a **heap game** Γ, all positions can be expressed as disjunctive sums drawn from a fixed sequence of components

$$H_0, H_1, H_2, \ldots,$$

known as **single-heap positions** or simply **Γ-heaps**. Moreover, every option of H_n has the form

$$H_n' = H_{a_1} + H_{a_2} + \cdots + H_{a_k} \quad (\text{each } a_i < n),$$

so that every move replaces a heap with a sum of strictly smaller heaps.

When Γ is a heap game, the nim value of any position

$$G = H_{a_1} + H_{a_2} + \cdots + H_{a_k}$$

can be obtained by calculating the nim value of each single-heap component H_{a_i} and applying the nim-addition rule. A solution to Γ can therefore be expressed in terms of the function

$$n \mapsto \mathscr{G}(H_n),$$

which is really just a mapping from $\mathbb{N} \to \mathbb{N}$. Moreover, individual values of this mapping can be calculated recursively (though not necessarily efficiently) by applying the nim-addition rule to each expression for H_n' and taking the mex. This is a uniquely simple framework that makes heap games particularly computationally friendly.

When the specific choice of Γ is clear from the context, we'll write $\mathscr{G}(n)$ as shorthand for $\mathscr{G}(H_n)$.

Example. NIM is a heap game, in which each H_n is taken to be a nim-heap of size n, with options given by

$$H_n = \{H_a : a < n\}.$$

Example. Let $\mathcal{S} \subset \mathbb{N}^+$. In the heap game $\Gamma = \text{SUBTRACTION}(\mathcal{S})$, the only legal moves are to remove exactly k tokens from a heap, for some $k \in \mathcal{S}$:

$$H_n = \{H_{n-k} : k \in \mathcal{S} \text{ and } k \leq n\}.$$

In particular, SUBTRACTION($\mathbb{N}^+$) is just another name for NIM, so subtraction games can be viewed as restrictions of NIM that constrain the number of tokens removed from a heap.

The set $\mathcal{S}$ is the **subtraction set** for Γ. If $\mathcal{S}$ is a singleton, say $\mathcal{S} = \{k\}$, then each H_n has at most one option, so that $\mathscr{G}(n) = 0$ or 1. A simple parity argument shows that $\mathscr{G}(n) = 0$ if $\lfloor n/k \rfloor$ is even; 1 otherwise.

Two-element subtraction sets are addressed in Exercise 2.1; three-element sets are an area of active research (see notes on page 196).

Example. DAWSON'S KAYLES was discussed in the introduction to this book; it was presented there as a game played with strips of boxes, in which a legal move is to delete any two adjacent boxes (see Figure I.1.2 on page 4).

We can also regard DAWSON'S KAYLES as a heap game: a legal move is to remove two tokens from any one heap, optionally splitting the remainder into two smaller heaps. Then

$$H_n = \{H_a + H_b : a + b = n - 2\}.$$

(The case $a = 0$ corresponds to leaving all the remaining tokens in a single heap, or equivalently, in the strip representation, removing boxes from the end of a strip.)

For example, from H_6, we can move to $H_2 + H_2$, $H_3 + H_1$, or H_4 (disregarding the heap of size zero, which is necessarily identically 0). Figure I.1.2(b) on page 4 depicts a move from H_{10} to $H_3 + H_5$.

Example. In the related game KAYLES, also played with strips of boxes, two types of moves are available. A player may either:

- remove *two adjacent* boxes, as in DAWSON'S KAYLES; or
- remove *any one* box.

In the heap game representation, we have

$$H_n = \{H_a + H_b : a + b = n - 2 \text{ or } n - 1\}.$$

Example. In GRUNDY'S GAME, a move consists of selecting a heap and splitting it into two nonempty heaps *of unequal size*. For example, the legal moves from H_6 are to $H_1 + H_5$ and $H_2 + H_4$, but not $H_3 + H_3$.

GRUNDY'S GAME was one of the first heap games to be studied and has remained a subject of intense focus. Its nim values $\mathscr{G}(n)$ have been computed up to $n = 2^{39}$, but no complete solution is known.

Periodicity

The successive $\mathscr{G}$-values of a given heap game Γ are easily computed using the mex and nim-addition rules. A typical option of H_n,

$$H'_n = H_{a_1} + H_{a_2} + \cdots + H_{a_k} \quad (\text{each } a_i < n)$$

has nim value

$$\mathscr{G}(a_1) \oplus \mathscr{G}(a_2) \oplus \cdots \oplus \mathscr{G}(a_k).$$

So assuming that each $\mathscr{G}(a)$ $(a < n)$ has been computed recursively, we can calculate $\mathscr{G}(n)$ by enumerating the options of H_n and mexing their $\mathscr{G}$-values.

	0										1										2										3			
	0	1	2	3	4	5	6	7	8	9	0	1	2	3	4	5	6	7	8	9	0	1	2	3	4	5	6	7	8	9	0	1	2	3
0+	0	0	1	1	2	0	3	1	1	0	3	3	2	2	4	0	5	2	2	3	3	0	1	1	3	0	2	1	1	0	4	5	2	7
34+	4	0	1	1	2	0	3	1	1	0	3	3	2	2	4	4	5	5	2	3	3	0	1	1	3	0	2	1	1	0	4	5	3	7
68+	4	8	1	1	2	0	3	1	1	0	3	3	2	2	4	4	5	5	9	3	3	0	1	1	3	0	2	1	1	0	4	5	3	7
102+	4	8	1	1	2	0	3	1	1	0	3	3	2	2	4	4	5	5	9	3	3	0	1	1	3	0	2	1	1	0	4	5	3	7
136+	4	8	1	1	2	0	3	1	1	0	3	3	2	2	4	4	5	5	9	3	3	0	1	1	3	0	2	1	1	0	4	5	3	7
170+	4	8	1	1	2	0	3	1	1	0	3	3	2	2	4	4	5	5	9	3	3	0	1	1	3	0	2	1	1	0	4	5	3	7

Figure 2.1. The first 204 nim values of DAWSON'S KAYLES.

Example. The first few values of SUBTRACTION$(2,3,5)$ are

$$0011223001122300112 23\ldots.$$

Each successive value $\mathscr{G}(n)$ is the mex of the values 2, 3, and 5 positions to the left. There is already a clear pattern: the sequence 0011223 repeats with period 7. It's easy to see that this pattern must continue indefinitely, since the mex calculation only "looks back" 5 steps in the sequence.

Example. Normal-play DAWSON'S KAYLES exhibits a similar phenomenon. The first few nim values are computed as follows:

$$\begin{array}{ll}
H_1 \to 0, & \mathscr{G}(1) = 0, \\
H_2 \to H_0, & \mathscr{G}(2) = \operatorname{mex}\{\mathscr{G}(0)\} = \operatorname{mex}\{0\} = 1, \\
H_3 \to H_1, & \mathscr{G}(3) = \operatorname{mex}\{\mathscr{G}(1)\} = \operatorname{mex}\{0\} = 1, \\
H_4 \to H_2 \text{ or } H_1 + H_1, & \mathscr{G}(4) = \operatorname{mex}\{\mathscr{G}(2), \mathscr{G}(1) \oplus \mathscr{G}(1)\} = \operatorname{mex}\{1, 0\} = 2, \\
H_5 \to H_3 \text{ or } H_2 + H_1, & \mathscr{G}(5) = \operatorname{mex}\{\mathscr{G}(3), \mathscr{G}(1) \oplus \mathscr{G}(2)\} = \operatorname{mex}\{1, 1\} = 0.
\end{array}$$

Continuing like this (using *cgsuite*, say) we obtain the table pictured in Figure 2.1. The figure illustrates the first 204 nim values of DAWSON'S KAYLES, grouped into six rows of 34. The reason for organizing them modulo 34 is immediately obvious: throughout the table we have $\mathscr{G}(n+34) = \mathscr{G}(n)$, with just six exceptions ($n = 0, 15, 17, 32, 35, 52$).

As it turns out, this mod-34 periodicity in DAWSON'S KAYLES continues indefinitely, with $n = 52$ the last exception, so that

$$\mathscr{G}(n+34) = \mathscr{G}(n) \quad \text{for all } n \geq 53.$$

Periodicity is an important phenomenon: by representing a heap game Γ as a periodic nim-sequence, we can compute the value of any single heap efficiently. The value of any *sum* of heaps—and therefore a perfect winning strategy—can then be computed using the nim-addition rule. Therefore, once Γ is known to be periodic, it is solved (in the sense of Chapter I). A major thrust of this section will be to identify conditions under which various classes of heap games can be proved periodic.

Before proceeding, we formalize the definitions of periodicity and some related notions.

Definition 2.1. Let $\mathbf{a} = (a_0, a_1, a_2, \ldots)$ be a sequence of integers. We say that $\mathbf{a}$ is **periodic with period** p **and preperiod** n_0 if, for all $n \geq n_0$, we have

$$a_{n+p} = a_n.$$

We say that $\mathbf{a}$ is **purely periodic** if it is periodic with preperiod 0.

Note that if $\mathbf{a}$ is periodic with period p and preperiod n_0, then it is also periodic with period kp and preperiod $n_0 + j$, for all j and k. The **minimal period** and **preperiod** are the least values of p and n_0 that suffice.

Definition 2.2. Let $\mathbf{a} = (a_0, a_1, a_2, \ldots)$ be a sequence of integers. We say that $\mathbf{a}$ is **arithmetic periodic with period** p, **saltus** s, and **preperiod** n_0 if, for all $n \geq n_0$, we have

$$a_{n+p} = a_n + s.$$

Definition 2.3. Let Γ be a heap game. We say that Γ is periodic (arithmetic periodic, purely periodic) if its sequence of nim values

$$\mathscr{G}(0), \mathscr{G}(1), \mathscr{G}(2), \ldots$$

is periodic (arithmetic periodic, purely periodic).

Finite Subtraction Games

SUBTRACTION($\mathcal{S}$) is a **finite subtraction game** if $\mathcal{S}$ is finite. Remarkably, *all* finite subtraction games are periodic.

Theorem 2.4. *SUBTRACTION($\mathcal{S}$) is periodic, for every finite $\mathcal{S} \subset \mathbb{N}^+$.*

Proof. Let m be the cardinality of $\mathcal{S}$. Then there are at most m distinct moves from any single heap, so that $\mathscr{G}(n) \leq m$ for all n.

Let $k = \max(\mathcal{S})$. Then there are just finitely many k-tuples of nonnegative integers $\leq m$. So there must exist integers $a < b$ such that

$$\mathscr{G}(a+i) = \mathscr{G}(b+i) \quad \text{for all } i \text{ with } 0 \leq i < k.$$

Since the value of $\mathscr{G}(n)$ depends only on the preceding k values of $\mathscr{G}$, a simple induction establishes that $\mathscr{G}$ is periodic, with period $b - a$ and preperiod a. □

Theorem 2.4 settles the ultimate periodicity of finite subtraction games, but the bound obtained on period lengths is enormous—on the order of $|\mathcal{S}|^{\max(\mathcal{S})}$. A more tractable analysis is given by the following theorem.

Theorem 2.5 (Subtraction Periodicity Theorem). *Let $\mathcal{S} \subset \mathbb{N}^+$ be finite and put $\Gamma = $ SUBTRACTION$(\mathcal{S})$. Let $k = \max(\mathcal{S})$. Suppose that there exist $n_0 \geq 0$ and $p \geq 1$ such that*

$$\mathscr{G}(n+p) = \mathscr{G}(n) \quad \text{for all } n \text{ such that } n_0 \leq n < n_0 + k.$$

Then it follows that

$$\mathscr{G}(n+p) = \mathscr{G}(n) \quad \text{for all } n \geq n_0,$$

so that Γ is periodic with period p and preperiod n_0.

Proof. By induction on n. Let $n \geq n_0 + k$. Then for each $i \in \mathcal{S}$ we have $i \leq n$ (by definition of k), so the options of H_n are exactly

$$\{H_{n-i} : i \in \mathcal{S}\}.$$

Likewise, the options of H_{n+p} are exactly $\{H_{n+p-i} : i \in \mathcal{S}\}$. But for each $i \in \mathcal{S}$, we have $n - i < n$, so by the inductive hypothesis

$$\mathscr{G}(n-i) = \mathscr{G}(n+p-i),$$

whereupon by the mex rule $\mathscr{G}(n) = \mathscr{G}(n+p)$. □

Octal Games

The **octal games** form a wide class that includes KAYLES, DAWSON'S KAYLES, SUBTRACTION$(\mathcal{S})$, and many other examples. In all octal games, a move consists of removing some number of tokens from a heap and then optionally splitting the remaining tokens into two heaps. For each $i \geq 0$, the rules specify the circumstances under which i tokens may be removed from a heap and whether the heap may subsequently be split.

The rules for a general octal game are described by a sequence of **code digits**

$$\mathbf{d}_0.\mathbf{d}_1\mathbf{d}_2\mathbf{d}_3\ldots$$

with $0 \leq \mathbf{d}_i < 8$. The ith digit $\mathbf{d}_i$ specifies the circumstances under which i tokens may be removed. The interpretations of each value for $\mathbf{d}_i$ are summarized in Figure 2.2 on the next page, and they can also be described by considering the binary representation of $\mathbf{d}_i$. Write

$$\mathbf{d}_i = e_0 + 2e_1 + 4e_2,$$

with each $e_j = 0$ or 1. Then:

- If $e_0 = 1$, then $H_i \to 0$ is a legal move: it is permissible to remove i tokens and leave 0 remaining heaps.
- If $e_1 = 1$, then $H_n \to H_{n-i}$ is a legal move, for all $n > i$: it is permissible to remove i tokens and leave exactly 1 remaining heap.

$\mathbf{d}_i$		It is legal to remove i tokens ...
0	000	Never
1	001	Provided that no tokens remain after they are removed
2	010	Provided that at least one token remains (and any remaining tokens are left in a single heap)
3	011	Always (but the remaining tokens must be left in a single heap)
4	100	Provided that the remaining tokens are split into exactly two nonempty heaps
5	101	Provided that *if* any tokens remain, *then* they are split into exactly two nonempty heaps
6	110	Provided that at least one token remains; the remaining tokens may optionally be split into two heaps
7	111	Always; the remaining tokens may optionally be split into two heaps

Figure 2.2. The various octal codes.

- If $e_2 = 1$, then $H_n \to H_a + H_b$ is a legal move, provided $a + b = n - i$ and $a, b \geq 1$. It is permissible to remove i tokens and leave exactly 2 remaining heaps.

Examples. Many examples in this section are octal games.

- NIM is $\mathbf{0.33333}\ldots$. This can be abbreviated $\mathbf{0.\overline{3}}$.
- Each SUBTRACTION($\mathcal{S}$) is an octal game: $\mathbf{d}_i = 3$ if $i \in \mathcal{S}$; 0 otherwise.
- KAYLES is **0.77** and DAWSON'S KAYLES is **0.07**.

The **length** of $\mathbf{d}_0.\mathbf{d}_1\mathbf{d}_2\mathbf{d}_3\ldots$ is the largest k such that $\mathbf{d}_k \neq 0$, or ∞ if no such k exists.

Note that the legal moves of an octal game are determined just by the *number of tokens removed* and the *number of heaps remaining.* In particular, the legal moves do not discriminate by heap size, so if Γ is a finite-length octal game, then all sufficiently large Γ-heaps start to look structurally alike.

Lemma 2.6. *Let H_n be a Γ-heap, for some octal game Γ. Then for all $a \geq 0$, $b \geq 1$, and $j \geq 0$,*

$$H_a + H_b \quad \textit{is an option of} \quad H_n$$

if and only if

$$H_a + H_{b+j} \quad \textit{is an option of} \quad H_{n+j}.$$

Proof. Let i be the number of tokens removed (in either case):

$$i = n - (a+b) = n + j - (a+b+j).$$

Let $\mathbf{d}_i$ be the corresponding digit of the octal code for Γ. If $a = 0$, then $H_a + H_b$ and $H_a + H_{b+j}$ both have exactly one heap (since $b \geq 1$). Therefore $H_a + H_b$ is an option of H_n if and only if $\mathbf{d}_i \in \{2, 3, 6, 7\}$, and likewise for H_{n+j}.

Similarly, if $a \geq 1$, then $H_a + H_b$ and $H_a + H_{b+j}$ both have exactly two heaps. Therefore $H_a + H_b$ is an option of H_n if and only if $\mathbf{d}_i \in \{4, 5, 6, 7\}$, and likewise for H_{n+j}. □

This basic observation underpins the following fundamental theorem.

Theorem 2.7 (Octal Periodicity Theorem). *Let* $\Gamma = \mathbf{d}_0.\mathbf{d}_1\mathbf{d}_2 \ldots \mathbf{d}_k$ *be an octal game of finite length* k. *Suppose that there exist* $n_0 \geq 1$ *and* $p \geq 1$ *such that*

$$\mathscr{G}(n+p) = \mathscr{G}(n) \quad \textit{for all } n \textit{ with } n_0 \leq n < 2n_0 + p + k.$$

Then it follows that

$$\mathscr{G}(n+p) = \mathscr{G}(n) \quad \textit{for all } n \geq n_0,$$

so that Γ *is periodic with period* p *and preperiod* n_0.

Proof. By induction on n. Let $n \geq 2n_0 + p + k$. Then each option of H_{n+p} has the form

$$H_a + H_b, \quad \text{with } a + b = n + p - i \text{ for some } i \leq k.$$

Since $n + p \geq 2n_0 + 2p + k$, we have $a + b \geq 2n_0 + 2p$. Without loss of generality, it follows that $b \geq n_0 + p$. Thus $b - p \geq n_0$ and

$$\mathscr{G}(b-p) = \mathscr{G}(b).$$

Therefore

$$\mathscr{G}(a) \oplus \mathscr{G}(b-p) = \mathscr{G}(a) \oplus \mathscr{G}(b)$$

and it follows that

$$H_a + H_{b-p} = H_a + H_b.$$

Now $b - p \geq 1$ (since $n_0 \geq 1$), so by Lemma 2.6, $H_a + H_{b-p}$ is an option of H_n. This shows that every option of H_{n+p} is equal to an option of H_n. A similar argument shows that every option of H_n is equal to an option of H_{n+p}, so we conclude that

$$\mathscr{G}(n+p) = \mathscr{G}(n).$$

□

	0										1	
	0	1	2	3	4	5	6	7	8	9	0	1
0+	0	1	2	3	1	4	3	2	1	4	2	6
12+	4	1	2	7	1	4	3	2	1	4	6	7
24+	4	1	2	8	5	4	7	2	1	8	6	7
36+	4	1	2	3	1	4	7	2	1	8	2	7
48+	4	1	2	8	1	4	7	2	1	4	2	7
60+	4	1	2	8	1	4	7	2	1	8	6	7
72+	4	1	2	8	1	4	7	2	1	8	2	7

Figure 2.3. The first 84 $\mathscr{G}$-values of KAYLES.

Suppose Γ is an octal game that is eventually periodic. Then Γ can be solved by calculating its first n $\mathscr{G}$-values, for some sufficiently large finite n, and then testing whether Γ satisfies the hypotheses of the Octal Periodicity Theorem. Both of these operations are easily automated, so that solutions are known for many octal games that would not be feasible to analyze by hand.

Example. To establish that DAWSON'S KAYLES has period 34 and preperiod 53, it suffices to compute the first

$$2 \cdot 53 + 2 \cdot 34 + 2 = 176$$

$\mathscr{G}$-values and verify periodicity among them. This we have done in Figure 2.1 on page 186, so in fact DAWSON'S KAYLES is periodic.

Example. Figure 2.3 shows the first 84 nim values of KAYLES. Further calculation reveals that the last row of this table repeats indefinitely, with $n = 70$ the last exceptional value. Therefore, after computing

$$2 \cdot 71 + 2 \cdot 12 + 2 = 168$$

values, we may conclude that KAYLES has period 12 and preperiod 71.

Example. Figure 2.4 on the next page summarizes the record-holders for longest known period, from 1956 to present. Remarkably, the current champion, **0.106**, exhibits no $\mathscr{G}$-value greater than 31, despite its period exceeding 3×10^{11}.

Example. The game **0.007** is played just like DAWSON'S KAYLES, except that *three* tokens must be removed from a heap (or three boxes deleted, in the strips-of-boxes representation) instead of two. Its $\mathscr{G}$-values are known up to $\mathscr{G}(2^{22})$, with no apparent periodicity. The contrast between **0.07** and **0.007** shows that the behavior of octal games is highly sensitive to small perturbations of the ruleset.

Γ	Publication	Period	Preperiod
0.07	1956, Guy & Smith	34	53
0.156	1967, J. Kenyon	349	3,479
0.165	1976, R. Austin	1,550	5,181
0.16	1989, Gangolli & Plambeck	149,459	105,351
0.454	2000, A. Flammenkamp	60,620,715	160,949,019
0.106	2002, A. Flammenkamp	328,226,140,474	465,384,263,797

Figure 2.4. Record-holders for the longest known period of an octal game, 1956–present.

Sparse Spaces

A close examination of the nim values for KAYLES (Figure 2.3) reveals a striking dichotomy. Call a natural number n **odious** if the binary expansion of n has an **od**d number of ones; **evil** if it has an **ev**en number of ones. Then the evil numbers

$$0,\ 3,\ 5,\ \text{and } 6$$

appear just a few times each in the KAYLES sequence, while the odious numbers

$$1,\ 2,\ 4,\ 7,\ \text{and } 8$$

occur infinitely often.

The evil numbers form a **sparse space** for KAYLES—a subgroup of $\mathbb{N}$ of index 2 containing just finitely many values in the KAYLES sequence. This pattern of behavior is intrinsic to octal games, and it is often possible to identify a "probable" sparse space well before periodicity becomes apparent. For reasons that will become clear, knowing the sparse space for Γ greatly improves the speed with which one can calculate its $\mathscr{G}$-values, and it would be quite impractical to determine the larger periods in Figure 2.4 without the sparse space machinery.

Definition 2.8. Let Γ be an octal game. A subgroup $\mathcal{S} \subset \mathbb{N}$ of index 2 is a **sparse space** for Γ if $\mathscr{G}(n) \in \mathcal{S}$ for just finitely many n. The complement $\mathcal{C} = \mathbb{N} \setminus \mathcal{S}$ is the corresponding **common coset**, and elements of $\mathcal{S}$ and $\mathcal{C}$ are called **rare** and **common** values of Γ, respectively.

Example. It remains unknown whether GRUNDY'S GAME is ultimately periodic, so we can't say for certain whether or not it has a sparse space. However there appears to be a well-established pattern among its known values. Let $\mathcal{S}$ be the group generated by

$$1,\ 6,\ 10,\ 18,\ 34,\ \ldots,\ 2^k \oplus 2,\ \ldots.$$

Here $n \in \mathcal{S}$ if and only if its binary expansion has an even number of ones, *disregarding the last digit.* (This is slightly different from the KAYLES space.) Remarkably, of the first 2^{39} $\mathscr{G}$-values of GRUNDY'S GAME, only 1,284 of them lie within $\mathcal{S}$. This is sufficient evidence to *presume* that $\mathcal{S}$ is a sparse space for GRUNDY'S GAME.

The structure of these 1,284 values is fascinating. The 1,271st occurs at heap size 82,860:

$$\mathscr{G}(82{,}860) = 108.$$

Thereafter there is a vast stretch during which no further rare values occur, until suddenly we see

$$\mathscr{G}(47{,}461{,}861) = 261$$

followed by twelve more rare values in (relatively) rapid succession. The last of these is

$$\mathscr{G}(48{,}399{,}022) = 259,$$

which is the last known rare value. See [**Fla**] for an exhaustive list.

GRUNDY'S GAME illustrates a fundamental feature of sparseness: once a bifurcation into rare and common values becomes established, it tends to persist. This tendency has a simple explanation. A typical excludent for H_n has the form

$$\mathscr{G}(a) \oplus \mathscr{G}(b), \quad a, b < n.$$

Now the nim-sum of two common values is always rare, so if $\mathscr{G}(a)$ is common for most $a < n$, then most of the excludents for H_n will be rare. So $\mathscr{G}(n)$ itself is quite likely to be common.

The existence of an apparent sparse space strongly suggests periodicity.

Theorem 2.9. *If Γ has a sparse space, then Γ is ultimately periodic.*

Proof. Suppose $\mathscr{G}(n)$ is common for all $n \geq n_0$. Then each such $\mathscr{G}(n)$ is equal to the least common value not among its excludents. But for $n \geq 2n_0 + k$ the common excludents all have the form

$$\mathscr{G}(a) \oplus \mathscr{G}(b), \quad \text{with } \mathscr{G}(a) \text{ rare and } b \geq n - n_0 - k.$$

(For moves leaving just one heap we can take $a = 0$, since 0 is always rare.) So the value of $\mathscr{G}(n)$ depends only on the preceding $n_0 + k$ values of $\mathscr{G}$. The rest of the argument is identical to the proof of Theorem 2.4. $\square$

We conclude with a final observation regarding sparse spaces. Denote by $2^{\mathbb{N}}$ the set of 2-powers

$$2^{\mathbb{N}} = \{2^k : k \in \mathbb{N}\}.$$

Then every sparse space $\mathcal{S}$ is characterized by $\mathcal{S} \cap 2^{\mathbb{N}}$.

Proposition 2.10. *There is a one-to-one correspondence between index-2 subgroups $\mathcal{S} \subset \mathbb{N}$ and proper subsets $\mathcal{A} \subsetneq 2^{\mathbb{N}}$, given by $\mathcal{S} \mapsto \mathcal{S} \cap 2^{\mathbb{N}}$.*

Proof. Certainly $\mathcal{S} \cap 2^{\mathbb{N}}$ must be a proper subset of $2^{\mathbb{N}}$, since $2^{\mathbb{N}}$ generates the entire group $\mathbb{N}$. To complete the proof we must show that every $\mathcal{A} \subsetneq 2^{\mathbb{N}}$ extends uniquely to an index-2 subgroup.

Given such $\mathcal{A}$, put $\mathcal{B} = 2^{\mathbb{N}} \setminus \mathcal{A}$ and let $\mathcal{S}$ be the group generated by

$$\mathcal{A} \cup \{a \oplus b : a, b \in \mathcal{B}\}.$$

Now $2^{\mathbb{N}}$ is a basis for $\mathbb{N}$, so $\mathcal{A}$ and $\mathcal{B}$ generate disjoint subgroups. It follows that $\mathcal{S} \cap 2^{\mathbb{N}} = \mathcal{A}$, and $\mathcal{S}$ has index 2 since $b \oplus \mathcal{S} = \mathbb{N} \setminus \mathcal{S}$ for any $b \in \mathcal{B}$.

Finally, let $\mathcal{S}'$ be any index-2 subgroup with $\mathcal{S}' \cap 2^{\mathbb{N}} = \mathcal{A}$. Then if $a, b \in \mathcal{B}$, we must have $a \oplus b \in \mathcal{S}'$, since otherwise $a \oplus \mathcal{S}'$ and $b \oplus \mathcal{S}'$ would be distinct cosets of $\mathcal{S}'$. It follows that $\mathcal{S}' = \mathcal{S}$. □

Examples. For the space of evil numbers we have $\mathcal{A} = \emptyset$, and for the (apparent) sparse space of GRUNDY'S GAME $\mathcal{A} = \{1\}$.

We conclude with a question from *Winning Ways* that is still unresolved.

Question. Is every finite octal game eventually periodic?

Hexadecimal Games

We can extend the definition of octal code to define an arbitrary **take-and-break game**

$$\Gamma = \mathbf{d}_0.\mathbf{d}_1\mathbf{d}_2\mathbf{d}_3\ldots$$

where $\mathbf{d}_i$ may be any nonnegative integer. Write $\mathbf{d}_i$ in binary, say,

$$\mathbf{d}_i = 2^0 e_0 + 2^1 e_1 + 2^2 e_2 + \cdots$$

with each $e_j = 0$ or 1. Then it is permissible to remove i tokens, splitting the remainder into exactly j heaps, precisely when $e_j = 1$.

A **hexadecimal game** is a take-and-break game Γ with all $\mathbf{d}_i < 16$. It is customary to abbreviate values of $\mathbf{d}_i = 10, 11, \ldots, 15$ by the hexadecimal digits $\mathbf{A}, \mathbf{B}, \ldots, \mathbf{F}$. Note that a hexadecimal game is just a take-and-break game in which it is never legal to split into more than three heaps.

Hexadecimal games exhibit new types of behavior not found among octal games. For example, finite-length octal games cannot exhibit arithmetic periodicity with nonzero saltus (see Exercise 2.12). By contrast, it is relatively easy to find arithmetic periodic hexadecimal games.

Example. Let $\Gamma = \mathbf{0.8}$. We will show that Γ is arithmetic periodic, with period 3, preperiod 1, and saltus 1. In particular, we will show that its nim sequence is

$$0000111222333444\ldots,$$

given for $n \geq 1$ by

$$\mathscr{G}(n) = \left\lfloor \frac{n-1}{3} \right\rfloor.$$

Now a legal move of Γ is to remove a single token and split the remainder into exactly three heaps. Therefore the moves from H_n are to

$$H_a + H_b + H_c, \quad \text{with } a, b, c \geq 1 \text{ and } a + b + c = n - 1.$$

But

$$\begin{aligned}\mathscr{G}(a) \oplus \mathscr{G}(b) \oplus \mathscr{G}(c) &\leq \mathscr{G}(a) + \mathscr{G}(b) + \mathscr{G}(c)\\ &\leq \frac{a-1}{3} + \frac{b-1}{3} + \frac{c-1}{3} = \frac{n-4}{3}.\end{aligned}$$

Since

$$\frac{n-4}{3} < \left\lfloor \frac{n-1}{3} \right\rfloor,$$

it follows that $\lfloor (n-1)/3 \rfloor$ is not an excludent of H_n. To complete the argument we must show that every smaller number *is* an excludent; that is, if $m < \lfloor (n-1)/3 \rfloor$, then H_n has an option of nim value m.

Fix such m. Since $\lfloor (n-1)/3 \rfloor > m$, we have $\lfloor (n-1)/3 \rfloor \geq m+1$, and hence $n - 3m \geq 4$. If $n - 3m$ is even, then put $a = (n - 3m - 2)/2$ and note that

$$a + a + (3m+1) = n - 1 \quad \text{and} \quad \mathscr{G}(a) \oplus \mathscr{G}(a) \oplus \mathscr{G}(3m+1) = 0 \oplus m = m.$$

Moreover $a \geq 1$, so H_n has the option $H_a + H_a + H_{3m+1}$ of value m.

Similarly, if $n - 3m$ is odd, then put $a = (n - 3m - 3)/2$ and note that

$$a + a + (3m+2) = n - 1 \quad \text{and} \quad \mathscr{G}(a) \oplus \mathscr{G}(a) \oplus \mathscr{G}(3m+2) = 0 \oplus m = m.$$

Moreover $a \geq 1$, so H_n has the option $H_a + H_a + H_{3m+2}$ of value m. This exhausts all cases.

Exercises

2.1 Let $\mathcal{S} = \{a, b\}$, with $a, b \geq 1$. Prove that SUBTRACTION($\mathcal{S}$) is purely periodic with period $a + b$.

2.2 *Ferguson's Pairing Property.* Let Γ be a subtraction game with subtraction set $\mathcal{S}$. Then for all $n \geq 0$,

$$\mathscr{G}(n) = 0 \quad \text{if and only if} \quad \mathscr{G}(n+s) = 1,$$

where s is the smallest element of $\mathcal{S}$.

2.3 Let Γ and Γ' be impartial rulesets. We say that Γ and Γ' have **identical positions** if every position of Γ is isomorphic to a position of Γ', and vice versa. Prove that for every impartial ruleset Γ that is closed under disjunctive sum, there is a heap game Γ' with identical positions.

2.4 Show that DAWSON'S CHESS (Exercise I.1.5 on page 14) is **0.137**.

2.5 Analyze the games $\mathbf{0.d}_1$, where $\mathbf{d}_1 = 2^k$. (Show that $\mathbf{0.4}$ is equivalent to DAWSON'S KAYLES. For $k \geq 3$, generalize the solution for $\mathbf{0.8}$.)

2.6 *Duplicate Kayles, etc.* Let $\Gamma = \mathbf{0.7}^k$ $(k \geq 2)$. The game

$$\Gamma' = \mathbf{0.0}^{t-1}\mathbf{7}^{k'} \qquad (t \geq 1),$$

with $k' = (k-2)t + 2$, is known as ***t*-plicate** Γ. Prove that

$$\mathscr{G}'(tn) = \mathscr{G}'(tn+1) = \cdots = \mathscr{G}'(tn + (n-1)) = \mathscr{G}(n) \text{ for all } n.$$

(Here $\mathscr{G}'(n)$ denotes the nim value of a Γ'-heap of size n.)

2.7 *Generalized Periodicity Theorem.* Let $\Gamma = \mathbf{d}_0.\mathbf{d}_1\mathbf{d}_2\ldots\mathbf{d}_k$ be a take-and-break game of finite length k. Let $t \geq 0$ be such that $\mathbf{d}_i < 2^{t+1}$ for all i. Suppose that there exist $n_0 \geq 0$ and $p \geq 0$ such that

$$\mathscr{G}(n+p) = \mathscr{G}(n) \quad \text{for all } n \text{ with } n_0 \leq n < tn_0 + (t-1)p + k.$$

Prove that

$$\mathscr{G}(n+p) = \mathscr{G}(n) \quad \text{for all } n \geq n_0.$$

2.8 *All-but subtraction games.* Let $\mathcal{S} \subset \mathbb{N}^+$ and suppose that $\mathbb{N}^+ \setminus \mathcal{S}$ is finite. Prove that SUBTRACTION($\mathcal{S}$) is arithmetic periodic.

2.9 Let $\Gamma = \mathbf{0.d}_1\mathbf{d}_2\mathbf{d}_3\ldots$. Suppose that there exist $i, j \geq 1$, with i odd, j even, and $\mathbf{d}_i, \mathbf{d}_j \geq 4$. Prove that Γ has just finitely many single-heap $\mathscr{P}$-positions.

2.10 Let $\Gamma = \mathbf{0.d}_1\mathbf{d}_2\mathbf{d}_3\ldots$. Suppose that there exist $i, j \geq 1$, with i odd, j even, and $\mathbf{d}_i, \mathbf{d}_j \geq 8$. Prove that $\lim_{n\to\infty} \mathscr{G}(n) = \infty$.

2.11 $\mathbf{0.3F}$ is arithmetic periodic with saltus 3.

2.12 *Austin's Theorem.* For $n \in \mathbb{N}$, denote by $f(n)$ the cardinality of the set

$$\{a \oplus b : a, b \in \mathbb{N} \text{ and } a + b = n\}.$$

Prove that:

(a) For all n, we have $f(2n+1) = f(n)$ and $f(2n) = f(n) + f(n-1)$.

(b) For all n, either $f(n) = f(a)$ for some $a \leq \frac{1}{2}n$ or else $f(n) = f(a) + f(b)$ for some $a \leq \frac{1}{2}n$ and $b \leq \frac{1}{4}n$.

(c) There is a real number $\alpha < 1$ such that $f(n) \leq \frac{5}{4}n^\alpha$ for all n.

(d) If Γ is an octal game with finite length, then Γ cannot be arithmetic periodic with nonzero saltus.

Notes

A general theory of subtraction games remains elusive, despite their apparent simplicity. There is an easy solution for two-element subtraction sets (Exercise 2.1), and three-element sets of the form $\{a, b, a+b\}$ were analyzed in *Winning Ways*. Several other special cases were recently solved by Ho [**Ho**].

Question. Does there exist a complete theory of finite subtraction games?

Octal games were introduced by Richard Guy and Cedric Smith in a seminal 1956 paper [**GS56**]. The sparse space theory is due to Berlekamp, who in 1973 discovered the bifurcation of GRUNDY'S GAME into rare and common values. Since then there's been little more to say about the normal-play *theory* of octal games, although advances in computing technology have gradually expanded its scope, at intervals of roughly ten years. At the time of this writing it's been just about that long since the last entry in Figure 2.4 was discovered, so the time is ripe for another:

Open Problem. Identify an octal game with period longer than **0.106**.

The existing entries in Figure 2.4 were first reported in [**GS56, Ken67, Aus76, GP89, Fla**].

Hexadecimal games have proven to be considerably more resistant to attack. Many years ago Jack Kenyon observed that **0.3F** has saltus 3 (Exercise 2.11), disproving a conjecture of Guy that the saltus must always be a power of 2.

Some time later, Richard Austin [**Aus76**] developed systematic methods for detecting arithmetic periodicity with saltus of the form 2^k. Austin's methods were extended by Howse and Nowakowski to games with odd saltus [**HN04**]; among other results, they discovered that **0.3F3** has saltus 5. Unfortunately all of these techniques apply only in certain restricted circumstances, and they often involve computing an enormous number of $\mathscr{G}$-values to verify relatively small periods. The longest known period for an arithmetic periodic hexadecimal game is 53, for **0.300F**, with saltus 16 [**HN04**].

Grossman and Nowakowski have recently discovered new types of regularity in hexadecimal games [**GN13**]. For example, **0.205200C** consists of an arithmetic periodic sequence "interleaved" with a periodic one; this is called **sapp regularity** in *Lessons in Play*. **0.2048** exhibits an even more striking form of regularity (**ruler regularity**), in which an infinite number of interrelated arithmetic periodic sequences are threaded together. These types of periodicity are generally established on a case-by-case basis, and few systematic methods exist.

In contrast to octal games, it's clear that hexadecimal games remain poorly understood, and much about them remains to be discovered. We close with a rather vague statement of this challenge.

Open Problem. Extend the analysis of hexadecimal games.

Exercise 2.2: Thomas Ferguson.

Exercise 2.6: Guy and Smith.

Exercise 2.8: Angela Siegel.

Exercise 2.11: Jack Kenyon.

Exercise 2.12: Richard Austin.

3. WYTHOFF

WYTHOFF is played with two heaps of tokens. On her turn a player may either:

- remove any number of tokens from a single heap (as in NIM); or else
- remove the *same* number of tokens from *both* heaps.

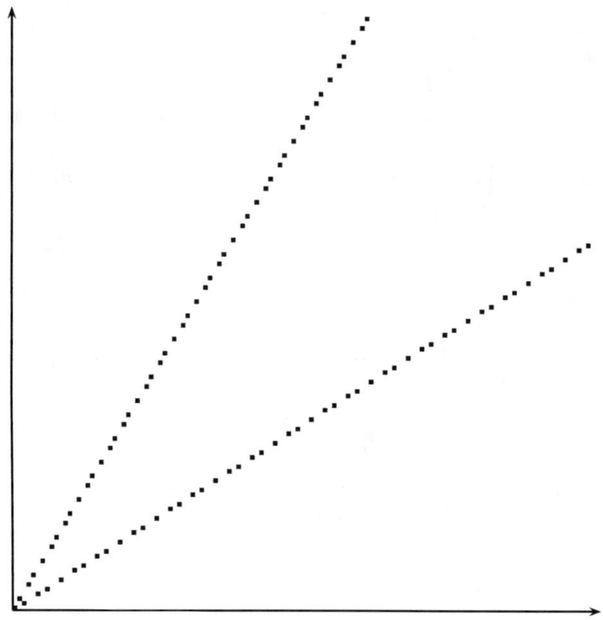

Figure 3.1. Initial $\mathscr{P}$-positions of WYTHOFF.

A position in WYTHOFF can be represented by an ordered pair $(a, b) \in \mathbb{N}^2$, and throughout this section we'll write $\mathscr{G}(a, b)$ for the nim value of (a, b). By symmetry it's clear that $\mathscr{G}(a, b) = \mathscr{G}(b, a)$ for all a and b.

Now $(0, 1)$ and $(1, 1)$ are both $\mathscr{N}$-positions, each having a move to $(0, 0)$; but $(1, 2)$ is a $\mathscr{P}$-position, since its only moves (up to symmetry) are to $(0, 1)$ and $(1, 1)$. Continuing like this, the first few $\mathscr{P}$-positions are

$$(0, 0), \quad (1, 2), \quad (3, 5), \quad (4, 7), \quad (6, 10), \quad (8, 13), \quad \ldots,$$

and they display a striking geometric regularity. Figure 3.1 shows all $\mathscr{P}$-positions with $a, b < 128$, plotted on the (a, b) coordinate plane. They cluster around the lines with slopes ϕ and $1/\phi$, where ϕ is the golden ratio

$$\phi = \frac{1 + \sqrt{5}}{2}.$$

The following lemma is key to proving this fact.

Lemma 3.1. *The* WYTHOFF *$\mathscr{P}$-positions are exactly those pairs of the form (a_n, b_n) or (b_n, a_n), where $(a_n, b_n) \in \mathbb{N}^2$ is defined recursively by*

$$a_n = \operatorname{mex}\{a_i, b_i : i < n\},$$
$$b_n = a_n + n.$$

(In particular, note that $a_0 = \operatorname{mex}(\emptyset) = 0$.)

Proof. We first show that there is no move from (a_n, b_n) to any (a_m, b_m) or (b_m, a_m). Such a move would necessarily have $m < n$. But by definition of a_n, we have $a_n \neq a_m$ and $a_n \neq b_m$. Moreover $a_n > a_m$, so $b_n > b_m$ and therefore $b_n \neq a_m$ and $b_n \neq b_m$. This leaves only the moves along the diagonal, but these are ruled out since $b_n - a_n = n$ while $b_m - a_m = m$.

To complete the proof we must show that if (a, b) is not of the form (a_n, b_n), then it has a move to some (a_n, b_n) or (b_n, a_n). We may assume that $a \leq b$.

First suppose $a = a_n$ for some n. If $b > b_n$, then there is a move directly to (a_n, b_n). Otherwise, $a_n \leq b < b_n$. Let $m = b - a$. Then $m < n$, so $a_m < a_n$ and there is a move to (a_m, b_m).

Finally, if $a \neq a_n$ for any n, then necessarily $a =$ some b_n. Therefore $b \geq a > a_n$, so there is a move to (b_n, a_n). □

Definition 3.2. Let $\mathcal{A}, \mathcal{B} \subset \mathbb{N}^+$. We say $\mathcal{A}$ and $\mathcal{B}$ are **complementary** if $\mathcal{B} = \mathbb{N}^+ \setminus \mathcal{A}$, that is, if $\mathcal{A} \cap \mathcal{B} = \emptyset$ and $\mathcal{A} \cup \mathcal{B} = \mathbb{N}^+$.

Lemma 3.3. *Fix irrational real numbers $\alpha, \beta > 1$ such that*

$$\frac{1}{\alpha} + \frac{1}{\beta} = 1.$$

Then the sets

$$\{\lfloor n\alpha \rfloor : n \in \mathbb{N}^+\} \quad \textit{and} \quad \{\lfloor n\beta \rfloor : n \in \mathbb{N}^+\}$$

are complementary.

Proof. Define $\mathcal{A}, \mathcal{B} \subset \mathbb{R}$ by

$$\mathcal{A} = \left\{\frac{i}{\alpha} : i \in \mathbb{N}^+\right\} \quad \text{and} \quad \mathcal{B} = \left\{\frac{j}{\beta} : j \in \mathbb{N}^+\right\}.$$

Since $1/\alpha + 1/\beta = 1$, we have

$$\frac{\alpha}{\beta} = \alpha - 1,$$

so that α/β is irrational. Therefore $\mathcal{A} \cap \mathcal{B} = \emptyset$, since $i/\alpha - j/\beta$ would imply $i/j = \alpha/\beta$.

Now put $\mathcal{X} = \mathcal{A} \cup \mathcal{B}$. For $x \in \mathcal{X}$, let $f(x)$ be the number of elements of $\mathcal{X}$ that are $\leq x$; that is,

$$f(x) = \big|\{y \in \mathcal{X} : y \leq x\}\big|.$$

Since $\mathcal{X}$ is a disjoint union of $\mathcal{A}$ and $\mathcal{B}$, it follows that $f[\mathcal{A}]$ and $f[\mathcal{B}]$ are complementary. We claim that

$$f[\mathcal{A}] = \{\lfloor i\beta \rfloor : i \in \mathbb{N}^+\} \quad \text{and} \quad f[\mathcal{B}] = \{\lfloor j\alpha \rfloor : j \in \mathbb{N}^+\}.$$

To see this, let $x \in \mathcal{A}$, and write $x = i/\alpha$. Then certainly

$$\big|\{y \in \mathcal{A} : y \le x\}\big| = i.$$

Moreover,

$$\begin{aligned}\big|\{y \in \mathcal{B} : y \le x\}\big| &= \big|\{j \in \mathbb{N}^+ : j/\beta \le x\}\big| \\ &= \big|\{j \in \mathbb{N}^+ : j \le x\beta\}\big| \\ &= \lfloor x\beta \rfloor.\end{aligned}$$

Now $x\beta = i\beta/\alpha = i(\beta - 1)$, so that

$$f(x) = i + \lfloor x\beta \rfloor = i + \lfloor i(\beta - 1)\rfloor = \lfloor i\beta \rfloor.$$

An identical argument shows that $f(j/\beta) = \lfloor j\alpha \rfloor$. Since $\mathcal{X}$ is a disjoint union of $\mathcal{A}$ and $\mathcal{B}$, we are done. □

Theorem 3.4. *The nth $\mathscr{P}$-position of* Wythoff *is given by*

$$(a_n, b_n) = \left(\lfloor n\phi \rfloor, \lfloor n\phi^2 \rfloor\right).$$

Proof. It is easily checked that $\phi^2 = \phi + 1$. Therefore $n\phi^2 = n\phi + n$, so

$$\lfloor n\phi^2 \rfloor = \lfloor n\phi + n \rfloor = \lfloor n\phi \rfloor + n.$$

Moreover, we have

$$\frac{1}{\phi} + \frac{1}{\phi^2} = 1$$

so by Lemma 3.3, $\lfloor n\phi \rfloor$ and $\lfloor n\phi^2 \rfloor$ are complementary. This proves that

$$\lfloor n\phi \rfloor = \operatorname{mex}\left\{\lfloor i\phi \rfloor, \lfloor i\phi^2 \rfloor : i < n\right\} \quad \text{and} \quad \lfloor n\phi^2 \rfloor = \lfloor n\phi \rfloor + n$$

and the theorem now follows from Lemma 3.1. □

The $\mathscr{G}$-Values of Wythoff

No general formula is known for computing arbitrary $\mathscr{G}$-values of Wythoff. In general they appear chaotic, though they exhibit a striking fractal-like pattern. Figure 3.2 illustrates the values $\mathscr{G}(a, b)$ for $a, b < 512$. Darker shades of gray indicate smaller $\mathscr{G}$-values, so that $\mathscr{G}(a, b) = 0$ is solid black and $\mathscr{G}(a, b) = 772$ (the maximum value attained on a 512×512 board) is solid white.

Despite this apparent chaos, the $\mathscr{G}$-values nonetheless have a high degree of geometric regularity. This regularity is visible both in the distribution of positions with bounded $\mathscr{G}$-value and in patterns of $\mathscr{G}$-values along restricted regions of the coordinate plane.

The first pattern is illustrated by Figure 3.3 on page 202, which shows the positions of $\mathscr{G}$-value exactly 30 on a 256×256 board. The lines X and Y of slopes ϕ and $1/\phi$ are superimposed to highlight the clustering

Figure 3.2. Intensity plot of the $\mathscr{G}$-values of WYTHOFF.

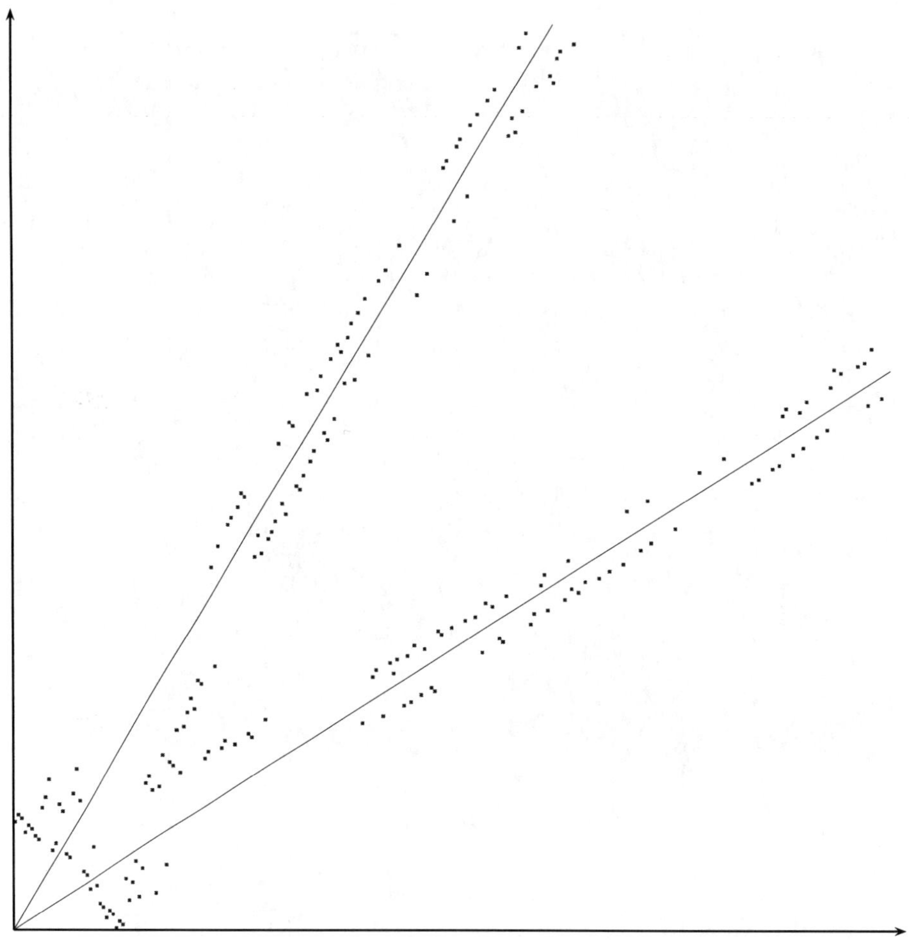

Figure 3.3. Initial WYTHOFF positions of $\mathscr{G}$-value exactly 30. The lines of slopes ϕ and $1/\phi$ are superimposed.

effect. Remarkably, the positions of $\mathscr{G}$-value m always lie within a bounded distance of X and Y, with the bound depending only on m. We won't prove this result here, but some further details are given in the notes on page 206.

A different type of regularity is established by considering the sequence of $\mathscr{G}$-values along a single row; that is, the sequence

$$b \mapsto \mathscr{G}(a,b)$$

for some fixed value of a. This sequence is always arithmetic periodic, though the best known bounds on the periods are quite large. We'll prove this result shortly, after establishing some elementary bounds on $\mathscr{G}(a,b)$.

Proposition 3.5. $\mathscr{G}(a,b) \le a+b$.

Proof. Every move from (a,b) is to some (a',b') with $a'+b' < a+b$, and the result follows by induction. □

Proposition 3.6. $\mathscr{G}(a,b) \ge -2a+b$.

Proof. Fix a, b, and $m < -2a+b$. We will show by a counting argument that $\mathscr{G}(a,b') = m$ for some $b' < b$, which certainly implies $\mathscr{G}(a,b) \ne m$. Let

$$\begin{aligned} \mathcal{X} &= \{b' < b : \mathscr{G}(a,b') < m\}, \\ \mathcal{Y} &= \{b' < b : \mathscr{G}(a-i,b') = m \text{ for some } i\}, \\ \mathcal{Z} &= \{b' < b : \mathscr{G}(a-i,b'-i) = m \text{ for some } i\}. \end{aligned}$$

If $\mathscr{G}(a,b') \ne m$, then either $\mathscr{G}(a,b') < m$, implying $b' \in \mathcal{X}$, or else (a,b') has a move to m. But if (a,b') has a move to m, then either $b' \in \mathcal{Y} \cup \mathcal{Z}$ or some $\mathscr{G}(a,b'-i) = m$. If any $\mathscr{G}(a,b'-i) = m$, then we are done, so to complete the proof it suffices to show that some $b' \notin \mathcal{X} \cup \mathcal{Y} \cup \mathcal{Z}$, i.e., that

$$|\mathcal{X} \cup \mathcal{Y} \cup \mathcal{Z}| < b.$$

Clearly $|\mathcal{X}| \le m$. Moreover, $|\mathcal{Y}| \le a$, since m can appear at most once in each row, and similarly $|\mathcal{Z}| \le a$. Therefore by choice of m

$$\begin{aligned} |\mathcal{X}| + |\mathcal{Y}| + |\mathcal{Z}| &\le m + a + a \\ &< b. \end{aligned}$$ □

Theorem 3.7. *For a fixed value of a, the sequence $b \mapsto \mathscr{G}(a,b)$ is arithmetic periodic.*

A generalization of the mex operator will be convenient in the proof of Theorem 3.7. For any integer $m \in \mathbb{Z}$, the expression

$$\operatorname*{mex}_{\ge m}(\mathcal{A})$$

denotes the least integer $\geq m$ not in the set $\mathcal{A}$. (This operator is defined even for negative values of m.) In particular, $\operatorname{mex}_{\geq 0}$ is just ordinary mex, and

$$\operatorname*{mex}_{\geq m}(\mathcal{A}) = m + \operatorname{mex}\{a - m : a \in \mathcal{A}\}.$$

Proof of Theorem 3.7. By Proposition 3.6, we have $\mathscr{G}(a,b) \geq -2a + b$ for all a and b. Therefore, whenever $b \geq 2a$, we can write

$$\mathscr{G}(a,b) = \operatorname*{mex}_{\geq -2a+b} \begin{pmatrix} \{\mathscr{G}(a-i,b) & : 1 \leq i \leq a\} \\ \cup\, \{\mathscr{G}(a-i,b-i) & : 1 \leq i \leq a\} \\ \cup\, \{\mathscr{G}(a,b-i) & : 1 \leq i \leq b\} \end{pmatrix}.$$

Moreover, for any $b' < -3a + b$, Proposition 3.5 implies $\mathscr{G}(a,b') < -2a + b$, so that $\mathscr{G}(a,b')$ does not contribute to this mex. It follows that for $b \geq 3a$,

$$\mathscr{G}(a,b) = \operatorname*{mex}_{\geq -2a+b} \begin{pmatrix} \{\mathscr{G}(a-i,b) & : 1 \leq i \leq a\ \} \\ \cup\, \{\mathscr{G}(a-i,b-i) & : 1 \leq i \leq a\ \} \\ \cup\, \{\mathscr{G}(a,b-i) & : 1 \leq i \leq 3a\} \end{pmatrix}.$$

Now let $f(a,b) = \mathscr{G}(a,b) - b$ for all a and b. We can rewrite the previous equation in terms of f:

$$f(a,b) = \operatorname*{mex}_{\geq -2a} \begin{pmatrix} \{f(a-i,b) & : 1 \leq i \leq a\ \} \\ \cup\, \{f(a-i,b-i) - i & : 1 \leq i \leq a\ \} \\ \cup\, \{f(a,b-i) - i & : 1 \leq i \leq 3a\} \end{pmatrix}.$$

This shows that $f(a,b)$ can be computed in terms of previous values $f(a',b')$, where (a',b') lies within the bounded rectangle from $(0, b-3a)$ to (a,b). To be precise, let $\mathcal{R}(a,b)$ be the $a \times (3a+1)$ matrix

$$\mathcal{R}(a,b) = \begin{pmatrix} f(0,b-3a) & f(0,b-3a+1) & \cdots & f(0,b-1) & f(0,b) \\ f(1,b-3a) & f(1,b-3a+1) & \cdots & f(1,b-1) & f(1,b) \\ \vdots & \vdots & & \vdots & \vdots \\ f(a,b-3a) & f(a,b-3a+1) & \cdots & f(a,b-1) & \text{---} \end{pmatrix}.$$

(The lower-right entry, corresponding to $f(a,b)$, is left undefined.) Then we have shown that the calculation of $f(a,b)$ depends only on $\mathcal{R}(a,b)$ and is independent of the specific value of b.

Now $\mathcal{R}(a,b)$ has exactly $a \cdot (3a+1)$ entries. Moreover, Propositions 3.5 and 3.6 imply that

$$-2a \leq f(a,b) \leq a.$$

Therefore there are at most $(3a+1)^{a(3a+1)}$ possible configurations for $\mathcal{R}(a,b)$. If a is fixed, then this number remains constant as b increases, so there must exist $b_0 > 0$ and $p > 0$ such that

$$\mathcal{R}(a, b_0) = \mathcal{R}(a, b_0 + p).$$

A straightforward induction now shows that $\mathcal{R}(a, b) = \mathcal{R}(a, b + p)$ for all $b \geq b_0$, so the sequence

$$b \mapsto f(a, b)$$

is periodic with period $\leq p$. But $\mathcal{G}(a, b) = f(a, b) + b$, so the sequence

$$b \mapsto \mathcal{G}(a, b)$$

is arithmetic periodic with period and saltus $\leq p$. □

Exercises

3.1 Prove the converse of Lemma 3.3: if $\lfloor n\alpha \rfloor$ and $\lfloor n\beta \rfloor$ are complementary, then α and β are irrational and $1/\alpha + 1/\beta = 1$.

3.2 Determine all misère $\mathcal{P}$-positions of WYTHOFF.

3.3 Every $\mathcal{G}$-value m appears exactly once on every row of the WYTHOFF plane.

3.4 For all m, there is exactly one a with $\mathcal{G}(a, a) = m$. Moreover, this value of a satisfies $m/2 \leq a \leq 2m$.

3.5 A position in EUCLID is an ordered pair (a, b) of positive integers. A move is to decrease the larger of a and b by any positive multiple of the smaller, provided that it remains ≥ 1. Prove that:
(a) (a, b) is a $\mathcal{P}$-position if and only if either $a/b < \phi$ or $b/a < \phi$ (where ϕ is the golden ratio).
(b) Every EUCLID position can be written as the sequential sum of nim-heaps (see Figure I.4.9 on page 41 for the definition of sequential sum).

3.6 Let $\mathcal{A} \subset \mathbb{N}^2$. Define a ruleset $\Gamma(\mathcal{A})$, played with two heaps of tokens, as follows. A player may either: (i) remove any number of tokens from a single heap; or (ii) remove a tokens from one heap and b from the other, where $(a, b) \in \mathcal{A}$. Prove that if $\mathcal{A}$ contains no elements of the form (a, a), then $\Gamma(\mathcal{A})$ has the same $\mathcal{P}$-positions as NIM.

3.7 Fix $r > 0$. r-WYTHOFF is played with two heaps of tokens. On her turn, a player may either: (i) remove any number of tokens from any one heap; or (ii) remove a tokens from one heap and b from the other, where $|a - b| < r$. Prove that the nth $\mathcal{P}$-position of r-WYTHOFF is given by

$$(a_n, b_n) = (\lfloor n\alpha \rfloor, \lfloor n\beta \rfloor)$$

where

$$\alpha = \frac{1}{2}\left(2 - r + \sqrt{r^2 + 4}\right) \quad \text{and} \quad \beta = \alpha + r.$$

3.8 Fix $k > 0$. 2^k-NIMHOFF is played with n heaps of tokens. On her turn, a player may either: (i) remove any number of tokens from any one heap; or (ii) remove exactly 2^k tokens from each of two distinct heaps. Prove that $\mathcal{G}(a_1, a_2, \ldots, a_n) = a_1 \circledast a_2 \circledast \cdots \circledast a_n$, where

$$a \circledast b = a \oplus b \oplus a^k b^k.$$

3.9 *Adjoining $\mathscr{P}$-positions as moves.* WYTHOFF^2 is played with two heaps of tokens. On her turn, a player may either: (i) move as in ordinary WYTHOFF; or (ii) remove a tokens from one heap and b from the other, where (a, b) is a WYTHOFF $\mathscr{P}$-position. Determine the $\mathscr{P}$-positions of WYTHOFF^2.

Notes

The $\mathscr{P}$-positions of WYTHOFF were first isolated in a 1907 paper by the Dutch mathematician Willem Wythoff [**Wyt07**]. More than a century after Wythoff's initial discovery, it remains unknown how to compute the $\mathscr{G}$-values of arbitrary coordinates efficiently. However, there have been promising advances in understanding the large-scale geometry of its $\mathscr{G}$-values.

Theorem 3.7, showing that every row of WYTHOFF is arithmetic periodic, is due to Norbert Pink [**Pin93, DFP99**]. The proof used here was given by Landman [**Lan02**].

Blass and Fraenkel showed that the positions of $\mathscr{G}$-value exactly 1 lie within a bounded distance of the lines with slope ϕ and $1/\phi$ [**BF90**]. Specifically, they showed that

$$8 - 6\phi < a_n - \phi n < 6 - 3\phi \qquad \text{and} \qquad -3\phi < b_n - \phi^2 n < 8 - 3\phi$$

where (a_n, b_n) is the nth position with $\mathscr{G}$-value 1. They also gave an algorithm for calculating (a_n, b_n) and proved that this algorithm is polynomial-time *provided that* a certain regularity hypothesis (the *convergence conjecture*) is true.

Miraculously, Gabriel Nivasch has recently generalized the Blass–Fraenkel results to positions of $\mathscr{G}$-value exactly m, for arbitrary m [**Niv09**]. Nivasch showed that all such positions lie within a bounded distance of the lines with slope ϕ and $1/\phi$, with the bound depending only on m, and generalized the Blass–Fraenkel algorithm to arbitrary m.

Nivasch has also shown that every m appears exactly once along every WYTHOFF diagonal (where a **diagonal** is a set of the form $\{(a, a + k) : a \in \mathbb{N}\}$). This generalizes Exercise 3.4, which is the case $k = 0$.

An intriguing new approach was recently proposed by Eric Friedman and Adam Landsberg, inspired by renormalization techniques from experimental physics [**FL09**]. Rather than try to compute the $\mathscr{G}$-values for WYTHOFF explicitly, Friedman and Landsberg studied their large-scale geometric structure empirically and showed that it can be explained by certain recurrence relations.

Generalizations of WYTHOFF. Aviezri Fraenkel has been an active force in the development of variant rulesets that generalize WYTHOFF. Exercises 3.7 through 3.9 are all due to Fraenkel [**Fra82, FL91, FO98**]; he has also analyzed the misère version of r-WYTHOFF (Exercise 3.7) and discovered some surprising connections to continued fractions and Fibonacci sequences [**Fra84**].

Fraenkel has also proposed the following ruleset, n-HEAP WYTHOFF. On her turn, a player may either:

- remove any number of tokens from any one heap; or
- remove a_i tokens from heap i, provided that $a_1 \oplus a_2 \oplus \cdots \oplus a_n = 0$.

Fraenkel has conjectured that for fixed values of $c_1 < \cdots < c_{n-2}$, the pairs (a, b) for which $(c_1, \ldots, c_{n-2}, a, b)$ is a $\mathscr{P}$-position have a geometric structure much like ordinary WYTHOFF. See **[Fra04]** for the details of Fraenkel's conjecture; Xinyu Sun has made some partial progress **[SZ04, Sun05]**.

Urban Larsson considered the **dual ruleset** $\Gamma^\star$, in which the $\mathscr{P}$-positions of Γ are the *only* moves available **[Lar]**. Larsson observed that repeated application of the $\star$-operator often gives rise to striking patterns, and he identified conditions under which a ruleset is **reflexive**, i.e., $\Gamma^{\star\star} = \Gamma$.

4. Generalized Sprague–Grundy Theory

In a **loopy impartial game**, the same position G might recur twice (or more) in a single play. If G is loopy, then it's possible for best play on G to continue indefinitely, as it might be to neither player's advantage to break out of a cycle.

A simple example is the game D shown in Figure 4.1. Here the subpositions of D are shown as vertices of a directed graph; the arcs of the graph indicate moves between them. So D has two options: a "pass move" back to D itself and a move to a loopfree position isomorphic to $*$.

Consider a typical play of D. If either player moves from D to $*$, her opponent will win immediately by reverting to 0. Therefore, competent players will repeatedly play the pass move from D, and play will continue indefinitely. Neither player can force a win, so D is neither an $\mathscr{N}$-position nor a $\mathscr{P}$-position; it belongs to a third outcome class $\mathscr{D}$ (the Drawn positions).

With $o(G)$ ranging over these three outcome classes, we have the usual equality relation given by the fundamental equivalence:

$$G = H \quad \text{if} \quad o(G + X) = o(H + X) \text{ for all finite loopy impartial games } X.$$

(We'll discuss formal definitions of $G+H$ and $o(G)$ shortly. In the meantime one can think of $G + H$ in the usual way: composed of side-by-side copies of G and H.) The equivalence classes modulo = give a new type of value, the **loopy nim value** of an impartial game G.

Sometimes all the loopiness of G reverses out, and G is equal to a loopfree game. In such cases we say that G has **finite rank**. If G is not equal to

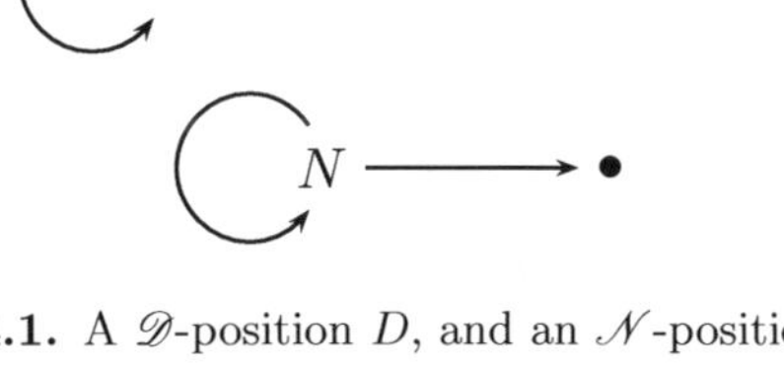

Figure 4.1. A $\mathscr{D}$-position D, and an $\mathscr{N}$-position N of rank ∞.

any loopfree game, then we say G has **rank** ∞. We'll see that

$$G \text{ is a } \mathscr{P}\text{-position} \quad \text{if and only if} \quad G = 0.$$

Moreover, if G is any game with finite rank, then $G = *m$ for some m, just as in the loopfree theory. In this case we can write $\mathscr{G}(G) = m$. However, the positions of rank ∞ have a more subtle structure: in addition to $\mathscr{D}$-positions (such as the game D discussed above), there are $\mathscr{N}$-positions of rank ∞. For example, consider the game N shown in Figure 4.1. N is a first-player win, having a move to 0, but any sum of the form $N + *m$ (for $m \geq 1$) behaves like the game D and so is a $\mathscr{D}$-position. Therefore N is not equal to any $*m$.

It's clear from this example that games with rank ∞ are *not* all equivalent, and the main result of this section is a classification theory for such positions. It turns out that the value of any such game G is characterized by its finite-rank options. For fixed G, let $\mathcal{A} \subset \mathbb{N}$ be given by

$$\mathcal{A} = \{a \in \mathbb{N} : \mathscr{G}(G') = a \text{ for some } G' \in G \text{ of finite rank}\}.$$

(Here, as in Sections 1 through 3, we write $G' \in G$ to mean that G' is an option of G.) Then we write

$$\mathscr{G}(G) = \infty(\mathcal{A}).$$

We'll show that

$$o(G + *m) = \begin{cases} \mathscr{N} & \text{if } m \in \mathcal{A}; \\ \mathscr{D} & \text{otherwise,} \end{cases}$$

and if H also has rank ∞, then

$$o(G + H) = \mathscr{D}$$

regardless of $\mathcal{A}$. Therefore the outcome $o(G + H)$ is completely determined by $\mathscr{G}(G)$ and $\mathscr{G}(H)$.

Formal Definitions

Definition 4.1. A **loopy impartial game** is a pair $G = (V, x)$ where V is a directed graph and x is a vertex of V. We say G is **finite** if V is finite. The **options** of G are games of the form $G' = (V, x')$, where there is an arc directed from x to x'.

Definition 4.2. Let $G = (V, x)$ and $H = (W, y)$ be loopy impartial games. We define $G + H = (V \times W, (x, y))$.

Definition 4.3. We say $G = (V, x)$ is **terminal** if x has no outedges in V.

Throughout the rest of this section, a **game** will be a finite loopy impartial game. It should be clear that games with *acyclic* graphs are exactly equivalent to the loopfree impartial games that we studied in Sections 1 through 3.

Outcomes

Definition 4.4. We define classes $\mathscr{P}_n$ as follows:

- $G \in \mathscr{P}_0$ if G is terminal.
- $G \in \mathscr{P}_{n+1}$ if for each $G' \in G$, there is some $G'' \in G'$ with $G'' \in \mathscr{P}_n$.

Note that $\mathscr{P}_n \subset \mathscr{P}_{n+1}$ for all n, by induction.

Definition 4.5. Let G be a finite loopy impartial game. We say G is

- a **$\mathscr{P}$-position** if $G \in \mathscr{P}_n$ for some n;
- an **$\mathscr{N}$-position** if some $G' \in G$ is a $\mathscr{P}$-position;
- a **$\mathscr{D}$-position** if G is neither a $\mathscr{P}$- nor an $\mathscr{N}$-position.

When G is a $\mathscr{P}$-position, then we write $\operatorname{rank}(G)$ for the least n such that $G \in \mathscr{P}_n$.

The Fundamental Theorem is obviously false in the loopy context, due to the existence of $\mathscr{D}$-positions. The following observation is its closest parallel.

Theorem 4.6. *Let G be a finite loopy impartial game.*

(a) *G is a $\mathscr{P}$-position iff every option of G is an $\mathscr{N}$-position.*

(b) *G is an $\mathscr{N}$-position iff some option of G is a $\mathscr{P}$-position.*

(c) *G is a $\mathscr{D}$-position iff no option of G is a $\mathscr{P}$-position and some option of G is a $\mathscr{D}$-position.*

Proof. (a) $\Rightarrow$ is straight from the definition of $\mathscr{P}_n$. For $\Leftarrow$, suppose that every $G' \in G$ is an $\mathscr{N}$-position. Since G has just finitely many options, there must be some n such that every such G' has some option $G'' \in \mathscr{P}_n$. It follows that $G \in \mathscr{P}_{n+1}$.

(b) is the definition of $\mathscr{N}$-position, and (c) follows immediately. □

In particular, this shows that no G can be both a $\mathscr{P}$-position and an $\mathscr{N}$-position, so that every G belongs to a unique **outcome class**:

Definition 4.7. We write $\mathscr{P}$, $\mathscr{N}$, and $\mathscr{D}$ for the classes of $\mathscr{P}$-, $\mathscr{N}$-, and $\mathscr{D}$-positions (the **outcome classes**), and $o(G)$ for the outcome class of G.

Theorem 4.8. *Let G be a finite loopy impartial game.*

(a) *If G is a $\mathscr{P}$-position, then second player can force a win.*

(b) *If G is an $\mathscr{N}$-position, then first player can force a win.*

(c) *If G is a $\mathscr{D}$-position, then either player can force a draw.*

Proof. (a) A straightforward induction shows that if $G \in \mathscr{P}_n$, then second player can force a win in at most $2n$ moves (i.e., n moves for each player).

(b) is immediate: some G' must be a $\mathscr{P}$-position (by definition), and first player can move to it.

(c) Either player can use the following strategy to force (at least) a draw: move to a $\mathscr{P}$-position if one is available; otherwise, move to a $\mathscr{D}$-position. Since every $G \in \mathscr{D}$ has an option $G' \in \mathscr{D}$, this guarantees she will never run out of moves. □

The following key theorem shows that $\mathscr{P}$-positions and zero positions coincide, just as in the loopfree case (cf. Theorem II.1.12 on page 56).

Theorem 4.9. *If $o(G) = \mathscr{P}$, then $G = 0$.*

Proof. We must show that

$$o(G+X) = o(X) \quad \text{for all } X.$$

The proof is by induction on rank(G). Fix G for the remainder of the proof, and assume that the full theorem is true for all games of rank $< G$.

We'll first prove that

$$\text{If} \quad o(X) = \mathscr{P}, \quad \text{then} \quad o(G+X) = \mathscr{P}, \quad \text{for all } X.$$

In this step, the proof is by induction on rank(X). For each $X' \in X$, there is some $X'' \in X'$ such that $o(X'') = \mathscr{P}$ and $\text{rank}(X'') < \text{rank}(X)$. By induction on rank($X$), we may assume that $o(G + X'') = \mathscr{P}$. Likewise, for each $G' \in G$, there is some $G'' \in G'$ such that $o(G'') = \mathscr{P}$ and $\text{rank}(G'') < \text{rank}(G)$. By induction on rank($G$), we may assume that $o(G'' + X) = \mathscr{P}$. It follows that every option of $G + X$ reverts to a known $\mathscr{P}$-position, so in fact $G + X$ is a $\mathscr{P}$-position.

We'll next show that

$$\text{if} \quad o(G+X) = \mathscr{P}, \quad \text{then} \quad o(X) = \mathscr{P}, \quad \text{for all } X.$$

The induction here is on rank($G + X$). For each $X' \in X$, there is an option $(G + X')' \in G + X'$ with $o((G + X')') = \mathscr{P}$ and $\text{rank}((G + X')') < \text{rank}(G + X)$. Now $(G + X')'$ must have the form $G + X''$ or $G' + X'$. If $o(G + X'') = \mathscr{P}$, then by induction on rank($G + X$) we can assume that $o(X'') = \mathscr{P}$. If instead $o(G' + X') = \mathscr{P}$, then choose $G'' \in G'$ with $o(G'') = \mathscr{P}$; then $o(G'' + X') = \mathscr{N}$. By induction on rank($G$), this implies $o(X') = \mathscr{N}$. This shows that every option of X is an $\mathscr{N}$-position.

We've established that

$$o(G+X) = \mathscr{P} \quad \text{if and only if} \quad o(X) = \mathscr{P}, \quad \text{for all } X,$$

and it follows immediately that

$$o(G+X) = \mathscr{N} \quad \text{if and only if} \quad o(X) = \mathscr{N}, \quad \text{for all } X,$$

which is sufficient to prove the theorem. □

Loopy Nim Values

The construction of the sets $\mathscr{P}_n$ can be generalized to give an inductive characterization of loopy nim value. Recall that $\mathscr{P}_n$ is the set of games that second player can win in $\leq n$ steps; likewise, we'll write $\mathscr{G}_n(G) = m$ to mean that the relation $G = *m$ can be verified in $\leq n$ stages. In all other cases we'll write $\mathscr{G}_n(G) = \infty$, so that $\mathscr{G}_n(G)$ is drawn from $\mathbb{N} \cup \{\infty\}$. It will be convenient to write $m < \infty$ for each $m \in \mathbb{N}$.

Definition 4.10. For each game G and integer $n \geq 0$, we define $\mathscr{G}_n(G) \in \mathbb{N} \cup \{\infty\}$ by

$$\mathscr{G}_0(G) = \begin{cases} 0 & \text{if } G \text{ is terminal;} \\ \infty & \text{otherwise.} \end{cases}$$

Then for fixed $n \geq 0$, let $m = \operatorname{mex}\{\mathscr{G}_n(G') : G' \in G\}$ and define

$$\mathscr{G}_{n+1}(G) = \begin{cases} m & \text{if for each } G' \in G \text{ with } \mathscr{G}_n(G') > m, \text{ there is} \\ & \quad \text{some } G'' \in G' \text{ with } \mathscr{G}_n(G'') = m; \\ \infty & \text{otherwise.} \end{cases}$$

So $\mathscr{G}_{n+1}(G) = m$ just if m is the mex of the "known value" options of G, and every *other* option reverts to a lower-rank position of value exactly m.

Proposition 4.11. *For all G and n:*

(a) *$\mathscr{G}_n(G) = 0$ if and only if $G \in \mathscr{P}_n$.*

(b) *If $\mathscr{G}_n(G) = m < \infty$, then also $\mathscr{G}_{n+1}(G) = m$.*

Proof. (a) This is clear for $n = 0$, so suppose $n > 0$. If $\mathscr{G}_n(G) = 0$, then necessarily every $\mathscr{G}_{n-1}(G') > 0$. So for every G', there is a G'' with $\mathscr{G}_{n-1}(G'') = 0$, which by induction implies $G'' \in \mathscr{P}_{n-1}$. The converse implication is similar.

(b) For each $a < m$, there is some G' with $\mathscr{G}_{n-1}(G') = a$. By induction on n, we can assume that $\mathscr{G}_n(G') = a$ as well.

Now consider any G' with $\mathscr{G}_n(G') \geq m$. By induction on n, we can assume that $\mathscr{G}_{n-1}(G') =$ either $\mathscr{G}_n(G')$ or ∞. Therefore $\mathscr{G}_{n-1}(G') > m$, so there is some G'' with $\mathscr{G}_{n-1}(G'') = m$. By induction on n, $\mathscr{G}_n(G'') = m$ as well. □

It follows from Proposition 4.11(b) that for a given game G, there are just two possibilities. Either:

- $\mathscr{G}_n(G) = \infty$ for all n; or else
- there is a unique value of m, and a minimal stage $n_0 \geq 0$, such that

$$\mathscr{G}_n(G) = \begin{cases} \infty & \text{if } n < n_0; \\ m & \text{if } n \geq n_0. \end{cases}$$

Definition 4.12. We write $\mathscr{G}(G) = m$ if $\mathscr{G}_n(G) = m$ for some (all) sufficiently large n. Otherwise, we write

$$\mathscr{G}(G) = \infty(\mathcal{A}),$$

where

$$\mathcal{A} = \{a \in \mathbb{N} : \mathscr{G}(G') = a \text{ for some option } G'\}.$$

The value $\mathscr{G}(G)$ is the **loopy nim value** of G.

The **rank** of G, denoted rank(G), is the least n such that $\mathscr{G}_n(G)$ is finite, or ∞ if no such n exists. (By Proposition 4.11(a), this is consistent with our earlier definition of rank(G) on page 209.)

The following theorem shows that $\mathscr{G}(G)$ determines $o(G)$.

Theorem 4.13. *Let G be a finite loopy impartial game.*

(a) *$\mathscr{G}(G) = 0$ if and only if G is a $\mathscr{P}$-position.*

(b) *If $\mathscr{G}(G)$ is an integer $m > 0$, then G is an $\mathscr{N}$-position.*

(c) *If $\mathscr{G}(G) = \infty(\mathcal{A})$ with $0 \in \mathcal{A}$, then G is an $\mathscr{N}$-position.*

(d) *If $\mathscr{G}(G) = \infty(\mathcal{A})$ with $0 \notin \mathcal{A}$, then G is a $\mathscr{D}$-position.*

Proof. (a) is a restatement of Proposition 4.11(a).

(b), (c) In both cases we have $\mathscr{G}(G') = 0$ for some G', so the conclusion follows from (a) and Theorem 4.6.

(d) G can't be a $\mathscr{P}$-position, since we proved "if and only if" in part (a). Nor can G be an $\mathscr{N}$-position, since no option of G is a $\mathscr{P}$-position. Therefore G must be a $\mathscr{D}$-position. □

Algebra of Loopy Nim Values

To complete the analysis of loopy nim values, we show that $\mathscr{G}(G + H)$ is determined by $\mathscr{G}(G)$ and $\mathscr{G}(H)$.

Theorem 4.14. *If $\mathscr{G}(G) = m$, then $G = *m$.*

Proof. If G is terminal, then necessarily $m = 0$ and the conclusion is immediate. Otherwise, let $n = \text{rank}(G)$. We'll show that $G + *m \in \mathscr{P}_{m+n}$, which suffices to prove the theorem by Theorem 4.9. Now

$$m = \text{mex}\{\mathscr{G}_{n-1}(G') : G' \in G\}.$$

Therefore each move to $G + *a$ reverts to some $G' + *a$ with $\mathscr{G}_{n-1}(G') = a$. By induction, it follows that $G' + *a \in \mathscr{P}_{a+n-1}$.

Now consider a move to $G' + *m$, and let $n' = \text{rank}(G')$. If $n' < n$ and $\mathscr{G}(G') = a < m$, then $G' + *m$ reverts to $G' + *a$, which by induction must be in $\mathscr{P}_{a+n'}$. If either $n' \geq n$ or $\mathscr{G}(G') > m$, then G' must have an option G'' with $\mathscr{G}(G'') = m$ and $\mathscr{G}_{n-1}(G'') = m$. So $G' + *m$ reverts to $G'' + *m$, which by induction must be in $\mathscr{P}_{m+n-1}$. □

Lemma 4.15. *If* $\text{rank}(G) = \infty$, *then* $\text{rank}(G + H) = \infty$ *for all* H.

Proof. Suppose (for contradiction) that the lemma is false, and choose a counterexample (G, H) with $\text{rank}(G + H)$ minimal. Let $n = \text{rank}(G + H)$ and $m = \mathscr{G}(G + H)$.

Observe that we must also have $\text{rank}(H) = \infty$, since $H = *a$ would imply $G = *m + *a$.

Now since G has rank ∞, it must have some option G' also of rank ∞, so by minimality of n we have $\text{rank}(G' + H) \geq n$. Therefore $G' + H$ must have an option X with $\mathscr{G}(X) = m$ and $\text{rank}(X) < n$. But X necessarily has the form $G'' + H$ or $G' + H'$; i.e., it has at least one component of rank ∞ (either H or G'). This contradicts minimality of n. □

Theorem 4.16. *For all G and H, we have*

$$\mathscr{G}(G + H) = \mathscr{G}(G) \oplus \mathscr{G}(H),$$

where $a \oplus b$ denotes ordinary nim-sum if $a, b \in \mathbb{N}$, and

$$\infty(\mathcal{A}) \oplus b = \infty(\mathcal{A} \oplus b), \qquad \infty(\mathcal{A}) \oplus \infty(\mathcal{B}) = \infty(\emptyset).$$

(Here $\mathcal{A} \oplus b = \{a \oplus b : a \in \mathcal{A}\}$.)

Proof. First suppose $\mathscr{G}(G) = a$ and $\mathscr{G}(H) = b$, both finite. Then by Theorem 4.14, $G = *a$ and $H = *b$, so in fact $G + H = *(a \oplus b)$.

Next suppose $\text{rank}(G) = \infty$ but $\text{rank}(H)$ is finite. Then Lemma 4.15 implies that $\text{rank}(G + H) = \infty$. Moreover, if $G' + H$ with $\mathscr{G}(G')$ finite, then $\mathscr{G}(G' + H) = \mathscr{G}(G') \oplus b$; and by Lemma 4.15, all other options of $G + H$ have rank ∞. This shows that $\mathscr{G}(G + H) = \infty(\mathcal{A} \oplus b)$.

Finally, if $\text{rank}(G) = \text{rank}(H) = \infty$, then by Lemma 4.15, $G + H$ and all its options have rank ∞. So $\mathscr{G}(G + H) = \infty(\emptyset)$. □

Exercises

4.1 If G is an option of itself, then $\operatorname{rank}(G) = \infty$.

4.2 Prove that rank is related to remoteness: if G has finite rank, then

$$\operatorname{rank}(G) = \tfrac{1}{2}\mathscr{R}(G + *m), \quad \text{where } m = \mathscr{G}(G).$$

(Remoteness is defined in Exercise I.4.5 on page 48.)

4.3 Describe the abstract structure of the monoid of loopy nim values.

4.4 *Subtraction games with pass moves.* For $\mathcal{S} \subset \mathbb{N}^+$, let $\Gamma(\mathcal{S})$ be the following modification of SUBTRACTION($\mathcal{S}$). A legal move is *either* to (i) remove n tokens from a heap, for some $n \in \mathcal{S}$; or else (ii) pass (remove 0 tokens), *provided that* the size of the heap is odd.

(a) Prove that if $\mathcal{S} \subset \mathbb{N}^+$ is finite, then the loopy nim values of $\Gamma(\mathcal{S})$ are ultimately periodic (cf. Theorem 2.4 on page 187).

(b) State and prove an analogue of Theorem 2.5.

(c) Show that if $\mathcal{S} = \{a, b\}$ is a two-element set, then the loopy nim values of $\Gamma(\mathcal{S})$ are purely periodic with period $a + b$ (cf. Exercise 2.1).

(d) Show that if $\mathcal{S}$ contains both an even integer and an odd integer, then $\mathscr{G}(n) = \infty$ for all sufficiently large n.

Notes

The theory of loopy nim values was first discovered by Cedric Smith in the 1960s [**Smi66**]. A decade later, it was independently discovered by Aviezri Fraenkel, who has been a sustained advocate of the theory [**FP73, FT75, FY86**].

Annihilation games are a particularly interesting class of loopy impartial games. They are played with tokens arranged on the vertices of a directed graph. A move consists of sliding any token along an arc of the graph; whenever two tokens occupy the same vertex, both tokens are immediately "annihilated" and removed from the game.

On an acyclic graph, the annihilation rule has no effect: if two tokens occupy the same vertex, then they must have the same value $*m$; and since $*m + *m = 0$, the token pair is irrelevant whether or not it is annihilated. In the cyclic case, however, the interaction between tokens imposed by the annihilation rule is genuine, and the disjunctive theory cannot easily be applied.

Annihilation games have been studied extensively by Fraenkel and Yesha, who showed that efficient strategies exist in normal play [**FY76, FY82**]. Ferguson has studied annihilation games in misère play [**Fer84**].

5. Nim Arithmetic

In this section we return to the study of ordinary nim values. Consider the recursive expression for nim-sum given by Proposition 1.4 on page 181. In order for $(\mathbb{N}, \oplus)$ to be a group, it must certainly satisfy

$$a \oplus b \neq a' \oplus b$$

whenever $a \neq a'$. So Proposition 1.4 asserts that $a \oplus b$ is the *smallest* integer consistent with all preceding nim-sums. In this sense $\oplus$ is the "simplest" operation that makes $\mathbb{N}$ into a group.

Now suppose we wish to define a new operation $\otimes$ that makes $(\mathbb{N}, \oplus, \otimes)$ into a field. Such an operation would certainly satisfy

$$(a \oplus a') \otimes (b \oplus b') \neq 0$$

whenever $a \neq a'$ and $b \neq b'$. Rewriting using the distributive law gives

$$(a \otimes b) \oplus (a' \otimes b) \oplus (a \otimes b') \oplus (a' \otimes b') \neq 0$$

so that

$$a \otimes b \neq (a' \otimes b) \oplus (a \otimes b') \oplus (a' \otimes b').$$

This yields the following "natural" genetic definition of multiplication.

Definition 5.1. For all $a, b \geq 0$, the **nim-product** $a \otimes b$ is given by

$$a \otimes b = \operatorname*{mex}_{\substack{a'<a \\ b'<b}} \big((a' \otimes b) \oplus (a \otimes b') \oplus (a' \otimes b')\big).$$

As it turns out, this construction gives rise to a field $(\mathbb{N}, \oplus, \otimes)$ with some remarkable properties. Recall that $\oplus$ is defined by its action on 2-powers (Theorem 1.6 on page 181). We can state an analogue for $\otimes$ in terms of **Fermat 2-powers**.

Definition 5.2. A **Fermat 2-power** is an integer of the form 2^{2^k} $(k \geq 0)$.

Theorem 5.3. $\otimes$ *is the unique operation such that:*

(i) $(\mathbb{N}, \oplus, \otimes)$ *is a field;*

(ii) $a \otimes a = \frac{3}{2}a$ *for every Fermat 2-power a; and*

(iii) $a \otimes b = ab$ *for distinct Fermat 2-powers a and b.*

We'll prove that $\otimes$ has these properties; uniqueness is left as an exercise. Note that the rules (ii) and (iii) can be used to calculate the value of any nim-product. In particular, if $n = 2^k$ is a 2-power, then by writing the exponent k as a sum of distinct 2-powers, we see that n can be expressed uniquely as the product of distinct Fermat 2-powers:

$$k = \sum_{a \in \mathcal{A}} 2^a, \quad \text{so that} \quad n = \prod_{a \in \mathcal{A}} 2^{2^a}.$$

(In all such expressions, we allow $\mathcal{A} = \emptyset$, in which case the sum resolves to 0 and the product resolves to 1.) It follows that every integer $n \geq 0$ can be expressed uniquely as the sum of products of Fermat 2-powers:

$$n = \sum_{\mathcal{A} \in \mathcal{F}} \prod_{a \in \mathcal{A}} 2^{2^a},$$

where $\mathcal{F}$ denotes a finite collection of finite subsets of $\mathbb{N}$. When an integer is written in this **normal form**, the nim-sum and nim-product operations are interchangeable with ordinary sum and product:

$$\sum_{\mathcal{A}\in\mathcal{F}} \prod_{a\in\mathcal{A}} 2^{2^a} = \bigoplus_{\mathcal{A}\in\mathcal{F}} \bigotimes_{a\in\mathcal{A}} 2^{2^a},$$

by Theorem 5.3(iii) and the usual nim-addition rule. To compute the nim-product of any two integers, we can write them in normal form and use the distributive law to express the result as a nim-sum of nim-products of Fermat 2-powers. Theorem 5.3(ii) can then be used to resolve any "duplicates" appearing in the nim-product expressions. Eventually we will obtain a normal form for the result, which can then be resolved by the ordinary rules of arithmetic.

Example. Let's compute $8 \otimes 8$. We have $8 = 4 \otimes 2$ in normal form, so

$$8 \otimes 8 = (4 \otimes 2) \otimes (4 \otimes 2) = (4 \otimes 4) \otimes (2 \otimes 2) = 6 \otimes 3.$$

Writing 6 and 3 in normal form and repeating:

$$\begin{aligned} 6 \otimes 3 &= (4 \oplus 2) \otimes (2 \oplus 1) \\ &= (4 \otimes 2) \oplus (4 \otimes 1) \oplus (2 \otimes 2) \oplus (2 \otimes 1) \\ &= 8 \oplus 4 \oplus 3 \oplus 2 \\ &= 13. \end{aligned}$$

We can proceed this way to calculate $a \otimes b$ for any a and b; the values for $0 \le a, b < 16$ are tabulated in Figure 5.1 on the facing page.

We proceed with the proof of Theorem 5.3.

Lemma 5.4. *$(\mathbb{N}, \oplus, \otimes)$ is a commutative ring with identity* 1.

Proof. Commutativity of $\otimes$ is transparent from the definition, and associativity and distributivity are routine inductions. Moreover, for all a we have

$$a \otimes 0 = \operatorname{mex}(\emptyset) = 0,$$

and therefore, inductively,

$$a \otimes 1 = \operatorname*{mex}_{a'<a} \big((a' \otimes 1) \oplus (a \otimes 0) \oplus (a' \otimes 0)\big) = \operatorname*{mex}_{a'<a} (a' \otimes 1) = a. \qquad \square$$

Lemma 5.5. *If $a \otimes b = 0$, then $a = 0$ or $b = 0$.*

Proof. If $a \neq 0$ and $b \neq 0$, then we can take $a' = b' = 0$ so that

$$(a' \otimes b) \oplus (a \otimes b') \oplus (a' \otimes b') = 0$$

is an excludent of $a \otimes b$. $\square$

$\otimes$	0	1	2	3	4	5	6	7	8	9	10	11	12	13	14	15
0	0	0	0	0	0	0	0	0	0	0	0	0	0	0	0	0
1	0	1	2	3	4	5	6	7	8	9	10	11	12	13	14	15
2	0	2	3	1	8	10	11	9	12	14	15	13	4	6	7	5
3	0	3	1	2	12	15	13	14	4	7	5	6	8	11	9	10
4	0	4	8	12	6	2	14	10	11	15	3	7	13	9	5	1
5	0	5	10	15	2	7	8	13	3	6	9	12	1	4	11	14
6	0	6	11	13	14	8	5	3	7	1	12	10	9	15	2	4
7	0	7	9	14	10	13	3	4	15	8	6	1	5	2	12	11
8	0	8	12	4	11	3	7	15	13	5	1	9	6	14	10	2
9	0	9	14	7	15	6	1	8	5	12	11	2	10	3	4	13
10	0	10	15	5	3	9	12	6	1	11	14	4	2	8	13	7
11	0	11	13	6	7	12	10	1	9	2	4	15	14	5	3	8
12	0	12	4	8	13	1	9	5	6	10	2	14	11	7	15	3
13	0	13	6	11	9	4	15	2	14	3	8	5	7	10	1	12
14	0	14	7	9	5	11	2	12	10	4	13	3	15	1	8	6
15	0	15	5	10	1	14	4	11	2	13	7	8	3	12	6	9

Figure 5.1. The nim-multiplication table for $\mathcal{P}_{16}$.

Now Theorem 5.3 is a direct consequence of the following lemma. Recall (from Section I.3) that $\text{lb}(n)$ denotes the base-2 logarithm of n, rounded down. For convenience, we also write

$$\mathcal{P}_n = \{a \in \mathbb{N} : a < n\},$$

the set of (strict) predecessors of n.

Lemma 5.6. *Suppose that n is a Fermat 2-power.*

(a) *If $a, b < n$, then $a \otimes b < n$.*

(b) *$(\mathcal{P}_n, \oplus, \otimes)$ is a field.*

(c) *If $a < n$, then $a \otimes n = an$.*

(d) *If $a < n$, then $\text{lb}(a \otimes a) = \text{lb}(a)$.*

(e) *$\{a \otimes a : a < n\} = \mathcal{P}_n$, and $\{(a \otimes a) \oplus a : a < n\} = \mathcal{P}_{n/2}$.*

(f) *$n \otimes n = \frac{3}{2}n$.*

Proof. By induction on n.

(a) Since $\otimes$ distributes over $\oplus$ and since $\mathcal{P}_n$ is closed under nim-sum, we may assume that a and b are 2-powers. Therefore we can write

$$a = ra' \quad \text{and} \quad b = sb',$$

where r and s are Fermat 2-powers, $a' < r$, and $b' < s$. By induction,

$$ra' \otimes sb' = (r \otimes a') \otimes (s \otimes b') = r \otimes s \otimes a' \otimes b'.$$

Without loss of generality, $s \le r$. If $s < r$, then a', b', and s are all $< r$, so by induction $s \otimes a' \otimes b' < r$, and

$$r \otimes s \otimes a' \otimes b' = r \cdot (s \otimes a' \otimes b') < r^2 \le n.$$

If $s = r$, then $r \otimes s = \frac{3}{2}r = r \oplus \frac{r}{2}$, by induction. Therefore

$$r \otimes s \otimes a' \otimes b' = \left(r \oplus \tfrac{r}{2}\right) \otimes a' \otimes b' = \left(r \cdot (a' \otimes b')\right) \oplus \left(\tfrac{r}{2} \otimes a' \otimes b'\right) < r^2 \le n.$$

(b) $\mathcal{P}_n$ is closed under $\oplus$ since n is a 2-power, and by (a) it is also closed under $\otimes$. Therefore $(\mathcal{P}_n, \oplus, \otimes)$ is a finite ring. By Lemma 5.5 it has no zero divisors, so it must be a field.

(c) We have

$$a \otimes n = \operatorname*{mex}_{\substack{a' < a \\ n' < n}} \left((a' \otimes n) \oplus (a \otimes n') \oplus (a' \otimes n')\right).$$

By induction on a, we may assume that $a' \otimes n = a'n$, so that

$$a \otimes n = \operatorname*{mex}_{\substack{a' < a \\ n' < n}} \Big((a'n) \oplus \big((a \oplus a') \otimes n'\big)\Big).$$

Now for a fixed value of $a' < a$, we have $a \oplus a' \in \mathcal{P}_n$. Since $\mathcal{P}_n$ is a field, multiplication by $a \oplus a'$ induces a permutation of $\mathcal{P}_n$, so

$$\mathcal{P}_n = \big\{(a \oplus a') \otimes n' : n' < n\big\}.$$

Therefore every integer i with $a'n \le i < (a'+1)n$ is an excludent of $a \otimes n$. This is true for all $a' < a$, so it follows that every integer $i < an$ is an excludent. Since an is not, we are done.

(d) Just as in (a), we can assume that a is a 2-power and

$$a \otimes a = \left(r \cdot (a' \otimes a')\right) \oplus \left(\tfrac{r}{2} \otimes a' \otimes a'\right)$$

with r a Fermat 2-power, $a' < r$, and $a = ra'$. But $\operatorname{lb}(a' \otimes a') = \operatorname{lb}(a')$ by induction, so

$$\operatorname{lb}\left(r \cdot (a' \otimes a')\right) = \operatorname{lb}(ra') = \operatorname{lb}(a).$$

Since also

$$\operatorname{lb}\left(\tfrac{r}{2} \otimes a' \otimes a'\right) < \operatorname{lb}(r) \le \operatorname{lb}(a),$$

this suffices.

(e) Certainly $a \otimes a < n$ whenever $a < n$, by (b). But if $a \otimes a = b \otimes b$, then $(a \oplus b) \otimes (a \oplus b) = 0$, so by Lemma 5.5, $a = b$. The conclusion

$$\{a \otimes a : a < n\} = \mathcal{P}_n$$

now follows by a cardinality argument.

Likewise, by (d) we have $\operatorname{lb}(a \otimes a) = \operatorname{lb}(a)$ whenever $a < n$, so

$$(a \otimes a) \oplus a < n/2.$$

Moreover, if

$$(a \otimes a) \oplus a = (b \otimes b) \oplus b,$$

then we have

$$(a \oplus b) \otimes (a \oplus b) = a \oplus b.$$

So $a \oplus b = 0$ or 1, i.e., $b = a$ or $a \oplus 1$. Again by a cardinality argument, we conclude that

$$\{(a \otimes a) \oplus a : a < n\} = \mathcal{P}_{n/2}.$$

(f) We have

$$\begin{aligned} n \otimes n &= \operatorname*{mex}_{a,b<n} \Big((a \otimes n) \oplus (n \otimes b) \oplus (a \otimes b)\Big) \\ &= \operatorname*{mex}_{a,b<n} \Big(\big(n \otimes (a \oplus b)\big) \oplus (a \otimes b)\Big). \end{aligned}$$

When $b = a$ this expression reduces to $a \otimes a$, so by (e) every element of $\mathcal{P}_n$ is an excludent. When $b = a \oplus 1$ it reduces to

$$(n \otimes 1) \oplus \big(a \otimes (a \oplus 1)\big) = n \oplus \big((a \otimes a) \oplus a\big),$$

so by (e) every element of $n + \mathcal{P}_{n/2}$ is an excludent, and hence every element of $\mathcal{P}_{3n/2}$, since

$$\mathcal{P}_{3n/2} = \mathcal{P}_n \cup (n + \mathcal{P}_{n/2}).$$

But $\frac{3}{2}n$ is *not* an excludent, since $n \otimes (a \oplus b) \geq 2n$ whenever $a \oplus b > 1$. This completes the proof. □

It follows immediately from Lemma 5.6(b) that $(\mathbb{N}, \oplus, \otimes)$ is a field. In fact we can say slightly more:

Theorem 5.7. $(\mathbb{N}, \oplus, \otimes)$ *is a quadratically closed field.*

Proof. Lemma 5.6(b) shows it to be a field, and Lemma 5.6(e) shows that square roots exist. Finally, Lemma 5.6(e) also shows that

$$\{(x \otimes x) \oplus x : x \in \mathbb{N}\} = \mathbb{N},$$

so that every polynomial of the form

$$(x \otimes x) \oplus x \oplus a$$

has a root in $\mathbb{N}$. This suffices, since arbitrary quadratics of the form

$$(x \otimes x) \oplus (b \otimes x) \oplus a$$

can now be solved by the transformation $x = b \oplus y$. □

We denote by $a \oslash b$ the unique integer c such that $c \otimes b = a$. Likewise, we denote by $a^{\boxed{n}}$ the nth power of a in $(\mathbb{N}, \oplus, \otimes)$:

$$a^{\boxed{n}} = \overbrace{a \otimes a \otimes \cdots \otimes a}^{n \text{ times}}.$$

Theorem 5.7 guarantees the existence of $a \oslash b$ and $a^{\boxed{1/2}}$, and it's also possible to describe them by explicit recursive constructions; see Exercise 5.7.

Exercises

5.1 The **Conway product** of G and H, denoted by $G \times H$, is defined by

$$G \times H = \{G' \times H + G \times H' + G' \times H' : G' \in G \text{ and } H' \in H\}.$$

(a) Prove the *nim-multiplication rule*: $*a \times *b = *c$, where $c = a \otimes b$.
(b) Prove that $\mathscr{G}(G \times H) = \mathscr{G}(G) \otimes \mathscr{G}(H)$ for all G and H.

5.2 Let $\Gamma = \text{Turning}(\mathcal{E})$ and $\Lambda = \text{Turning}(\mathcal{F})$ (cf. Exercise 1.7 on page 182). The **cross product** $\Gamma \times \Lambda$ is played with a two-dimensional array of $m \times n$ coins, labeled $C_{i,j}$ for $1 \le i \le m$ and $1 \le j \le n$. A legal move is to select any $\mathcal{A} \in \mathcal{E}$ and $\mathcal{B} \in \mathcal{F}$ and turn over coins $C_{i,j}$ for all $i \in \mathcal{A}$ and $j \in \mathcal{B}$, provided that the *northeast-most* coin is turned heads-to-tails. Prove that:
(a) Every position in $\Gamma \times \Lambda$ can be expressed as a disjunctive sum of single-heads positions.
(b) If coin C_i in Γ has nim value a and coin C_j in Λ has nim value b, then coin $C_{i,j}$ in $\Gamma \times \Lambda$ has nim value $a \otimes b$.
(c) In particular, the nim values of Nim $\times$ Nim are exactly given by the nim-multiplication table (Figure 5.1 on page 217).

5.3 If $k \ge 2$, then $x^{\boxed{3}} = 2^{2^k}$ has exactly three solutions in $\mathbb{N}$.

5.4 If a is a 2-power, then the sequence

$$\langle a^{\boxed{n}} : n \text{ is a Fermat 2-power}\rangle$$

is monotonically nonincreasing with limit a.

5.5 If n is a Fermat 2-power, then

$$n^{\boxed{n}} = n \oplus 1 \quad \text{and} \quad (n^2)^{\boxed{n}} = n^2 \oplus n \oplus e, \text{ for some } e \in \{0, 1\}.$$

(It is an open problem to determine e as a function of n.)

5.6 If n is a Fermat 2-power, then the $(n+1)^{\text{st}}$ nim-roots of unity are given by

$$\left(a^{\boxed{2}} \oplus n \oplus \tfrac{n}{2}\right) \oslash \left(a^{\boxed{2}} \oplus a \oplus \tfrac{n}{2}\right) \quad \text{for} \quad 0 \le a < n.$$

5.7 *Explicit inverses and square roots.* For all $a \in \mathbb{N}$:
(a) $1 \oslash a = \text{mex}(\mathcal{B})$, where $\mathcal{B}$ is the closure of $\{0\}$ under the mapping

$$b \mapsto (1 \oplus (a \oplus a') \otimes b) \oslash a',$$

with a' ranging over all integers $< a$.
(b) $a^{\boxed{1/2}} = \text{mex}(\mathcal{C})$, where $\mathcal{C}$ is the closure of $\left\{(a')^{\boxed{1/2}} : a' < a\right\}$ under the mapping

$$(b_1, b_2) \mapsto (b_1 \otimes b_2 \oplus a) \oslash (b_1 \oplus b_2),$$

in which (b_1, b_2) ranges over pairs of *unequal* elements of $\mathcal{C}$.

5.8 For every 2-power a, write $q(a) = (a \otimes a)/a$. Lemma 5.6(f) shows that if a is a Fermat 2-power, then $q(a) = 3/2$. Prove the following for every 2-power a:

(a) If a is *not* a Fermat 2-power, then $q(a) > 3/2$.

(b) If a has the form 2^{2^k+1}, then $q(a) = 13/8$.

(c) If $a = 2^{2^k+j}$ and $a' = 2^{2^{k'}+j}$, with $j < 2^k, 2^{k'}$, then $q(a) = q(a')$. This defines a sequence indexed by j (and independent of choices of k) whose first two values are $3/2$ and $13/8$. Determine the next few values in this sequence.

5.9 If s is a Fermat 2-power, then every base-s lexicode $\mathcal{L}$ is necessarily closed under componentwise nim-product. Therefore $\mathcal{L}$ is a vector space over the finite field of order s. (See Exercises 1.8 and 1.9 for the definition of lexicode.)

Notes

Nim-product was discovered by Conway in the context of the transfinite Field $\mathbf{ON}_2$ (cf. Section VIII.4). It has been studied extensively by Lenstra [**Len78**].

Much of Lemma 5.6 emerges as a consequence of the more general theory presented in Section VIII.4. The proof given in this section is due to Knuth and is instructive for its elementary nature.

Exercises 5.2 through 5.6: Lenstra.

Exercise 5.7(b): Clive Bach.

Exercise 5.9: Conway and Sloane.

Chapter V

Misère Play

Throughout Chapters II through IV, the theory assumed the **normal-play convention**: whoever makes the last move wins. Under the **misère-play convention**, the outcomes are inverted and the player who makes the last move loses.

For a given combinatorial game G, the play convention (normal or misère) determines the outcome $o(G)$ and the relations $G = H$, but it has no bearing on the combinatorial structure of G itself or on the meaning of the sum $G + H$. Therefore we immediately have misère variants for all of the rulesets discussed in previous chapters. As we'll gradually come to appreciate, misère play is rife with complications, and the success of the normal-play theory is often difficult to replicate.

At the root of these complications is the treatment of 0. Denote by $o^+(G)$ the normal outcome of G (what we've called simply $o(G)$ in previous chapters) and by $o^-(G)$ its misère outcome. The normal outcome can be defined by a succinct recursion (here assuming G is short impartial):

$$o^+(G) = \mathscr{N} \text{ if and only if } o^+(G') = \mathscr{P} \text{ for some } G' \in G.$$

The corresponding recursion for misère play unavoidably requires an additional clause:

$$o^-(G) = \mathscr{N} \text{ if and only if either } \begin{cases} G \cong 0; \text{ or else} \\ o^-(G') = \mathscr{P} \text{ for some } G' \in G. \end{cases}$$

In this sense, normal play is the "natural" convention, requiring the fewest additional constraints. And while this additional constraint appears innocuous, it complicates the disjunctive theory to an extraordinary degree.

For a sense of the difficulties, consider DAWSON'S KAYLES. The normal-play solution (described in Section IV.2) was published by Guy and Smith in the mid-1950s. The largest single-heap nim value is 8 (Figure IV.2.1 on page 186), so the largest nim value that can appear in any sum of heaps is 15. So there are essentially 16 distinct types of positions—one $\mathscr{P}$-position type (nim value 0) and 15 $\mathscr{N}$-position types (values 1 through 15). Their combinatorial structure is described by the group $\mathbb{Z}_2^4$, with generators corresponding to nim value 1, 2, 4, and 8.

By contrast, the misère solution—originally proposed by Dawson more than 75 years ago—remains an open problem. We *do* know how to find the outcome of a position G, provided all heaps in G have size at most 33. Even in this restricted context, there are 638 distinct position types, including 109 $\mathscr{P}$-position types, and their combinatorial structure is quite complicated. Extending this solution to include heaps of size 34 is beyond the reach of current theoretical and computational resources.

Despite this apparent runaway combinatorial complexity in the general case, there are certain special cases for which the misère theory is tractable. The misère-play solution to NIM is essentially the same as the normal-play solution, with just minor modifications. This was observed by Charles Bouton (in the same paper that introduced the normal-play solution), and subsequent authors hoped to find a reduction of arbitrary misère games to misère NIM, analogous to the Sprague–Grundy theory for normal play. Eventually Conway showed that this is only possible for a special class of games known as **tame games**. Systematic methods for studying the remaining **wild** games appear to require novel algebraic techniques.

Sections 1 and 2 review the theory of misère NIM and its generalization to tame games. Section 3 gives Conway's construction of misère canonical forms. The misère quotient construction, a recent technique for analyzing wild games, is introduced in Section 4 and explored in more detail in Section 5. Finally, Section 6 describes a generalization of the theory to partizan games.

1. Misère NIM

Let's recall the normal-play solution to NIM (Theorem I.1.1 on page 3).

Theorem (Bouton's Theorem). *The* NIM *position with heaps $a_1, a_2, \ldots, a_k$ is a normal $\mathscr{P}$-position if and only if*

$$a_1 \oplus a_2 \oplus \cdots \oplus a_k = 0.$$

The misère solution is essentially the same, with a slight modification:

Theorem 1.1 (Bouton's Theorem, Misère Version). *The* NIM *position G with heaps $a_1, a_2, \ldots, a_k$ is a misère $\mathscr{P}$-position if and only if*

$$a_1 \oplus a_2 \oplus \cdots \oplus a_k = 0,$$

unless *every $a_i = 0$ or 1. In that case, G is a $\mathscr{P}$-position if and only if*

$$a_1 \oplus a_2 \oplus \cdots \oplus a_k = 1.$$

In particular, if G has at least one heap of size ≥ 2, then its normal and misère outcomes coincide. So the strategy for misère NIM can be summarized like so:

Play just as you would in normal NIM unless your move would leave a position consisting entirely of heaps of size 0 or 1. In that case, play to leave an odd number of heaps of size 1.

Proof of Theorem 1.1. It suffices to prove the following, for every NIM position G:

(a) If G contains no heaps of size ≥ 2, say $G \cong n \cdot *$, then $o^-(G) = \mathscr{P}$ if and only if n is odd.

(b) Otherwise, $o^-(G) = o^+(G)$.

The proof is by induction on G. For (a), the only move from G is to remove a copy of $*$, inverting the parity of n. The conclusion follows trivially.

Consider now case (b). Suppose first that G has just a single heap of size ≥ 2, so that $G \cong *m + (n \cdot *)$ for some $m \geq 2$ and $n \geq 0$. Now $*m$ has moves to both 0 and $*$, so G has moves to both

$$n \cdot * \quad \text{and} \quad (n+1) \cdot *,$$

one of which must be a $\mathscr{P}$-position. Therefore $o^-(G) = \mathscr{N} = o^+(G)$.

If G has more than one heap of size ≥ 2, then every $G' \in G$ has at least one such heap. By induction, $o^-(G') = o^+(G')$ for all such G', and it follows immediately that $o^-(G) = o^+(G)$. □

Simplification

The next goal is to identify situations in which misère games can be reduced to NIM. As always, we'll define misère **equality** of impartial games via the fundamental equivalence:

$$G = H \quad \text{iff} \quad o^-(G + X) = o^-(H + X) \quad \text{for all } X.$$

Then the **misère game value** of G is its equivalence class modulo $=$.

Note that we formerly used the same symbol "$G = H$" to denote normal-play equality. Throughout this chapter, the misère interpretation will be the default when nothing is said to the contrary, and we'll say "$G = H$ in normal play" or "$G = H$ in misère play" in the few cases when it's important to distinguish between the two.

Likewise, throughout this chapter $o(G)$ will mean $o^-(G)$, and the latter will be used only when necessary to disambiguate.

Lemma 1.2 (Replacement Lemma). *Suppose that $G \cong \{G'_1, \ldots, G'_k\}$ and $H \cong \{H'_1, \ldots, H'_k\}$ and that $G'_i = H'_i$ for all i. Then $G = H$.*

Proof. We must show that

$$o(G + X) = \mathscr{N} \quad \text{if and only if} \quad o(H + X) = \mathscr{N}, \quad \text{for all } X.$$

The proof is by induction on X. First suppose that $o(G + X) = \mathscr{N}$; there are three possibilities.

- If $o(G+X') = \mathscr{P}$ for some $X' \in X$, then by induction $o(H+X') = \mathscr{P}$ as well.
- If $o(G'_i + X) = \mathscr{P}$ for some G'_i, then by assumption $o(H'_i + X) = \mathscr{P}$ as well.
- Finally, if $G + X \cong 0$, then we must also have $H + X \cong 0$, since G and H have equally many options.

In all three cases, $o(H+X) = \mathscr{N}$. The argument for the converse implication is identical. □

Definition 1.3. Let $G \cong \{G'_1, \ldots, G'_k\}$. Let H be a game whose options include those of G:

$$H \cong \{G'_1, \ldots, G'_k, H'_1, \ldots, H'_l\}.$$

We say that H **simplifies to** G provided that:

(i) each new option H'_j includes a move back to G itself (a **reverting** move); and

(ii) if $G \cong 0$, then $o(H) = \mathscr{N}$.

The strange-looking clause (ii) in Definition 1.3 is known as the **proviso**. The idea is that reverting moves ordinarily do not change the value of G: any time a player can win on $G + X$, she can win on $H + X$ by following the same strategy, "reverting" to G should her opponent move to any H'_j. However, this strategy fails in the specific case $G \cong X \cong 0$ and $o(H) = \mathscr{P}$: here first player wins on $G + X$, by virtue of having no option at all, but on $H + X$ she is nonetheless compelled to move. The additional clause is necessary to guarantee the existence of a suitable alternative move on H.

Note that exactly the same argument works in normal play, *without* the proviso. This will emerge as a consistent theme throughout this chapter:

arguments in the misère theory closely resemble those that are familiar from normal play, but with additional complications in the base case owing to the exceptional treatment of 0. Unfortunately, as Conway wrote in ONAG,

> ...the complications so produced persist indefinitely, and make the misère play theory much more complicated than the normal one. [**Con01**]

Theorem 1.4. *If H simplifies to G, then $G = H$.*

Proof. We must show that

$$o(G+X) = \mathscr{N} \quad \text{if and only if} \quad o(H+X) = \mathscr{N}, \quad \text{for all } X.$$

The proof is by induction on X. First suppose that $o(G+X) = \mathscr{N}$. Then there are three possibilities:

- If $o(G+X') = \mathscr{P}$ for some $X' \in X$, then by induction $o(H+X') = \mathscr{P}$ as well, so that $o(H+X) = \mathscr{N}$.
- If $o(G'+X) = \mathscr{P}$ for some $G' \in G$, then since H simplifies to G, we also have $G' \in H$. Therefore $o(H+X) = \mathscr{N}$.
- Finally, it might be the case that no option of $G+X$ is a $\mathscr{P}$-position, but $o(G+X) = \mathscr{N}$ because $G+X \cong 0$. In this case, we must have $G \cong 0$ and $X \cong 0$, so that $H+X \cong H$. Here the proviso ensures that $o(H+X) = o(H) = \mathscr{N}$.

Next suppose that $o(H+X) = \mathscr{N}$. Here there are two cases:

- If $o(H+X') = \mathscr{P}$ for some $X' \in X$, then by induction $o(G+X') = \mathscr{P}$ as well, so that $o(G+X) = \mathscr{N}$.
- If $o(H'+X) = \mathscr{P}$ for some $H' \in H$, then either $H' \in G$ or else H' has a reverting move back to G (so that $G \in H'$). So either $H'+X$ is an option of $G+X$ or else $H'+X$ has $G+X$ as an option; and in both cases $G+X$ must be an $\mathscr{N}$-position. □

The Misère Mex Rule

The normal-play theory of Chapter IV is driven by two key results: the *mex rule* (Theorem IV.1.2 on page 180) and the *nim-addition rule* (Theorem II.4.7 on page 85). We can state misère analogues of these rules, but in each case the proviso sharply limits the scope of the result.

First recall the normal-play mex rule:

Theorem (Normal Mex Rule). *Let $a_1, a_2, \ldots, a_k \in \mathbb{N}$, and suppose that*

$$G \cong \{*a_1, *a_2, \ldots, *a_k\}.$$

Then $G = *m$ in normal play, where $m = \operatorname{mex}\{a_1, a_2, \ldots, a_k\}$.

Now the misère version:

Theorem 1.5 (Misère Mex Rule). *Let $a_1, a_2, \ldots, a_k \in \mathbb{N}$, and suppose that*

$$G \cong \{*a_1, *a_2, \ldots, *a_k\}.$$

Then $G = *m$ in misère play, where $m = \operatorname{mex}\{a_1, a_2, \ldots, a_k\}$, provided that *at least one a_i is 0 or 1.*

Proof. We show that G simplifies to $*m$. Each $*a_i$ with $a_i > m$ has a reverting move back to $*m$, so it suffices to verify the proviso. If G has 0 as an option, then $m > 0$, so $*m \not\cong 0$ and there is nothing to prove. Otherwise, by assumption G must have $*$ as an option. Since $*$ is a $\mathscr{P}$-position, this means G is an $\mathscr{N}$-position, so the proviso is satisfied. □

Example. Let $G = \{*2\}$. Then G is a $\mathscr{P}$-position, so that $G \neq 0$ (in contrast to normal play).

G can't simplify to $*$, since it doesn't have 0 as an option; and it can't simplify to 0, since it fails the proviso. In fact G isn't equal to *any* Nim position, though we won't be ready to prove this for some time (see Exercise 3.1 on page 247).

We'll give G the special name $*2_\#$, pronounced *star-two-sharp*.

Theorem (Normal Nim-Addition Rule). *For any $a, b \in \mathbb{N}$, we have*

$$*a + *b = *c \quad \textit{in normal play,} \quad \textit{where} \quad c = a \oplus b.$$

Theorem 1.6 (Misère Nim-Addition Rule). *For any $a, b \in \mathbb{N}$, we have*

$$*a + *b = *c \quad \textit{in misère play,} \quad \textit{where} \quad c = a \oplus b,$$

provided that *at least one of a or b is 0 or 1.*

Proof. This is immediate when $a = 0$ or $b = 0$, so assume (without loss of generality) that $a > 0$ and $b = 1$. Then the options of $*a + *b$ are

$$*a + 0 \quad \text{and} \quad *a' + * \quad (a' < a).$$

By induction on a, we have $*a' + * = *(a' \oplus 1)$ for each such a'. Therefore

$$*a + *b = \bigl\{*a, *(0 \oplus 1), *(1 \oplus 1), \ldots, *\bigl((a-1) \oplus 1\bigr)\bigr\}.$$

Therefore, by the misère mex rule it suffices to prove that

$$a \oplus 1 = \operatorname{mex}(\mathcal{A}), \quad \text{where} \quad \mathcal{A} = \bigl\{a, 0 \oplus 1, 1 \oplus 1, \ldots, (a-1) \oplus 1\bigr\}.$$

It is clear that $a \oplus 1 \notin \mathcal{A}$. To complete the proof, we must show that if $a' < a \oplus 1$, then $a' \in \mathcal{A}$. If $a' = a$, then we are done; otherwise $a' \oplus 1 < a$. But this implies $a' = (a' \oplus 1) \oplus 1 \in \mathcal{A}$. □

Example. $*m + *$ simplifies to the nim-heap $*(m \oplus 1)$, for all m. The "next simplest" NIM position is

$$G = *2 + *2.$$

In contrast to normal play we have $G \neq 0$, since G is a $\mathscr{P}$-position and 0 an $\mathscr{N}$-position. Moreover, $G \neq *$, since Theorem 1.1 implies that

$$o(G + G) = \mathscr{P}, \qquad \text{whereas} \qquad o(* + G) = \mathscr{N}.$$

Since larger nim-heaps are $\mathscr{N}$-positions, we conclude that G is not equal to any $*m$.

What happens if we try to simplify G according to the misère mex rule? By the nim-addition rule (and the Replacement Lemma) we have

$$G = \{*2 + 0,\ *2 + *\} = \{*2,\ *3\}.$$

But that's as far as it goes! In fact it turns out that no further simplifications are possible, though we won't be prepared to prove this until Section 3.

Henceforth we'll refer to G by its abbreviated name $*2_2$.

Misère Nim Value

Recall (from page 181) the recursive characterization of normal nim value:

$$\mathscr{G}^+(G) = \operatorname*{mex}_{G' \in G} \mathscr{G}^+(G').$$

This was denoted by $\mathscr{G}(G)$ in Chapter IV; here we use the notation $\mathscr{G}^+(G)$ to disambiguate from misère play. We have the following analogous definition.

Definition 1.7. Let G be impartial. The **misère nim value** of G, denoted by $\mathscr{G}^-(G)$, is defined by

$$\mathscr{G}^-(G) = \begin{cases} 1 & \text{if } G \cong 0; \\ \operatorname*{mex}_{G' \in G} \mathscr{G}^-(G') & \text{otherwise.} \end{cases}$$

Just as in normal play, nim value tells the outcome.

Proposition 1.8. *$\mathscr{G}^-(G)$ is the unique value of m such that $G + *m$ is a $\mathscr{P}$-position. (In particular, G is a $\mathscr{P}$-position if and only if $\mathscr{G}^-(G) = 0$.)*

Proof. The proof is by induction on G. The base case $G \cong 0$ is clear, since $\mathscr{G}^-(0) = 1$ and $0 + *$ is a $\mathscr{P}$-position.

So assume $G \not\cong 0$. Then for each $a < m$, there is some $G' \in G$ with $\mathscr{G}^-(G') = a$. By induction, $G' + *a$ is a $\mathscr{P}$-position, so that $G + *a$ is an $\mathscr{N}$-position. Moreover, each $G' + *m$ is also an $\mathscr{N}$-position, since $\mathscr{G}^-(G') \neq m$. This shows that every option of $G + *m$ is an $\mathscr{N}$-position, and since $G + *m$ has at least one option, it must be a $\mathscr{P}$-position.

Uniqueness follows immediately, since $G + *a$ is an option of $G + *m$ when $a < m$, and it has $G + *m$ as an option when $a > m$. □

Recall also that the normal nim value of a sum is determined by the nim values of its components:

Theorem. $\mathscr{G}^+(G + H) = \mathscr{G}^+(G) \oplus \mathscr{G}^+(H)$, *for all* G *and* H.

Once again, there is a restricted misère analogue.

Proposition 1.9. $\mathscr{G}^-(G + *) = \mathscr{G}^-(G) \oplus 1$, *for all* G.

Proof. Similar to the proof of Theorem 1.6. We have $\mathscr{G}^-(0) = 1$ and $\mathscr{G}^-(*) = 0$, so if $G \cong 0$, then we are done. Otherwise, by induction

$$\begin{aligned}\mathscr{G}^-(G + *) &= \operatorname*{mex}_{G' \in G} \left(\mathscr{G}^-(G' + *),\ \mathscr{G}^-(G)\right) \\ &= \operatorname*{mex}_{G' \in G} \left(\mathscr{G}^-(G') \oplus 1,\ \mathscr{G}^-(G)\right).\end{aligned}$$

Let $a = \mathscr{G}^-(G)$. We must show that

$$a \oplus 1 = \operatorname{mex}(\mathcal{A}), \quad \text{where} \quad \mathcal{A} = \{\mathscr{G}^-(G') \oplus 1 : G' \in G\} \cup \{a\}.$$

It is clear that $a \oplus 1 \notin \mathcal{A}$ (since no $G' \in G$ has $\mathscr{G}^-(G') = a$). To complete the proof, we must show that if $a' < a \oplus 1$, then $a' \in \mathcal{A}$. If $a' = a$, then we are done; otherwise $a' \oplus 1 < a$, so there is some $G' \in G$ with $\mathscr{G}^-(G') = a' \oplus 1$. Then $\mathscr{G}^-(G') \oplus 1 = a$. □

Unfortunately, this is about all we can say in general about misère nim values of sums, as the following example demonstrates.

Example. Let $G = *$ and $H = *2 + *2$. Then G and H are both $\mathscr{P}$-positions (cf. Theorem 1.1), so that

$$\mathscr{G}^-(G) = 0 = \mathscr{G}^-(H).$$

However,

$$\mathscr{G}^-(G + *2) = 1 \oplus 2 = 3$$

by Proposition 1.9, whereas

$$\begin{aligned}\mathscr{G}^-(H + *2) &= \operatorname{mex}\{\mathscr{G}^-(H),\ \mathscr{G}^-(H + *)\} \\ &= \operatorname{mex}\{\quad 0 \quad,\quad 0 \oplus 1 \quad\} \\ &= 2.\end{aligned}$$

Therefore $\mathscr{G}^-(G) = \mathscr{G}^-(H)$, but $\mathscr{G}^-(G + *2) \neq \mathscr{G}^-(H + *2)$. This shows that the value of $\mathscr{G}^-(G + *2)$ cannot in general be computed from $\mathscr{G}^-(G)$.

The following theorem characterizes the misère nim values of NIM positions.

Theorem 1.10. *Let G be a* NIM *position.*

(a) *If G contains no heaps of size ≥ 2, say $G \cong n \cdot *$, then*

$$\mathscr{G}^-(G) = \begin{cases} 1 & \text{if } n \text{ is even;} \\ 0 & \text{if } n \text{ is odd.} \end{cases}$$

(b) *Otherwise, $\mathscr{G}^-(G) = \mathscr{G}^+(G)$.*

Proof. (a) is a trivial induction on n. For (b), put $m = \mathscr{G}^+(G)$. Then Bouton's Theorem gives $o^+(G + *m) = \mathscr{P}$, so by Theorem 1.1 we have $o^-(G + *m) = \mathscr{P}$. By Proposition 1.8, $\mathscr{G}^-(G) = m$. □

Notes

The solution to misère NIM was given in 1901 by Bouton [**Bou01**]. The simplification rule (Definition 1.3) was given by Grundy and Smith [**GrS56**]; most of the remaining material in this section is due to Conway [**Con01**].

2. Genus Theory

A more detailed look at the values of NIM suggests two flavors of positions:

- **fickle** positions with no heaps of size ≥ 2; and
- **firm** positions with at least one heap of size ≥ 2,

corresponding to the two distinct cases in the proof of Theorem 1.1.

We've seen that $\mathscr{G}^-(G + H)$ and $\mathscr{G}^-(G) \oplus \mathscr{G}^-(H)$ are not necessarily equal, even if G and H are NIM positions. But Theorem 1.10 shows that they *are* equal for NIM positions of the same flavor (firm or fickle). Moreover, if we know both $\mathscr{G}^+(G)$ and $\mathscr{G}^-(G)$, we can tell whether G is firm or fickle, since

- $\mathscr{G}^-(G) = \mathscr{G}^+(G)$ if G is firm, but
- $\mathscr{G}^-(G) \neq \mathscr{G}^+(G)$ if G is fickle.

Therefore, given *both* the normal and misère nim values for NIM positions G and H, we can determine $\mathscr{G}^+(G + H)$ and also whether $G + H$ is firm or fickle. From this information we can deduce the value $\mathscr{G}^-(G + H)$, and hence the outcome $o^-(G + H)$. This suggests a new type of value, which we call the **genus** of G, obtained by conjoining its normal and misère nim values.

Definition 2.1. The **genus** of G (plural **genera**), denoted by $\mathscr{G}^\pm(G)$, is the ordered pair

$$\mathscr{G}^\pm(G) = (a, b),$$

where $a = \mathscr{G}^+(G)$ and $b = \mathscr{G}^-(G)$. We usually write $\mathscr{G}^\pm(G)$ using the abbreviated notation

$$\mathscr{G}^\pm(G) = a^b.$$

Examples. From Theorem 1.10 we have

$$\mathscr{G}^\pm(*a) = \begin{cases} 0^1 & \text{if } a = 0; \\ 1^0 & \text{if } a = 1; \\ a^a & \text{if } a \geq 2. \end{cases}$$

Two further genera are easily obtained:

$$\begin{aligned} \mathscr{G}^\pm(*2 + *2) \quad\quad &= 0^0; \\ \mathscr{G}^\pm(*2 + *2 + *) &= 1^1. \end{aligned}$$

These are the only genera that can arise in NIM, since Theorem 1.10 gives

$$\mathscr{G}^\pm(G) = 0^1 \text{ or } 1^0$$

when $G \cong n \cdot *$, depending on the parity of n, and

$$\mathscr{G}^\pm(G) = a^a$$

when G is firm, with $a = \mathscr{G}^+(G)$.

So the only genus values that arise in NIM are 0^1, 1^0, and a^a for $a \geq 0$. Remarkably, the analysis of misère NIM generalizes verbatim to *any* game whose genus values are drawn from this ensemble. Such games are said to be **tame**. Formally:

Definition 2.2. G is **tame** if for every subposition H of G,

$$\mathscr{G}^\pm(H) = 0^1 \text{ or } 1^0 \text{ or } a^a \text{ for some } a.$$

We say G is **wild** if G is not tame.

Definition 2.3. Let G be tame. We say G is **fickle** if $\mathscr{G}^\pm(G) = 0^1$ or 1^0; **firm** otherwise.

The crucial fact about tameness is the following.

Theorem 2.4. *If G and H are tame, then so is $G+H$. Furthermore, $G+H$ is fickle if and only if G and H are both fickle.*

We'll prove Theorem 2.4 in a moment. A key consequence is that if G and H are tame, then $\mathscr{G}^\pm(G + H)$ can be computed from $\mathscr{G}^\pm(G)$ and $\mathscr{G}^\pm(H)$: one can easily calculate

$$\mathscr{G}^+(G + H) = \mathscr{G}^+(G) \oplus \mathscr{G}^+(H)$$

using the nim-addition rule, and (since $G + H$ is tame) this uniquely determines $\mathscr{G}^\pm(G+H)$ except when $\mathscr{G}^+(G+H)$ is 0 or 1. But in that case, there

$$\begin{aligned} 0^1 \oplus 0^1 &= 0^1, & a^a \oplus 0^1 &= a^a, \\ 0^1 \oplus 1^0 &= 1^0, & a^a \oplus 1^0 &= (a \oplus 1)^{a\oplus 1}, \\ 1^0 \oplus 1^0 &= 0^1, & a^a \oplus b^b &= (a \oplus b)^{a\oplus b}. \end{aligned}$$

Figure 2.1. Addition table for genus values of tame games.

are just two possibilities for $\mathscr{G}^{\pm}(G+H)$, and exactly one of them is fickle. So in all cases the genus of $G+H$ is determined by those of G and H. The addition rules are summarized in Figure 2.1.

Example. Let $G = *2_{\#}$ (defined on page 228). Then the only option of G is $*2$, which is tame of genus 2^2. So G is tame of genus 0^0.

Since $*2_2$ (defined on page 229) is also tame of genus 0^0, this shows that $*2_2$ and $*2_{\#}$ cannot be distinguished by any tame game:

$$\begin{aligned} \mathscr{G}^{\pm}(*2_2 + X) &= \mathscr{G}^{\pm}(*2_{\#} + X), && \text{and therefore} \\ o(*2_2 + X) &= \quad o(*2_{\#} + X), && \text{whenever } X \text{ is tame.} \end{aligned}$$

However, $*2_2$ and $*2_{\#}$ *can* be distinguished by a suitable wild game; we'll see this in Section 3.

We'll need one further tool in the proof of Theorem 2.4.

Definition 2.5. A **unit** is one of the four genera 0^1, 1^0, 0^0, or 1^1.

Definition 2.6. Suppose that all options of G are tame. The **discriminant** of G, denoted by $\delta(G)$, is the set of units represented by options of G:

$$\delta(G) = \{0^1, 1^0, 0^0, 1^1\} \cap \mathscr{G}^{\pm}[G].$$

(Recall that $\mathscr{G}^{\pm}[G] = \{\mathscr{G}^{\pm}(G') : G' \in G\}$.)

Theorem 2.7. *Suppose that $G \not\cong 0$ and all options of G are tame.*

(a) *If $\delta(G)$ contains* exactly *one fickle unit and* exactly *one firm unit, then G is wild.*

(b) *If $\delta(G)$ contains* exactly *one fickle unit and* no *firm units, then G is tame and fickle.*

(c) *Otherwise, G is tame and firm.*

Proof. (a) If $\delta = \{0^1, 0^0\}$ or $\{1^0, 1^1\}$, then $\mathscr{G}^{+}(G) \le 1$ but $\mathscr{G}^{-}(G) \ge 2$. Conversely, if $\delta = \{0^1, 1^1\}$ or $\{1^0, 0^0\}$, then $\mathscr{G}^{+}(G) \ge 2$ but $\mathscr{G}^{-}(G) \le 1$. In either case G is wild.

(b) If $\delta = \{0^1\}$, then by the mex rule $\mathscr{G}^{\pm}(G) = 1^0$. Likewise, if $\delta = \{1^0\}$, then $\mathscr{G}^{\pm}(G) = 0^1$.

(c) If δ contains both fickle units or both firm units, then $\mathscr{G}^{\pm}(G) = a^a$ for some $a \geq 2$. Otherwise, the options of G must all be firm. Therefore $\mathscr{G}^{\pm}(G')$ has the form $(a')^{a'}$ for all $G' \in G$, so $\mathscr{G}^{\pm}(G) = a^a$, with $a = \text{mex}(a')$. $\square$

Proof of Theorem 2.4. Suppose G and H are tame. If either $G \cong 0$ or $H \cong 0$, then the conclusions are trivial (noting that 0 is fickle), so assume $G \not\cong 0$ and $H \not\cong 0$. By induction, we may assume that the theorem holds for every $G' + H$ and every $G + H'$. In particular, since every subposition of G and H is tame, we have that $G' + H$ and $G + H'$ are tame.

Case 1: Both G and H are firm. Then by induction, every $G' + H$ and $G + H'$ is tame and firm, so by Theorem 2.7(c) so is $G + H$.

Case 2: Both G and H are fickle. If $\mathscr{G}^{\pm}(G) = \mathscr{G}^{\pm}(H) = 0^1$, then by Theorem 2.7, $\delta(G) = \delta(H) = \{1^0\}$, so $\delta(G + H) = \{1^0\}$. Similarly, if $\mathscr{G}^{\pm}(G) = \mathscr{G}^{\pm}(H) = 1^0$, then $\delta(G + H) = \{1^0\}$; and if $\mathscr{G}^{\pm}(G) = 1^0$ while $\mathscr{G}^{\pm}(H) = 0^1$ (or vice versa), then $\delta(G + H) = \{0^1\}$. In all cases $G + H$ is tame and fickle.

Case 3: Otherwise, without loss of generality we may assume that G is fickle and H firm. So $\mathscr{G}^{\pm}(H) = a^a$ for some a. If $a \geq 2$, then (by the mex rule) $\delta(H)$ contains either both fickle units or both firm units, so $\delta(G + H)$ must also. By Theorem 2.7(c), this implies $G + H$ is tame and firm.

If $a = 0$ or 1, then $\delta(H)$ cannot contain any fickle units, so every option of H is firm. Therefore every $G + H'$ is firm, so every option of $G + H$ is firm. By Theorem 2.7(c) so is $G + H$. $\square$

Strategies for Tame Sums

Recall the strategy for misère NIM given at the beginning of this chapter (on page 225):

> Play just as you would in normal NIM unless your move would leave a position consisting entirely of heaps of size 0 or 1. In that case, play to leave an odd number of heaps of size 1.

This strategy generalizes in a straightforward way to arbitrary sums of tame games. Given a sum

$$G = G_1 + G_2 + \cdots + G_k,$$

with all G_i tame, we can state the misère strategy as follows:

> Play just as you would in normal G unless your move would leave a position consisting entirely of fickle components. In that case, play to leave an odd number of components with genus 1^0.

Periodicity

The Octal Periodicity Theorem (Theorem IV.2.7 on page 190) generalizes in the following way to tame misère games.

Theorem 2.8 (Tame Periodicity Theorem). *Let* $\Gamma = \mathbf{d}_0.\mathbf{d}_1\mathbf{d}_2\ldots\mathbf{d}_k$ *be an octal game of finite length* k. *Suppose that there exist* $n_0 \geq 1$ *and* $p \geq 1$ *such that*

$$H_n \textit{ is tame} \quad \textit{for all } n < 2n_0 + 2p + k$$

and

$$\mathscr{G}^{\pm}(n+p) = \mathscr{G}^{\pm}(n) \quad \textit{for all } n \textit{ with } n_0 \leq n < 2n_0 + p + k.$$

Then it follows that

$$H_n \textit{ is tame} \quad \textit{for all } n$$

and

$$\mathscr{G}^{\pm}(n+p) = \mathscr{G}^{\pm}(n) \quad \textit{for all } n \geq n_0.$$

Proof. By induction on n. Let $n \geq 2n_0 + p + k$. By an argument identical to the proof of Theorem IV.2.7, every option of H_{n+p} has the same genus as some option of H_n, and vice versa. Since H_n is tame, so is H_{n+p}, and we have $\mathscr{G}^{\pm}(n+p) = \mathscr{G}^{\pm}(n)$. □

Some Notation

We've seen several examples of games that are not equal to nim-heaps, including

$$*2_{\#} = \{*2\} \quad \text{and} \quad *2_2 = *2 + *2.$$

An extension of this notation is useful for describing further examples. Given existing games $*G$, $*H$, $*J$, we write

$$*GHJ \quad \text{as shorthand for} \quad \{*G, *H, *J\}.$$

Subscripts are used to denote sums; thus

$$*G_H \quad \text{abbreviates} \quad *G + *H.$$

Finally, a # symbol denotes a singleton option, so that

$$*G_{\#} \quad \text{abbreviates} \quad \{*G\}.$$

For example:

$$\begin{aligned} *2_{\#}320 &= \{\{*2\}, *3, *2, 0\}, \\ *2_{\#\#2} &= \{\{*2\}\} + *2, \\ *2_{2\#\#} &= \{\{*2 + *2\}\}. \end{aligned}$$

$\delta(G)$	Restive Genera	$\delta(G)$	Restless Genera
$\{0^1, 0^0\}$	1^2, 1^3, 1^4, ...	$\{0^1, 1^1\}$	2^0, 3^0, 4^0, ...
$\{1^0, 1^1\}$	0^2, 0^3, 0^4, ...	$\{1^0, 0^0\}$	2^1, 3^1, 4^1, ...

Figure 2.2. Genera of restive and restless games.

Restive and Restless Games

Suppose that G is wild but every option of G is tame. By Theorem 2.7(a), we know that there are exactly four possible discriminants for G. Each choice of discriminant gives rise to a different family of genera, as summarized in Figure 2.2.

It turns out that those genera with $\mathscr{G}^+(G) = 0$ or 1 (in the left-hand column of Figure 2.2) behave quite differently from those with $\mathscr{G}^-(G) = 0$ or 1 (in the right-hand column). These are called **restive** and **restless** games, respectively.

Examples. Here are some simple examples exhibiting all four discriminant types:

- $*2_{\#}0$ has $\delta = \{0^0, 0^1\}$ and genus 1^2;
- $*2_3 1$ has $\delta = \{1^1, 1^0\}$ and genus 0^2;
- $*2_{\#}1$ has $\delta = \{0^0, 1^0\}$ and genus 2^1;
- $*2_3 0$ has $\delta = \{1^1, 0^1\}$ and genus 2^0.

The next theorem shows that restive games are well behaved when played together with Nim positions.

Theorem 2.9. *Suppose R is restive of genus a^b. Let G be a* Nim *position. Then $R + G$ is a misère $\mathscr{P}$-position if and only if*

$$\mathscr{G}^+(G) = a,$$

unless every heap in G has size 0, 1, b, or $b \oplus 1$. In that case, $R + G$ is a misère $\mathscr{P}$-position if and only if

$$\mathscr{G}^+(G) = b.$$

Note that Theorem 2.9 is also true when R is tame of genus 0^1 or 1^0. Bouton's Theorem (Theorem 1.1 on page 225) can therefore be viewed as a special case, with $R \cong 0$ of genus 0^1. In this sense restive games behave like variants of fickle tame games.

Theorem 2.9 can't be generalized to allow arbitrary tame G (Exercise 2.6), so this appears to be about as far as one can push this type of argument. Restless games appear totally inscrutable, even when all other components of the sum are nim-heaps.

Proof of Theorem 2.9. Note that since R is restive of genus a^b, it must have tame options of genera

$$(a \oplus 1)^0,\ (a \oplus 1)^1,\ 2^2,\ 3^3,\ \ldots,\ (b-1)^{b-1}.$$

The proof proceeds by induction on G.

Case 1: Every heap in G has size 0, 1, b, or $b \oplus 1$. Then $\mathscr{G}^+(G) = 0$, 1, b, or $b \oplus 1$. We must show that $o(R+G) = \mathscr{P}$ when $\mathscr{G}^+(G) = b$, and $\mathscr{N}$ otherwise. There are several subcases.

- If $\mathscr{G}^+(G) = b \oplus 1$ and $b \oplus 1 > b$, then G has an option G' with $\mathscr{G}^+(G') = b$. By induction, $o(R+G') = \mathscr{P}$, so $o(R+G) = \mathscr{N}$.
- If $\mathscr{G}^+(G) = b \oplus 1$ and $b \oplus 1 < b$, then R must have an option R' with $\mathscr{G}^\pm(R') = (b \oplus 1)^{b \oplus 1}$. Since R' is tame, we have $\mathscr{G}^\pm(R'+G) = 0^0$, so $o(R'+G) = \mathscr{P}$.
- If $\mathscr{G}^\pm(G) = 0^1$ (resp. 1^0), then $o(R'+G) = \mathscr{P}$ with $\mathscr{G}^\pm(R') = (a \oplus 1)^0$ (resp. $(a \oplus 1)^1$).
- If $\mathscr{G}^\pm(G) = 0^0$ or 1^1, then G is firm, so it must contain at least two heaps of size b or $b+1$. Each such heap has moves to both 0 and $*$, so G must have a move to some G' with $\mathscr{G}^+(G') = b$. Therefore $o(R+G') = \mathscr{P}$ by induction.
- Finally, if $\mathscr{G}^+(G) = b$, then every G' has $\mathscr{G}^+(G') \neq b$, so $o(R+G') = \mathscr{N}$. Likewise every $R'+G$ is tame and firm, and since $\mathscr{G}^\pm(R') \neq b^b$, we have $\mathscr{G}^\pm(R'+G) \neq 0^0$. Therefore $o(R'+G) = \mathscr{N}$. Since $R+G \not\cong 0$, we conclude that $o(R+G) = \mathscr{P}$.

Case 2: G has exactly one heap of size $m \notin \{0, 1, b, b \oplus 1\}$ (and all other heaps have size 0, 1, b, or $b \oplus 1$). We must show that $o(R+G) = \mathscr{N}$ in all cases.

- If $m > b$, then m has moves to 0, 1, b, and $b \oplus 1$. One of these moves must yield a G' with $\mathscr{G}^+(G') = b$, so that $o(R+G') = \mathscr{P}$ by induction.
- Suppose $m < b$ and G has an *odd* number of heaps of size b or $b \oplus 1$. Then G has an option G' with $\mathscr{G}^+(G') = a$, obtained by reducing one of the b-heaps to m or $m \oplus 1$ (depending on parity). Since G' has heaps of size m, we have $o(R+G') = \mathscr{P}$ by induction.
- Suppose $m < b$ and G has an *even* number of heaps of size b or $b \oplus 1$. Then $\mathscr{G}^+(G) = m$ or $m \oplus 1$. Since $b > m$ and $b > m \oplus 1$, this shows that R' has an option with $\mathscr{G}^+(R') = \mathscr{G}^+(G)$. Since R' and G are tame, and G is firm, we have $\mathscr{G}^\pm(R'+G) = 0^0$.

Case 3: G has at least two heaps of size $\notin \{0, 1, b, b \oplus 1\}$.

- If $\mathscr{G}^+(G) = a$, then $\mathscr{G}^+(G') \neq a$ for all G'. By induction, $o(R + G') = \mathscr{N}$ for all G'. Moreover, since $\mathscr{G}^+(R) = a$, we also have $\mathscr{G}^+(R') \neq a$ for any R'. Therefore $\mathscr{G}^+(R' + G) \neq 0$, and since $R' + G$ is tame and firm, $o(R' + G) = \mathscr{N}$. This shows that $o(R + G) = \mathscr{P}$.
- Suppose $\mathscr{G}^+(G) \neq a$. If $a < \mathscr{G}^+(G)$, then certainly $\mathscr{G}^+(G') = a$ for some G', so by induction $o(R + G') = \mathscr{P}$. But if $a > \mathscr{G}^+(G)$, then necessarily $a = 1$ and $\mathscr{G}^+(G) = 0$. Since $a = 1$, we have $\delta(R) = \{0^1, 0^0\}$ so there is some option R' of genus 0^0. Therefore $\mathscr{G}^\pm(R' + G) = 0^0$ and therefore $o(R' + G) = \mathscr{P}$. In all cases, $o(R + G) = \mathscr{N}$. □

General Reversibility

The theory of tame games can be extended slightly by allowing wild options of G, provided every such G' reverts back to a tame game with the same genus as G. In the following discussion, a **tame genus** is a genus of the form 0^1, 1^0, or a^a; and a **restive genus** is a genus of the form 0^b or 1^b, with $b \geq 2$.

Definition 2.10. Fix a game G of tame genus a^b. We say that G is **generally tame** if:

(i) removing wild options from G does not change its genus; and

(ii) every wild $G' \in G$ has generally tame options $G_1'', G_2'' \in G'$ with

$$\mathscr{G}^\pm(G_1'') = a^x, \quad \mathscr{G}^\pm(G_2'') = y^b$$

for some x and y. (We allow for the possibility that $G_1'' = G_2''$.)

Example. Let $G = \{*4, *2_\#0\}$. G is not tame, since it has the wild option $*2_\#0$. However $\mathscr{G}^\pm(G) = 0^0$, and the option $*2_\#0$ does not contribute to this genus, since $\mathscr{G}^\pm(\{*4\}) = 0^0$ as well. So condition (i) in Definition 2.10 is satisfied. Moreover $*2_\#0$ has the option $*2_\#$ also of genus 0^0, so $*2_\#0$ meets condition (ii). This shows that G is generally tame.

The strategy for sums of tame games works for generally tame games as well, with the following modifications:

- Always move another position that is generally tame.
- If your opponent moves to a wild position, always respond in the *same* component with an appropriate reverting move.

Restive games can be generalized too.

Definition 2.11. Fix a game G of restive genus a^b. We say that G is **generally restive** if:

(i) removing wild options from G does not change its genus; and

(ii) every wild $G' \in G$ has options $G''_1, G''_2 \in G'$ that are generally tame *or* generally restive, and satisfy
$$\mathscr{G}^{\pm}(G''_1) = a^x, \quad \mathscr{G}^{\pm}(G''_2) = y^b$$
for some $x, y \in \{0, 1, b, b \oplus 1\}$.

Properties of generally tame and generally restive games are explored in the exercises.

Extended Genus

Given $\mathscr{G}^{\pm}(G)$, one can compute $\mathscr{G}^{\pm}(G+*)$, and therefore $o(G+*)$. However, unless G is tame, it is not possible to determine $o(G + *2)$ from $\mathscr{G}^{\pm}(G)$ alone. Moreover, an easy argument shows that $*2$ is a subposition of every nontrivial G:

Theorem 2.12. *If $G \neq 0$ and $G \neq *$, then G has a subposition equal to $*2$.*

Proof. If $G' \neq 0$ and $G' \neq *$ for some $G' \in G$, then by induction G' has a subposition equal to $*2$, and therefore so does G. But if every option of G is equal to 0 or $*$, then the Replacement Lemma implies G is equal to
$$\{\,\} \quad \text{or} \quad \{0\} \quad \text{or} \quad \{*\} \quad \text{or} \quad \{0, *\},$$
which are equal to 0, $*$, 0, and $*2$ respectively. □

In particular, any nontrivial, hereditarily closed set of games must contain $*2$ (or equivalent). So if Γ is any nontrivial heap game, then $*2$ must appear as a Γ-heap. In order to develop a strategy for sums of Γ-heaps, it is therefore necessary (as a first step) to understand sums of the form
$$G + *2, \qquad G + *2 + *2, \qquad G + *2 + *2 + *2, \qquad \ldots$$
regardless of the precise structure of Γ.

Fortunately such sums are remarkably well behaved: there is a straightforward extension of $\mathscr{G}^{\pm}$ that captures the outcomes of all games
$$G + n \cdot *2.$$

Definition 2.13. The **extended genus** of G, denoted by $\mathscr{G}^*(G)$, is the sequence
$$\mathscr{G}^*(G) = (a, b_0, b_1, b_2, \ldots),$$
where $a = \mathscr{G}^+(G)$ and $b_n = \mathscr{G}^-(G + n \cdot *2)$. We usually write
$$\mathscr{G}^*(G) = a^{b_0 b_1 b_2 \cdots}.$$

Theorem 2.14. *Suppose that* $G \not\cong 0$ *and* $\mathscr{G}^*(G) = a^{b_0 b_1 b_2 \cdots}$. *Then*

$$a = \text{mex}\{a'\}, \quad b_0 = \text{mex}\{b_0'\}, \quad b_{n+1} = \text{mex}\{b_{n+1}', b_n, b_n \oplus 1\},$$

where $a'^{b_0' b_1' b_2' \cdots}$ *denotes the extended genus of a typical option* $G' \in G$.

Proof. The first two assertions are trivial. For the third, note that the options of $G + (n+1) \cdot *2$ are to

$$G' + (n+1) \cdot *2, \quad G + n \cdot *2, \quad \text{and} \quad G + n \cdot *2 + *.$$

Now

$$\mathscr{G}^-\big(G' + (n+1) \cdot *2\big) = b_{n+1}' \quad \text{and} \quad \mathscr{G}^-\big(G + n \cdot *2\big) = b_n$$

by definition of $\mathscr{G}^*$, and

$$\mathscr{G}^-\big(G + n \cdot *2 + *\big) = b_n \oplus 1$$

by Proposition 1.9 (page 230). □

Theorem 2.15. *Let* $\mathscr{G}^*(G) = a^{b_0 b_1 b_2 \cdots}$. *Then* $b_{n+1} = b_n \oplus 2$ *for all sufficiently large* n.

Proof. By Theorem 2.14, we have

$$b_{n+1} = \text{mex}\{b_n, b_n \oplus 1, b_{n+1}'\}$$

for all $n \geq 0$ (where b_n' ranges over the values of $\mathscr{G}^-(G' + n \cdot *2)$, for $G' \in G$). By induction, we can assume that there exists an n_0 such that $b_{n+1}' = b_n' \oplus 2$ for all $n \geq n_0$. For such n we can write

$$b_{n+1} = \text{mex}\{b_n, b_n \oplus 1, b_n' \oplus 2\}.$$

Now the values b_n are bounded (since G has just finitely many options), so there must be some $n \geq n_0 + 1$ with $b_{n+1} < b_n$. For such n we have $b_{n+1} \leq b_n \oplus 1$, and therefore $b_{n+1} < b_n \oplus 1$ (since $b_n \oplus 1$ is an excludent), and we can write

$$b_{n+1} = \text{mex}\{b_n' \oplus 2\}.$$

The only possible values for this mex are b_n, $b_n \oplus 1$, and $b_n \oplus 2$, but the first two have already been ruled out. This shows that $b_{n+1} = b_n \oplus 2$, and the result now follows by a straightforward recurrence. □

Thus every extended genus is eventually periodic with period 2, and it is customary to write simply

$$\mathscr{G}^*(G) = a^{b_0 b_1 b_2 \cdots b_N}$$

when it is known that $b_{n+1} = b_n \oplus 2$ for all $n \geq N$.

Example. Let G be tame of genus a^b. Since $*2$ is tame and firm, the addition rules for tame games give (for all $n \geq 1$)

$$\mathscr{G}^{\pm}(G + n \cdot *2) = \begin{cases} a^a & \text{if } n \text{ is even;} \\ (a \oplus 2)^{a \oplus 2} & \text{if } n \text{ is odd.} \end{cases}$$

Therefore if G is tame, then $\mathscr{G}^*(G)$ is completely determined by $\mathscr{G}^{\pm}(G)$ and we have

$$\mathscr{G}^*(G) = 0^{12} \text{ or } 1^{03} \text{ or } a^a$$

when

$$\mathscr{G}^{\pm}(G) = 0^1 \text{ or } 1^0 \text{ or } a^a,$$

respectively.

Exercises

2.1 For any integers $a, b \geq 0$, there is a game G with $\mathscr{G}^{\pm}(G) = a^b$.

2.2 Every position in **0.56** is tame.

2.3 (a) If G is restive, then $\mathscr{G}^*(G)$ is completely determined by $\mathscr{G}^{\pm}(G)$. Determine all possible extended genera for G.
(b) Give examples of restless games G and H with $\mathscr{G}^{\pm}(G) = \mathscr{G}^{\pm}(H)$, but $\mathscr{G}^*(G) \neq \mathscr{G}^*(H)$.
(c) Determine all possible extended genera for a restless game G.

2.4 If G and H are generally tame, then so is $G + H$.

2.5 If G is generally restive, then $G + G$ is generally tame with genus 0^0. Also, $G + G + G$ is generally tame with genus a^a, where $a = \mathscr{G}^+(G)$.

2.6 Theorem 2.9 is false if G is permitted to be an arbitrary tame game (rather than specifically a NIM position).

2.7 Theorem 2.9 remains true if R is generally restive.

2.8 Suppose R is generally restive of genus a^b, and let m be an integer with $1 < m < \min\{b, b \oplus 1\}$. Prove that $R + *m$ is generally tame, and determine its genus.

2.9 *The Noah's Ark Theorem.* Let G be restless. Assume that:
(i) $G + G$ is a misère $\mathscr{N}$-position; and
(ii) for every $a \geq 2$, if G has an option of genus a^a, then G also has an option of genus $(a \oplus 1)^{(a \oplus 1)}$.
Prove that $o(2n \cdot G + T) = o(T)$ for every $n \geq 0$ and tame game T. Prove furthermore that $o\big((2n+1) \cdot G + T\big) = o(G + T)$.

Notes

Almost all of the material in this section is drawn from ONAG and *Winning Ways*, but our terminology is slightly different. Conway defined the **genus** of G to be $\mathscr{G}^*(G)$, rather than $\mathscr{G}^{\pm}(G)$. This has the unfortunate effect of disguising the central point of the genus theory, which is that $\mathscr{G}^{\pm}(G)$ (rather than $\mathscr{G}^*(G)$) exactly determines the value of G in relation to the universe of tame games. The symbol $\mathscr{G}^*(G)$ appears to have less theoretical relevance than $\mathscr{G}^\pm(G)$, although its practical use is substantial when calculating genus values by hand. For this reason we have invented the term **extended genus** for $\mathscr{G}^*(G)$.

The meaning of **tame** is also inconsistent in the literature; the definition in ONAG agrees with ours, while the *Winning Ways* definition coincides with our **generally tame**. The ONAG definition is more consistent with later developments (in particular, the quotient theory of Sections 4 and 5), so we have chosen to adopt it here.

Dean Allemang [**All84, All01**] has further generalized the extended genus construction. Allemang's theory replaces the "saltus" $*2$ with an arbitrary list of games $X_1, X_2, \ldots, X_k$. The **generalized genus** of G (relative to $X_1, \ldots, X_k$) is the multidimensional array given by

$$\mathscr{G}^-(G + n_1 \cdot X_1 + n_2 \cdot X_2 + \cdots + n_k \cdot X_k)$$

for $(n_1, n_2, \ldots, n_k) \in \mathbb{N}^k$. Unfortunately this generalization is not always well behaved, and in particular there is no known analogue of Theorem 2.15. The generalized genus theory has been largely supplanted by the misère quotient construction discussed later in this chapter.

3. Misère Canonical Form

Genus calculations provide a complete theory for tame games in misère play, but they have little to say about wild games, as the following example demonstrates.

Example. Let $G = *2$, $H = *2_{2}$, and $X = *2_{3}0$. Then G and H are both Nim positions of genus 2^2, so that $\mathscr{G}^\pm(G) = \mathscr{G}^\pm(H)$.

However, $o(G + X) \neq o(H + X)$:

- $G + X$ is an $\mathscr{N}$-position, having a winning move to X.
- $H + X$ is a $\mathscr{P}$-position, as follows. Using Theorem 1.6, we have

$$H = \{*2_2,\ *2_3\}.$$

Then the four possible opening moves on $H + X$ are to

$$*2_2 + X, \quad *2_3 + X, \qquad H + *2_3, \qquad H + 0.$$

These can be countered, respectively, by moving to

$$*2_2 + 0, \quad *2_3 + *2_3, \quad *2_3 + *2_3, \quad *2_2 + 0,$$

each of which is a Nim position of genus 0^0.

In this section we'll show that the reduction given by Definition 1.3 (on page 226) is the *only* available simplification in misère play. A game with no available simplifications is in canonical form, in a sense exactly analogous to the partizan theory from Chapter II: nonisomorphic games in canonical form are unequal.

The Mate of G

Definition 3.1. The **mate** of G, denoted by G^-, is defined by

$$G^- = \begin{cases} * & \text{if } G \cong 0; \\ \{(G')^- : G' \in G\} & \text{otherwise.} \end{cases}$$

If we visualize G as a tree, then G^- is obtained by "hanging" an additional move from every leaf node of G (see Figure 3.1).

Lemma 3.2. *$G + G^-$ is a (misère) $\mathscr{P}$-position.*

Proof. If $G \cong 0$, then $G + G^- = *$, a $\mathscr{P}$-position. Otherwise, every option of $G + G^-$ reverts to a position of the form

$$G' + (G')^-, \quad \text{for some } G' \in G,$$

which we may assume is a $\mathscr{P}$-position by induction. □

The normal-play analogue of Lemma 3.2 asserts that $G+G$ is a (normal) $\mathscr{P}$-position, and the proof is just the mirror-image argument: every option reverts to a position of the form $G' + G'$. So Lemma 3.2 can be viewed as a modified mirror-image argument in which the base case is $*$ instead of 0.

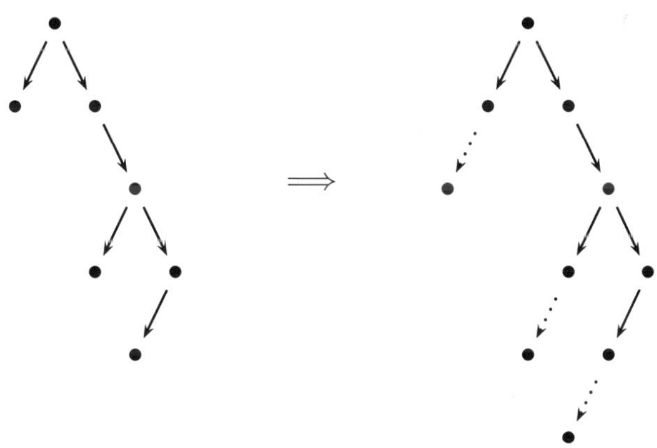

$*2_{\#}0$ $\qquad\qquad$ $(*2_{\#}0)^-$

Figure 3.1. From G to G^-. The new options are shown with dotted arrows in the figure on the right. (These are *impartial* game trees in which either player may use any edge.)

Now if $G \neq H$, then there must exist some T with $o(G+T) \neq o(H+T)$. Such T is said to **distinguish** G from H. The following lemma shows that we can always choose the distinguishing game T so that $o(G+T) = \mathscr{P}$.

Lemma 3.3. *Suppose that $G \neq H$. Then there is some T such that*

$$o(G+T) = \mathscr{P} \quad \textit{and} \quad o(H+T) = \mathscr{N}.$$

Proof. Since $G \neq H$, there is some U such that *either*

$$(\dagger) \qquad o(G+U) = \mathscr{P} \quad \text{and} \quad o(H+U) = \mathscr{N}$$

or

$$(\ddagger) \qquad o(G+U) = \mathscr{N} \quad \text{and} \quad o(H+U) = \mathscr{P}.$$

In case (†) we simply take $T = U$. In case (‡), put

$$T = \{G_1^-, G_2^-, \ldots, G_k^-, U\}$$

where $G_1, G_2, \ldots, G_k$ are the options of G. Then $H+T$ is an $\mathscr{N}$-position, since it has a move to $H+U$. Consider $G+T$. The option $G+U$ is an $\mathscr{N}$-position by hypothesis, but all other options revert to positions of the form

$$G_i + G_i^-$$

which are $\mathscr{P}$-positions by Lemma 3.2. Therefore $o(G+T) = \mathscr{P}$. □

The Simplest Form Theorem

We can now prove the Simplest Form Theorem. The following idea is key.

Definition 3.4. We say G is **linked to** H (**by** T), and we write $G \bowtie H$, if

$$o(G+T) = o(H+T) = \mathscr{P} \text{ for some impartial game } T.$$

Theorem 3.5. $G \bowtie H$ *if and only if* $G =$ *no* H' *and no* $G' = H$.

Proof. If $G \cong H \cong 0$, then $*$ links G to H, and we are done. So for the remainder of the proof we may assume that $G \ncong 0$ or $H \ncong 0$.

First suppose that $G \bowtie H$, and fix T so that $o(G+T) = o(H+T) = \mathscr{P}$. Then $o(G'+T) = \mathscr{N}$ for every G', and $o(H'+T) = \mathscr{N}$ for every H'. So T distinguishes G from every H' and every G' from H.

Conversely, suppose $G =$ no H' and no $G' = H$. Then for each $G_i \in G$, there is (by Lemma 3.3) some X_i such that

$$o(G_i + X_i) = \mathscr{P} \quad \text{and} \quad o(H + X_i) = \mathscr{N}.$$

Likewise, for each $H_j \in H$ there is some Y_j with

$$o(H_j + Y_j) = \mathscr{P} \quad \text{and} \quad o(G + Y_j) = \mathscr{N}.$$

Put

$$T = \{X_i, Y_j\},$$

ranging over all i and j, and consider $G + T$. The options to $G + Y_j$ are $\mathscr{N}$-positions *a priori*, and all other options revert to positions of the form

$$G_i + X_i,$$

which are $\mathscr{P}$-positions. Therefore every option of $G + T$ is an $\mathscr{N}$-position. Since $G \ncong 0$ or $H \ncong 0$, we have $T \ncong 0$, and hence $G + T \ncong 0$. Thus $G + T$ is a $\mathscr{P}$-position. The same argument shows that $H + T$ is also a $\mathscr{P}$-position, so that T links G to H. □

Theorem 3.6. $G = H$ *if and only if the following four conditions hold:*

(i) *G is linked to no H'.*

(ii) *No G' is linked to H.*

(iii) *If $G \cong 0$, then H is an $\mathscr{N}$-position.*

(iv) *If $H \cong 0$, then G is an $\mathscr{N}$-position.*

Proof. If T links G to H', then $o(G + T) = \mathscr{P}$ but $o(H + T) = \mathscr{N}$, so $G \neq H$. Thus $G = H$ implies (i) and, by symmetry, (ii). Likewise, if $G = H$, then certainly $o(G) = o(H)$; since 0 is an $\mathscr{N}$-position, (iii) and (iv) must also hold.

Conversely, suppose that $G \neq H$. Let T be a game of minimal birthday that distinguishes G from H. Without loss of generality, we have

$$o(G + T) = \mathscr{P} \quad \text{and} \quad o(H + T) = \mathscr{N}.$$

If $H + T \cong 0$, then $H \cong 0$ and $T \cong 0$. Therefore $G + T \cong G$ and it follows that G is a $\mathscr{P}$-position, violating (iv).

If $H + T \ncong 0$, then some option of $H + T$ must be a $\mathscr{P}$-position. It cannot be $H + T'$, for this would imply $o(G+T') \neq o(H+T')$, contradicting minimality of T. Therefore some $H' + T$ is a $\mathscr{P}$-position, whereupon T links G to H'. □

Theorems 3.5 and 3.6 show that a game is in canonical form if it has no **reversible** moves, in the following sense.

Definition 3.7. Let G be an impartial game. An option $G' \in G$ is said to be **(misère) reversible (through** G''**)** if $G'' = G$ for some $G'' \in G'$.

Definition 3.8. We say that G is in **canonical form** (or **simplest form**) if no subposition of G has any reversible options.

Lemma 3.9. *Suppose that neither G nor H has any reversible options, and assume that $G = H$. Then for every $G' \in G$, there is an $H' \in H$ such that $G' = H'$, and vice versa.*

Proof. Fix $G' \in G$. Since $G = H$, we have $G' \not\bowtie H$ by Theorem 3.6. Then by Theorem 3.5, either $G'' = H$ for some $G'' \in G'$ or else $G' = H'$ for some $H' \in H$. But $G'' = H$ would imply $G'' = G$, contradicting the assumption that G has no reversible options. □

Theorem 3.10 (Simplest Form Theorem). *Suppose that G and H are in canonical form, and assume that $G = H$. Then $G \cong H$.*

Proof. Lemma 3.9 shows that for every $G' \in G$, there is an $H' \in H$ with $G' = H'$. By induction, this implies that $G' \cong H'$. Therefore $G \subset H$, and likewise $H \subset G$, so that $G \cong H$. □

A final theorem connects this notion of canonical form with the simplification techniques of Section 1.

Theorem 3.11. *Suppose that every option of H is in canonical form and that some option of H is reversible through G. Then H simplifies to G.*

Proof. It suffices to show that:

(i) every option of G is an option of H; and

(ii) every *other* option $H' \in H$ has $G \in H'$.

Since $G = H$, we have $G' \not\bowtie H$ for every $G' \in G$. Thus either $G'' = H$ for some $G'' \in G'$ or else $G' = H'$ for some $H' \in H$. But $G'' = H$ is impossible, since G is presumed to be in canonical form. Therefore, since G' and H' are both in canonical form, it follows that $G' \in H$. This proves (i).

For (ii), fix an option $H' \in H$. Then $G \not\bowtie H'$, so either $G' = H'$ for some $G' \in G$, or else $G = H''$ for some $H'' \in H'$. These imply, in turn, that either $H' \in G$ or $G \in H'$. □

Games Born by Day 4

The Simplest Form Theorem can be used to classify the games born by day n.

Theorem 3.12. *There are exactly* $(1, 1, 1, 2, 17)$ *distinct misère impartial games with birthday* $(0, 1, 2, 3, 4)$.

Proof. Clearly

$$* = \{0\}$$

is the only game with birthday 1. On day 2, we have additionally $\{*\}$ and $*2 = \{0, *\}$, but $\{*\} = 0$ by the misère mex rule. Therefore

$$*2 = \{0, *\}$$

is the unique game with birthday 2.

0	$*2_\#$	$*2_\#0$	$*2_\#20$	$*2_\#210$
$*$	$*3_\#$	$*2_\#1$	$*2_\#21$	$*2_\#320$
$*2$	$*32$	$*2_\#2$	$*2_\#30$	$*2_\#321$
$*3$	$*2_{\#\#}$	$*2_\#3$	$*2_\#31$	$*2_\#3210$
$*4$			$*2_\#32$	

Figure 3.2. The 22 distinct misère impartial games born by day 4.

Next suppose that G has birthday 3. Then G must have $*2$ as an option, and its remaining options must be among 0 and $*$. Furthermore, by the misère mex rule, the options of G must contain *neither* or *both* of 0 and $*$. This leaves just two possibilities

$$*3 = \{0, *, *2\} \quad \text{and} \quad *2_\# = \{*2\}.$$

Finally, suppose that G has birthday 4. Then G must have either $*3$ or $*2_\#$ as an option (or both). If $*2_\#$ is not an option of G, then by the misère mex rule the only possibilities are

$$*4 = \{0, *, *2, *3\}, \quad *3_\# = \{*3\}, \quad \text{and} \quad *32 = \{*2, *3\}.$$

Sixteen possibilities remain: those in which $*2_\#$ is an option of G, and G's remaining options comprise one of the 16 subsets of $\{0, *, *2, *3\}$.

Which of these are canonical? Suppose that such a G simplifies to H. Then at least one option of G reverses through H, so that $\text{b}(H) \le 2$. In particular, since $*2_\#$ has birthday 3, it cannot be an option of H and so must have H as an option. This implies $H = *2$.

Now G's options are a superset of H's, so 0 and $*$ are options of G. Moreover, $*2$ cannot be an option of G, since it is neither an option of H nor contains H as an option. We conclude that among the 16 possibilities, the only *non*-canonical games are $*2_\#10$ and $*2_\#310$. Thus there are 14 birthday-4 games with $*2_\#$ as an option. □

The full list of games born by day 4 is summarized in Figure 3.2.

Exercises

3.1 (a) Prove that $*2_\#$ is not equal to any NIM position.
(b) Exhibit a specific game X such that $o(*2_\# + X) \neq o(*2_2 + X)$.

3.2 (a) Give an example to show that $G = H$ need *not* imply $G^- = H^-$.
(b) Show that for *every* G, there is an H with $G = H$ but $G^- \neq H^-$.

3.3 True or false: If $G =^- H$, then necessarily $G =^+ H$.

3.4 Determine which games born by day 4 are tame, restive, and restless.

3.5 There are precisely 4,171,780 distinct games born by day 5. (There are 2^{22} games to consider. Subtract from 2^{22} one term for each of the five games H born by day 3. Carefully observe the proviso in the $H = 0$ case.)

3.6 If $G + H = 0$, then either $G = H = 0$ or $G = H = *$.

3.7 The *Misère Periodicity Theorem.* Let $\Gamma = \mathbf{d}_0.\mathbf{d}_1\mathbf{d}_2 \ldots \mathbf{d}_k$ be an octal game of finite length k. Suppose that there exist $n_0 \geq 1$ and $p \geq 1$ such that

$$H_{n+p} = H_n \quad \text{for all } n \text{ with } n_0 \leq n < 2n_0 + p + k.$$

Then it follows that

$$H_{n+p} = H_n \quad \text{for all } n \geq n_0.$$

Notes

Grundy and Smith undertook the first systematic effort to develop a misère theory in a 1956 paper [**GrS56**]. They understood the proviso and knew about reversible moves, and they correctly calculated the number of games born by day 5 (4,171,780). Without the Simplest Form Theorem, however, they were unable to prove this answer correct.

In the 1970s Conway finally proved the Simplest Form Theorem. In *On Numbers and Games*, he wrote of this theorem:

> It was formerly known to some people as *Grundy's conjecture*, although Professor Smith informs me that in fact Grundy conjectured no such thing, and firmly believed the opposite! [**Con01**]

Conway's breakthrough verified the Grundy–Smith count of day 5 games, and using similar techniques he calculated the number of games born by day 6. The original count published in ONAG was slightly inaccurate; the corrected answer was given by Chris Thompson in 1999 [**Tho99**] and independently verified by Hoey and Siegel. For the record, there are exactly

$$\begin{aligned} 2^{4171780} &- 2^{2096640} - 2^{2095104} - 2^{2094593} - 2^{2094080} - 2^{2091523} - 2^{2091522} \\ &- 2^{2088960} - 2^{2088705} - 2^{2088448} - 2^{2088193} - 2^{2086912} - 2^{2086657} \\ &- 2^{2086401} - 2^{2086145} - 2^{2085888} - 2^{2079234} + 2^{1960962} + 21 \end{aligned}$$

games born by day 6. (A derivation of this result can be found in [**CS**].)

The Simplest Form Theorem also makes possible an abstract structure theory for misère impartial games (analogous to the theory of normal-play partizan games in Chapter III). Conway proved several key results in this direction. Denote by $\mathcal{M}$ the monoid of misère impartial game values; then:

Theorem (Conway's Cancellation Theorem). *$\mathcal{M}$ is cancellative: $G + J = H + J$ implies $G = H$, for all $G, H, J \in \mathcal{M}$.*

He also showed that for each G, there are just finitely many games H and J with $G = H + J$. Proofs of these results are surprisingly tricky; they can be found in Allemang's thesis [**All84**] and in [**CS**].

In [**CS**] it is shown that if $n \cdot G = n \cdot H$, then either $G = H$ or $G = H + *$. This in turn implies that $*$ is the unique torsion element of the group of fractions of $\mathcal{M}$. This *almost* characterizes the structure of $\mathcal{M}$, but not quite: it might still be possible that $m \cdot G = n \cdot H$, for distinct m and n.

Open Problem. Characterize the abstract structure of the monoid $\mathcal{M}$ of misère impartial games. (Conjecture: $\mathcal{M}$ is isomorphic to the free commutative monoid on a countable set of generators, modulo a factor of $\mathbb{Z}_2$ corresponding to $*$.)

Exercise 3.2: Dan Hoey.

Exercise 3.5: Grundy and Smith.

4. Misère Quotients

The central question in misère impartial games can be phrased like so:

What is the misère analogue of the Sprague–Grundy theory?

Misère canonical forms are one answer to this question. In a sense it is the "complete" answer: the resulting **misère game value** of G encodes all the information needed to determine $o(G + X)$, for any impartial X.

The problem, of course, is that this is rather a lot of information, and calculations involving misère canonical forms are often nearly intractable. For example, *cgsuite* gives the following canonical form for a KAYLES heap of size 9:

$$K_9 = *(X_1X2_32_230)_1(X_1X2_32_230)X_3X_22_320 \quad (X = *2_2321)$$

and the values K_n become exponentially more complicated as n increases. In this sense, misère canonical forms are a poor substitute for the Sprague–Grundy theory, which reduces normal-play calculations to a simple arithmetic computation.

Genus values provide a different kind of answer: they *do* succeed in reducing the theory to a simple arithmetic computation, but *only* for a restricted class of games, the tame games. Neither approach works terribly well for wild games: misère canonical forms retain too much information; genus values retain too little.

Fortunately, there is an intermediate approach that often works in cases where both classical methods fail. The key idea is to localize the fundamental equivalence to a specific set of games $\mathscr{A}$—for example, the set of all KAYLES positions.

Definition 4.1. Let $\mathscr{A}$ be a set of impartial games. Then for $G, H \in \mathscr{A}$, we define

$$G \equiv H \pmod{\mathscr{A}} \quad \text{if} \quad o(G + X) = o(H + X) \text{ for all } X \in \mathscr{A}.$$

Examples. If $\mathscr{T}$ is the set of tame games, then the genus theory implies

$$G \equiv H \pmod{\mathscr{T}} \quad \text{if and only if} \quad \mathscr{G}^{\pm}(G) = \mathscr{G}^{\pm}(H).$$

The corresponding equivalence class of G retains just the information needed to determine $o(G+X)$ for *tame* X.

If $\mathscr{I}$ is the set of *all* short impartial games, then

$$G \equiv H \pmod{\mathscr{I}} \quad \text{if and only if} \quad G = H.$$

The corresponding equivalence class of G is its misère game value, from which we can recover the outcome of $o(G+X)$ for *any* impartial X.

If $\mathscr{A}$ is the set of KAYLES positions, then the equivalence class of G modulo $\mathscr{A}$ is a new type of value. It retains just the information needed to determine $o(G+X)$, where X is another KAYLES position. Note that this is exactly what one needs for perfect play of KAYLES.

If $\mathscr{A}$ is closed under addition, then the resulting equivalence classes inherit its structure:

Proposition 4.2. *Let $\mathscr{A}$ be closed under addition. If $G \equiv H \pmod{\mathscr{A}}$ and $J \in \mathscr{A}$, then $G + J \equiv H + J \pmod{\mathscr{A}}$.*

Proof. For all $X \in \mathscr{A}$, we have $J + X \in \mathscr{A}$, so that

$$(G+J)+X \equiv G+(J+X) \equiv H+(J+X) \equiv (H+J)+X \pmod{\mathscr{A}}. \quad \square$$

Furthermore, if $0 \in \mathscr{A}$, then $G \equiv H$ implies $o(G) = o(H)$, so the outcome of G can be recovered from its equivalence class modulo $\equiv$.

Let $\mathcal{Q}$ be the set of equivalence classes modulo $\mathscr{A}$; Proposition 4.2 shows that $\mathcal{Q}$ has the structure of a commutative monoid. Also let $\Phi(G)$ denote the equivalence class of G, so that

$$\Phi : \mathscr{A} \to \mathcal{Q}$$

is the quotient map. In order to avoid confusion, we'll use lowercase letters

$$x,\ y,\ z,\ \ldots$$

for elements of $\mathcal{Q}$, and we'll write the operation on $\mathcal{Q}$ multiplicatively:

$$\text{if} \quad x = \Phi(G) \quad \text{and} \quad y = \Phi(H), \quad \text{then} \quad xy = \Phi(G+H).$$

Finally, we denote by $\mathcal{P}$ the **$\mathscr{P}$-portion** of $\mathcal{Q}$:

$$\mathcal{P} = \Phi[\mathscr{A} \cap \mathscr{P}] = \{\Phi(G) : G \in \mathscr{A} \text{ and } G \text{ is a } \mathscr{P}\text{-position}\}.$$

The structure $(\mathcal{Q}, \mathcal{P})$ is the **misère quotient** of $\mathscr{A}$, denoted by $\mathcal{Q}(\mathscr{A})$. It serves as a *local* misère analogue of the Sprague–Grundy theory in the following sense. The Sprague–Grundy theory reduces normal-play sums to calculations in the arithmetic of the group

$$\mathbb{Z}_2^{\omega}.$$

Likewise, the quotient $\mathcal{Q}(\mathscr{A})$ reduces misère-play sums to calculations in the arithmetic of $\mathcal{Q}$, *provided that* all the components of the sum are elements of $\mathscr{A}$.

Given the misère quotient $\mathcal{Q}(\mathscr{A})$ and the Φ-values on a set of generators for $\mathscr{A}$, we can recover the outcomes for arbitrary elements of $\mathscr{A}$. For suppose we are studying some game $G \in \mathscr{A}$ that decomposes in $\mathscr{A}$:

$$G = G_1 + G_2 + \cdots + G_k, \quad \text{each } G_i \in \mathscr{A}.$$

If we know the Φ-values of each G_i:

$$x_i = \Phi(G_i),$$

then we can simply multiply them out in the arithmetic of $\mathcal{Q}$ to determine $\Phi(G)$:

$$\Phi(G) = x = x_1 x_2 \cdots x_k$$

and then check whether $x \in \mathcal{P}$. In the typical case, $\mathscr{A}$ will be the set of all positions in a heap game (such as KAYLES), and the set of generators for $\mathscr{A}$ will consist of individual heaps. Thus our strategy is precisely analogous to the theory of Section IV.2, in which compound heap games were solved (in normal play) by adding up single-heap nim values. In many cases, including KAYLES, the monoid $\mathcal{Q}$ turns out to be finite, and the problem of determining the outcome of G reduces to a small number of operations on the finite multiplication table $\mathcal{Q}$.

Example: KAYLES

A presentation for the KAYLES quotient is given in Figure 4.1 on the following page. Here $\mathcal{Q}$ is described by seven generators, corresponding to heaps of sizes 1, 2, 5, 9, 12, 25, and 27. Every other position in KAYLES is equivalent to some combination of these seven heaps. The quotient has 40 total elements, including the nine $\mathscr{P}$-positions shown in Figure 4.1. The single-heap Φ-values are periodic with period 12, just as the $\mathscr{G}$-values are periodic in normal play.

Example. Let's compute the outcome of

$$G = K_{17} + K_{25} + K_{51} + K_{57}.$$

We first look up the Φ-values of each heap in Figure 4.1:

$$\Phi(K_{17}) = d^2e, \quad \Phi(K_{25}) = f, \quad \Phi(K_{51}) = g, \quad \Phi(K_{57}) = b^2c.$$

Multiplying these together in the commutative monoid $\mathcal{Q}$ gives the value

$$\Phi(G) = b^2cd^2efg.$$

$$\begin{aligned}\mathcal{Q} \cong \langle a,b,c,d,e,f,g \mid\ & a^2 = 1,\ b^3 = b,\ bc^2 = b,\ c^3 = c,\ bd = bc,\\ & cd = b^2,\ d^3 = d,\ be = bc,\ ce = b^2,\ e^2 = de,\\ & bf = ab,\ cf = ab^2c,\ d^2f = f,\ f^2 = b^2,\\ & b^2g = g,\ c^2g = g,\ dg = cg,\ eg = cg,\\ & fg = ag,\ g^2 = b^2\rangle\end{aligned}$$

$$\mathcal{P} = \{a, b^2, ac, ac^2, d, ad^2, e, ade, adf\}$$

	1	2	3	4	5	6	7	8	9	10	11	12
0+	a	b	ab	a	c	ab	b	ab^2	d	b	bc	e
12+	ab^2	b	abc	ab^2	d^2e	ab	b	ade	b^2c	bc	abc	b^2c
24+	f	b	g	ab^2c	b^2c	abc	b	ab^2	g	bc	abc	b^2c
36+	ab^2	b	ab	ab^2	b^2c	abc	b	ab^2	g	b	abc	b^2c
48+	ab^2	b	g	ab^2	b^2c	abc	b	ab^2	b^2c	b	abc	b^2c
60+	ab^2	b	g	ab^2	b^2c	abc	b	ab^2	g	bc	abc	b^2c
72+	ab^2	b	g	ab^2	b^2c	abc	b	ab^2	g	b	abc	b^2c
84+	ab^2	b	g	ab^2	b^2c	abc	b	ab^2	g	b	$\cdots$	

Figure 4.1. The full solution to KAYLES (misère quotient and Φ-values).

Applying successively the relations $d^2f = f$, $be = bc$, $bf = ab$, and $bc^2 = b$ (again from Figure 4.1) gives

$$b^2cd^2efg = b^2cefg = b^2c^2fg = ab^2c^2g = ab^2g.$$

Since $b^2g = g$, this is equal to ag, after which no further reductions are possible. Since $ag \notin \mathcal{P}$, we conclude that G is an $\mathscr{N}$-position. A typical winning move is from K_{51} to K_{49}, replacing a heap of value g with one of value ab^2. This leaves a composite position G' with value a^2b^4, which is equal to b^2, a $\mathscr{P}$-position.

A similar calculation can be used to determine the outcome of any KAYLES position—so the presentation in Figure 4.1 is an efficient strategy for KAYLES, in a rigorous sense.

Closure

In all of our examples so far, the set of games $\mathscr{A}$ is **closed** in the following sense:

Definition 4.3. A set of impartial games $\mathscr{A}$ is **closed** provided that:

(i) if $G \in \mathscr{A}$ and $H \in \mathscr{A}$, then $G + H \in \mathscr{A}$ (**additive closure**); and

(ii) if $G \in \mathscr{A}$ and $G' \in G$, then $G' \in \mathscr{A}$ (**hereditary closure**).

In particular, if $\mathscr{A}$ is the set of positions that arise in some heap game Γ, then $\mathscr{A}$ will automatically be closed. Throughout most of this section and the next, we'll accept closure as a basic assumption on the sets of games under consideration. Although the definition of misère quotient requires only condition (i) in order to work, much of the structure theory depends on (ii) as well.

If $\mathscr{A}$ is any set of games, we write $\mathrm{cl}(\mathscr{A})$ for the closure of $\mathscr{A}$ (i.e., the smallest closed set containing $\mathscr{A}$). If G is an individual game, then $\mathrm{cl}(G)$ is shorthand for $\mathrm{cl}(\{G\})$; likewise $\mathrm{cl}(G, H)$ is shorthand for $\mathrm{cl}(\{G, H\})$, and so on. We also write $\mathscr{A}[G]$ to mean $\mathrm{cl}(\mathscr{A} \cup \{G\})$, i.e., the closed extension obtained when we adjoin G as a new element of $\mathscr{A}$.

If Γ is a ruleset, then $\mathcal{Q}(\Gamma)$ is shorthand for $\mathcal{Q}(\mathscr{A})$, where $\mathscr{A}$ is the set of positions of Γ. If Γ is a heap ruleset with heaps H_n, then $\mathcal{Q}_n(\Gamma)$ denotes the **partial quotient** $\mathcal{Q}(H_1, H_2, \ldots, H_n)$.

Tame Quotients

The KAYLES quotient in Figure 4.1 was obtained by computer calculation and is quite tedious to verify by hand. Let's investigate some simpler examples: the misère quotients of tame games. Such **tame quotients** don't tell us anything new, since genus values already give a complete theory for tame games, but it's helpful to see how the genus theory maps onto the new language.

Example. Let $\mathscr{A}$ be the set of NIM positions with all heaps of size 1 or 2, so that $\mathscr{A} = \mathrm{cl}(*, *2)$. Let

$$(\mathcal{Q}, \mathcal{P}) = \mathcal{Q}(\mathscr{A})$$

be the misère quotient of $\mathscr{A}$. What is its structure?

Certainly $\mathcal{Q}$ is generated by two elements

$$a = \Phi(*) \quad \text{and} \quad b = \Phi(*2).$$

From the theory of misère NIM, we have

$$*2 + *2 + *2 \equiv *2 \pmod{\mathscr{A}}$$

since both are tame of genus 2^2. Therefore $b^3 = b$. We also know that

$$* + * \equiv 0 \pmod{\mathscr{A}}$$

since in fact $* + * = 0$ (cf. Theorem 1.6 on page 228). This shows that $a^2 = 1$. These two relations, in fact, characterize the structure of $\mathcal{Q}$: the six elements

$$1,\ a,\ b,\ ab,\ b^2,\ ab^2$$

are easily seen to be distinct, as they correspond to games

$$0,\ *,\ *2,\ *3,\ *2+*2,\ *2+*3$$

with genera

$$0^1,\ 1^0,\ 2^2,\ 3^3,\ 0^0,\ 1^1.$$

Genera 1^0 and 0^0 are the only $\mathscr{P}$-positions, so we can write

$$\mathcal{Q} \cong \langle a, b : a^2 = 1,\ b^3 = b\rangle \quad \text{and} \quad \mathcal{P} = \{a, b^2\}.$$

The four elements of $\mathcal{Q}$ corresponding to firm genera,

$$b,\ ab,\ b^2,\ ab^2,$$

form a subgroup of $\mathcal{Q}$ isomorphic to $\mathbb{Z}_2 \times \mathbb{Z}_2$ (since firm genus values behave like nim values) with identity b^2. The fickle elements $\{1, a\}$ form a separate copy of $\mathbb{Z}_2$, with identity 1.

Henceforth we'll refer to $(\mathcal{Q}, \mathcal{P})$ by the name $\mathcal{T}_2$, the **tame quotient** on two generators.

Example. Let $\mathscr{A} = \mathrm{cl}(*, *2, *4)$. Then $\mathcal{Q}$ is generated by three elements

$$a = \Phi(*),\ b = \Phi(*2),\ c = \Phi(*4).$$

Just as in the previous example, the theory of NIM implies $a^2 = 1$, $b^3 = b$, and $c^3 = c$. But we also have

$$b^2 = c^2$$

since b^2 and c^2 correspond to games $*2 + *2$ and $*4 + *4$, both of genus 0^0.

These relations characterize $\mathcal{Q}$, since the ten values

$$1,\ a,\ b^2,\ ab^2,\ b,\ ab,\ c,\ ac,\ bc,\ abc$$

represent distinct genera 0^1, 1^0, 0^0, ..., 7^7. We see that

$$\mathcal{Q} \cong \langle a, b, c : a^2 = 1,\ b^3 = b,\ c^3 = c,\ b^2 = c^2\rangle \quad \text{and} \quad \mathcal{P} = \{a, b^2\}.$$

This type of argument generalizes easily to show that

$$\mathcal{T}_n = \mathcal{Q}(*, *2^1, *2^2, \ldots, *2^{n-1})$$

has order $2^n + 2$, for all $n \geq 2$.

Example. Let $\mathscr{A}$ be the set of all NIM positions. Then $\mathcal{Q}$ has the generators

$$a = \Phi(*),\ b_1 = \Phi(*2^1),\ b_2 = \Phi(*2^2),\ b_3 = \Phi(*2^3),\ \ldots$$

corresponding to genera

$$0^1,\ 2^2,\ 4^4,\ 8^8,\ \ldots.$$

The firm genera form a subgroup $\mathcal{K}$ isomorphic to $\mathbb{Z}_2^{\omega}$, corresponding to the "normal case" in Bouton's Theorem. The fickle genera form a separate copy of $\mathbb{Z}_2$, corresponding to the parity argument when all heaps have size 1.

We'll write $\mathcal{T}_\infty = \mathcal{Q}(\mathscr{A})$, the full quotient of NIM. If $\mathscr{T}$ is the set of all tame games, then we have

$$\mathcal{Q}(\mathscr{T}) \cong \mathcal{T}_\infty,$$

and this is an equivalent way of stating the genus theory.

Here is a full presentation:

$$\mathcal{Q} \cong \langle a, b_1, b_2, \ldots : a^2 = 1,\ b_1^3 = b_1,\ b_2^3 = b_2,\ \ldots,\ b_1^2 = b_2^2 = b_3^2 = \cdots \rangle,$$

$$\mathcal{P} = \{a, b^2\}.$$

Example: 0.75

Consider the octal game $\Gamma = \mathbf{0.75}$. Recall (from Section IV.2) that the options from a Γ-heap are:

- remove one token, optionally splitting the remainder into two heaps;
- remove any heap of size exactly 2; or
- remove any two tokens from a heap of size ≥ 4, splitting the remainder into exactly two heaps.

The first few heaps are easily calculated using the misère mex rule:

- $H_1 = \{H_0\} = *$,
- $H_2 = \{H_0, H_1\} = *2$,
- $H_3 = \{H_2, H_1 + H_1\} = *$,
- $H_4 = \{H_3, H_2 + H_1, H_1 + H_1\} = \{0, *, *3\} = *2$.

But at heap 5 we find a restive game of genus 1^4:

- $H_5 = \{H_4, H_3 + H_1, H_2 + H_2, H_2 + H_1\} = \{*2, 0, *2_2, *3\} = *2_2320$.

$\mathcal{Q}_5(\Gamma)$ is not isomorphic to any $\mathcal{T}_n$. It turns out to be a quotient of order 8, which we call $\mathcal{R}_8$ (the $\mathcal{R}$ is for "restive"):

$$\mathcal{R}_8 = \langle a, b, c \mid a^2 - 1,\ b^3 = b,\ bc = ab,\ c^2 = b^2 \rangle \quad \text{and} \quad \mathcal{P} = \{a, b^2\}.$$

It will be easiest to show this after we've developed more of the structure theory, so we'll defer proof until the next section. We'll also show that at heap 5, the quotient stabilizes (every larger heap is equivalent to a combination of prior heaps) and periodicity quickly emerges. The full Φ-values are

$$a\ b\ a\ b\ c\ b\ c\ b\ ab^2\ b\ ab^2\ b\ ab^2\ \cdots$$

with period 2 and preperiod 8.

Periodicity

The Octal Periodicity Theorem (Theorem IV.2.7 on page 190) can be generalized to arbitrary misère quotients. The proof is similar to the normal-play version, with one significant complication. The classical proof relies upon the fact that

$$\mathscr{G}[G] = \mathscr{G}[H] \quad \text{implies} \quad \mathscr{G}(G) = \mathscr{G}(H),$$

which follows directly from the mex rule. In the quotient context there is no mex rule, so we'll need to prove explicitly that

$$\Phi[G] = \Phi[H] \quad \text{implies} \quad \Phi(G) = \Phi(H).$$

(Recall from Section I.3 that $\mathscr{G}[G] = \{\mathscr{G}(G') : G' \in G\}$, and likewise $\Phi[G] = \{\Phi(G') : G' \in G\}$.)

Theorem 4.4. *Let $\mathscr{A}$ be a closed set of games. Suppose that $G \subset \mathscr{A}$, and suppose further that $\Phi[G] = \Phi[H]$ for some $H \in \mathscr{A}$. Then*

$$G \equiv H \pmod{\mathscr{A}[G]}.$$

(Remember that G is identified with the set of its options, so "$G \subset \mathscr{A}$" means that every option of G is an element of $\mathscr{A}$.)

Proof. We first prove that

$$o(nG + X) = o(nH + X)$$

for each $n \geq 0$ and $X \in \mathscr{A}$. The proof is by induction on n and X. It suffices to show that

$$(\dagger) \qquad o(nG + X) = \mathscr{N} \quad \text{if and only if} \quad o(nH + X) = \mathscr{N}.$$

First suppose that $o(nG+X) = \mathscr{N}$. For $n = 0$ there is nothing to prove, so assume that $n \geq 1$. There are three cases.

Case 1: $o(nG + X') = \mathscr{P}$ for some $X' \in X$. Then $o(nH + X') = \mathscr{P}$ by induction on X, and we are done.

Case 2: $o\big((n-1)G + G' + X\big) = \mathscr{P}$ for some $G' \in G$. Then certainly $G' + X \in \mathscr{A}$, so by induction on n we have

$$o\big((n-1)H + G' + X\big) = \mathscr{P}.$$

Now $\Phi[G] = \Phi[H]$, so there is some H' with $\Phi(G') = \Phi(H')$. In particular $G' \equiv H' \pmod{\mathscr{A}}$, and since $(n-1)H + X \in \mathscr{A}$, this implies

$$o\big((n-1)H + H' + X\big) = \mathscr{P}.$$

Case 3 (the proviso): $nG + X \cong 0$. Then $G \cong 0$ (since $n \geq 1$), so

$$\Phi[H] = \Phi[G] = \emptyset$$

and it follows that $H \cong 0$. Since also $X \cong 0$, we have $nH + X \cong 0$.

The same argument in reverse shows that if $o(nH + X) = \mathscr{N}$, then also $o(nG + X) = \mathscr{N}$. This proves (†).

Now if $Y \in \mathscr{A}[G]$, then $Y = nG + X$ for some $n \geq 0$ and $X \in \mathscr{A}$. Therefore applying (†) twice and using the fact that $H + X \in \mathscr{A}$, we have

$$o(G+Y) = o(G+nG+X) = o(H+nH+X) = o(H+nG+X) = o(H+Y).$$

Therefore $G \equiv H \pmod{\mathscr{A}[G]}$. □

Corollary 4.5. *Let $\mathscr{A}$ be a closed set of games. Suppose that $G \subset \mathscr{A}$, and suppose further that $\Phi[G] = \Phi[H]$ for some $H \in \mathscr{A}$. Then*

$$\mathcal{Q}(\mathscr{A}[G]) \cong \mathcal{Q}(\mathscr{A})$$

with $\Phi(G) = \Phi(H)$.

Proof. Every element of $\mathscr{A}[G]$ has the form $nG + X$ for some $n \geq 0$ and $X \in \mathscr{A}$, and Theorem 4.4 shows that

$$nG + X \equiv nH + X \pmod{\mathscr{A}[G]}.$$

Moreover, for each $A, B \in \mathscr{A}$ we have

$$A \equiv B \pmod{\mathscr{A}} \quad \text{if and only if} \quad A \equiv B \pmod{\mathscr{A}[G]},$$

since $o(A+nG+X) \neq o(B+nG+X)$ implies $o(A+nH+X) \neq o(B+nH+X)$.

It follows that the equivalence classes of $\mathscr{A}[G]$ are in one-to-one correspondence with those of $\mathscr{A}$. □

Theorem 4.6 (Quotient Periodicity Theorem). *Let $\Gamma = \mathbf{d}_0.\mathbf{d}_1\mathbf{d}_2\ldots\mathbf{d}_k$ be an octal game of finite length k. Let $\mathscr{A}$ be the set of all Γ-positions, and let $(\mathcal{Q}, \mathcal{P}) = \mathcal{Q}(\mathscr{A})$, with quotient map $\Phi : \mathscr{A} \to \mathcal{Q}$. Suppose that there exist $n_0 \geq 1$ and $p \geq 1$ such that*

$$\Phi(H_{n+p}) = \Phi(H_n) \quad \textit{for all } n \textit{ with } n_0 \leq n < 2n_0 + p + k.$$

Then

$$\Phi(H_{n+p}) = \Phi(H_n) \quad \textit{for all } n \geq n_0.$$

Proof. Let $n \geq 2n_0 + p + k$. To show that $\Phi(H_{n+p}) = \Phi(H_n)$, it suffices to prove that $\Phi[H_{n+p}] = \Phi[H_n]$, by Theorem 4.4.

If $x \in \Phi[H_{n+p}]$, then

$$x = \Phi(H_a)\Phi(H_b), \qquad \text{with } a + b = n + p - i \text{ for some } i \leq k.$$

Since $n + p \geq 2n_0 + 2p + k$, we have $a + b \geq 2n_0 + 2p$. Without loss of generality, it follows that $b \geq n_0 + p$. Thus $b - p \geq n_0$ and we have

$$\Phi(b - p) = \Phi(b).$$

Therefore

$$\Phi(a)\Phi(b-p) = \Phi(a)\Phi(b).$$

Now $b-p \neq 0$ (since $n_0 \geq 1$), so $H_a + H_{b-p}$ is an option of H_n, and it follows that $x \in \Phi[H_n]$. A similar argument works for the converse direction. □

Exercises

4.1 Determine the outcome of $K_9 + K_{15} + 3 \cdot K_{25} + K_{118}$ (where K_n denotes a KAYLES heap of size n).

4.2 A **bipartite monoid** is a pair $(\mathcal{Q}, \mathcal{P})$, where $\mathcal{Q}$ is a commutative monoid and $\mathcal{P} \subset \mathcal{Q}$ is an arbitrary subset of $\mathcal{Q}$. A **homomorphism** $f : (\mathcal{Q}, \mathcal{P}) \to (\mathcal{Q}', \mathcal{P}')$ is a monoid homomorphism $f : \mathcal{Q} \to \mathcal{Q}'$ such that $x \in \mathcal{P}$ iff $f(x) \in \mathcal{P}'$.

If there exists a surjective homomorphism $f : (\mathcal{Q}, \mathcal{P}) \to (\mathcal{Q}', \mathcal{P}')$, then we say $(\mathcal{Q}', \mathcal{P}')$ is a **quotient** of $(\mathcal{Q}, \mathcal{P})$. A bipartite monoid $(\mathcal{Q}, \mathcal{P})$ is **reduced** if it has no proper quotients. We write r.b.m. as shorthand for reduced bipartite monoid. Prove that:

(a) Every bipartite monoid $(\mathcal{Q}, \mathcal{P})$ has a unique reduced quotient (up to isomorphism). This r.b.m. is called the **reduction** of $(\mathcal{Q}, \mathcal{P})$.

(b) If $\mathscr{A}$ is a set of games, then $(\mathscr{A}, \mathscr{A} \cap \mathscr{P})$ is a bipartite monoid whose reduction is $\mathcal{Q}(\mathscr{A})$.

(c) If $(\mathcal{Q}, \mathcal{P}) = \mathcal{Q}(\mathscr{A})$ and $\mathscr{B} \subset \mathscr{A}$, then there is a submonoid $\mathcal{R} < \mathcal{Q}$ such that $\mathcal{Q}(\mathscr{B})$ is the reduction of $(\mathcal{R}, \mathcal{R} \cap \mathcal{P})$.

(d) If $\mathscr{A}$ is a closed set of games and $(\mathcal{Q}, \mathcal{P})$ is an r.b.m., then the following are equivalent: (i) $(\mathcal{Q}, \mathcal{P}) = \mathcal{Q}(\mathscr{A})$; (ii) there exists a surjective monoid homomorphism $\Phi : \mathscr{A} \to \mathcal{Q}$ such that for all $G \in \mathscr{A}$,

$$\Phi(G) \in \mathcal{P} \quad \text{if and only if} \quad G \not\cong 0 \text{ and } \Phi(G') \notin \mathcal{P} \text{ for all } G' \in G.$$

Notes

Conway's work on misère theory in the 1970s led to easy solutions for several tame octal games, including **0.56**. At the same time, other results seemed only to confirm the intrinsic complications of the general case. In ONAG Conway undertook an analysis of misère GRUNDY'S GAME, computing the outcome $o(G_n)$ up to $n = 50$ by brute-force application of the genus theory. His conclusion was rather discouraging:

> It would become intolerably tedious to push this sort of analysis much further, and I think there is no practicable way of finding the outcome of G_n for much larger n. [**Con01**]

Nonetheless others continued to press forward. Dean Allemang, in conjunction with his *generalized genus theory* (see Notes to Section 2 on page 241), formally introduced the localized misère equivalence relation (Definition 4.1) and proved a form of the Quotient Periodicity Theorem (Theorem 4.6) [**All84, All01**]. Using these and other techniques, he found solutions to several wild misère octal games, including **0.26**, **0.53**, and **4.7**.

The solution to misère KAYLES was discovered by William Sibert in 1973 by brute-force analysis. Sibert didn't make his solution public until the late 1980s, so the *Winning Ways* authors were unaware of it when the first edition was published. Eventually he contacted Conway and in 1992, the *Sibert–Conway Decomposition* finally saw the light of day [**CS92**]. However it failed to resolve the controversy over the ultimate tractability of the misère theory:

> A referee of a draft of the Sibert–Conway paper wrote "the actual solution will have no bearing on other problems," while another wrote "the ideas are likely to be applicable to some other games." [**BCG01**]

Fortunately Thane Plambeck was counted among the optimists. He extended the Sibert–Conway results to several other octal games [**Pla92**] and sought to subsume all such results into a general theory. Plambeck's vision wasn't fully realized until 2004, when in a flash of insight he discovered the misère quotient concept. This was published in a 2005 paper [**Pla05**], which also gave the KAYLES solution in the succinct form of Figure 4.1. This nicely illustrates the power of the quotient theory: Sibert's original solution was 43 pages long!

Plambeck and Siegel later extended the quotient construction into a systematic method for calculating finite misère quotients and discovered misère solutions for many new octal games [**PS08a**], including **0.15**.

5. The Structure of Finite Misère Quotients

The misère quotient of a closed set $\mathscr{A}$ can be viewed as a concise description of its misère-play combinatorics. In this section we'll investigate the general structure theory of finite misère quotients, including a deep relationship between normal and misère play that applies to a wide range of games.

We begin with some preliminary observations. The simplest quotient is the **trivial** quotient $\mathcal{Q}(0)$, which has $\mathcal{Q} = \{1\}$ and $\mathcal{P} = \emptyset$. All other quotients are **nontrivial**.

If $\mathscr{A}$ is closed and contains any nonempty G, then necessarily $* \in \mathscr{A}$, so that $\Phi(*) \in \mathcal{P}$. Therefore every nontrivial misère quotient has a nonempty $\mathcal{P}$-portion. We can say slightly more:

Proposition 5.1. *Every nontrivial misère quotient* $(\mathcal{Q}, \mathcal{P})$ *contains an element* $a \in \mathcal{P}$ *with* $a^2 = 1$.

Proof. Let $(\mathcal{Q}, \mathcal{P}) = \mathcal{Q}(\mathscr{A})$ with quotient map $\Phi : \mathscr{A} \to \mathcal{Q}$. If $\mathscr{A} \neq \{0\}$, then since $\mathscr{A}$ is hereditarily closed, it must be the case that $* \in \mathscr{A}$. Let $a = \Phi(*)$; then since $* + * = 0$, we have $a^2 = 1$. Moreover, $a \in \mathcal{P}$ and $1 \notin \mathcal{P}$, so that $a \neq 1$. □

The following proposition can be viewed as a localized analogue of Lemma 3.2 on page 243.

Proposition 5.2. *Let* $(\mathcal{Q}, \mathcal{P}) = \mathcal{Q}(\mathscr{A})$ *be a nontrivial misère quotient. For each* $x \in \mathcal{Q}$*, there is some* $y \in \mathcal{Q}$ *with* $xy \in \mathcal{P}$*.*

Proof. If $x = 1$, then $x\Phi(*) = \Phi(*) \in \mathcal{P}$. Otherwise, fix $G \in \mathscr{A}$ with $\Phi(G) = x$, and consider $G + G$. If $\Phi(G + G) \in \mathcal{P}$, then we can simply put $y = x$. Otherwise, $G + G$ is an $\mathscr{N}$-position, and since $x \neq 1$ we must have $G + G \not\cong 0$. Therefore there is some option $G + G'$ with $\Phi(G + G') \in \mathcal{P}$, and we can put $y = \Phi(G')$. □

Corollary 5.3. *Let* $(\mathcal{Q}, \mathcal{P}) = \mathcal{Q}(\mathscr{A})$ *and fix* $G \in \mathscr{A}$*. If* G' *is any option of* G*, then* $\Phi(G') \neq \Phi(G)$*.*

Proof. Put $x = \Phi(G)$ and choose $y \in \mathcal{Q}$ with $xy \in \mathcal{P}$. Let $H \in \mathscr{A}$ be such that $\Phi(H) = y$. Then $G + H$ is a $\mathscr{P}$-position, so $G' + H$ must be an $\mathscr{N}$-position. Therefore $\Phi(G')y \notin \mathcal{P}$, so that y distinguishes $\Phi(G')$ from x. □

Quotients of Small Order

The preceding results can be used to classify the misère quotients of small order. We'll show fairly easily that:

- every finite nontrivial misère quotient has even order;
- $\mathcal{T}_1$ is the only quotient of order 2 (up to isomorphism);
- there are no quotients of order 4; and
- $\mathcal{T}_2$ is the only quotient of order 6 (up to isomorphism).

Theorem 5.4. *Let* $(\mathcal{Q}, \mathcal{P}) = \mathcal{Q}(\mathscr{A})$ *be finite and nontrivial. Then the cardinality of* $\mathcal{Q}$ *is even.*

Proof. Since $\mathscr{A}$ is nontrivial, we have $* \in \mathscr{A}$. Put $a = \Phi(*)$, and define $\sigma : \mathcal{Q} \to \mathcal{Q}$ by $\sigma(x) = ax$. We'll show that σ defines a one-to-one pairing of elements of $\mathcal{Q}$.

Given $x \in \mathcal{Q}$, fix $G \in \mathscr{A}$ with $\Phi(G) = x$. Then G is an option of $G + *$, so Corollary 5.3 gives

$$ax = \Phi(G + *) \neq \Phi(G).$$

This shows that $\sigma(x) \neq x$, and $* + * = 0$ implies that $\sigma(\sigma(x)) = x$. □

Theorem 5.5. *Let* $\mathscr{A}$ *be a nonempty closed set of games. Then either:*

(i) $\mathcal{Q}(\mathscr{A}) \cong \mathcal{T}_0$*; or*

(ii) $\mathcal{Q}(\mathscr{A}) \cong \mathcal{T}_1$*; or*

(iii) *there exists a closed subset* $\mathscr{B} \subset \mathscr{A}$ *with* $\mathcal{Q}(\mathscr{B}) \cong \mathcal{T}_2$*.*

Proof. There are two cases.

Case 1: Every $G \in \mathscr{A}$ is equal to 0 or $*$. If $\mathscr{A} = \{0\}$, then $\mathcal{Q}(\mathscr{A}) = \mathcal{T}_0$. Otherwise, it must be the case that $* \in \mathscr{A}$, so that

$$\mathcal{Q}(\mathscr{A}) \cong \mathcal{Q}(*) = \mathcal{T}_1.$$

Case 2: Otherwise, Theorem 2.12 (page 239) implies that there is a $G \in \mathscr{A}$ with $G = *2$. Choose such G with minimal birthday and let $\mathscr{B} = \mathrm{cl}(G)$. By minimality of G, every element of $\mathscr{B}$ is equal to some element of $\mathrm{cl}(*, *2)$. Therefore

$$\mathcal{Q}(\mathscr{B}) \cong \mathcal{Q}(*2) = \mathcal{T}_2. \qquad \square$$

Theorem 5.6. *There are no misère quotients of order* 4 *and exactly one each of orders* 1, 2, *and* 6 *(up to isomorphism).*

Proof. Let $(\mathcal{Q}, \mathcal{P}) = \mathcal{Q}(\mathscr{A})$ and consider the three cases in Theorem 5.5. Cases (i) and (ii) imply that $|\mathcal{Q}| = 1$ or 2, respectively. In case (iii), there is some $\mathscr{B} \subset \mathscr{A}$ with $\mathcal{Q}(\mathscr{B}) \cong \mathcal{T}_2$. Thus there is a submonoid $\mathcal{R} < \mathcal{Q}$ whose reduction is $\mathcal{T}_2$, i.e., there is a surjective homomorphism

$$\rho : \mathcal{R} \to \mathcal{T}_2.$$

Therefore $|\mathcal{Q}| \geq |\mathcal{R}| \geq |\mathcal{T}_2|$. This shows that $|\mathcal{Q}| \geq 6$. Moreover, if $|\mathcal{Q}| = 6$, then necessarily $\mathcal{R} = \mathcal{Q}$ and ρ is an isomorphism, so

$$(\mathcal{Q}, \mathcal{P}) \cong \mathcal{T}_2. \qquad \square$$

The Kernel and Normal Play

If $(\mathcal{Q}, \mathcal{P}) = \mathcal{Q}(\mathscr{A})$, then certain combinatorial properties of $\mathscr{A}$ can be deduced from the abstract algebraic structure of $\mathcal{Q}$. Notably, there are many situations in which the strategy for $\mathscr{A}$ reduces to normal play—much as the strategy for misère NIM reduces to normal play whenever the position contains heaps of size ≥ 2.

Recall (Definition B.4.2 on page 470) that an element $x \in \mathcal{Q}$ is an **idempotent** if $x^2 = x$. Let

$$z_1, z_2, \ldots, z_k$$

enumerate all the idempotents of a finite commutative monoid $\mathcal{Q}$. Then their product

$$z = z_1 z_2 \cdots z_k$$

must also be an idempotent. z is a sort of "master idempotent" that absorbs all the others, in the sense that $z \cdot z_i = z$ for all i. The **kernel** of $\mathcal{Q}$ is the set

$$\mathcal{K} = \{zx : x \in \mathcal{Q}\}.$$

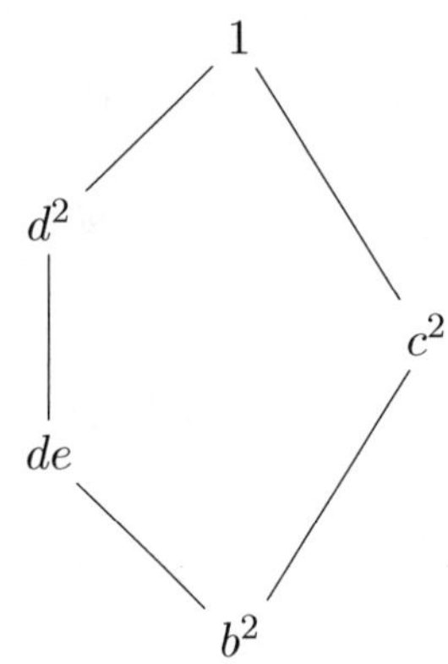

Figure 5.1. A partial order of the idempotents in the KAYLES quotient, where $x \leq y$ if and only if $xy = x$.

Now if $x \in \mathcal{K}$, then since $\mathcal{Q}$ is finite, there must be some n with x^n an idempotent (this is given by Lemma B.4.9 on page 472). Since z is the unique idempotent in $\mathcal{K}$, it follows that $x^n = z$. It follows that $\mathcal{K}$ is an Abelian group with identity z.

Example. Recall that the tame quotient $\mathcal{T}_2$ is presented by

$$\mathcal{Q} \cong \langle a, b : a^2 = 1,\ b^3 = b \rangle.$$

$\mathcal{T}_2$ has two idempotents: the identity 1 and the kernel identity $z = b^2$. The kernel is given by

$$\mathcal{K}_2 = \{z, az, b, ab\},$$

which is isomorphic to $\mathbb{Z}_2 \times \mathbb{Z}_2$. The remaining two elements $\{1, a\}$ form a separate, disjoint copy of $\mathbb{Z}_2$ that is mapped into the kernel by multiplication by z.

Likewise, for any $n \geq 2$, the kernel $\mathcal{K}_n$ of $\mathcal{T}_n$ is isomorphic to $\mathbb{Z}_2^n$, and

$$\mathcal{T}_n = \mathcal{K}_n \cup \{1, a\}.$$

$\mathcal{K}_n$ corresponds exactly to the "normal-play component" of the strategy for misère NIM (equivalently, the firm genus values); the group $\{1, a\}$ corresponds to the exceptional case (the fickle genus values).

Example. The KAYLES quotient (Figure 4.1 on page 252) has five idempotents:

$$1,\ b^2,\ c^2,\ d^2,\ de.$$

Their product is equal to b^2, which is the kernel identity. A partial order of these idempotents is shown in Figure 5.1, with $x \leq y$ if and only if $xy = x$.

There is a remarkable correspondence between the structure of $\mathcal{K}$ and the normal-play solution to KAYLES. The kernel has 16 elements and is generated by

$$z, \ za, \ zb, \ zc, \ zg.$$

Now any KAYLES position G can be mapped to an element of the kernel via the projection

$$G \mapsto z\Phi(G).$$

If we make the correspondence

$$\bigl(z, za, zb, zc, zg\bigr) \mapsto \bigl(0, 1, 2, 4, 8\bigr),$$

then it turns out that $z\Phi(G)$ aligns perfectly with the normal-play nim value of G. For example, consider K_{12}, an individual heap of size 12. From Figure 4.1 we see that $\Phi(K_{12}) = e$, and we have

$$ze = zc$$

(applying the relation $be = bc$ in the presentation for $\mathcal{Q}$). Now c maps to nim value 4 under our correspondence and, sure enough, we see from Figure IV.2.3 on page 191 that $\mathscr{G}(K_{12}) = 4$.

Therefore the structure of $\mathcal{K}$ exactly matches the normal-play strategy for KAYLES, and we can use this fact to deduce lots of information about its misère strategy. For example, an inspection of Figure 4.1 reveals that $\Phi(K_n) \in \mathcal{K}$ for all $n \geq 26$. It follows that as long as there is at least one heap of size ≥ 26, the strategies for normal- and misère-play KAYLES coincide. Note the similarity to the misère strategy for NIM.

Closer scrutiny reveals an even stronger statement: $\Phi(K_n) \in \mathcal{K}$ except when $n \in \{1, 4, 9, 12, 17, 20, 25\}$. So in fact misère KAYLES reduces to normal KAYLES unless the position consists entirely of heaps drawn from this seven-element set!

Now KAYLES is sufficiently simple that this type of reasoning can be worked out by "brute force," without the quotient theory. However we'll soon show that both NIM and KAYLES are special cases of a more general principle that applies to a wide class of games.

Definition 5.7. Let $(\mathcal{Q}, \mathcal{P})$ be a misère quotient. Let $\mathcal{K}$ be the kernel of $\mathcal{Q}$, with identity z. We say $(\mathcal{Q}, \mathcal{P})$ is **normal** if $\mathcal{K} \cap \mathcal{P} = \{z\}$.

Theorem 5.8. *Let* $(\mathcal{Q}, \mathcal{P}) = \mathcal{Q}(\mathscr{A})$, *with quotient map* $\Phi : \mathscr{A} \to \mathcal{Q}$. *If* $(\mathcal{Q}, \mathcal{P})$ *is normal, then for all* $G, H \in \mathscr{A}$, *we have*

$$\mathscr{G}(G) = \mathscr{G}(H) \quad \Longrightarrow \quad z\Phi(G) = z\Phi(H).$$

Proof. Let $\mathcal{K}$ be the kernel of $\mathcal{Q}$, let z be the identity of $\mathcal{K}$, and fix $Z \in \mathscr{A}$ with $\Phi(Z) = z$.

We'll first show that if $G \in \mathscr{A}$ and $\mathscr{G}(G) = 0$, then $z\Phi(G) = z$. Consider $X = G + Z + Z$. Since Z is a $\mathscr{P}$-position, we have $Z \not\cong 0$, so $X \not\cong 0$. Now the options of X all have the form

$$X' \cong G' + Z + Z \quad \text{or} \quad X' \cong G + Z' + Z,$$

so that $\Phi[X] \subset \mathcal{K}$. Moreover, since $z = \Phi(X)$, it follows from Corollary 5.3 that $z \notin \Phi[X]$. Since z is the unique element of $\mathcal{K} \cap \mathcal{P}$, this implies

$$\Phi[X] \cap \mathcal{P} = \emptyset.$$

Thus every X' is an $\mathscr{N}$-position. But $X \not\cong 0$, so X must be a $\mathscr{P}$-position. Therefore $\Phi(X) \in \mathcal{K} \cap \mathcal{P}$, and so $\Phi(X) = z$.

Now if $\mathscr{G}(G) = \mathscr{G}(H)$, then

$$\mathscr{G}(G + G) = \mathscr{G}(G + H) = 0,$$

so that by the preceding argument,

$$z\Phi(G + G) = z\Phi(G + H) = z.$$

This shows that $zx^2 = zxy$, where $x = \Phi(G)$ and $y = \Phi(H)$. But $\mathcal{K}$ is a group, so x is invertible in $\mathcal{K}$, and we conclude that $zx = zy$. $\square$

Theorem 5.9. *Let $(\mathcal{Q}, \mathcal{P}) = \mathcal{Q}(\mathscr{A})$, and let $\mathbb{Z}_2^n$ be the normal-play quotient of $\mathscr{A}$ (i.e., the group of nim values of $\mathscr{A}$). Then there is a surjective group homomorphism $\rho : \mathbb{Z}_2^n \to \mathcal{K}$.*

Proof. If $\Phi(G), \Phi(H) \in \mathcal{K}$ and $\mathscr{G}(G) = \mathscr{G}(H)$, then Theorem 5.8 implies $\Phi(G) = \Phi(H)$. Therefore we can define ρ by

$$\rho(a) = \Phi(G), \quad \text{for any } G \in \Phi^{-1}[\mathcal{K}] \text{ with } \mathscr{G}(G) = a.$$

This map is clearly surjective. To see that it's a homomorphism, suppose $\mathscr{G}(G) = a$, $\mathscr{G}(H) = b$, and $\Phi(G) = x$, $\Phi(H) = y$, with $x, y \in \mathcal{K}$. Then $\mathscr{G}(G + H) = a \oplus b$, and $\Phi(G + H) = xy$, so it follows that

$$\rho(a \oplus b) = \Phi(G)\Phi(H). \qquad \square$$

In particular, if $(\mathcal{Q}, \mathcal{P})$ is normal, then its kernel $\mathcal{K}$ is isomorphic to $\mathbb{Z}_2^n$, for some n. Moreover, if $\Phi(G) \in \mathcal{K}$ and G is a normal $\mathscr{P}$-position, then G is also a misère $\mathscr{P}$-position. Therefore misère play of $\mathscr{A}$ reduces to normal play, so long as the position remains in the kernel.

Recall the strategy for misère NIM (cf. pages 225 and 234):

> Play just as you would in normal NIM unless your move would leave a position consisting entirely of heaps of size 0 or 1. In that case, play to leave an odd number of heaps of size 1.

We can now generalize to arbitrary rulesets Γ with normal misère quotients:

> Play just as you would in normal Γ unless your move would leave a position whose Φ-value lies outside of $\mathcal{K}$. In that case, pay attention to the fine structure of the misère quotient.

This is perhaps the strongest evidence that the quotient theory is the "right" way to generalize the theory of NIM. Unfortunately it isn't fully general. Although the vast majority of quotients encountered in practice have normal kernels, there do exist abnormal examples. The smallest known of these is

$$G = *(2_{\#}1)(2_{\#}0)4310$$

whose quotient of order 120,

$$\begin{aligned}\mathcal{Q}(G) \cong \langle a, b, c, d, e, f \mid\ & a^2 = 1,\ b^4 = b^2,\ b^2c = c,\ c^3 = ac^2, \\ & b^2d = b^3,\ cd = bc,\ d^2 = 1,\ ce = ab^2e, \\ & b^2e^2 = c^2,\ be^3 = b^3e,\ e^4 = c^2,\ b^2f = ab^2e, \\ & cf = b^2e,\ e^2f = ae^3,\ bef^5 = bef,\ f^6 = f^2\rangle\end{aligned}$$

has kernel isomorphic to $\mathbb{Z}_2 \times \mathbb{Z}_4$. There also exist quotients with elements of odd order; an example can be found in [**PS08a**]. The rabbit hole of the misère theory, it seems, extends deeper still.

The Mex Function

Misère quotients—at least those with normal kernels—provide a reasonably satisfactory generalization of the theory of NIM. A related question is:

> What is the misère analogue of the mex rule?

We'll show that for every set $\mathscr{A}$ with misère quotient $(\mathcal{Q}, \mathcal{P})$, there is a unique partial function

$$F : \mathrm{Pow}(\mathcal{Q}) \to \mathcal{Q}$$

that generalizes the mex rule in the following sense.

Definition 5.10. Let $F : \mathrm{Pow}(\mathcal{Q}) \to \mathcal{Q}$ be a partial function. Suppose that for all games G with $G \subset \mathscr{A}$, we have the following:

(i) $\mathcal{Q}(\mathscr{A}[G]) \cong \mathcal{Q}(\mathscr{A})$ if and only if $F(\Phi[G])$ is defined; and

(ii) if $F(\Phi[G])$ is defined, then $\Phi(G) = F(\Phi[G])$.

Then we say F is a **mex function** for $\mathscr{A}$.

A closely related structure is the **transition algebra** of $\mathscr{A}$.

Definition 5.11. Let $\mathscr{A}$ be a closed set of games with quotient $(\mathcal{Q}, \mathcal{P})$ and quotient map $\Phi : \mathscr{A} \to \mathcal{Q}$. The **transition algebra** of $\mathscr{A}$ is the set of pairs

$$T(\mathscr{A}) = \{(\Phi(G), \Phi[G]) : G \in \mathscr{A}\}.$$

The map $\Psi : \mathscr{A} \to T(\mathscr{A})$ is defined by $\Psi(G) = (\Phi(G), \Phi[G])$.

The transition algebra has a commutative monoid structure with identity given by $\Psi(0) = (1, \emptyset)$:

Lemma 5.12. *If $(x, \mathcal{E}) \in T(\mathscr{A})$ and $(y, \mathcal{F}) \in T(\mathscr{A})$, then so is $(xy, y\mathcal{E} \cup x\mathcal{F})$.*

Proof. If $\Psi(G) = (x, \mathcal{E})$ and $\Psi(H) = (y, \mathcal{F})$, then

$$\Psi(G + H) = (xy, y\mathcal{E} \cup x\mathcal{F}). \qquad \square$$

The projection map $T(\mathscr{A}) \to \mathcal{Q}(\mathscr{A})$, given by $(x, \mathcal{E}) \mapsto x$, gives the following commutative diagram:

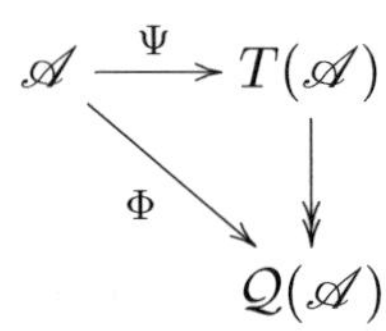

Now suppose that $\Psi(G) = (x, \mathcal{E})$ and $y \in \mathcal{E}$. Then for any $z \in \mathcal{Q}$, it's necessarily true that either $xz \notin \mathcal{P}$ or $yz \notin \mathcal{P}$. To see this, fix an H with $\Phi(H) = z$ and a $G' \in G$ with $\Phi(G') = y$; then $G + H$ and $G' + H$ cannot both be $\mathscr{P}$-positions.

This observation motivates the following definition, which specifies a sort of "largest possible" choice of $\mathcal{E}$ for a given x:

Definition 5.13. Let $x \in \mathcal{Q}$. The **meximal set** of x, denoted by $\mathcal{M}_x$, is given by

$$\mathcal{M}_x = \{y \in \mathcal{Q} : \text{there is no } z \in \mathcal{Q} \text{ such that both } xz, yz \in \mathcal{P}\}.$$

Lemma 5.14. *If $\Phi(G) = x$, then $\Phi[G] \subset \mathcal{M}_x$.*

Proof. Let $y \in \Phi[G]$, so that $\Phi(G') = y$ for some $G' \in G$. Then for any $Z \in \mathscr{A}$, one of $G+Z$ or $G'+Z$ must be an $\mathscr{N}$-position, so either $x\Phi(Z) \notin \mathcal{P}$ or $y\Phi(Z) \notin \mathcal{P}$. Therefore $y \in \mathcal{M}_x$. $\square$

Theorem 5.15 (Generalized Mex Rule)**.** *Let $T = T(\mathscr{A})$. Suppose that $G \subset \mathscr{A}$, with $G \not\cong 0$. Then the following are equivalent, for $x \in \mathcal{Q}$:*

(i) *$\mathcal{Q}(\mathscr{A}[G]) \cong \mathcal{Q}(\mathscr{A})$ and $\Phi(G) = x$.*

(ii) *The following two conditions are satisfied:*

 (a) *$\Phi[G] \subset \mathcal{M}_x$; and*

(b) *for each* $(y, \mathcal{E}) \in T$ *and each* $n \geq 1$ *such that* $x^n y \notin \mathcal{P}$*, either:* $x^n y' \in \mathcal{P}$ *for some* $y' \in \mathcal{E}$*; or else* $x^{n-1}x'y \in \mathcal{P}$ *for some* $x' \in \Phi[G]$.

Proof. (i) $\Rightarrow$ (ii): Condition (a) is just Lemma 5.14. For (b), fix $Y \in \mathscr{A}$ with $\Phi(Y) = y$ and $\Phi[Y] = \mathcal{E}$. If $x^{n+1}y \notin \mathcal{P}$ for some $n \geq 0$, then $(n+1) \cdot G + Y$ is an $\mathscr{N}$-position. Since $G \not\equiv 0$, it follows that

$$\text{either} \quad \text{some } (n+1) \cdot G + Y' \quad \text{or} \quad \text{some } n \cdot G + G' + Y \quad \text{is a } \mathscr{P}\text{-position.}$$

Therefore $x^{n+1}\Phi(Y') \in \mathcal{P}$ or $x^n \Phi(G')y \in \mathcal{P}$.

(ii) $\Rightarrow$ (i): Fix $H \in \mathscr{A}$ with $\Phi(H) = x$. We will show that

$$G \equiv H \pmod{\mathscr{A}[G]}.$$

As in Theorem 4.4 on page 256, it suffices to show that

$$(\dagger) \qquad o(nG + X) = \mathscr{N} \quad \text{if and only if} \quad o(nH + X) = \mathscr{N}$$

for each $n \geq 0$ and $X \in \mathscr{A}$. The $\Rightarrow$ direction of (†) is similar to Theorem 4.4. For the analogue of Case 2, suppose that

$$o\big((n-1)G + G' + X\big) = \mathscr{P}$$

for some $G' \in G$. Then by induction on n, we can assume that

$$o\big((n-1)H + G' + X\big) = \mathscr{P}.$$

Put $K = (n-1)H + X$. Then $\Phi(G')\Phi(K) \in \mathcal{P}$, so by condition (a), $x\Phi(K) \notin \mathcal{P}$. Therefore $o(nH + X) = \mathscr{N}$.

For the $\Leftarrow$ direction of (†), suppose that $o(nH + X) = \mathscr{N}$, with $n \geq 1$. Put $(y, \mathcal{E}) = \Psi(X)$. Then $x^n y \notin \mathscr{P}$, so by condition (b), either $x^n y' \in \mathscr{P}$ for some $y' \in \mathcal{E}$ or else $x^{n-1}x'y \in \mathscr{P}$ for some $x' \in \Phi[G]$. But the former implies $o(nH + X') = \mathscr{P}$ for some $X' \in X$, so that by induction $o(nG + X') = \mathscr{P}$, while the latter implies

$$o\big((n-1)H + G' + X\big) = \mathscr{P}$$

for some $G' \in G$, so by induction on n

$$o\big((n-1)G + G' + X\big) = \mathscr{P}.$$

In both cases, we have $o(nG + X) = \mathscr{N}$. □

Corollary 5.16. *Every set of games* $\mathscr{A}$ *admits a unique mex function, depending only on* $T(\mathscr{A})$.

Proof. Condition (ii) of the generalized mex rule depends only on $\Phi[G]$ and $T(\mathscr{A})$; therefore the same is true for condition (i). □

Exercises

5.1 Prove that if $\mathcal{K}$ is the kernel of $(\mathcal{Q}, \mathcal{P})$, then $\mathcal{K} \cap \mathcal{P} \neq \emptyset$.

5.2 $(\mathcal{Q}, \mathcal{P})$ is **regular** if $\mathcal{K} \cap \mathcal{P}$ is a singleton. Prove that Theorems 5.8 and 5.9 remain true if "normal" is weakened to "regular."

5.3 Describe the transition algebras and mex functions for $\mathrm{cl}(*2)$ and $\mathrm{cl}(*2_{\#})$.

5.4 If $G \subset \mathscr{A}$ and $\Phi[G] = \mathcal{M}_x$ for some $x \in \mathscr{A}$, then $\mathcal{Q}(\mathscr{A}[G]) \cong \mathcal{Q}(\mathscr{A})$ and $\Phi(G) = x$.

5.5 Suppose that $(\mathcal{Q}, \mathcal{P})$ is normal with kernel $\mathcal{K}$. If $G \subset \mathscr{A}$ and $\Phi[G] \subsetneq \mathcal{K}$, then $\mathcal{Q}(\mathscr{A}[G]) \cong \mathcal{Q}(\mathscr{A})$ and $\Phi(G) \in \mathcal{K}$.

5.6 The *Mex Interpolation Principle.* Let $G \subset \mathscr{A}$, and suppose that there is some $H \in \mathscr{A}$ such that $\Phi[H] \subset \Phi[G] \subset \mathcal{M}_x$, where $x = \Phi(H)$. Prove that $\mathcal{Q}(\mathscr{A}[G]) \cong \mathcal{Q}(\mathscr{A})$ and $\Phi(G) = x$. Show furthermore that $\mathscr{A}$ and $\mathscr{A}[G]$ have identical mex functions.

Notes

This section largely follows the developments in [**PS08a**], with some material drawn from [**Sie13b**].

The classification theory has been extended by Siegel. With some effort, it is possible to show that $\mathcal{R}_8$ is the only quotient of order 8 and $\mathcal{T}_3$ is the only quotient of order 10. There are multiple quotients of order 12, and thereafter the quotients proliferate rapidly. The details of these and other results are found in [**Sie13b**].

Many questions about finite misère quotients remain open, including:

Question. Let $(\mathcal{Q}, \mathcal{P}) = \mathcal{Q}(\mathscr{A})$, with quotient map $\Phi : \mathscr{A} \to \mathcal{Q}$, and let $G, H \in \mathscr{A}$. Does $\Phi(G) = \Phi(H)$ necessarily imply $\mathscr{G}(G) = \mathscr{G}(H)$?

Question. If $(\mathcal{Q}, \mathcal{P})$ is a misère quotient and $\mathcal{A}$ is a maximal subgroup of $\mathcal{Q}$, must $\mathcal{A} \cap \mathcal{P}$ be nonempty?

Although the theory of finite quotients remains incomplete, still less is known about the structure of infinite misère quotients. Such quotients are commonplace: for example, a proof that $\mathcal{Q}(*(2_{\#}0)0)$ is infinite can be found in [**PS08a**]. Now every finitely generated monoid is finitely presented (cf. Rédei's Theorem, Theorem B.4.8 on page 472), so every finitely generated set of games $\mathscr{A}$ necessarily has a finitely presented misère quotient. In particular, this is true for every partial quotient $\mathcal{Q}_n(\Gamma)$ of every heap ruleset Γ. However, a more general structure theory for such quotients remains elusive. Even a computational finiteness test remains out of reach:

Open Problem. Specify an algorithm to determine whether or not $\mathcal{Q}(\mathscr{A})$ is infinite, given a finitely generated set of impartial games $\mathscr{A}$.

(The input to such an algorithm is a finite set of generators for $\mathscr{A}$.) The existing algorithms for calculating misère quotients [**PS08b**] *assume* that $\mathcal{Q}(\mathscr{A})$ is finite, terminating gracefully when a finite quotient is found but going into an infinite loop when $\mathcal{Q}(\mathscr{A})$ is infinite. Here's a more difficult problem:

Open Problem. Specify an algorithm to calculate a presentation for $\mathcal{Q}(\mathscr{A})$, given a finitely generated set of impartial games $\mathscr{A}$.

Against this bleak landscape are a handful of positive results. Plambeck and Siegel [**PS08b**] isolated the infinite quotients of **0.26** and **4.7**, building on earlier work of Allemang [**All84**]. These examples are not finitely generated and exhibit a little-understood phenomenon known as **algebraic periodicity**. Weimerskirch [**Wei07**] isolated the infinite partial quotient $\mathcal{Q}_7(\mathbf{0.3122})$ but left open the full solution to misère **0.3122**.

An exciting new direction of research was recently proposed by Alan Guo and Ezra Miller [**GM11, GM12**], who defined a superset of octal games known as **lattice games**. Using techniques from combinatorial commutative algebra [**MS05, KM**], they showed that certain lattice games admit a particular type of algebraic decomposition known as **affine stratifications**, from which a finitely representable winning strategy can be recovered. In particular, affine stratifications provide succinct encodings for certain infinite misère quotients.

Guo and Miller initially conjectured that *all* lattice games admit affine stratifications, but this conjecture was quickly shown to be false by Alex Fink [**Fin12**]. Very recently, Miller proved that lattice games with finite misère quotient *do* admit affine stratifications [**Mil13**]. These results suggest that the Guo–Miller theory holds promise for bridging this gap: either by yielding concise solutions for octal games with infinite quotients or by clarifying their structure and pointing the way towards hardness results.

6. Partizan Misère Canonical Form

In this section we show that canonical forms exist for short partizan games in misère play. The proof is an elegant synthesis of the corresponding proofs for normal-play partizan games (Section II.2) and misère-play impartial ones (Section 3 of this chapter).

As in normal play, there are four outcome classes, denoted in this section by $\mathscr{L}$, $\mathscr{N}$, $\mathscr{P}$, and $\mathscr{R}$. We also write $o(G)$ for the misère outcome class of G. Note that this is the same notation used in Chapter II to discuss normal play. This overlap should cause no confusion; in the rare cases when it's necessary to discuss normal and misère outcomes together, we disambiguate by writing $o^+(G)$ for normal-play outcome class and $o^-(G)$ for misère-play outcome class.

As always, write

$$G = H \quad \text{if} \quad o(G + X) = o(H + X) \text{ for all } X,$$
$$G \geq H \quad \text{if} \quad o(G + X) \geq o(H + X) \text{ for all } X,$$

and remember that throughout this section, $o(G+X)$ and $o(H+X)$ denote misère outcomes, and X ranges over all partizan games.

We begin with a useful test for the inequality $G \geq H$.

Proposition 6.1. *$G \geq H$ if and only if, for all short games X:*

(i) *if $o(H+X) \geq \mathscr{P}$, then $o(G+X) \geq \mathscr{P}$; and*

(ii) *if $o(H+X) \geq \mathscr{N}$, then $o(G+X) \geq \mathscr{N}$.*

Note that $o(G) \geq \mathscr{P}$ is equivalent to "Left can win playing second on G," and $o(G) \geq \mathscr{N}$ is equivalent to "Left can win playing first on G."

Proof of Proposition 6.1. $\Rightarrow$ is immediate. For the converse, we must show that $o(G+X) \geq o(H+X)$, for all X. If $o(H+X) = \mathscr{R}$, then there is nothing to prove; if $o(H+X) = \mathscr{P}$ or $\mathscr{N}$, it is immediate from (i) or (ii), respectively. Finally, if $o(H+X) = \mathscr{L}$, then by (i) and (ii) we have $o(G+X) \geq$ both $\mathscr{P}$ and $\mathscr{N}$, whence $o(G+X) = \mathscr{L}$. □

Ends and Adjoints

The **adjoint** G° can be viewed variously as a partizan analogue of the mate G^- and a misère analogue of the negative $-G$.

Definition 6.2. G is a **Left end** (resp. **Right end**) if G has no Left (resp. Right) option.

Definition 6.3. The **adjoint** of G, denoted by G°, is defined by

$$G^\circ = \begin{cases} * & \text{if } G \cong 0; \\ \{(G^R)^\circ \mid 0 \} & \text{if } G \not\cong 0 \text{ and } G \text{ is a Left end;} \\ \{ 0 \mid (G^L)^\circ\} & \text{if } G \not\cong 0 \text{ and } G \text{ is a Right end;} \\ \{(G^R)^\circ \mid (G^L)^\circ\} & \text{otherwise.} \end{cases}$$

Proposition 6.4. *$G + G^\circ$ is a $\mathscr{P}$-position, for all G.*

Proof. By symmetry, it suffices to show that Left can win $G + G^\circ$ playing second. Now $G + G^\circ$ cannot be a Right end, since the definition of G° ensures that it has at least one Right option. So it suffices to show that Left has a winning response to every move by Right. There are two cases:

- If Right moves to $G^R + G^\circ$ or $G + (G^L)^\circ$, Left has a mirror image move on the other component, which wins by induction on G.
- If G is a Left end and Right moves to $G + 0$, then Left has no move and so wins *a priori*. □

Theorem 6.5. *If $G \not\geq H$, then:*

(a) *There is some T such that $o(G+T) \leq \mathscr{P}$ and $o(H+T) \geq \mathscr{N}$.*

(b) *There is some U such that $o(G+U) \leq \mathscr{N}$ and $o(H+U) \geq \mathscr{P}$.*

Proof. Certainly one of (a) or (b) must hold, so it suffices to show that (a) $\Rightarrow$ (b) and (b) $\Rightarrow$ (a). The arguments are identical, so we will show that (a) $\Rightarrow$ (b).

Fix T as in (a), and put

$$U = \{(H^R)^\circ \mid T\}.$$

Now from $G + U$, Right has a winning move to $G + T$, so $o(G + U) \leq \mathscr{N}$. Next consider $H + U$. It is certainly not a Right end, since U has T as a Right option. Now if Right moves to $H^R + U$, Left has a winning response to $H^R + (H^R)^\circ$. If instead Right moves to $H + T$, then Left wins *a priori* since $o(H + T) \geq \mathscr{N}$. Therefore $o(H + U) \geq \mathscr{P}$. □

Theorem 6.6. *If H is a Left end and G is not, then $G \not\geq H$.*

Proof. Put

$$T = \Big\{(H^R)^\circ \;\Big\|\; \{\ \mid (G^L)^\circ\}\Big\}.$$

First consider $H + T$. If Right moves to $H^R + T$, then Left has a winning response to $H^R + (H^R)^\circ$. If instead Right moves to $H + \{\ \mid (G^L)^\circ\}$, then Left wins outright, since (by the assumption on H) both components are Left ends. Therefore $o(H + T) \geq \mathscr{P}$.

Next consider $G + T$. Right has a move to $G + \{\ \mid (G^L)^\circ\}$, from which Left's only options have the form $G^L + \{\ \mid (G^L)^\circ\}$. There must be at least one such option, by the assumption on G; and each such option has a mirror-image response by Right, to $G^L + (G^L)^\circ$. This shows that $o(G + T) \leq \mathscr{N}$.

Therefore $o(G + T) \not\geq o(H + T)$, so in fact $G \not\geq H$. □

A startling corollary of this theorem shows that *every* nonempty game is unequal to 0:

Corollary 6.7. *If $G \not\cong 0$, then $G \neq 0$, for all G.*

Proof. If $G \not\cong 0$, then either G is not a Left end or else G is not a Right end. But if G is not a Left end, then $G \not\geq 0$, while if G is not a Right end, then $G \not\leq 0$. □

Dominated and Reversible Options

The following definition is exactly the same as in the normal-play case (Definition II.2.3 on page 64). Here, all inequalities represent misère comparison.

Definition 6.8. Let G be a short game.

(a) A Left option G^L is **dominated (by $G^{L'}$)** if $G^{L'} \geq G^L$ for some other Left option $G^{L'}$.

(b) A Right option G^R is **dominated (by $G^{R'}$)** if $G^{R'} \leq G^R$ for some other Right option $G^{R'}$.

(c) A Left option G^L is **reversible (through G^{LR})** if $G^{LR} \leq G$ for some Right option G^{LR}.

(d) A Right option G^R is **reversible (through G^{RL})** if $G^{RL} \geq G$ for some Left option G^{RL}.

Theorem 6.9 (cf. Theorem II.2.4 on page 65). *Let G be a short game and suppose G' is obtained from G by removing some dominated Left option G^{L_1}. Then $G' = G$.*

Proof. Since the Left options of G' are a subset of those of G and since G' still has at least one Left option (namely, the dominating option G^{L_2}), we certainly have $G' \leq G$. Thus it suffices to show that $G' \geq G$.

By Proposition 6.1, it suffices to show that for all X:

(a) If $o(G+X) \geq \mathscr{P}$, then $o(G'+X) \geq \mathscr{P}$.

(b) If $o(G+X) \geq \mathscr{N}$, then $o(G'+X) \geq \mathscr{N}$.

The proof is by induction on X.

(a) Suppose $o(G+X) \geq \mathscr{P}$. Then every $o(G^R+X) \geq \mathscr{N}$ and every $o(G+X^R) \geq \mathscr{N}$. Now G and G' have exactly the same Right options, so every $o\big((G')^R+X\big) \geq \mathscr{N}$. Also, by induction on X, we have

$$o(G'+X^R) = o(G+X^R) \geq \mathscr{N}$$

for all X^R. This shows that $o(G'+X) \geq \mathscr{P}$, except possibly in the case where $G'+X$ is a Right end. But since G and G' have the same Right options, this would imply that $G+X$ is a Right end, contradicting the assumption $o(G+X) \geq \mathscr{P}$.

(b) Suppose $o(G+X) \geq \mathscr{N}$. We know that $G+X$ is not a Left end (since G is assumed to have a Left option), so either some $o(G+X^L) \geq \mathscr{P}$ or some $o(G^L+X) \geq \mathscr{P}$. If $o(G+X^L) \geq \mathscr{P}$, then by induction $o(G'+X^L) \geq \mathscr{P}$, so we are done. If $o(G^L+X) \geq \mathscr{P}$ and G^L is a Left option of G', then the conclusion is immediate. So the only remaining case is when $G^L = G^{L_1}$. But since $G^{L_2} \geq G^{L_1}$, we have

$$o(G^{L_2}+X) \geq o(G^{L_1}+X) \geq \mathscr{P}.$$

In all cases, $o(G'+X) \geq \mathscr{N}$. □

Theorem 6.10 (cf. Theorem II.2.5 on page 65). *Let G be a short game, and suppose some Left option G^{L_1} is reversible through $G^{L_1R_1}$. Let*

$$G' = \big\{G^{L_1R_1L}, G^{L'} \bigm| G^R\big\},$$

where $G^{L_1R_1L}$ *ranges over all Left options of* $G^{L_1R_1}$ *and* $G^{L'}$ *ranges over all Left options of* G except *for* G^{L_1}*. Then* $G' = G$*.*

Proof. By Proposition 6.1, it suffices to show that for all X:

(a) $o(G+X) \geq \mathscr{P}$ if and only if $o(G'+X) \geq \mathscr{P}$.

(b) $o(G+X) \geq \mathscr{N}$ if and only if $o(G'+X) \geq \mathscr{N}$.

The proof is by induction on X.

(a) If either $G+X$ or $G'+X$ is a Right end, then both must be, since G and G' have the same Right options. If both are Right ends, then necessarily $o(G+X) \leq \mathscr{N}$ and $o(G'+X) \leq \mathscr{N}$, so (a) is satisfied. So assume neither is a Right end; then by induction on X,

$$\begin{aligned} o(G+X) \geq \mathscr{P} &\text{ iff every } o(G^R+X) \geq \mathscr{N} \text{ and every } o(G+X^R) \geq \mathscr{N} \\ &\text{ iff every } o\big((G')^R+X\big) \geq \mathscr{N} \text{ and every } o(G'+X^R) \geq \mathscr{N} \\ &\text{ iff } o(G'+X) \geq \mathscr{P}. \end{aligned}$$

(b) First suppose $o(G+X) \geq \mathscr{N}$. We know that G is not a Left end (since it has G^{L_1} as a Left option), so either some $o(G+X^L) \geq \mathscr{P}$ or else some $o(G^L+X) \geq \mathscr{P}$. There are three subcases.

- If $o(G+X^L) \geq \mathscr{P}$, then $o(G'+X^L) \geq \mathscr{P}$, by induction on X.
- If $o(G^L+X) \geq \mathscr{P}$ for $G^L \neq G^{L_1}$, then G^L is also a Left option of G' and the conclusion is immediate.
- Suppose instead that $o(G^{L_1}+X) \geq \mathscr{P}$. Then in particular,

  $$o(G^{L_1R_1}+X) \geq \mathscr{N}.$$

  So either $o(G^{L_1R_1L}+X) \geq \mathscr{P}$ or $o(G^{L_1R_1}+X^L) \geq \mathscr{P}$. In the first case we are done, since $G^{L_1R_1L}$ is a Left option of G'. But in the second case, we have $G \geq G^{L_1R_1}$ (by hypothesis) so

  $$o(G+X^L) \geq o(G^{L_1R_1}+X^L) \geq \mathscr{P}$$

  and therefore $o(G'+X^L) \geq \mathscr{P}$, by induction on X.

Finally, suppose $o(G'+X) \geq \mathscr{N}$. Then again we have three subcases: either $o(G^L+X) \geq \mathscr{P}$ or $o(G'+X^L) \geq \mathscr{P}$ or $o(G^{L_1R_1L}+X) \geq \mathscr{P}$. The first two subcases are just the same as before, so suppose $o(G^{L_1R_1L}+X) \geq \mathscr{P}$. Then in particular, $o(G^{L_1R_1}+X) \geq \mathscr{N}$. But $G \geq G^{L_1R_1}$, so

$$o(G+X) \geq o(G^{L_1R_1}+X) \geq \mathscr{N}.$$

□

Misère Canonical Form

Definition 6.11 (cf. Definition II.2.6 on page 66). A short game K is in **misère canonical form** if no subposition of K has any (misère) dominated or reversible options.

Theorem 6.12. *For every short game G, there exists a game $K = G$ in misère canonical form.*

Proof. Identical to the corresponding proof for normal play (Theorem II.2.7 on page 66). □

Lemma 6.13 (cf. Lemma II.2.8 on page 67 and Lemma 3.9 on page 245). *Suppose that neither G nor H has any dominated or reversible options, and assume $G = H$. Then for every G^L, there is an H^L such that $G^L = H^L$, and vice versa; and likewise for Right options.*

The proof of Lemma 6.13 is somewhat more involved than in normal play, and it relies on the following generalization of the impartial theory.

Definition 6.14 (cf. Definition 3.4 on page 244). We say G is **downlinked** to H (by T), and we write $G \ltimes H$, if

$$o(G+T) \leq \mathscr{P} \text{ and } o(H+T) \geq \mathscr{P} \text{ for some short game } T.$$

Likewise, we say G is **uplinked** to H (by T), and we write $G \rtimes H$, if

$$o(G+T) \geq \mathscr{P} \text{ and } o(H+T) \leq \mathscr{P} \text{ for some short game } T.$$

Theorem 6.15 (cf. Theorem 3.5 on page 244). *$G \ltimes H$ if and only if no $G^L \geq H$ and $G \geq$ no H^R.*

Proof. Suppose T downlinks G to H, so $o(G+T) \leq \mathscr{P}$ and $o(H+T) \geq \mathscr{P}$. Then necessarily $o(G^L+T) \leq \mathscr{N}$ and $o(H^R+T) \geq \mathscr{N}$, so T witnesses both $G^L \not\geq H$ and $G \not\geq H^R$.

Conversely, suppose that no $G^L \geq H$ and $G \geq$ no H^R. Then for each G_i^L, Theorem 6.5 yields an X_i such that

$$o(G_i^L + X_i) \leq \mathscr{P} \text{ and } o(H + X_i) \geq \mathscr{N}.$$

Likewise, for each H_j^R, there is some Y_j such that

$$o(G + Y_j) \leq \mathscr{N} \text{ and } o(H_j^R + Y_j) \geq \mathscr{P}.$$

Put

$$T = \begin{cases} * & \text{if } G \cong H \cong 0; \\ \{0 \mid (H^L)^\circ\} & \text{if } G \cong 0 \text{ and } H \text{ is a nonzero Right end;} \\ \{(G^R)^\circ \mid 0\} & \text{if } H \cong 0 \text{ and } G \text{ is a nonzero Left end;} \\ \{Y_j, (G^R)^\circ \mid X_i, (H^L)^\circ\} & \text{otherwise.} \end{cases}$$

We claim that G is downlinked to H by T. We will show that $o(G+T) \leq \mathscr{P}$; the proof that $o(H+T) \geq \mathscr{P}$ is identical.

We first show that $G+T$ has a Left option. If G has a Left option, this is automatic. If G or H has a Right option, then T necessarily has a Left option. This exhausts every case except when $G \cong 0$ and H is a Right end; but then Left's move to 0 is built into the definition of T.

Thus $G+T$ is not a Left end, and it therefore suffices to show that every Left option is losing. If Left moves to $G_i^L + T$, Right can respond to $G_i^L + X_i$, which wins by choice of X_i. If Left moves to $G + (G^R)^\circ$, Right can respond to $G^R + (G^R)^\circ$, winning. Left's move to $G + Y_j$ loses automatically, by choice of Y_j. The only remaining possibility is Left's additional move to 0 in the first two cases of the definition of T. But that move is only available when $G \cong 0$, so it ends the game immediately. □

Theorem 6.16 (cf. Theorem 3.6 on page 245). *$G \geq H$ if and only if the following four conditions hold:*

(i) *G is downlinked to no H^L.*

(ii) *No G^R is downlinked to H.*

(iii) *If H is a Left end, then so is G.*

(iv) *If G is a Right end, then so is H.*

Proof. For $\Rightarrow$ (i), fix any game T. If $o(G+T) \leq \mathscr{P}$, then by hypothesis, $o(H+T) \leq \mathscr{P}$ as well. Therefore $o(H^L+T) \leq \mathscr{N}$, so T cannot downlink G to H^L. $\Rightarrow$ (ii) is similar, and $\Rightarrow$ (iii) and (iv) are just restatements of Theorem 6.6 (and its mirror image).

We now prove $\Leftarrow$. By Proposition 6.1, we must show that for all T:

- if $o(H+T) \geq \mathscr{P}$, then $o(G+T) \geq \mathscr{P}$; and
- if $o(H+T) \geq \mathscr{N}$, then $o(G+T) \geq \mathscr{N}$.

The proof proceeds by induction on T. First suppose (for contradiction) that $o(H+T) \geq \mathscr{P}$ but $o(G+T) \leq \mathscr{N}$. Then either $o(G+T^R) \leq \mathscr{P}$ or $o(G^R+T) \leq \mathscr{P}$ or else $G+T$ is a Right end. If $o(G+T^R) \leq \mathscr{P}$, then by induction on T we have $o(H+T^R) \leq \mathscr{P}$, contradicting the assumption that $o(H+T) \geq \mathscr{P}$. If $o(G^R+T) \leq \mathscr{P}$, then T downlinks G^R to H, contradicting (ii). Finally, if $G+T$ is a Right end, then in particular G is a Right end, so by (iv), H is a Right end. Therefore $H+T$ is a Right end, contradicting the assumption that $o(H+T) \geq \mathscr{P}$.

This shows that $o(H+T) \geq \mathscr{P}$ implies $o(G+T) \geq \mathscr{P}$. The proof that $o(H+T) \geq \mathscr{N}$ implies $o(G+T) \geq N$ is identical, with (i) and (iii) used in place of (ii) and (iv). □

Proof of Lemma 6.13. Fix G^L. Since $G \le H$, Theorem 6.16 implies that G^L is not uplinked to H. By Theorem 6.15, either $G^L \le H^L$ for some H^L or else $G^{LR} \le H$ for some G^{LR}. But the latter would imply

$$G^{LR} \le H \le G,$$

contradicting the assumption that G has no reversible options. Therefore $G^L \le H^L$ for some H^L. By exactly the same argument, there is some $G^{L'}$ such that $H^L \le G^{L'}$. But now we have

$$G^L \le H^L \le G^{L'}.$$

Since G has no dominated options, this implies $G^L = H^L = G^{L'}$. The other cases follow by symmetry. □

Theorem 6.17 (Simplest Form Theorem)**.** *Suppose that G and H are in misère canonical form, and assume $G = H$. Then $G \cong H$.*

Proof. Identical to the proof of Theorem II.2.9 on page 67. □

Partizan Misère Quotients

The misère quotient construction (described in Sections 4 and 5 for impartial games) also has a partizan analogue. Given a closed set $\mathscr{A}$ of short games, we define

$$G \equiv H \pmod{\mathscr{A}} \quad \text{if} \quad o(G+X) = o(H+X) \text{ for all } X \in \mathscr{A}$$

just as in the impartial case. Writing $\mathscr{Q}$ for the corresponding quotient monoid and $\Phi : \mathscr{A} \to \mathscr{Q}$ for the quotient map, we have an induced **outcome tetrapartition**

$$\Pi : \mathscr{Q} \to \{\mathscr{L}, \mathscr{N}, \mathscr{P}, \mathscr{R}\}$$

given by

$$\Pi(\Phi(G)) = o(G)$$

for all G. (This is well-defined, since $\Phi(G) = \Phi(H)$ implies $o(G) = o(H)$.) This quotient has the structure of a partially ordered monoid, with the ordering given by

$$x \ge y \quad \text{if} \quad \Pi(xz) \ge \Pi(yz) \text{ for all } z \in \mathscr{Q}.$$

In an intriguing twist, the monoid often has order relations that are not present in the full universe of all partizan games. Some examples are explored in the exercises.

The theory of partizan misère quotients is ripe for further exploration:

Open Problem. Develop a structure theory for partizan misère quotients.

Significant partial progress has been made toward this goal, which is discussed in the notes below.

Exercises

6.1 *Games born by day n.* Prove that:
   (a) All four games born by day 1 are pairwise incomparable.
   (b) If G and H are born by day 2 and $G \not\geq H$, then $G \neq H$. Conclude that there are 256 pairwise unequal games born by day 2. (Contrast with normal play: Theorem III.1.2 on page 154.)
   (c) There exist nonisomorphic games born on day 3 that are equal.

6.2 True or false: If $G \geq H$ in misère play, then $G \geq H$ in normal play.

6.3 Write $G \geq_S H$ if the following four conditions hold: (i) every H^L is equal to some G^L; (ii) every G^L is equal to some H^L; (iii) if H is a Left end, then so is G; (iv) if G is a Right end, then so is H. Prove that for all G and H:
   (a) If $G \geq_S H$, then $G \geq H$.
   (b) If G and H are born by day 2 and $G \geq H$, then $G \geq_S H$.
   (c) There exist G and H born on day 3 with $G \geq H$, but $G \not\geq_S H$.

6.4 Let x and y be numbers in normal-play canonical form. If $x > y$ in normal play, then $* : x > * : y$ in misère play.

6.5 Let $\mathscr{A} = \mathrm{cl}(1, \overline{1})$. Show that $\mathcal{Q}(\mathscr{A}) \cong \mathbb{Z}$ as a partially ordered monoid.

6.6 *Dicotic games in misère play.* Let $\mathscr{A} = \tilde{\mathbb{G}}^0$, the set of all short dicotic games (as defined in Section II.1).
   (a) $* + * \equiv 0 \pmod{\mathscr{A}}$.
   (b) If G is dicotic and $H + \overline{H}$ is a misère $\mathscr{N}$-position for every subposition H of G, then $G + \overline{G} \equiv 0 \pmod{\mathscr{A}}$.
   (c) A game G is **binary** if every nonempty subposition of G has *exactly* one move for each player. Suppose G is binary, and assume every alternating run of every subposition of G has length at most 3. Show that $G + \overline{G} \equiv 0$ $\pmod{\mathscr{A}}$. (Use (b).)

6.7 G is said to be a **dead end** if either: every subposition of G is a Left end; or every subposition of G is a Right end; or G is not an end and every proper subposition of G is a dead end. Let $\mathscr{A}$ be the set of all dead ends. Prove that:
   (a) $\mathscr{A}$ is closed under disjunctive sum.
   (b) The following are dead ends: every dicotic game; every HACKENBUSH position; every DOMINEERING position.
   (c) If $G \in \mathscr{A}$ and G is an end, then $G + \overline{G} \equiv 0 \pmod{\mathscr{A}}$.

Notes

For decades, partizan games in misère play were considered essentially intractable. Then in 2007, Mesdal and Ottaway [**MO07**] showed that every nonempty game is distinct from 0 (our Corollary 6.7), suggesting that a coherent theory is possible. Later the same year, the full theory was isolated by Aaron Siegel [**Sie13a**] in essentially the form presented here. Siegel also obtained an exact count of 256 partizan misère games born by day 2 (Exercise 6.1) and an upper bound of 2^{183} games born by day 3.

These breakthroughs ushered in a flurry of further advances. In the same paper that introduced the full canonical theory, Siegel included a brief note suggesting the partizan misère quotient construction described on page 276. The study of such quotients was enthusiastically pursued by Meghan Allen [**All09, All13**], who introduced a number of examples. Allen placed a particular emphasis on the special case $\mathscr{A} = \tilde{\mathbb{G}}^0$, the set of all dicotic games, and showed that the quotient theory of such games is intermediate between the impartial and partizan cases and has many nice properties (see Exercise 6.6).

Several years later, Rebecca Milley [**Mill13**] carried these efforts still further. Milley explored the quotient of $\tilde{\mathbb{G}}^0$ in more detail and introduced a new class of games, the **dead ends** (Exercise 6.7), which were further investigated by Milley and Renault [**MR13**]. Some additional results on dicotic games were also obtained by McKay, Milley, and Nowakowski [**MMN13**], including a complete theory of sums of games $* : x$, where x is a number in (normal-play) canonical form.

Similarly, Milley, Nowakowski, and Ottaway studied the class of **alternating** games, in which every G^L is a Left end and every G^R is a Right end [**MNO12**]. These games were originally introduced by Ottaway in the normal-play context (cf. Exercise II.3.16 on page 82) and are also discussed in Milley's thesis [**Mill13**].

It is clear from the range of these discoveries that the misère theory of partizan games has been blown wide open. There is a whole spectrum of misère theories obtained by restricting the class of partizan games in various ways, each richer than the curiously impenetrable canonical theory. One can safely presume that much about them remains to be discovered.

Open Problem. Extend the theory of partizan games in misère play.

Exercise 6.4: McKay, Milley, and Nowakowski.

Exercise 6.6: Meghan Allen.

Exercise 6.7: Rebecca Milley.

Chapter VI

Loopy Games

Most of the theory we've developed so far is specific to **short games**, in which play is guaranteed to end after just finitely many moves. However, many games permit repetition, including well-known examples such as CHESS, GO, and FOX AND GEESE. In this chapter the focus is on extending the theory to such **loopy games**.

We briefly studied impartial loopy games in Section IV.4; there they succumbed to a remarkably simple generalization of the Sprague–Grundy theory. It turns out that there's a similarly straightforward extension of the *partizan* theory to a special class of loopy games known as **stoppers**. The general case appears more difficult, and a fully satisfactory theory has yet to be worked out.

Section 1 of this chapter is an overview of the general theory of loopy games. Stoppers are studied in detail in Sections 2 and 3. In Section 4, we introduce a broader class of games, the **stopper-sided games**, to which much of the theory can still be applied. Finally, Section 5 describes a structure theory for certain classes of stoppers.

1. Coping with Cycles

Let's take a closer look at FOX AND GEESE and some of its endgame situations. (It might be helpful to review the introduction to FOX AND GEESE on pages 26–28.)

Recall that FOX AND GEESE is played on an 8×8 checkerboard, with Left commanding several **geese** against Right's single **fox**. The geese move

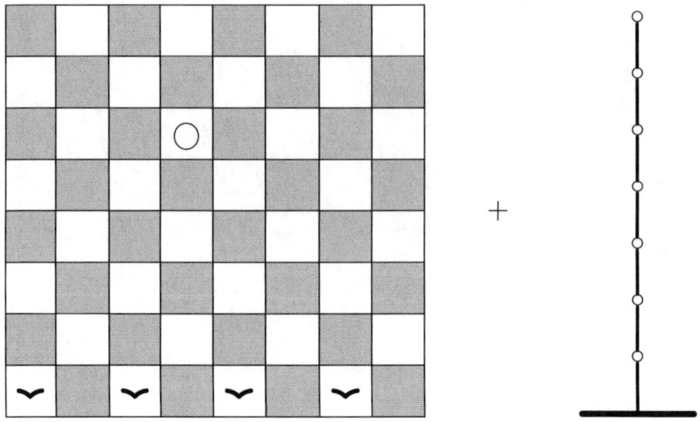

Figure 1.1. What is the value of an escaped fox?

one square diagonally southeast or southwest; the fox moves one square diagonally in any direction. The usual starting position is shown in Figure I.4.1 on page 27.

Figure 1.1 shows a typical FOX AND GEESE endgame in a sum with a large HACKENBUSH stalk. In this position the fox has escaped. The geese are unable to move, so Left's only option is to play on the HACKENBUSH component and remove one of her edges. But she can only do this a finite number of times, while Right is free to continue moving the fox around indefinitely. Therefore Left will eventually run out of moves and lose (no matter who moves first). The same argument works with an arbitrarily long HACKENBUSH stalk, which suggests that

$$(\text{escaped fox}) < n$$

for any integer n, and consequently

$$(\text{escaped fox}) < G$$

for any loopfree G.

Another endgame position is shown in Figure 1.2 on the facing page. The fox is trapped against the corner of the board, and Left has the option to end the game in one move at any time. Meanwhile Right is constrained to shuttle the fox indefinitely between the two accessible squares.

Let's call this position T (for a *T*rapped fox). What is the value of T? Certainly $T > 0$ since Left can win any time she chooses. But $T < 1$, since on $T - 1$ Right can postpone moving on -1 until after Left traps the fox. The same argument shows that $T < 2^{-n}$ for any n, so in fact T is an infinitesimal.

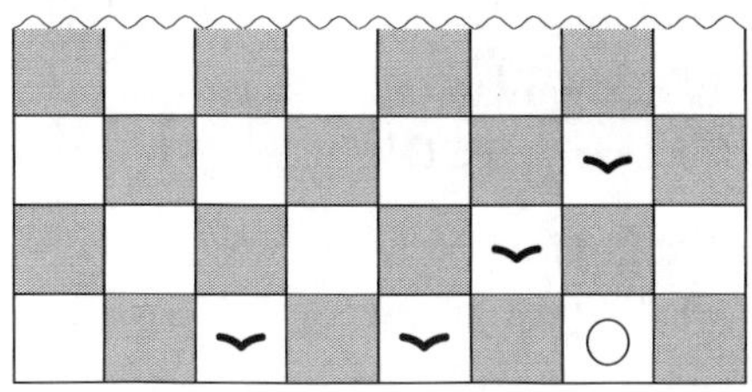

Figure 1.2. A trapped fox is a rather large infinitesimal.

As far as infinitesimals go, though, it's a rather large one, for if G is any dicotic game, then

$$T > G$$

essentially by a Lawnmower argument (cf. Theorem II.4.2 on page 83). On $T-G$, Left will continue to play on $-G$ for as long as possible and will trap the fox only after $-G$ has ended. So T is a sort of super-infinitesimal that exceeds every loopfree infinitesimal.

Formal Definitions

The following formalization is a bit dry but will help to make the theory precise. It's natural to model loopy games as directed graphs, whose vertices represent positions and whose edges represent legal moves. This was exactly the formalization we used in the impartial theory of Section IV.4. In the partizan case, it's necessary to distinguish between two types of edges, the **Left edges** and **Right edges** of the graph V.

Definition 1.1. A **bigraph** is a structure $V = (V, E^L, E^R)$, where V is any set (the **vertices** of the bigraph), and E^L and E^R are sets of ordered pairs drawn from V (the **Left edges** and **Right edges** of V, respectively).

The **negative bigraph** $\overline{V}$ has the same set of vertices as V, but with E^L and E^R interchanged: $\overline{V} = (V, E^R, E^L)$.

A **cycle** of V is a cycle of the ordinary directed graph $(V, E^L \cup E^R)$.

A **monochromatic cycle** of V is a cycle of (V, E^L) or (V, E^R).

In particular note that cycles may admit edges of either type, in any combination.

Definition 1.2. A **loopy game** is a pair $G = (V, x)$, where V is a bigraph and x a vertex of V (the **start vertex** of G). We say G is:

- **finite** if V is finite;
- **impartial** if $E^L = E^R$;

- **loopfree** if V is acyclic (i.e., contains no cycles).

The **Left options** of G are games of the form $G^L = (V, x^L)$ where there is a Left edge directed from x to x^L; **Right options** are defined analogously.

Definition 1.3. The **negative** of a loopy game $G = (V, x)$ is given by interchanging its Left and Right edges:

$$-G = (\overline{V}, x).$$

When convenient, we will write $\overline{G}$ as a synonym for $-G$.

Definition 1.4. Let $G = (V, x)$ and $H = (W, y)$ be loopy games. We define

$$G + H = \big(V \times W, (x, y)\big)$$

just as in the impartial case (cf. Definition IV.4.2 on page 208). Here the direct product $V \times W$ is obtained by taking direct products independently on the Left and Right edge sets.

For the rest of this chapter, our attention will be restricted to *finite* loopy games. Little is known about the transfinite case, and we won't have much to say about it, save for a few exercises in Section VIII.1.

It should be clear that finite loop*free* games (using Definition 1.2) are exactly equivalent to short games, as defined in Chapter II.

Outcomes and Values

A play of a loopy game G has three possible outcomes. Left (resp. Right) might **win** the game outright, if at some point Right (resp. Left) finds himself unable to move; but it is also possible for play to continue indefinitely, in which case we say the game is **drawn** and no one wins.

Since the outcome might depend on who moves first, we actually have *nine* distinct outcome classes, summarized in Figure 1.3. There are the four

		Left moves first		
		Left wins	Draw	Right wins
Right moves first	Left wins	$\mathscr{L}$	$\hat{\mathscr{P}}$	$\mathscr{P}$
	Draw	$\hat{\mathscr{N}}$	$\mathscr{D}$	$\check{\mathscr{P}}$
	Right wins	$\mathscr{N}$	$\check{\mathscr{N}}$	$\mathscr{R}$

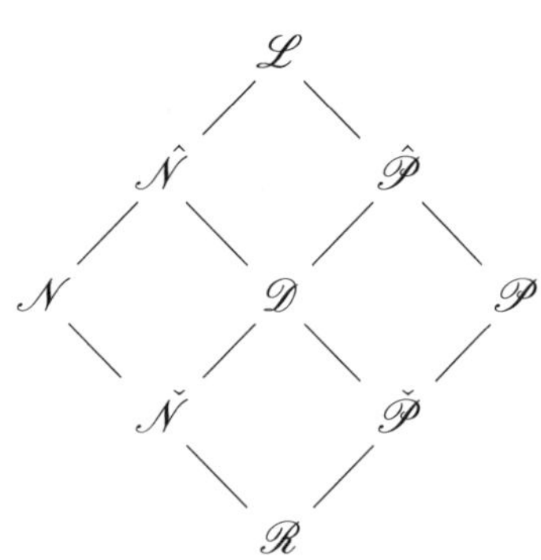

Figure 1.3. Nine outcome classes of loopy games.

traditional classes $\mathscr{L}$, $\mathscr{N}$, $\mathscr{P}$, $\mathscr{R}$; the class $\mathscr{D}$ of *D*rawn positions that's familiar from the impartial theory; and four new ones: $\hat{\mathscr{P}}$, $\check{\mathscr{P}}$, $\hat{\mathscr{N}}$, $\check{\mathscr{N}}$ (prounced *P-hat*, *P-check*, *N-hat*, *N-check*). $\hat{\mathscr{P}}$ is a more favorable version of $\mathscr{P}$ in which Left wins moving second but manages a draw if she plays first. The other three are similar modifications of $\mathscr{P}$ and $\mathscr{N}$.

Just as in the loopfree theory, the outcome classes are naturally partially ordered by favorability to Left, as shown in Figure 1.3.

Formally, we can define sets of games from which Left or Right can force a win in $\leq n$ moves, and we can use these to define outcome classes:

Definition 1.5. We define $\mathscr{P}_n^L$ (games that Left can win in n moves as second player) and $\mathscr{P}_n^R$ (likewise for Right) inductively by:

$$\begin{aligned}
G &\in \mathscr{P}_0^L && \text{if} \quad G \text{ has no Right option;}\\
G &\in \mathscr{P}_0^R && \text{if} \quad G \text{ has no Left option;}\\
G &\in \mathscr{P}_{n+1}^L && \text{if} \quad \text{for every } G^R\text{, there is a } G^{RL} \in \mathscr{P}_n^L;\\
G &\in \mathscr{P}_{n+1}^R && \text{if} \quad \text{for every } G^L\text{, there is a } G^{LR} \in \mathscr{P}_n^R.
\end{aligned}$$

Now we define $\mathscr{P}^L$ (games that Left can win as second player) and $\mathscr{P}^R$:

$$\mathscr{P}^L = \bigcup_n \mathscr{P}_n^L; \qquad \mathscr{P}^R = \bigcup_n \mathscr{P}_n^R.$$

Likewise, we define $\mathscr{N}^L$ (games that Left can win as first player) and $\mathscr{N}^R$:

$$\begin{aligned}
G &\in \mathscr{N}^L && \text{if} \quad \text{some } G^L \in \mathscr{P}^L;\\
G &\in \mathscr{N}^R && \text{if} \quad \text{some } G^R \in \mathscr{P}^R.
\end{aligned}$$

Now $\mathscr{P}^L$ and $\mathscr{N}^R$ are clearly disjoint, so there are three possibilities for a given G: either $G \in \mathscr{P}^L$ or $G \in \mathscr{N}^R$ or else $G \notin \mathscr{P}^L \cup \mathscr{N}^R$. These translate to "Left wins," "Right wins," and "Draw," respectively, in the rows of Figure 1.3. Likewise, $\mathscr{N}^L$ and $\mathscr{P}^R$ can be used to define the columns in Figure 1.3, yielding a formal definition of the nine outcome classes.

As always, we denote by $o(G)$ the outcome class of G. Now we can invoke the fundamental equivalence:

Definition 1.6. Let G and H be loopy games. Then we write

$$\begin{aligned}
G = H &\quad \text{if} \quad o(G+X) = o(H+X) \text{ for every loopy game } X,\\
G \geq H &\quad \text{if} \quad o(G+X) \geq o(H+X) \text{ for every loopy game } X.
\end{aligned}$$

Definition 1.7. The (**normal-play finite loopy partizan**) **game value** of a loopy game G is its equivalence class modulo $=$.

The following terminology is convenient.

Definition 1.8. Let G be a loopy game.

(a) Left **wins** G **playing first (second)** if $G \in \mathscr{N}^L$ ($G \in \mathscr{P}^L$).

(b) Left **survives** G **playing first (second)** if $G \notin \mathscr{P}^R$ ($G \notin \mathscr{N}^R$).

(c) A Left option G^L is a **winning move** if $G^L \in \mathscr{P}^L$.

(d) A Left option G^L is a **survival move** if $G^L \notin \mathscr{N}^R$.

The corresponding definitions for Right are analogous.

It is clear from the definitions that Left can win (resp. survive) G playing first if and only if she has a winning (resp. survival) move on G.

Example: $\mathbf{on} + \mathbf{off} = \mathbf{dud}$

There are four loopy games with just a single position, summarized in Figure 1.4. In the game graphs in Figure 1.4, Left edges are drawn with arrows that protrude to the Left; Right edges with arrows that protrude to the Right. This will be our default convention throughout this chapter.

on is a sort of "pass move" that's only available to Left. It's easy to see that for any loopy game G, Left can survive

$$\mathbf{on} + G$$

playing first or second, since she'll never run out of opportunities to pass. In fact we have $\mathbf{on} \geq G$ for all G: if Left can win (or draw) $G+X$, then she can certainly win (or draw) $\mathbf{on} + X$ just by replicating her strategy for $G + X$. The **on** component guarantees her a pass move any time the strategy calls for a move on G.

In particular, this implies $\mathbf{on} > G$ for all loopfree G (since $\mathbf{on} \geq G + 1$), so **on** is much like the escaped fox of Figure 1.1. In fact it's not hard to see that the escaped fox has value **off**: it doesn't matter where on the board the fox is; all that matters is that he always has *some* move available (and Left never has any).

$$0 = \{\,|\,\} =$$

$$\mathbf{on} = \{\mathbf{on}\,|\,\} =$$

$$\mathbf{off} = \{\,|\,\mathbf{off}\} =$$

$$\mathbf{dud} = \{\mathbf{dud}\,|\,\mathbf{dud}\} =$$

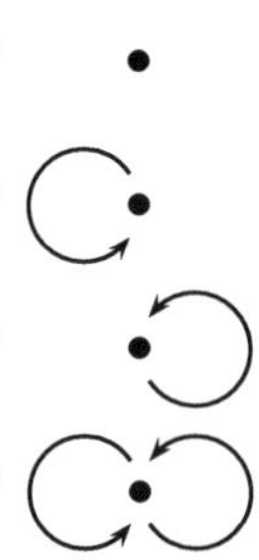

Figure 1.4. Single-position loopy games: 0, **on**, **off**, **dud**.

What about the sum **on** + **off**, from which both players are permitted to pass? It's clear that the outcome is a draw; moreover,

$$\mathbf{on} + \mathbf{off} + X$$

is a $\mathscr{D}$-position for *any* X: neither player will ever run out of moves, regardless of the structure of X. So we might say that **on** + **off** is a sort of

deathless **u**niversal **d**raw,

or **dud** for short.

We conclude that

$$\mathbf{dud} + G = \mathbf{dud}$$

for any G, since $\mathbf{dud} + G + X$ and $\mathbf{dud} + X$ are necessarily both draws!

Strategies

In the treatment of short games in Chapter II, many proofs were by induction. Generalizing such inductive arguments to loopy games is sometimes awkward, and it's often clearer to reason in terms of **strategies** for Left and Right, as we did (informally) in the previous example. We'll now introduce machinery for making such arguments precise.

Definition 1.9. Let G be a loopy game and let $\mathscr{A}$ be the set of all subpositions of G. A **Left strategy** for G is a partial function $\sigma : \mathscr{A} \to \mathscr{A}$ such that whenever $X \in \mathscr{A}$ and X has at least one Left option, then $\sigma(X)$ is defined and is a Left option of X. Right strategies are defined analogously.

Definition 1.10. Let σ be a Left strategy for G. A **σ-play** is a (finite or infinite) play

$$\vec{G} = (G_0, G_1, G_2, \ldots)$$

of G, such that $G_i = \sigma(G_{i-1})$ for every even integer $i > 0$. We say that $\vec{G}$ is **played according to σ**.

So a σ-play is a play in which Left is second player and makes every move according to σ.

Definition 1.11. Let σ be a Left strategy for G. We say σ is a **survival** (resp. **winning**) **strategy** if Left survives (resp. wins) every σ-play.

The following result shows (essentially) that any argument stated in terms of strategies can be converted to a proof by induction, and vice versa.

Theorem 1.12. *Let G be a loopy game.*

(a) *Left wins G playing second if and only if there exists a Left winning strategy for G.*

(b) *Left survives G playing second if and only if there exists a Left survival strategy for G.*

Proof. (a) First suppose that $G \in \mathscr{P}^L$. For each subposition X of G, define the **rank** of X to be the least n such that $X \in \mathscr{P}^L_n$ (or ∞ if no such n exists). Let $\sigma(X)$ be any X^L of minimal rank.

Then if $\vec{G}$ is a σ-play, the successive even-index positions

$$G_0, G_2, G_4, \ldots$$

have strictly decreasing rank, so eventually some $G_{2i} \in \mathscr{P}^L_0$, and therefore has no Right option.

Next suppose that $G \notin \mathscr{P}^L$. Let $\sigma : \mathscr{A} \to \mathscr{A}$ be any strategy for Left; we'll define a σ-play $\vec{G}$ that is not a win for Left. $\vec{G}$ will satisfy $G_i \notin \mathscr{P}^L$ for all even i.

Fix an even integer i and suppose we are given $G_i \notin \mathscr{P}^L$. Then there is necessarily some G_i^R with $G_i^R \notin \mathscr{N}^L$. Put $G_{i+1} = G_i^R$. If G_{i+1} has no Left option, then we stop, and $\vec{G}$ has length $i+1$. Otherwise, put $G_{i+2} = \sigma(G_{i+1})$. Since $G_{i+1} \notin \mathscr{N}^L$, we necessarily have $G_{i+2} \notin \mathscr{P}^L$.

This defines the σ-play $\vec{G}$, and it's clear from the definition that Left does not win $\vec{G}$. Therefore σ is not a winning strategy.

(b) If there is a Left survival strategy for G, then no G^R can have a Right winning strategy. By (a), Right cannot win any G^R playing second; therefore Left can survive any G^R playing first, and hence G playing second. The converse implication is given by the same argument in reverse. □

As an immediate application, we have the following simple result.

Theorem 1.13 (cf. Theorem II.1.13 on page 56). *For any loopy game G, we have either $o(G - G) = \mathscr{P}$ or $o(G - G) = \mathscr{D}$.*

Proof. The mirror-image strategy is a survival strategy for $G - G$ for either player. Since $G - G$ is its own negative, the only possibilities are $o(G - G) = \mathscr{P}$ and $o(G - G) = \mathscr{D}$. □

The Swivel Chair

Recall the following key result about short games: if G and H are short, then $G \geq H$ if and only if Left can win $G - H$ playing second (Theorem II.1.19 on page 58). The proof of this theorem depends crucially on the fact that if $G \geq H$, then $G - H \geq 0$. For loopy games G and H, however, $G \geq H$ does *not* necessarily imply $G - H \geq 0$. For example, **on** $\geq$ **on**; but **on** $-$ **on** $=$ **dud**, while **dud** $\not\geq 0$.

$$\begin{matrix} G+X \\ -H+H \end{matrix} \to \begin{matrix} G+X^R \\ -H+H \end{matrix} \to \begin{matrix} G+X^R \\ -H+H^L \end{matrix} \to \begin{matrix} G+X^R \\ -H^L+H^L \end{matrix} \to \begin{matrix} G+X^R \\ -H^{LR}+H^L \end{matrix} \to$$

$$\to \begin{matrix} G+X^R \\ -H^{LR}+H^{LR} \end{matrix} \to \cdots \to \begin{matrix} G+X^R \\ -H^{LRL\cdots L}+H^{LRL\cdots L} \end{matrix} \to \begin{matrix} G^L+X^R \\ -H^{LRL\cdots L}+H^{LRL\cdots L} \end{matrix}$$

Figure 1.5. A typical opening sequence for the swivel chair strategy. Because Left's strategy for $G-H$ is a winning strategy, the long alternating sequence must eventually end with a move on G or X^R.

Various generalizations of Theorem II.1.19 play a central role in the theory of loopy games. The following theorem is the most straightforward version.

Theorem 1.14. *Let G and H be loopy games.*

(a) *If $G \geq H$, then Left can survive $G-H$ playing second.*

(b) *If Left can win $G-H$ playing second, then $G \geq H$.*

Proof of Theorem 1.14(a). We have $o(G-H) \geq o(H-H)$, and from Theorem 1.13 we know that $o(H-H) = \mathscr{P}$ or $\mathscr{D}$. □

The proof of Theorem 1.14(b) is more involved and employs a *swivel chair argument*, the first of many that we will see. In the typical case, we suppose Left can survive both $G-H$ and $H+X$ playing second, and we wish to show she can also survive $G+X$. We envision Left sitting down to play a rousing game of $G+X$ against Right. Secretly, she sets up a dummy copy of $-H+H$ alongside $G+X$:

$$\begin{matrix} G+X \\ -H+H \end{matrix}$$

One can regard the *columns* of this diagram as copies of $G-H$ and $X+H$. Whenever Right plays on $G+X$, Left simply follows her known survival strategies on $G-H$ and $X+H$, transferring each move on H or $-H$ to its mirror image component. A typical sequence is shown in Figure 1.5. In order for this strategy to work, we must arrange that play can never get "stuck" on $-H+H$; otherwise Left can never guarantee that she will make another move on $G+X$ at all! In Theorem 1.14(b), this guarantee comes from the fact that $G-H$ is a *winning* strategy, so the corresponding run of $G-H$ is necessarily finite.

Definition 1.15. Let σ and τ be Left strategies for $G-H$ and $H+X$, respectively. The **swivel run** starting from $H+X$ is the run of H,

$$\vec{H} = (H_0, H_1, H_2, \ldots),$$

defined as follows. Put $H_0 = H$, and for all $i \geq 0$,

$$\tau(H_{2i} + X) = H_{2i+1} + X \quad \text{and} \quad \sigma(G - H_{2i+1}) = G - H_{2i+2}.$$

If for some i, $\tau(H_{2i} + X)$ is undefined or has the form $H_{2i} + X^L$, then we stop and the run ends; likewise for $\sigma(G - H_{2i+1})$. Otherwise, the swivel run is infinite.

Lemma 1.16. *Suppose that Left can win $G - H$ playing second. Then if she can survive $H + X$ playing first (second), she can also survive $G + X$ playing first (second).*

Proof. First assume that Left is second player. Consider a subposition $\dot{G} + \dot{X}$ of $G + X$. Call $\dot{G} + \dot{X}$ **safe** (through $\dot{H}$) if there is a subposition $\dot{H}$ of H such that Left can win $\dot{G} - \dot{H}$ playing second and can survive $\dot{H} + \dot{X}$ playing second. Now $G + X$ itself is safe (by hypothesis), so it suffices to prove that if Right ever moves away from a safe position, then Left can return to one.

Suppose that Right moves to $\dot{G} + \dot{X}^R$ (the proof is the same if he moves to $\dot{G}^R + \dot{X}$). Let σ be a winning strategy for Left for $\dot{G} - \dot{H}$, and τ a survival strategy for $\dot{H} + \dot{X}$, and let $\vec{H}$ be the swivel run starting from $\dot{H} + \dot{X}^R$. Since σ is a winning strategy, $\vec{H}$ must have finite length, say n. If n is even, then $\tau(H_n + \dot{X}^R) = H_n + \dot{X}^{RL}$, so $\dot{G} + \dot{X}^{RL}$ is safe through H_n. Likewise if n is odd, then some $\dot{G}^L + \dot{X}^R$ is safe through H_n.

If Left is first player, then the same argument shows that Left has an opening move to a safe position. □

Proof of Theorem 1.14(b). Fix any loopy game X. If Left can survive $H + X$ playing first (second), she can also survive $G + X$ playing first (second), by Lemma 1.16. Now we know that Right can win $H - G$ playing second, so a symmetric argument shows that if Right can survive $G + X$ playing first (second), then he can also survive $H + X$ playing first (second). This establishes that $o(G + X) \geq o(H + X)$. □

If Left can merely *survive* $G - H$ playing second, then the proof of Lemma 1.16 breaks down: the swivel run from $\dot{H} + \dot{X}^R$ might be infinite, so that from $\dot{G} + \dot{X}^R$ Left can find no good response.

Example. Let $G = 0$ and $H = \mathbf{dud}$. Then certainly Left can survive $G - H$ playing second, but $G \not\geq H$. In particular, if $X = -1$, then Left can survive $H + X$ but not $G + X$.

Examples of this sort suggest that the asymmetry in Theorem 1.14 is fundamental. This is the central difficulty of the loopy theory, and finding ways to resolve this asymmetry will be a major theme of this chapter.

Exercises

1.1 If $G \geq 0$, then Left can win G playing second (resolving the asymmetry in Theorem 1.14 in the case $H = 0$).

1.2 If G is not equal to any loopfree game, then either some $G^L \geq G$ or else some $G^R \leq G$. (Therefore Theorem II.1.30 holds if and only if G is loopfree.)

1.3 A Left strategy σ for G is a **complete survival strategy** if, whenever Left can survive a subposition X of G playing first, then $\sigma(X)$ is a survival move and σ is a survival strategy for $\sigma(X)$. Prove that every game G admits a complete survival strategy.

1.4 For loopy games G and H, define

$$G \geq^\sharp H \quad \text{if} \quad \text{Left can win } G - H \text{ playing second,}$$

$$G \geq^\flat H \quad \text{if} \quad \text{Left can survive } G - H \text{ playing second.}$$

(Ordinary $\geq$ could be called $\geq^\natural$ in this context.) Show that:

(a) $\geq^\sharp$ is transitive but not reflexive, and $\geq^\flat$ is reflexive but not transitive.

(b) If G is loopfree and $G \geq^\flat H \geq^\sharp J$, then $G \geq^\sharp J$.

(c) G^{L_1} is **strongly dominated** if $G^{L_2} \geq^\sharp G^{L_1}$ for some G^{L_2}. If G' is obtained from G by removing a strongly dominated option, then $G' =^\sharp G$.

(d) G^{L_1} is **strongly reversible** if $G^{L_1 R_1} \leq^\sharp G$ for some $G^{L_1 R_1}$. If G' is obtained from G by bypassing a strongly reversible option, then $G' =^\sharp G$.

(e) $G \geq^\sharp G$ if and only if G is equal to a loopfree game. (Apply (c) and (d).)

Notes

Almost immediately after Conway introduced the axiomatic theory of loopfree partizan games, efforts were underway to extend it to loopy games. The asymmetry in Theorem 1.14 was recognized early on. Fraenkel and Tassa [**FT82**] addressed this problem by defining two separate relations, as in Exercise 1.4. The Fraenkel–Tassa approach yields some useful theorems (such as Exercise 1.4(e)), but its applicability is limited.

The first hints of a more general theory appeared in a groundbreaking paper by Robert Li [**Li76**]. Li's paper inspired Conway to develop a more general theory, with the help of his students, Simon Norton and Clive Bach. These developments will be discussed in subsequent sections.

2. Stoppers

A loopy game G is a **stopper** if no subposition of G admits an infinite alternating run. For example, every position in Fox and Geese is necessarily a stopper: the geese can only make a bounded number of moves throughout the game, constraining the length of any alternating run.

Some elementary facts about stoppers are worth emphasizing:

- If G is a stopper, then G admits no infinite plays, so every play of G must eventually **stop**. Therefore $o(G)$ must be one of the four traditional outcome classes $\mathscr{N}$, $\mathscr{P}$, $\mathscr{L}$, $\mathscr{R}$.
- If G is a stopper and H is loopfree, then $G+H$ is necessarily a stopper. *However*:
- If H is also a stopper (not necessarily loopfree), then $G+H$ might *not* be a stopper. We showed this for **on** + **off** in the previous section.

It turns out that stoppers are free of many complications that afflict arbitrary loopy games. In particular we'll prove the following (somewhat less elementary) results, for all stoppers G and H:

- $G \geq H$ if and only if Left can survive $G - H$, playing second. (The asymmetry in Theorem 1.14 disappears, yielding a more successful generalization of the loopfree theory.)
- We can eliminate dominated options and bypass reversible ones without changing the value of G, just as for loopfree games.
- Every stopper has a simplest form, and if G and H are in simplest form and $G = H$, then necessarily $G \cong H$.

All of these facts are, essentially, consequences of the following theorem, which improves Theorem 1.14 in the case where G and H are stoppers.

Theorem 2.1. *Let G and H be stoppers. Then $G \geq H$ if and only if Left can survive $G - H$ playing second.*

The proof uses a swivel chair argument, much like Theorem 1.14. As before, we break out the key argument into a lemma.

Lemma 2.2. *Suppose that Left can survive $G - H$ playing second, with H a stopper. Then if she can survive $H + X$ playing first (second), she can also survive $G + X$ playing first (second).*

Proof. Consider a subposition $\dot{G}+\dot{X}$ of $G+X$. We define $\dot{G}+\dot{X}$ to be **safe** if there is some subposition $\dot{H}$ of H such that Left can survive both $\dot{G}-\dot{H}$ and $\dot{H}+\dot{X}$ playing second. The rest of the proof is just like Lemma 1.16. This time, the strategy σ for $\dot{G}-\dot{H}$ is merely assumed to be a survival strategy; the fact that H is a stopper guarantees that every associated swivel run is finite. □

Proof of Theorem 2.1. Fix any loopy game X. If Left can survive $H+X$ playing first (second), then Lemma 2.2 shows that she can also survive $G+X$, since H is a stopper. Likewise if Right can survive $G + X$ playing first (second), then he can also survive $H + X$, since G is a stopper. This proves that $o(G + X) \geq o(H + X)$. $\square$

Stoppers as Limits

In Section 1 we showed that $\mathbf{on} \geq G$ for any game G. We can say something even stronger: the sequence

$$0,\ 1,\ 2,\ 3,\ \ldots$$

"tends to **on**" in the sense that if G is a finite loopy game that exceeds all the integers, then $G = \mathbf{on}$. This example illustrates one of the most fascinating aspects of loopy games: they often witness precise limits for natural sequences of loopfree games. These limits have many interesting properties, which we'll explore in this section and in Section 5.

Definition 2.3. Let $\mathcal{A}$ be a set of game values. We say that G is the **supremum** of $\mathcal{A}$, and we write $G = \sup(\mathcal{A})$, provided that:

(i) $G \geq X$ for all $X \in \mathcal{A}$; and

(ii) if $H \geq X$ for all $X \in \mathcal{A}$, then also $H \geq G$.

We say that G is the **infimum** of $\mathcal{A}$, and we write $G = \inf(\mathcal{A})$, if (i) and (ii) hold with all inequalities reversed.

Throughout this chapter, the quantifiers in Definition 2.3 are understood to range over all *finite* loopy games. Some of these results can be generalized to the transfinite case; this is addressed briefly in the exercises to Chapter VIII (for example, see Exercise VIII.1.12 on page 411).

Proposition 2.4. $\mathbf{on} = \sup(\mathbb{N})$.

Proof. If $G \mathrel{\lhd\!\!\mid} \mathbf{on}$, then Right can win $G - \mathbf{on}$ playing first. Choose n so that Right can win in at most n moves. Then he can win $G - n$ using the same strategy, so that $G \mathrel{\lhd\!\!\mid} n$. $\square$

Note that by the Archimedean Principle (Theorem II.3.7 on page 71), we have in fact $\mathbf{on} = \sup(\mathbb{G})$.

Just as **on** is the supremum of all loopfree games, another stopper, **over**, is the supremum of all infinitesimals. **over** is defined by

$$\mathbf{over} = \{0 \mid \mathbf{pass}\} =$$

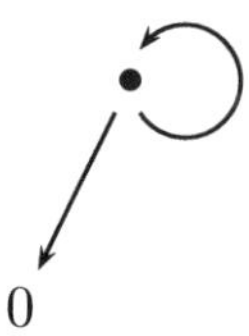

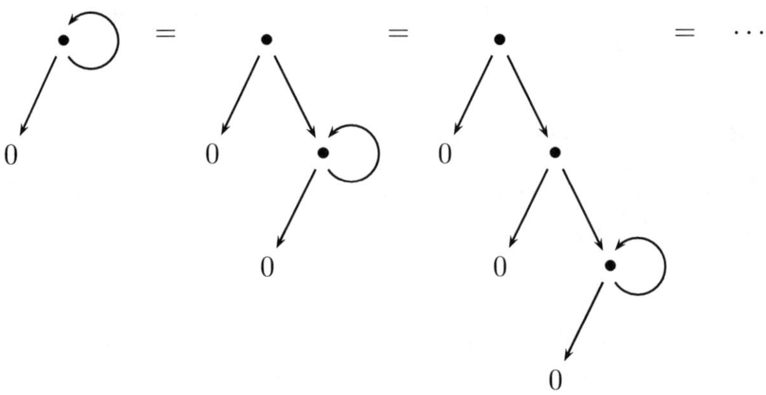

Figure 2.1. The loop in **over** can be expanded indefinitely without changing its value.

We also write **under** $=$ $-$**over**. Note that **over** is equal to the trapped fox T from Figure 1.2 (page 281).

Our intent is to prove that **over** $= \sup_n(\uparrow n)$ and **over** $= \inf_n(2^{-n})$. These facts reflect a structural similarity between **over** and the loopfree games $\uparrow n$ and 2^{-n}. We can expand the game graph of **over** by adding extra vertices to make the first few passes explicit, as in Figure 2.1. All of these representations clearly have the same value. Comparing the expanded game graph to the game trees of $\uparrow n$ and 2^{-n}, as in Figure 2.2, illustrates the similarity: **over**, $\uparrow n$, and 2^{-n} appear identical for their first n moves.

The fact that **over** $= \sup_n(\uparrow n)$ is not hard to prove directly, but the following **sidling** technique will make it trivial.

Sidling

Sidling is an iterative process for producing successive approximations to an arbitrary loopy game G.

Definition 2.5. Let G be any loopy game and let X be a stopper. The nth **sidling approximation** to G (**starting from** X) is defined by

$$\Sigma_0^X(G) = X \quad \text{and} \quad \Sigma_{n+1}^X(G) = \{\Sigma_n^X(G^L) \mid \Sigma_n^X(G^R)\}.$$

Example. Let $G =$ **over** $= \{0 \mid \textbf{pass}\}$. Then

$$\Sigma_1^{\textbf{on}}(G) = \{\textbf{on} \mid \textbf{on}\} = \textbf{on},$$
$$\Sigma_2^{\textbf{on}}(G) = \{0 \mid \textbf{on}\} = 1,$$
$$\Sigma_3^{\textbf{on}}(G) = \{0 \mid 1\} = \tfrac{1}{2},$$
$$\Sigma_4^{\textbf{on}}(G) = \{0 \mid \tfrac{1}{2}\} = \tfrac{1}{4},$$

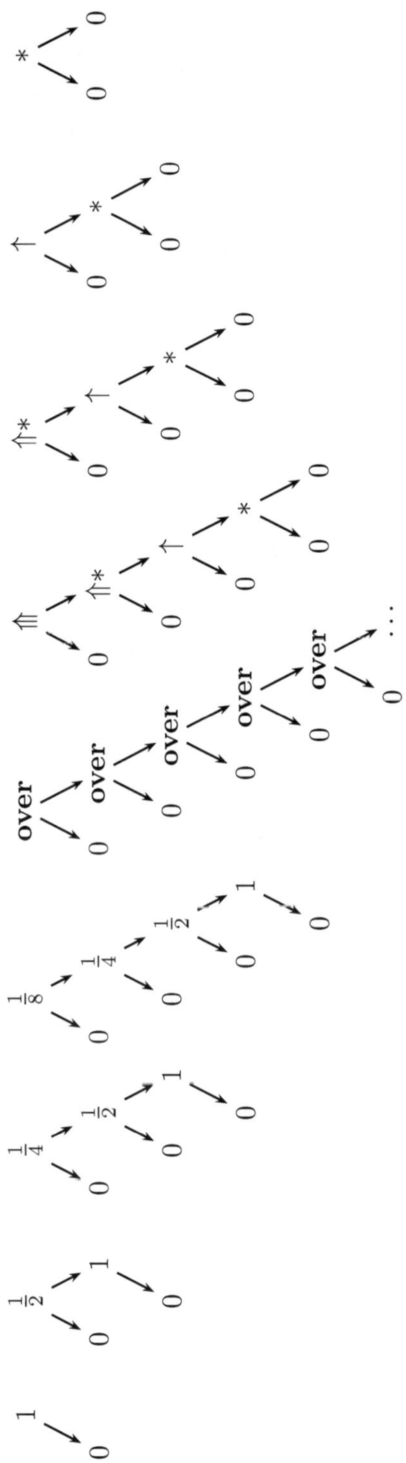

Figure 2.2. 2^{-n}, **over**, and $\Uparrow n$ are structurally similar for large values of n.

and so on, yielding the familiar sequence of approximations

$$\mathbf{on},\ 1,\ \tfrac{1}{2},\ \tfrac{1}{4},\ \ldots.$$

(Here we're using the obvious fact that $\Sigma_n^{\mathbf{on}}(0) = 0$ for all $n \geq 1$.) Sidling from the other direction, starting with $X = 0$, we have

$$\begin{aligned}\Sigma_1^0(G) &= \{0 \mid 0\} = *,\\ \Sigma_2^0(G) &= \{0 \mid *\} = \uparrow,\\ \Sigma_3^0(G) &= \{0 \mid \uparrow\} = \Uparrow*,\\ \Sigma_4^0(G) &= \{0 \mid \Uparrow*\} = ⤊.\end{aligned}$$

yielding the sequence

$$*,\ \uparrow,\ \Uparrow*,\ ⤊,\ \ldots.$$

These are exactly the sequences indicated by Figure 2.2.

This example illustrates a more general phenomenon: for sufficiently large X, the sidling approximations necessarily form a sequence of upper bounds for G.

Proposition 2.6. *If G and X are stoppers and $X \geq H$ for every subposition H of G, then*

$$\Sigma_n^X(G) \geq G \quad \textit{for all } n.$$

Proof. Let H be a subposition of G. Then on the difference

$$\Sigma_{n+1}^X(H) - H,$$

Right's moves all have the form

$$\Sigma_n^X(H^R) - H \quad \text{or} \quad \Sigma_{n+1}^X(H) - H^L,$$

which Left can revert to

$$\Sigma_n^X(H^R) - H^R \quad \text{and} \quad \Sigma_n^X(H^L) - H^L,$$

respectively. By induction on n, we may assume that Left can survive each $\Sigma_n^X(H^L) - H^L$ and $\Sigma_n^X(H^R) - H^R$ playing second, so this shows she can survive $\Sigma_{n+1}^X(H) - H$ as well. □

In particular, the **onside approximations** $\Sigma_n^{\mathbf{on}}(G)$ always give a sequence of upper bounds for G, and the **offside approximations** $\Sigma_n^{\mathbf{off}}(G)$ give a sequence of lower bounds. Here's an example where the bounds converge.

Example. Let G be the following game:

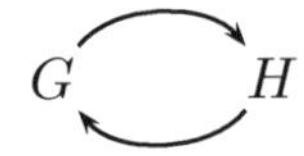

Here $G = \{\ |\ H\}$ and $H = \{G\ |\ \}$. Then the onside approximations $\Sigma_n^{\mathbf{on}}(G)$ and $\Sigma_n^{\mathbf{on}}(H)$ are given by

$$\begin{array}{rcccccccc} n & = & 0\ , & 1 & , & 2\ , & 3\ , & \ldots \\ \Sigma_n^{\mathbf{on}}(G) & = & \mathbf{on}\ , & \{\ |\ \mathbf{on}\} = 0 & , & \{\ |\ \mathbf{on}\} = 0\ , & \{\ |\ \mathbf{on}\} = 0\ , & \ldots \\ \Sigma_n^{\mathbf{on}}(H) & = & \mathbf{on}\ , & \{\mathbf{on}\ |\ \} = \mathbf{on} & , & \{0\ |\ \} = 1\ , & \{0\ |\ \} = 1\ , & \ldots \end{array}$$

Since $\Sigma_3^{\mathbf{on}}(G) = \Sigma_2^{\mathbf{on}}(G)$, and likewise for H, the approximations have converged, and therefore $\Sigma_n^{\mathbf{on}}(G) = 0$ and $\Sigma_n^{\mathbf{on}}(H) = 1$ for all $n \geq 2$.

Definition 2.7. We define relations $G \sim_n H$, for $n \geq 0$, inductively:

- $G \sim_0 H$ is true for all G and H;
- $G \sim_{n+1} H$ if for each G^L there is some H^L with $G^L \sim_n H^L$, and vice versa; and likewise for Right options.

We say G and H are **similar for n moves** (or just **n-similar**) if $G \sim_n H$. If $G \sim_n H$ for all n, then we say G and H are **similar** and write $G \simeq H$.

Certainly $\Sigma_n^X(G) \sim_n G$ for all X and n, straight from the definitions.

Theorem 2.8. *Let $\mathcal{A}$ be a set of stoppers and let G be a stopper. Suppose that:*

(i) *$G \geq X$ for all $X \in \mathcal{A}$; and*

(ii) *for each $n \geq 0$, there is an $X \in \mathcal{A}$ with $G \sim_n X$ (for some form of X).*

Then $G = \sup(\mathcal{A})$.

Proof. It suffices to show that if $H \geq X$ for every $X \in \mathcal{A}$, then $H \geq G$.

Suppose (for contrapositive) that $H \lhd\!\!\mid G$. Then Left can win on $G - H$ playing first. Choose n so that Left can win on $G - H$ in at most $n - 1$ moves. By assumption (ii), there is some $X \in \mathcal{A}$ with $G \sim_n X$. Then Left can win on $X - H$ by following exactly the same strategy she used on $G - H$. This implies $H \lhd\!\!\mid X$. □

Corollary 2.9. *If $X \leq H$ for every subposition H of G, then*

$$G - \sup_n \left(\Sigma_n^X(G)\right).$$

In particular, it is always true that

$$G = \sup_n \left(\Sigma_n^{\mathbf{off}}(G)\right) \quad \textit{and} \quad G = \inf_n \left(\Sigma_n^{\mathbf{on}}(G)\right).$$

Proof. Immediate from Proposition 2.6 and Theorem 2.8. □

The limits of **over** are now easily established:

Proposition 2.10. $\mathbf{over} = \sup_n(\uparrow n) = \inf_n(2^{-n})$.

Proof. Follows from the sidling approximations to **over** on page 292. □

The limit **over** $= \inf_n(2^{-n})$ can also be visualized in terms of ordinal sums. Norton's Lemma (Theorem II.4.18 on page 91) generalizes verbatim to loopy games, with many useful applications. Observe that

$$\mathbf{over} = 1 : \mathbf{off},$$

straight from the definition. But also

$$\tfrac{1}{2} = (1 : -1), \quad \tfrac{1}{4} = (1 : -2), \quad \tfrac{1}{8} = (1 : -3), \quad \ldots .$$

Now for all n, we have $-n : \mathbf{off} = \mathbf{off}$, so that

$$\mathbf{over} = \big(1 : (-n : \mathbf{off})\big) = \big((1 : -n) : \mathbf{off}\big) = (2^{-n} : \mathbf{off}).$$

By Norton's Lemma, every game G has the same order-relations with **over** as with 2^{-n} for some sufficiently large n. It follows that **over** $= \inf_n(2^{-n})$.

More Loopy Infinitesimals

The supremum of the games $\uparrow^{[n]}$ is defined like so:

$$\uparrow^{[\mathbf{on}]} = \{\mathbf{pass} \mid *\} =$$

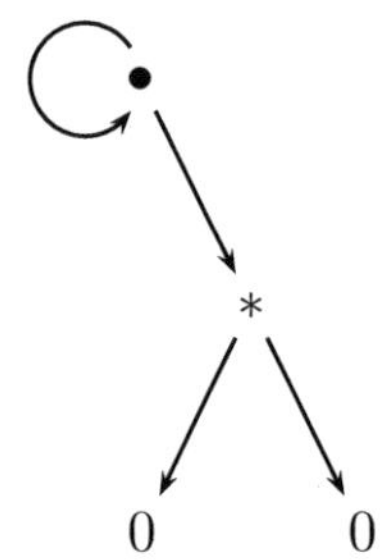

($\uparrow^{[\mathbf{on}]}$ is sometimes denoted by **upon** in the literature.) We also write

$$\uparrow^{[\mathbf{on}]}* = \{0, \mathbf{pass} \mid 0\} =$$

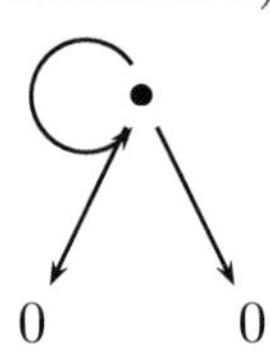

It's easily checked that $\uparrow^{[\mathbf{on}]}* = \uparrow^{[\mathbf{on}]} + *$ by playing the appropriate differences. There are also loopy analogues of $\uparrow^n$ and $\uparrow^n*$:

$$\uparrow^{\mathbf{on}} = \big\{0 \mid -\uparrow^{[\mathbf{on}]}*\big\},$$

$$\uparrow^{\mathbf{on}}* = \big\{0, * \mid -\uparrow^{[\mathbf{on}]}\big\} = \uparrow^{\mathbf{on}} + *.$$

Proposition 2.11. $\uparrow^{[\mathbf{on}]} = \sup_n(\uparrow^{[n]})$ *and* $\uparrow^{\mathbf{on}} = \inf_n(\uparrow^n)$.

Proof. $\Sigma_n^{\mathbf{off}}(\uparrow^{[\mathbf{on}]}) = \uparrow^{[n-1]}$ and $\Sigma_n^{\mathbf{on}}(\uparrow^{\mathbf{on}}) = \uparrow^{n-1}$. The picture is shown in Figure 2.3 on the facing page. □

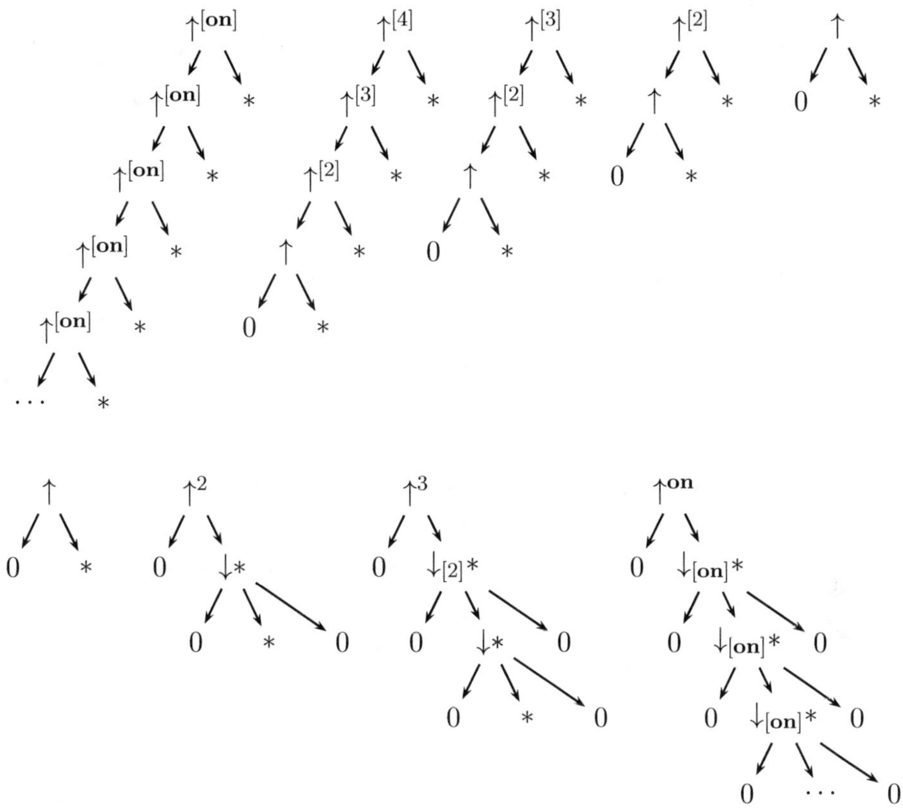

Figure 2.3. $\uparrow^{[\mathbf{on}]}$ is similar to $\uparrow^{[n]}$, and $\uparrow^{\mathbf{on}}$ is similar to $\uparrow^{n}$.

So we can think of $\uparrow^{[\mathbf{on}]}$ (informally) as having the infinite uptimal expansion $0.11111\ldots$. In terms of ordinal sums:

$$\uparrow* = (* : 1), \quad \uparrow^{[2]}* = (* : 2), \quad \uparrow^{[3]}* = (* : 3), \quad \ldots, \quad \uparrow^{[\mathbf{on}]}* = (* : \mathbf{on}),$$

further justifying our notation.

A similar example is

$$+_{\mathbf{on}} = \{0 \,\|\, 0 \mid \mathbf{off}\},$$

the smallest positive game of all!

Proposition 2.12. *If $G > 0$, then $G \geq +_{\mathbf{on}}$.*

Proof. We show that Left can survive playing second on $G - +_{\mathbf{on}}$. If Right plays to $G - 0$, then Left wins *a priori*, so suppose Right plays to some $G^R - +_{\mathbf{on}}$. Then Left can move to

$$G^R + \{\mathbf{on} \mid 0\}.$$

Stopper	Definition	$\sup_n$	$\inf_n$	Negative
on	$\{\mathbf{pass} \mid \cdot\}$	n	—	**off**
over	$\{0 \mid \mathbf{pass}\}$	2^{-n}	$\uparrow n$	**under**
$\uparrow^{[\mathbf{on}]}$	$\{\mathbf{pass} \mid *\}$	$\uparrow^{[n]}$	†	$\downarrow_{[\mathbf{on}]}$
$\uparrow^{[\mathbf{on}]}*$	$\{0, \mathbf{pass} \mid 0\}$	$\uparrow^{[n]}*$	†	$\downarrow_{[\mathbf{on}]}*$
$\uparrow^{\mathbf{on}}$	$\{0 \mid -\uparrow^{[\mathbf{on}]}*\}$	†	$\uparrow^{n}$	$\downarrow_{\mathbf{on}}$
$\uparrow^{\mathbf{on}}*$	$\{0, * \mid -\uparrow^{[\mathbf{on}]}\}$	†	$\uparrow^{n}*$	$\downarrow_{\mathbf{on}}*$
$+_{\mathbf{over}}$	$\{0 \parallel 0 \mid \mathbf{under}\}$	$+_{2^{-n}}$	$\{0 \parallel 0 \mid \downarrow n\}$	$-_{\mathbf{over}}$
$+_{\mathbf{on}}$	$\{0 \parallel 0 \mid \mathbf{off}\}$	—	$+_{n}$	$-_{\mathbf{on}}$

†See Exercise 2.4.

Figure 2.4. Several of the most important stoppers.

Now if Right plays to G^R+0, Left simply continues with her winning strategy from G^R. If instead Right plays to $G^{RR} + \{\mathbf{on} \mid 0\}$, then Left reverts to $G^{RR} + \mathbf{on}$, which she can survive since $\mathbf{on} \geq -G^{RR}$. □

We also have $+_{\mathbf{on}} = \inf_n(+_n)$, since $\mathbf{on} = \sup(\mathbb{N})$. These and several other stoppers are summarized in Figure 2.4.

Atomic Weights of Stoppers

The atomic weight relation from Section II.7 can be extended verbatim to loopy games:

$$G \sim_{\star} H \quad \text{if} \quad o(G + X + \star) = o(H + X + \star) \quad \text{for all } X.$$

Here, as before, $o(G + \star)$ is shorthand for $o(G + *m)$, for any $*m$ that does not occur as the value of a subposition of G.

It's easy to see that $\operatorname{aw}(\uparrow^{[\mathbf{on}]}) = 1$, as expected, since

$$0.11 < \uparrow^{[\mathbf{on}]} < 0.12$$

(with the bounds given in uptimal notation).

Also of interest are the games

$$\mathbf{ace} = \{0 \mid +_{\mathbf{on}}\}, \quad \mathbf{deuce} = \{0 \mid \mathbf{ace}\}, \quad \mathbf{trey} = \{0 \mid \mathbf{deuce}\},$$

whose graphs are shown in Figure 2.5 on the next page, and in general

$$\mathbf{pip}_0 = +_{\mathbf{on}}, \quad \mathbf{pip}_{n+1} = \{0 \mid \mathbf{pip}_n\}.$$

These games satisfy the relations

$$\mathbf{ace} + \mathbf{ace} = \mathbf{deuce}, \quad \mathbf{deuce} + \mathbf{ace} = \mathbf{trey}, \quad \ldots, \quad n \cdot \mathbf{ace} = \mathbf{pip}_n \ (n \geq 1),$$

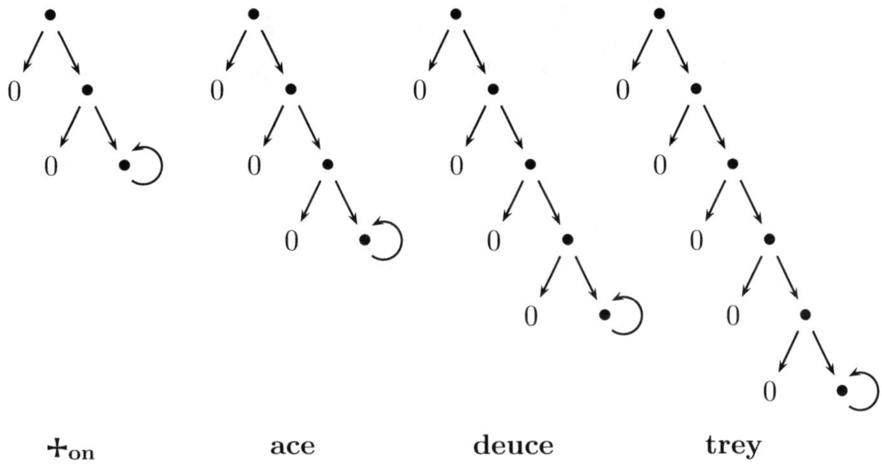

Figure 2.5. The games **ace**, **deuce**, **trey**, ...

and

$$*.11 < \mathbf{ace} < *.12,$$

so that $\mathrm{aw}(\mathbf{ace}) = 1$ and therefore $\mathrm{aw}(\mathbf{pip}_n) = n$. In fact it's not hard to show that

$$\mathbf{ace} - \uparrow^{[\mathbf{on}]}* = \uparrow^{\mathbf{on}},$$

so that **ace** and $\uparrow^{[\mathbf{on}]}*$ are quite close indeed.

These sorts of one-off results are easy enough to obtain, and they suggest a more general theory.

Open Problem. Extend the atomic weight calculus to (an appropriate subclass of) stoppers.

Exercises

2.1 Give examples to show that if G and H are stoppers, then $G + H$ might be any of the nine outcome classes in Figure 1.3 (page 282).

2.2 Draw game trees (as in Figure 2.3 on page 297) to illustrate the limits $\uparrow^{[\mathbf{on}]}* = \sup_n(\uparrow^{[n]}*)$ and $\uparrow^{\mathbf{on}}* = \inf_n(\uparrow^n*)$.

2.3 If $G > 0$ and G is dicotic, then $G \geq +_{\mathbf{over}}$.

2.4 Exhibit a decreasing sequence G_n of loopfree games with $\uparrow^{[\mathbf{on}]} = \inf_n G_n$, and an increasing sequence H_n with $\uparrow^{\mathbf{on}} = \sup_n H_n$.

2.5 Exhibit a stopper G such that $G = \sup_n(*2 : n)$. Express G as the inf of another sequence.

2.6 If G and H are loopfree and $G \simeq H$, then $G \cong H$.

2.7 True or false: There exists a stopper G such that

$$G = \sup\{0.1,\ 0.101,\ 0.10101,\ 0.1010101,\ \ldots\}.$$

2.8 A stopper P is a **pseudonumber** if for every subposition Q of P and every Q^L and Q^R, we have $Q^L < Q^R$.

(a) Assuming P to be in canonical form, prove that P is a pseudonumber if and only if $P^L \le P \le P^R$ for every P^L and P^R.

(b) Prove the *Pseudonumber Avoidance Theorem*: Suppose that P is equal to a pseudonumber and G is not. If Left has a winning (resp. survival) move on $G+P$, then she has a winning (resp. survival) move of the form $G^L + P$.

(c) Show that every pseudonumber is either a number, **on**, **off**, or a game of the form $x + \textbf{over}$ or $x + \textbf{under}$, where x is a number. Conclude that the pseudonumbers are totally ordered.

2.9 *Stops.* For G a finite stopper, define

$$L(G) = \begin{cases} G & \text{if } G \text{ is equal to a pseudonumber;} \\ \max\limits_{G^L}\big(R(G^L)\big) & \text{otherwise;} \end{cases}$$

$$R(G) = \begin{cases} G & \text{if } G \text{ is equal to a pseudonumber;} \\ \min\limits_{G^R}\big(L(G^R)\big) & \text{otherwise.} \end{cases}$$

Prove that:

(a) $L(G)$ and $R(G)$ are well-defined (the recursion necessarily terminates) and are independent of the form of G.

(b) $L(G) \ge R(G)$ for all G.

(c) If $L(G) = \textbf{off}$, then $G = \textbf{off}$; if $R(G) = \textbf{on}$, then $G = \textbf{on}$.

(d) The *Pseudonumber Translation Theorem*: If P is equal to a pseudonumber and G is not, then $G + P = \big\{G^L + P \bigm| G^R + P\big\}$.

2.10 *Pseuduptimals.* For all positive pseudonumbers P (cf. Exercise 2.8), we define

$$\uparrow^{[P]} = (* : P) - * \quad \text{and} \quad \uparrow^{1+P} = \big\{0 \bigm| \downarrow_{[P]}*\big\}.$$

These form a common generalization of $\uparrow^{[\textbf{on}]}$ and the fractional uptimals from Exercise II.4.24 on page 99.

(a) Determine the canonical forms of $\uparrow^{[\textbf{over}]}$ and $\uparrow^{1+\textbf{over}}$.

(b) Express $\uparrow^{[\textbf{over}]}$ and $\uparrow^{1+\textbf{over}}$ as the sup and inf of sequences of loopfree games. (See Exercise II.4.24 on page 99.)

(c) Prove that if $P > Q \ge 1$, then $\uparrow^P \ll \uparrow^Q$.

(d) Prove that $\uparrow^{[1+P]} - \uparrow^{[P]} = \uparrow^{1+P}$, except when P is loopfree.

(e) Which pseuduptimals are absorbed by $\uparrow^P$?

Notes

Stoppers were introduced by Conway in a 1978 paper titled simply *Loopy games* [**Con78**]. The theory was expanded in *Winning Ways*, and most of the material in this section derives from one of those two sources. Note that $\uparrow^{[\textbf{on}]}$ and $+_{\textbf{on}}$ were called **upon** and **tiny**, respectively, in *Winning Ways*.

Exercise 2.8(c): Robert Li.

3. Simplification of Stoppers

Stoppers can be simplified just like ordinary loopfree games, and remarkably this gives a version of the Simplest Form Theorem. In this section, as throughout this chapter, we continue to assume that all games under consideration are finite.

The definitions of dominated and reversible moves from Section II.2 extend verbatim to loopy games. This gives the following generalizations of Theorems II.2.4 and II.2.5:

Theorem 3.1 (cf. Theorem II.2.4 on page 65). *Let G be a stopper, and suppose that G' is obtained from G by removing a dominated Left option from some subposition of G. Then G' is also a stopper, and $G' = G$.*

Proof. It is clear that G' is a stopper and $G' \leq G$, since G' is obtained from G by removing a Left option. So we must show that $G' \geq G$, i.e., that Left can survive $G' - G$ playing second.

The game graph of G' has the same vertices as G, so there is a one-to-one correspondence between subpositions X of G and subpositions X' of G'. Consider a subposition

$$X' - Y$$

of $G' - G$. We say $X' - Y$ is **safe** if the corresponding subposition, $X - Y$, of $G - G$ satisfies $X \geq Y$.

Certainly $G' - G$ is safe, so it suffices to prove that any time Right moves from a safe position $X' - Y$, then Left can return to another safe position.

So suppose Right moves from some safe position, say $X'_0 - Y_0$, to a new position $X'_1 - Y_1$. Then since $X_0 \geq Y_0$ and since $X_1 - Y_1$ is a Right option of $X_0 - Y_0$, Left must have a move from $X_1 - Y_1$ to a position $X_2 - Y_2$ with

$$X_2 \geq Y_2.$$

This means $X'_2 - Y_2$ is safe. Moreover Left has a safe reply from $X'_1 - Y_1$ to $X'_2 - Y_2$, *unless* $X_2 = X_1^{L_1}$, the dominated Left option. But in that case there is some $X_1^{L_2}$ with

$$X_1^{L_2} > X_1^{L_1} \geq Y_2$$

so that $\left(X_1^{L_2}\right)' - Y_2$ is a safe reply. □

Theorem 3.2 (cf. Theorem II.2.5 on page 65). *Let G be a stopper, and suppose that G' is obtained from G by bypassing a reversible Left option from some subposition of G. Then G' is also a stopper and $G' = G$.*

Proof. *Step 1*: We first show that G' is a stopper. Suppose (for contradiction) that there is an alternating cycle

$$X'_1, X'_2, \ldots, X'_k$$

originating from a subposition of G'. Since G is a stopper, this cycle must involve one of the new options. Without loss of generality, we can assume this option is from X_1' to X_2', so that $X_2 = X_1^{LRL}$. But then

$$X_1, X_1^L, X_1^{LR}, X_2, \ldots, X_k$$

is an alternating cycle originating from a subposition of G, a contradiction.

Step 2: We next show that Left can survive $G' - G$ playing second. As in Theorem 3.1, call a subposition $X' - Y$ **safe** if the corresponding subposition $X - Y$ of $G - G$ satisfies $X \geq Y$. It suffices to show that if Right moves away from a safe position, then Left can return to one.

Suppose Right moves from a safe position $X_0' - Y_0$ to $X_1' - Y_1$. Then there must be some $X_2 - Y_2$, a Left option of $X_1 - Y_1$, with $X_2 \geq Y_2$. So Left has a safe reply to $X_2' - Y_2$ *unless* $X_2 = X_1^{L_1}$, the reversible Left option. But in that case, Left must have a survival move on

$$X_1^{L_1 R_1} - Y_2.$$

So either some $X_1^{L_1 R_1 L} \geq Y_2$ or else $X_1^{L_1 R_1} \geq$ some Y_2^R. If $X_1^{L_1 R_1 L} \geq Y_2$, then Left has a safe reply to

$$(X_1^{L_1 R_1 L})' - Y_2'.$$

Otherwise,

$$X_1 \geq X_1^{L_1 R_1} \geq Y_2^R,$$

so that $X_1' - Y_2^R$ is a safe reply.

Step 3: Finally, we show that Left can survive $G - G'$ playing second. This time, we say $X - Y'$ is safe if $X \geq Y$.

Suppose Right moves from $X_0 - Y_0'$ to $X_1 - Y_1'$. Then there must be some $X_2 - Y_2$, a Left option of $X_1 - Y_1$, with $X_2 \geq Y_2$, *unless* $Y_1 = Y_0^{L_1 R_1 L}$ is an option introduced through the bypass. But in that case, Left certainly has a survival move from $Y_0^{L_1 R_1} - Y_0^{L_1 R_1 L}$ (the mirror-image move). Since

$$X_0 \geq Y_0 \geq Y_0^{L_1 R_1},$$

Left must also have a survival move from $X_0 - Y_0^{L_1 R_1 L}$, which translates to a safe reply on $X_1 - Y_1'$. □

Theorem 3.3. *For every stopper G, there exists a stopper $K = G$ such that no subposition of K has any dominated or reversible options.*

Proof. By Theorems 3.1 and 3.2, we can eliminate dominated options and bypass reversible options from any subposition of G, without affecting its value.

Eliminating a dominated option strictly decreases the number of edges in the graph of G, so only finitely many such eliminations are possible. To complete the proof, we must show that there is no infinite sequence

$$G_1, G_2, G_3, \ldots$$

with each G_{i+1} obtained from G_i by bypassing a reversible option. Assume (for contradiction) that such a sequence exists. By passing to an appropriate subsequence, we can assume that each G_{i+1} is obtained from G_i by bypassing a reversible Left option H_i^L of a fixed subposition H_i. We may furthermore assume that each H_{i+1}^L was newly introduced by bypassing H_i^L. (Since there are just finitely many edges in the graph of G, there are necessarily infinitely many i for which H_i^L was introduced during a previous bypass.)

But this implies that each H_{i+1}^L has the form H_i^{LRL}, and so we obtain arbitrarily long alternating runs of H, contradicting the fact that G is a stopper. □

Recall that any *short* games $G = H$ with no dominated or reversible moves must be isomorphic. (This is Theorem II.2.9 on page 67.) An analogous result holds when G and H are stoppers, but with a slightly weaker conclusion: in place of isomorphism, we have *similarity* in the sense of Definition 2.7 (page 295).

Theorem 3.4. *Let G and H be stoppers. Suppose that no subposition of G or H has any dominated or reversible options, and assume that $G = H$. Then $G \simeq H$.*

Proof. Assume that $G \sim_n H$ for all pairs of stoppers G and H with these properties; we'll show that $G \sim_{n+1} H$. The theorem then follows by induction. (The base case $n = 0$ is immediate from the definition of $\sim_0$.)

Fix G^L. Right can survive $G - H$, so he has a survival move on $G^L - H$. So either $G^L \le H^L$ for some H^L or else $G^{LR} \le H$ for some G^{LR}. But the latter would imply

$$G^{LR} \le H < G,$$

contradicting the assumption that G has no reversible options. Therefore $G^L \le H^L$ for some H^L. By exactly the same argument, there is some $G^{L'}$ such that $H^L \le G^{L'}$. But now we have

$$G^L \le H^L \le G^{L'}.$$

Since G has no dominated options, this implies $G^L = H^L = G^{L'}$. By induction on n, we have $G^L \sim_n H^L$. The other cases follow by symmetry. □

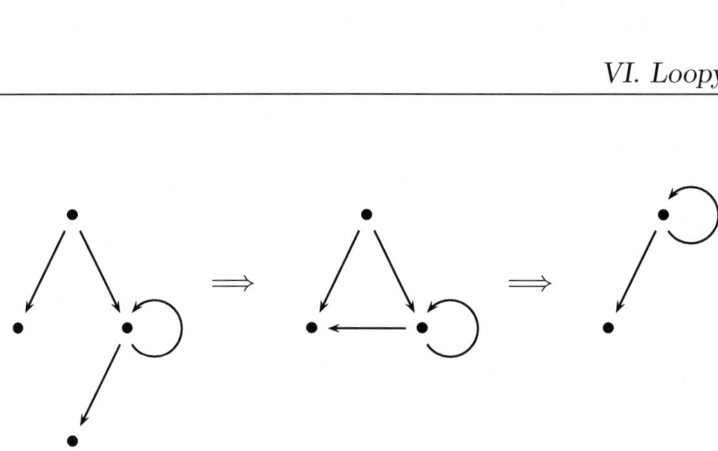

Figure 3.1. Fusion simplifies **over**.

Fusion

We've now established that for every stopper G, there is a $K = G$ with no dominated or reversible options, and this K is unique *up to similarity*. But similarity is not quite isomorphism, as shown in Figure 2.1 on page 292. Different forms of **over** (for example) might have different numbers of subpositions, even though no subpositions of either form have any dominated or reversible options. (The theory of Chapter II shows that this situation isn't possible for short games.)

Similarity can be strengthened to isomorphism with the help of a new type of simplification: fusion of equivalent subpositions. Figure 3.1 shows how **over** might be condensed in two stages. First, the two terminal positions (which are obviously equal) are fused into a single vertex; then the two positions of value **over** are fused in a separate step. Fusion can be seen as a loopy analogue of the Replacement Lemma (Lemma II.2.2 on page 64).

Theorem 3.5 (Fusion Lemma). *Let G be a stopper, and assume that no subposition of G has any dominated or reversible options. Suppose there are distinct subpositions H and J of G with $H = J$. Let G' be obtained from G by replacing all moves to J (throughout all subpositions of G) with moves to H. Then G' is also a stopper and $G' = G$.*

Proof. *Step 1*: We first show that G' is a stopper. Suppose (for contradiction) that there is an infinite alternating run

$$X'_0, X'_1, X'_2, \ldots$$

starting from some subposition X'_0 of G'. Let

$$X_0, X_1, X_2, \ldots$$

be the corresponding sequence of subpositions of G. To obtain a contradiction, we'll construct an alternating run

$$Y_0, Y_1, Y_2, \ldots$$

of a subposition of G. The positions Y_i will all satisfy $Y_i = X_i$.

Put $Y_0 \cong X_0$. Given $Y_i = X_i$, we describe how to construct Y_{i+1}. Suppose that $X'_{i+1} = (X'_i)^L$ (the argument is the same when $X'_{i+1} = (X'_i)^R$). If X_{i+1} is a Left option of X_i, then by Theorem 3.4, there is some Y_i^L with $Y_i^L = X_{i+1}$, and we put $Y_{i+1} \cong Y_i^L$.

Otherwise, it must be the case that J is a Left option of X_i, and also $X_{i+1} \cong H$ (so that X'_{i+1} is reached from X'_i via a replaced move). In that case (also by Theorem 3.4) there is some Y_i^L with $Y_i^L = J$, and we put $Y_{i+1} \cong Y_i^L$. Since $J = H$, this satisfies $Y_{i+1} = X_{i+1}$.

This shows that there is an infinite alternating run of Y_0, contradicting the assumption that G is a stopper.

Step 2: Next we show that $G' = G$. We'll show Left can survive $G' - G$ moving second; an identical proof shows that Right can do the same.

Call a subposition $X' - Y$ of $G' - G$ **safe** if the corresponding subposition $X - Y$ of $G - G$ satisfies $X \geq Y$. It suffices to show that if Right ever moves away from a safe position, then Left can return to one.

So suppose Right moves from $X'_0 - Y_0$ to $X'_1 - Y_1$, with $X_0 \geq Y_0$. Now $X_0 - Y_0$ has a move to $X_1 - Y_1$, except possibly when $X'_1 = H'$ was reached by a replaced option. But in that case $X_0 - Y_0$ has a move to $J - Y_1$, which is equal to $X_1 - Y_1$. So in all cases, Left can survive $X_1 - Y_1$ moving first.

Let $X_2 - Y_2$ be her survival response, so that $X_2 \geq Y_2$. Then $X'_2 - Y_2$ is safe. Left can move to this position from $X'_1 - Y_1$, except when $X_2 = J$. But in that case Left can move to $H' - Y_2$, and since $H = X_2 \geq Y_2$, this is also safe. □

In Theorem 3.5, it is essential that G be free of dominated and reversible options. Otherwise, the proof of Step 1 breaks down and G' might fail to be a stopper (see Exercise 3.1 on page 307).

Canonical Form

We can now prove an analogue of the Simplest Form Theorem.

Definition 3.6. A stopper K is in **canonical form** provided that:

(i) no subposition of K has any dominated or reversible options; and

(ii) if X and Y are distinct subpositions of K, then $X \neq Y$.

Theorem 3.7. *For every stopper G, there exists a stopper $K = G$ in canonical form.*

Proof. By Theorem 3.3 there exists a $J = G$ with no dominated or reversible options. Now K can be obtained from J by repeated application of the Fusion Lemma. □

Theorem 3.8 (Simplest Form Theorem). *Suppose that G and H are stoppers in canonical form, and assume that $G = H$. Then $G \cong H$.*

Proof. Let $\mathscr{A}$ and $\mathscr{B}$ be the sets of subpositions of G and H, respectively. Consider some $X \in \mathscr{A}$. By repeated application of Theorem 3.4, there must exist an $f(X) \in \mathscr{B}$ with $X = f(X)$. This defines a mapping $f : \mathscr{A} \to \mathscr{B}$, and condition (ii) in Definition 3.6 implies that f is one-to-one.

Likewise, to each $Y \in \mathscr{B}$ there must be some $X \in \mathscr{A}$ with $X = Y$, and condition (ii) in Definition 3.6 implies that $Y = f(X)$. Therefore f is a bijection, and Theorem 3.4 shows it to be an isomorphism. □

Longer Cycles

The cycles in all of our previous examples are 1-cycles, which can be regarded as simple pass moves (cf. Figures 2.4 and 2.5). We'll now prove results showing that 2- and 3-cycles are always noncanonical, so that the "next simplest" canonical stoppers necessarily exhibit 4-cycles.

Theorem 3.9. *Suppose that G is a stopper in canonical form. Then G contains no monochromatic cycles of length greater than 1.*

Proof. Suppose G has a monochromatic n-cycle, say,

$$G_0, G_1, G_2, \ldots, G_n \quad (n \geq 1),$$

with each $G_{i+1} = G_i^L$ and $G_0 = G_n$. We'll show that, in fact,

$$G_0 = G_1 = G_2 = \cdots = G_n,$$

so that condition (ii) of Definition 3.6 implies that $n = 1$.

Since $G_0 = G_n$, it suffices to prove that

$$G_0 \leq G_1 \leq G_2 \leq \cdots \leq G_n,$$

i.e., that Left can survive any $G_{i+1} - G_i$ playing second. If Right moves to $G_{i+1} - G_{i+1}$, Left responds to $G_{i+2} - G_{i+1}$, another position of the same form (if $i + 1 = n$, then since $G_n \cong G_0$, Left has a response to $G_1 - G_0$).

If Right steps off the cycle, it must be to some position of the form

$$G_{i+1} - G_i^L \quad \text{or} \quad G_{i+1}^R - G_i.$$

But it cannot be the case that $G_{i+1} \leq G_i^L$, since this would imply that G_{i+1} is dominated by G_i^L (as a Left option of G_i). So Left must have a winning move from $G_{i+1} - G_i^L$.

Likewise, it cannot be the case that $G_{i+1}^R \le G_i$, since this would imply that G_{i+1} is reversible through G_{i+1}^R (as a Left option of G_i). So Left must also have a winning move from $G_{i+1}^R - G_i^L$. □

Theorem 3.9 can be extended to cycles that are "almost monochromatic," having just a single move for Right.

Theorem 3.10. *Suppose that G is a stopper in canonical form. Then G contains no bichromatic cycles with just a single move for Right.*

Proof. Suppose (for contradiction) that G has such a cycle, say,

$$G_0, G_1, G_2, \ldots, G_n \quad (n \ge 1),$$

with each $G_{i+1} = G_i^L$ and $G_0 = G_n^R$. We first show that

$$G_0 \le G_1 \le G_2 \le \cdots \le G_n.$$

The proof is identical to the proof of Theorem 3.9, except when Right moves from $G_n - G_{n-1}$ to $G_n - G_n$, in which case Left must respond to $G_n - G_0$. Now if Right moves to $G_0 - G_0$, Left continues around the cycle just as before; any other move for Right loses outright, since G_n has no dominated or reversible moves.

This shows, in particular, that $G_0 \le G_{n-1}$. But $G_0 = G_{n-1}^{LR}$, contradicting the assumption that G_{n-1} has no reversible moves. □

Theorem 3.11. *If G is a stopper in canonical form, then every cycle of G is either a 1-cycle or has length ≥ 4.*

Proof. By Theorems 3.9 and 3.10, every cycle of G is either a 1-cycle or contains at least two moves for each player. □

Stoppers with canonical 4-cycles exist; an example is shown in Figure 3.2. (In this diagram, the arrows are carefully arranged so that moves for Left always point to the Left, but the edges are additionally labeled with "L" or "R" for clarity.) With more effort it's possible to construct examples with more elaborate cycles (see Exercise 3.5, Figure 3.3, and the notes for this section).

Exercises

3.1 In the statement of the Fusion Lemma, it is assumed that no subposition of G has any dominated or reversible options. Show by example that both hypotheses are necessary.

3.2 Prove that fusion does not introduce any dominated or reversible options.

3.3 Determine the atomic weight of each position in Figure 3.2 on the next page.

3.4 Express each position in Figure 3.2 as the sup and inf of loopfree games.

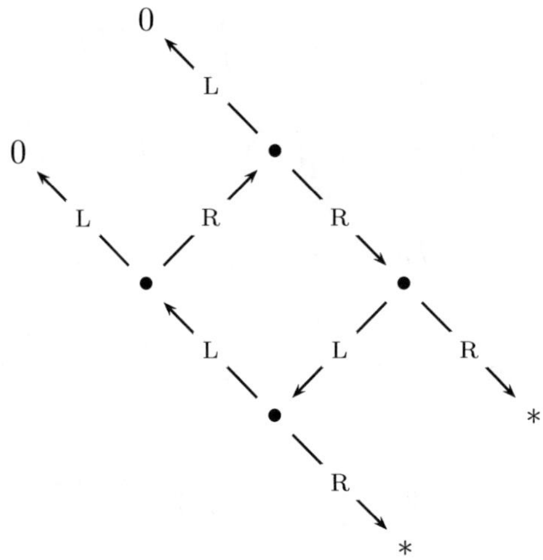

Figure 3.2. A stopper with a 4-cycle in canonical form.

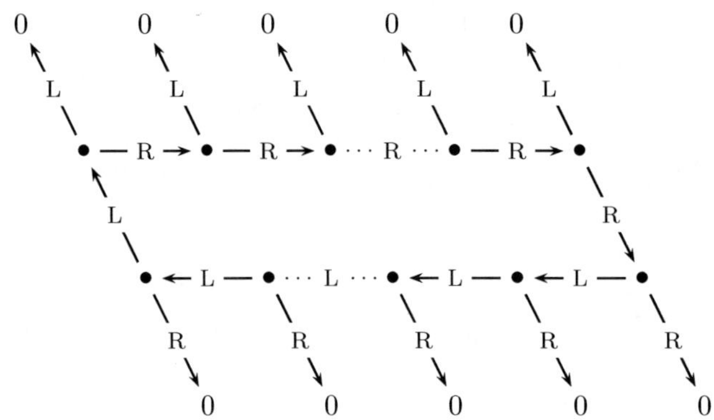

Figure 3.3. A schema for arbitrarily long canonical cycles (cf. Exercise 3.5).

3.5 Suppose that G contains a cycle with m consecutive Left edges, followed by n consecutive Right edges (so that the cycle alternates just once between Left and Right). Theorems 3.9 and 3.10 show that if G is canonical, then necessarily $m \geq 2$ and $n \geq 2$. Figure 3.2 gives an example with $m = n = 2$.

(a) Figure 3.3 illustrates a schema of such G. The dotted edges indicate variable numbers of edges, allowing for arbitrary choices of m and n. Prove that Figure 3.3 is canonical if and only if $m \geq 3$ and $n \geq 3$.

(b) Give an example of a canonical stopper exhibiting a cycle with $m = 3$ and $n = 2$.

Notes

Theorems 3.3 and 3.4 are central results in the loopy theory. They were first observed by Conway; partial proofs appear in [**Con78**] and in *Winning Ways*.

Theorems 3.5 and 3.7–3.11 are due to Siegel [**Sie05, Sie09c**]. These techniques were also used to study **long irreducible cycles** (cycles of length > 1 in canonical stoppers); examples of arbitrarily long cycles are given in Exercise 3.5.

Let $\mathcal{C}$ be a cycle appearing in a canonical stopper G. The **cycle pattern** P of $\mathcal{C}$ is the corresponding sequence of L's and R's as one traverses the cycle; for example, Figure 3.2 has cycle pattern LLRR. There are some obvious restrictions on P: since G is a stopper, P cannot be strictly alternating; and by Theorems 3.9 and 3.10, P must contain at least two L's and at least two R's.

Exercise 3.5 shows that if a cycle alternates just once between Left and Right, then these are the only restrictions on P. It is an open problem whether this is true for cycles with multiple alternations. In [**Sie09c**] it is remarked that $P =$ LLLRLRRRLR is the shortest unresolved pattern.

Open Problem. Characterize the cycle patterns that can occur in a canonical stopper. In particular, determine whether LLLRLRRRLR is a possible pattern.

Shortly after the fusion technique was introduced, long irreducible cycles with rich structure were discovered in one-dimensional PHUTBALL by Nowakowski, Ottaway, and Siegel [**Sie09b**].

All known cycle patterns in PHUTBALL alternate just once between L and R. Call a cycle **tame** if its cycle pattern has this property; **wild** otherwise. Then:

Open Problem. Exhibit a ruleset (with reasonably "natural" rules) containing a position G, such that the canonical form of G has a wild cycle.

4. Sides

Theorem 2.1 (page 290) is a key result in the theory of stoppers: $G \geq H$ if and only if Left can survive $G - H$, playing second. There is no clear analogue of this theorem when G or H is not a stopper. Consequently, in the general case most of the machinery in Sections 2 and 3 breaks down, and the whole theory becomes much more complicated.

However, in many cases, an arbitrary loopy game G can be decomposed into a pair of stoppers, the **sides** of G. When this is possible, then calculations involving G can be carried out in terms of its sides, and the complications of the general theory can be avoided.

The motivating observation is straightforward: if all we want to know is whether Left can survive G (playing first or second), then we might as well assume that draws are wins for Left. Likewise, if we just want to know whether Right can survive G, then we might assume that draws are wins for Right. So write $\hat{o}(G)$ and $\check{o}(G)$ for the **biased outcomes** with draws

		$\hat{o}(G)$			
		$\mathscr{L}$	$\mathscr{P}$	$\mathscr{N}$	$\mathscr{R}$
$\check{o}(G)$	$\mathscr{L}$	$\mathscr{L}$	—	—	—
	$\mathscr{N}$	$\hat{\mathscr{N}}$	—	$\mathscr{N}$	—
	$\mathscr{P}$	$\hat{\mathscr{P}}$	$\mathscr{P}$	—	—
	$\mathscr{R}$	$\mathscr{D}$	$\check{\mathscr{P}}$	$\check{\mathscr{N}}$	$\mathscr{R}$

Figure 4.1. The true outcome of G is determined by its biased outcomes.

redefined as wins for Left or Right, respectively. That is,

$$\hat{o}(G) = \begin{cases} \mathscr{L} & \text{if } o(G) = \mathscr{L},\ \hat{\mathscr{P}},\ \hat{\mathscr{N}}, \text{ or } \mathscr{D}; \\ \mathscr{P} & \text{if } o(G) = \mathscr{P} \text{ or } \check{\mathscr{P}}; \\ \mathscr{N} & \text{if } o(G) = \mathscr{N} \text{ or } \check{\mathscr{N}}; \\ \mathscr{R} & \text{if } o(G) = \mathscr{R}, \end{cases}$$

with $\check{o}(G)$ specified analogously. The true outcome $o(G)$ is completely determined by the biased outcomes $\hat{o}(G)$ and $\check{o}(G)$. This is shown in Figure 4.1: of the sixteen possible pairs $(\hat{o}(G), \check{o}(G))$, seven are prohibited by the restriction $\hat{o}(G) \geq \check{o}(G)$; the remaining nine map one-to-one onto the nine loopy outcome classes.

A biased variant of the fundamental equivalence can now be invoked:

Definition 4.1. Let G and H be loopy games. Then we write

$$\begin{aligned} G \mathrel{\hat{=}} H \quad &\text{if} \quad \hat{o}(G+X) = \hat{o}(H+X) \quad \text{for all } X, \\ G \mathrel{\check{=}} H \quad &\text{if} \quad \check{o}(G+X) = \check{o}(H+X) \quad \text{for all } X, \\ G \mathrel{\hat{\geq}} H \quad &\text{if} \quad \hat{o}(G+X) \geq \hat{o}(H+X) \quad \text{for all } X, \\ G \mathrel{\check{\geq}} H \quad &\text{if} \quad \check{o}(G+X) \geq \check{o}(H+X) \quad \text{for all } X. \end{aligned}$$

The advantage of this approach is that $\hat{o}(G)$ and $\check{o}(G)$ are restricted to the four traditional outcome classes $\mathscr{L}$, $\mathscr{P}$, $\mathscr{N}$, and $\mathscr{R}$, so it's plausible that there exist stoppers S and T with

$$G \mathrel{\hat{=}} S \quad \text{and} \quad G \mathrel{\check{=}} T.$$

In such cases we say that S and T are the **sides** of G (the **onside** and **offside**, respectively), and we write

$$G = S \,\&\, T.$$

To determine the outcome of a sum $G+H$, we can variously replace G by S and T (and likewise for H) and use the theory of Sections 2 and 3 to compute the biased outcomes. Since the full outcome of $G + H$ is determined by its biased outcomes, the problem of computing $o(G + H)$ is effectively reduced to the theory of stoppers.

It turns out that S and T are unique (when they exist), so that writing them in simplest form gives a genuine canonical representation of G. Moreover, in many cases S and T will be ordinary loopfree games, and the full theory of Chapter II can be used to investigate the structure of G.

This is really all we need to get a theory going! For instance:

Example. We see that

$$\hat{o}(\mathbf{on} + X) = \hat{o}(\mathbf{dud} + X) = \mathscr{L} \quad \text{for all } X,$$

since Left can never run out of moves in either case. Therefore **dud** has onside **on**, and by symmetry, offside **off**. This shows that

$$\mathbf{dud} = \mathbf{on} \,\&\, \mathbf{off}.$$

In order for the sides to be *useful*, we need to have some method for calculating them. That will be the subject of the remainder of this section.

Note that the sided relations $\hat{\geq}$ and $\check{\geq}$ have the following simple interpretation in terms of outcomes.

Proposition 4.2. *For all loopy games G and H:*

(a) *$G \hat{\geq} H$ iff whenever Left can survive $H + X$ playing first (resp. second), then she can also survive $G + X$ playing first (resp. second).*

(b) *$G \check{\geq} H$ iff whenever Left can win $H + X$ playing first (resp. second), then she can also win $G + X$ playing first (resp. second).*

Proof. (a) Directly from the definition, we see that $G \hat{\geq} H$ if and only if

$$\begin{aligned} \hat{o}(H + X) \geq \mathscr{P} \quad &\Longrightarrow \quad \hat{o}(G + X) \geq \mathscr{P} \quad \text{and} \\ \hat{o}(H + X) \geq \mathscr{N} \quad &\Longrightarrow \quad \hat{o}(G + X) \geq \mathscr{N}, \quad \text{for all } X. \end{aligned}$$

Now $\hat{o}(H + X) \geq \mathscr{P}$ if and only if $o(H + X) \geq \check{\mathscr{P}}$. Similar identities apply to each clause of the previous equation, so in fact $G \hat{\geq} H$ if and only if

$$\begin{aligned} o(H + X) \geq \check{\mathscr{P}} \quad &\Longrightarrow \quad o(G + X) \geq \check{\mathscr{P}} \quad \text{and} \\ o(H + X) \geq \check{\mathscr{N}} \quad &\Longrightarrow \quad o(H + X) \geq \check{\mathscr{N}}, \quad \text{for all } X, \end{aligned}$$

which is equivalent to the conclusion of the proposition.

(b) is proved the same way, using the fact that $\check{o}(H + X) \geq \mathscr{P}$ if and only if $o(H + X) \geq \hat{\mathscr{P}}$. □

Concentrating Strategies

Recall from Section 1 the asymmetry in Theorem 1.14:

(a) if $G \geq H$, then Left can survive $G - H$ playing second; whereas

(b) if Left can win $G - H$ playing second, then $G \geq H$.

When G and H are stoppers, this asymmetry nicely resolves itself, because we can replace "win" by "survive" in clause (b) to obtain an equivalent condition. In a more general context, this isn't possible, but a partial converse can be achieved through suitable restrictions on Left's survival strategy for $G - H$.

Definition 4.3. Let σ be a strategy for $G+H$. We say that σ **concentrates on** G if every infinite σ-run involves infinitely many moves on G.

Theorem 4.4. *If Left has a survival strategy for $G - H$ that concentrates on G, then $G \mathrel{\hat{\geq}} H$. Likewise, if Left has a survival strategy for $G - H$ that concentrates on $-H$, then $G \mathrel{\check{\geq}} H$.*

Note that any winning strategy σ for $G - H$ vacuously concentrates on *both* G and $-H$ (since there can be *no* infinite σ-runs), so this result generalizes Theorem 1.14(b). Likewise, if G and H are stoppers, then any survival strategy σ must concentrate on both G and $-H$, since no infinite run can be confined to either. So this result also generalizes Theorem 2.1.

The method of proof is exactly like Theorems 1.14 and 2.1. As before, we extract the main idea out into a lemma.

Lemma 4.5. *Suppose that Left has a survival strategy for $G - H$ that concentrates on G. Then if she can survive $H + X$ playing first (second), she can also survive $G + X$ playing first (second).*

Proof. Consider a subposition $\dot{G} + \dot{X}$ of $G + X$. We define $\dot{G} + \dot{X}$ to be **safe** if there is some subposition $\dot{H}$ of H such that:

(i) Left has a survival strategy for $\dot{G} - \dot{H}$ that concentrates on $\dot{G}$; and

(ii) Left has a survival strategy for $\dot{H} + \dot{X}$ (without restriction).

The rest of the proof is just like Lemmas 1.16 and 2.2. The fact that σ concentrates on $\dot{G}$ guarantees that every associated swivel run is finite. □

Proof of Theorem 4.4. Suppose Left has a survival strategy for $G - H$ that concentrates on G, and fix any loopy game X. If Left can survive $H+X$ playing first (second), then she can also survive $G+X$, by Lemma 4.5. This is sufficient to establish $\hat{o}(G+X) \geq \hat{o}(H+X)$. The second assertion follows by symmetry. □

Example. In Section 2 we considered the following pair of games:

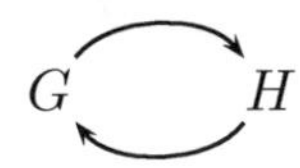

There we showed that $\Sigma_n^{\mathbf{on}}(G) = 0$ and $\Sigma_n^{\mathbf{on}}(H) = 1$ for all sufficiently large n, so in particular

$$0 \geq G \quad \text{and} \quad 1 \geq H.$$

One can also see that

$$G \hat{\geq} 0 \quad \text{and} \quad H \hat{\geq} 1,$$

since on $G - 0$ Right has no moves at all; and on $H - 1$ Right's only move is to $H - 0$, after which infinite play concentrates on H. Therefore $G \hat{=} 0$ and $H \hat{=} 1$, and by symmetry

$$G = 0 \,\&\, {-1} \quad \text{and} \quad H = 1 \,\&\, 0.$$

Proposition 4.6. *If S is a stopper and G is any loopy game, then*

$$S \hat{\geq} G \quad \textit{if and only if} \quad S \geq G$$

and

$$S \check{\leq} G \quad \textit{if and only if} \quad S \leq G.$$

Proof. Suppose that $S \hat{\geq} G$. Left can certainly survive $G - G$ playing second (by playing the mirror-image strategy), so by Proposition 4.2 she can survive $S - G$ as well. But since S is a stopper, any Left survival strategy for $S - G$ necessarily concentrates on $-G$, so that $S \check{\geq} G$. Therefore $S \geq G$.

The converse implication is clear from the definitions, and the second claim follows by symmetry. □

In particular, note that if S and T are both stoppers, then

$$S \hat{\geq} T \quad \text{if and only if} \quad S \geq T \quad \text{if and only if} \quad S \check{\geq} T.$$

Several basic facts about the relationship between G and its sides now follow:

Proposition 4.7. *Let G be a loopy game, and suppose that $G = S \,\&\, T$. Then:*

(a) $S \geq G \geq T$.

(b) *(Uniqueness of sides) If also $G = S' \,\&\, T'$, then $S = S'$ and $T = T'$.*

Proof. Both (a) and (b) follow directly from Proposition 4.6. (For (b), we use the facts that $S \hat{=} S'$ and $T \check{=} T'$.) □

Proposition 4.8. *If S is a stopper, then $S = S \,\&\, S$.*

Proof. Since $S = S$, we certainly have $S \hat{=} S$ and $S \check{=} S$. □

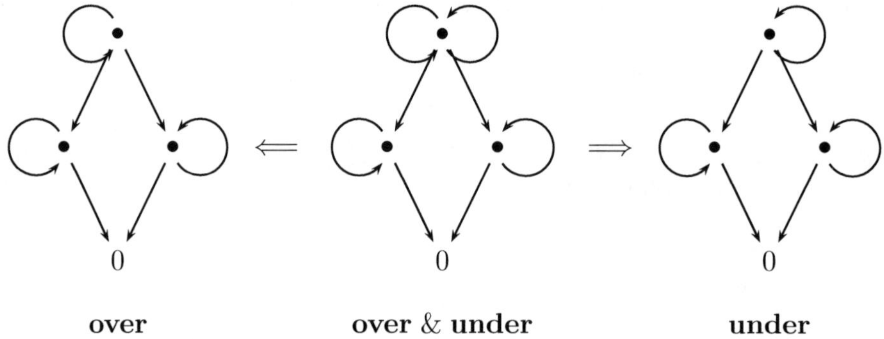

Figure 4.2. Picking plums produces the sides.

Plumtrees

When the only cycles in the graph of G are 1-cycles, then the sides of G are particularly easy to determine. This type of game is important enough to deserve a special name.

Definition 4.9. A loopy game G is a **plumtree** if the only cycles in its graph are 1-cycles.

Most of the examples we've considered so far are plumtrees, including all the games in Figure 2.4 on page 298. The name "plumtree" reflects the appearance of their game graphs: they can always be drawn as trees, variously decorated with pass moves (the **plums**).

Computing the sides of a game G is in general a hard problem, but when G is a plumtree it becomes trivial. We can obtain the onside S by removing Right's plum from every subposition that has plums for *both* players. Here's an example:

Example. Consider the sum

$$G = \mathbf{over} + \mathbf{under}.$$

Its game graph is drawn in Figure 4.2. In this example, only a single position (G itself) has plums for both players, so the onside and offside can be obtained by removing them, as shown in the figure.

The sides can then be simplified using the techniques from Section 2. Both Left options from the onside S are reversible: the pass move reverses through S^R, to 0; and Left's other option reverses out entirely through 0. This establishes that $S = \mathbf{over}$, and by symmetry the offside is **under**, so that

$$\mathbf{over} + \mathbf{under} = \mathbf{over} \mathbin{\&} \mathbf{under}.$$

Theorem 4.10. *Let G be a plumtree. Let S be obtained by removing Right's plum from every subposition that has plums for both players, and let T be obtained by removing Left's plums from such subpositions. Then S and T are stoppers, and*

$$G = S \mathbin{\&} T.$$

Proof. Suppose that H is a subposition of S and let $\vec{H} = (H_0, H_1, \ldots)$ be an infinite run of H. Since S is a plumtree, no position of $\vec{H}$ can be visited more than once by a nonpass move. It follows that for some n, every move beyond H_n is a pass move. But H_n has a plum for at most one player, so $\vec{H}$ cannot be alternating. This shows that S is a stopper.

Now clearly $S \geq G$, since S is obtained by removing Right options from subpositions of G. To show that S is the onside of G, we must also establish that $G \mathrel{\hat{\geq}} S$. Since S is a stopper, we need only show that Left can survive $G - S$, playing second.

To survive $G - S$, Left follows the mirror-image strategy, *except* when Right makes a pass move on G that has no counterpart on S. By definition of S, this can only happen if Left *also* has a pass move on G, so she can respond with her own pass move, preserving mirror-image symmetry.

This shows that S is the onside of G, and an identical argument shows that T is the offside. □

Bach's Carousel

When $G = S \mathbin{\&} T$ for stoppers S and T, then we can put S and T into simplest form, obtaining a true canonical representation of G. We've just shown that this is always possible when G is a plumtree, and it is natural to ask whether *every* loopy game is similarly **stopper-sided**. Somewhat surprisingly, the answer appears to be no!

Example. Bach's Carousel is shown in Figure 4.3 on the next page (named in honor of Clive Bach, who discovered it). In this diagram we've dropped the convention that Left options always point to the left and simply labeled each edge with an L, R, or E as appropriate (options labeled with E are available to *E*ither player).

Let G be the southwest vertex of the Carousel (from which both players have a move to 0), and let S be a stopper. We'll show that there cannot exist survival strategies for Left for $G - S$ and $S - G$ that concentrate on G and S, respectively. Assume (for contradiction) that such strategies exist, say σ and τ. Without loss of generality, assume that S is given in canonical form.

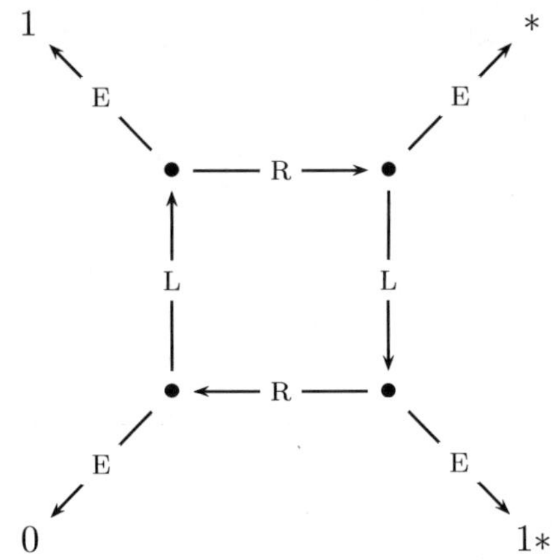

Figure 4.3. Bach's Carousel.

Let G^L, G^{LR}, and G^{LRL} be successive moves around the Carousel proceeding from G. Note the following key properties of G:

- Moves around the Carousel fail to be reversible: $G - G^{LR}$ is a first-player win, with Left's winning move to $G - *$, and Right's to $0 - G^{LR}$. Similarly, $G^L - G^{LRL}$ is a first-player win.
- Moves around the Carousel also fail to be dominated: Right has winning moves from $0 - G^L$ and $* - G^{LRL}$, and Left has winning moves from $1* - G$ and $1 - G^{LR}$.

Now $\tau(S - G^L)$ must be a survival move for Left (where G^L is the move around the Carousel). Since $G - G^{LR}$ is a first-player win and $S \mathrel{\hat{=}} G$, it cannot have the form $S - G^{LR}$, so $\tau(S - G^L) = S^L - G^L$. Likewise $\sigma(G - S^L)$ must be a survival move for Left, and since S has no reversible options, it must have the form $G^{L'} - S^L$.

Since Left can survive $G^{L'} - S^L$ and $S^L - G^L$ and since Right can win $0 - G^L$, it cannot be the case that $G^{L'} = 0$. Therefore $G^{L'} = G^L$. It follows that σ and τ are concentrating strategies for $G^L - S^L$ and $S^L - G^L$.

Iterating this argument, we find that there exists an S^{LR} such that $G^{LR} - S^{LR}$ and $S^{LR} - G^{LR}$ both admit concentrating strategies; likewise an S^{LRL}, and so on around the Carousel. This yields an infinite alternating run of S, contradicting the assumption that S is a stopper.

This strongly suggests that the Carousel is not stopper-sided, leaving the whole theory curiously incomplete. More than 30 years ago, Conway asked how one might resolve this, and today we appear no closer to a solution. Quoting verbatim from *Winning Ways*:

Question. Is there an alternative notion of simplest form that works for *all* finite loopy games (in particular, for the Carousel)?

To this we add two additional questions:

Question. Is the sum of finite stoppers necessarily stopper-sided?

Question. Given stoppers S and T with $S \geq T$, must there necessarily exist a game G with $G = S \,\&\, T$?

This last question has an affirmative answer when S and T are plumtrees (Exercise 4.7), but the general case remains unresolved.

The Sidling Theorem

When G is not a plumtree, more sophisticated techniques are needed to calculate the sides of G. The sidling approximations $\Sigma_n^{\mathbf{on}}(G)$ and $\Sigma_n^{\mathbf{off}}(G)$, introduced in Definition 2.5 on page 292, are sequences of stoppers that approach G from above and below, respectively. We've seen that in some cases, these approximations converge to a constant value for all sufficiently large n. The following theorem shows that whenever this happens, then their limits are equal to the sides of G.

Theorem 4.11 (Sidling Theorem). *Let G be a loopy game, and suppose that for some $n_0 \geq 0$, we have*

$$\Sigma_{n_0+1}^{\mathbf{off}}(H) = \Sigma_{n_0}^{\mathbf{off}}(H)$$

for every subposition H of G. Then $G \doteq \Sigma_{n_0}^{\mathbf{off}}(G)$.

To prove Theorem 4.11, we must first generalize some of the machinery from Section 2.

Proposition 4.12 (cf. Proposition 2.6). *For all loopy games G and $n \geq 0$,*

$$\Sigma_n^{\mathbf{on}}(G) \geq G \quad \textit{and} \quad \Sigma_n^{\mathbf{off}}(G) \leq G.$$

Proof. On the difference $\Sigma_n^{\mathbf{on}}(G) - G$, Left can follow the mirror-image strategy as in the proof of Proposition 2.6. After a finite number of moves, the position will reach $\mathbf{on} - H$ for some subposition H of G, after which Left can confine her play completely to the **on** component. This is a survival strategy, and it clearly concentrates on $\Sigma_n^{\mathbf{on}}(G)$. This shows that $\Sigma_n^{\mathbf{on}}(G) \mathrel{\hat{\geq}} G$, which suffices since $\Sigma_n^{\mathbf{on}}(G)$ is a stopper. □

In fact Proposition 4.12 can be proved for any sidling superscript, but the proof is more involved (see Exercise 4.6).

Theorem 4.13 (cf. Theorem 2.8 on page 295). *Let $\mathcal{A}$ be a set of stoppers and let G be an arbitrary loopy game. Suppose that:*

(i) *$G \geq X$ for all $X \in \mathcal{A}$; and*

(ii) *for each $n \geq 0$, there is an $X \in \mathcal{A}$ with $G \sim_n X$ (for some form of X).*

Then for any stopper *S with $S \geq X$ for every $X \in \mathcal{A}$, we have $S \check{\gtrsim} G$.*

Proof. Suppose (for contrapositive) that $S \check{\not\gtrsim} G$. Then on $G - S$ Right has no survival strategy that concentrates on G. But since S is a stopper, *every* survival strategy necessarily concentrates on G, and it follows that Right has no survival strategy at all on $G - S$.

Therefore Left can win on $G - S$ playing first. Choose n so that Left can win on $G - S$ in at most $n - 1$ moves. By assumption (ii), there is some $X \in \mathcal{A}$ with $G \sim_n X$. Then Left can win on $X - S$ by following exactly the same strategy used for $G - S$, and since X and S are both stoppers, this implies $S \lhd\!\!| X$. □

Proof of Theorem 4.11. Let $S = \Sigma^{\mathbf{off}}_{n_0}(G)$. Since $\Sigma^{\mathbf{off}}_{n_0+1}(H) = \Sigma^{\mathbf{off}}_{n_0}(H)$ for every subposition H of G and since each $\Sigma^{\mathbf{off}}_{n+1}(G)$ depends only on the values of $\Sigma^{\mathbf{off}}_{n}(H)$, it follows that $\Sigma^{\mathbf{off}}_{n}(G) = S$ for all $n \geq n_0$. Put

$$\mathcal{A} = \{\Sigma^{\mathbf{off}}_{n}(G) : n \geq 0\};$$

we have just shown that S is the greatest element of $\mathcal{A}$.

Now $G \sim_n \Sigma^{\mathbf{on}}_{n}(G)$, by definition of sidling, so the hypotheses of Theorem 4.13 are satisfied. Therefore $S \check{\gtrsim} G$. Since also $G \geq S$ (Proposition 4.12), this shows that $G \doteq S$. □

The Sidling Theorem nicely ties together many ideas from the theory of loopy games. Sadly, however, its practical relevance is limited, as there are many stopper-sided games whose sidling approximations fail to converge. Several other techniques for computing sides are discussed in the notes on page 321.

Upsum and Downsum

The decomposition $G = S \,\&\, T$ dissects G into a pair of stoppers; we can likewise dissect the sum operation $G + H$ into a pair of operations on stoppers. The resulting **upsum** and **downsum** operations have some quite natural properties.

Definition 4.14. If $G + H = S \,\&\, T$ for stoppers S and T, then we say that S is the **upsum** of G and H, and T the **downsum**, and we write

$$S = G \curlywedge H \quad \text{and} \quad T = G \curlyvee H.$$

Of course, upsum and downsum are only well-defined when $G + H$ is stopper-sided. This is something of a hole in the theory, since there exist games (such as Bach's Carousel) that don't appear to be stopper-sided; and we don't even know whether the sum of *stoppers* is necessarily stopper-sided. In fact the plumtrees are the largest known class of stopper-sided games that is closed under upsum and downsum.

The remaining theory in this chapter makes quite heavy use of upsum and downsum. Rather than continually restating assumptions about their existence, we will assume that this theory only applies to classes of games over which upsum and downsum are defined and that are closed under them. The theory is already quite powerful in the case of plumtrees; a fuller understanding of its extensions will have to await later theoretical developments.

Proposition 4.15. *The operations $\curlywedge$ and $\curlyvee$ are commutative and associative.*

Proof. Commutativity is immediate from the definition. To see that $\curlywedge$ is associative (say), suppose that

$$G + H = S \mathbin{\&} T \quad \text{and} \quad H + J = U \mathbin{\&} V.$$

Then for all X,

$$\hat{o}(S + J + X) = \hat{o}(G + H + J + X) = \hat{o}(G + U + X)$$

so that $S \curlywedge J = G \curlywedge U$. □

Proposition 4.16. *For all G, H, and J:*

(a) $\overline{G \curlywedge H} = \overline{G} \curlyvee \overline{H}$ *and* $\overline{G \curlyvee H} = \overline{G} \curlywedge \overline{H}$.

(b) $G \mathrel{\hat{\gtrsim}} H$ *implies* $G \curlywedge J \geq H \curlywedge J$ *and* $G \mathrel{\check{\gtrsim}} H$ *implies* $G \curlyvee J \geq H \curlyvee J$.

(c) $G \curlywedge H \geq G \curlyvee H$.

(Recall that $\overline{G}$ is a synonym for the negative $-G$. The notation $\overline{G}$ is somewhat more convenient in expressions involving upsum and downsum.)

Proof. Write $G + H = S \mathbin{\&} T$.

(a) Here $\overline{G} + \overline{H} = \overline{T} \mathbin{\&} \overline{S}$, since whenever $\hat{o}(G + X) = \hat{o}(S + X)$, say, then certainly $\check{o}(\overline{G} + \overline{X}) = \check{o}(\overline{S} + \overline{X})$.

(b) If $G \mathrel{\hat{\gtrsim}} H$, then $\hat{o}(G + X) \geq \hat{o}(H + X)$ for all X, so in particular $\hat{o}(G + J + X) \geq \hat{o}(H + J + X)$ for all X. It follows that $G \curlywedge J \geq H \curlywedge J$. The second assertion is proved the same way.

(c) This is straight from Proposition 4.7. □

Proposition 4.17 (A, B, C Property)**.** *For all stoppers A, B, and C,*

$$A \curlywedge B \geq C \quad \textit{if and only if} \quad A \geq \overline{B} \curlyvee C.$$

Proof. If $A \curlywedge B \geq C$, then certainly $A + B \mathrel{\hat{\gtrsim}} C$. So

$$\hat{o}(A + B + \overline{C}) \geq \hat{o}(C + \overline{C})$$

and it follows that Left can survive $A + B + \overline{C}$, playing second. Since A is a stopper, her play must concentrate on $B + \overline{C}$, so by Theorem 4.4

$$A \mathrel{\check{\gtrsim}} \overline{B} + C.$$

Therefore $A \geq \overline{B} \curlyvee C$. The converse is given by the same argument in reverse, using the fact that $\overline{C}$ is a stopper. $\square$

Exercises

4.1 Determine the onside and offside of the following games G and H. Express all answers in canonical form.

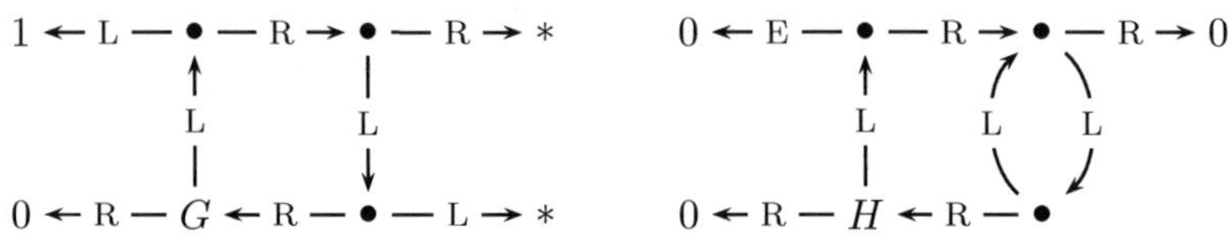

4.2 The **height** of a plumtree G is the longest sequence of nonpass moves proceeding from G. Determine the number of distinct plumtrees of height ≤ 2.

4.3 If G and H are plumtrees, then $G + H$ is stopper-sided.

4.4 G is a **zugzwang game** if $G^L < G < G^R$ for every G^L and G^R. Prove that if G is a zugzwang game, then $G = x \,\&\, y$ for some $x, y \in \mathbb{D}$.

4.5 G is a **weak zugzwang game** if $G^L \leq G \leq G^R$ for every G^L and G^R. Prove that if G is a weak zugzwang game, then $G = S \,\&\, T$ for some pseudonumbers S and T (cf. Exercise 2.8 on page 300).

4.6 (Cf. Propositions 2.6 and 4.12) Let G be an arbitrary game and let T be a stopper. Suppose that $T \geq H$ for every subposition H of G. Show that $\Sigma_n^T(G) \geq G$ for all n. (Show that any strategy for $G + X$ can be converted into one for $\Sigma_n^T(G) + X$.)

4.7 *Grafting plumtrees.* Let S and T be stopper plumtrees, with $S \geq T$. Let

$$R = \left\{ S^L,\ \hat{d}(T^L) \,\middle|\, \check{d}(S^R),\ T^R \right\}$$

where for all G,

$$\hat{d}(G) = \{G^L \mid G^R, \mathbf{dud}\} \quad \text{and} \quad \check{d}(G) = \{G^L, \mathbf{dud} \mid G^R\}.$$

Prove that $R = S \,\&\, T$.

4.8 *Onside simplification.* A Left option G^{L_1} is **onside-dominated** if $G^{L_2} \mathrel{\hat{\gtrsim}} G^{L_1}$ for some other Left option G^{L_2}. Likewise, a Right option G^{R_1} is **onside-dominated** if $G^{R_2} \mathrel{\hat{\lesssim}} G^{R_1}$ for some other Right option G^{R_2}, *provided that* no alternating cycle contains the move $G \to G^{R_2}$.

G^{R_1} is **onside-reversible** if $G^{R_1L_1} \hat{\geq} G$ for some $G^{R_1L_1}$. Likewise, G^{L_1} is **onside-reversible** if $G^{L_1R_1} \hat{\leq} G$ for some $G^{L_1R_1}$, *provided that* no alternating cycle contains the sequence of moves $G \to G^{L_1} \to G^{L_1R_1}$.

(a) If G' is obtained from G by eliminating an onside-dominated option from some subposition of G, then $G' \hat{=} G$.

(b) If G' is obtained from G by bypassing an onside-reversible option from some subposition of G, then $G' \hat{=} G$.

(c) State and prove the corresponding definitions and results for offside simplification.

(d) Suppose G' is obtained from G by fusing Y to X, where $X \hat{=} Y$ and there is no even-length alternating path from X to Y. Show that $G' \hat{=} G$.

Notes

Decomposition into sides was first recognized by Robert Li [**Li76**]. Li studied only **zugzwang games** in which it is a disadvantage to move (Exercises 4.4 and 4.5), generalizing ordinary numbers to the loopy context.

On reading Li's paper, Conway suspected a more general theory. Together with Norton and Bach, he developed the theory of stoppers and showed that they naturally represent the sides of many games. These results were published in a groundbreaking paper titled simply *Loopy games* [**Con78**] and also in *Winning Ways*. Later, Moews generalized the sidling technique and applied it to GO [**Moe93, Moe96b**].

The proof of the Sidling Theorem given here is previously unpublished (as far as the author knows) and is somewhat simpler than the proof in *Winning Ways*. In many cases, however, the Sidling Theorem fails to isolate the sides of G, even when G is stopper-sided. There are several additional techniques that are more successful in practice, such as *onside simplification* (Exercise 4.8) and *unraveling* [**Sie05, Sie09c**], but none is known to be fully general.

Open Problem. Give an algorithm to determine whether an arbitrary loopy game G is stopper-sided and to calculate its sides if it is.

Games that have been successfully analyzed using this theory include BACKSLIDING TOADS AND FROGS [**Sie09a**] (a loopy variant of TOADS AND FROGS described in *Winning Ways*) and HARE AND HOUNDS [**Sieb**]. Albert and Siegel developed an effective implementation of sided simplification [**Sie05, Sie09c**] that works in *most* cases and is incorporated into *cgsuite*.

Conway's version of the theory is somewhat different from ours. In Conway's framework, a game G comes equipped with a mapping p that associates to each infinite play $\vec{G}$ an outcome $p(\vec{G}) \in \{\mathscr{L}, \mathscr{D}, \mathscr{R}\}$. Therefore infinite plays are not necessarily draws, but depend on the annotation p. An (annotated) game is **free** if $p(\vec{G}) = \mathscr{D}$ for every infinite play $\vec{G}$, and **fixed** if $p(\vec{G}) \neq \mathscr{D}$ for every such $\vec{G}$. Then we define

$$G = H \quad \text{if} \quad o(G + X) = o(H + X) \text{ for all } X,$$

as usual, but G, H, and X range over all *annotated* games, and $o(G)$ observes the annotation of G in the case of infinite plays. We define G^+ (resp. G^-) to be the fixed game with the same structure as G, but with all draws on G redefined as wins for Left (resp. Right).

The advantage of this approach is that Left can survive $G^+ - H^+$ *if and only if* $G \mathrel{\hat{\gtrsim}} H$, so that Theorem 4.4 becomes an equivalent condition, simplifying some of the proofs. However, the use of annotated games in the fundamental equivalence is a bit unsettling: it destroys the careful separation of the structure of games from their outcomes; and it distracts from the primary emphasis on games in which infinite plays are draws. For these reasons, we've departed from the literature and emphasized biased outcomes $\hat{o}$ and $\check{o}$ rather than annotated games G^+ and G^-.

Note that if S is a stopper, then $G^+ = S^+$ if and only if S is the onside of G (in our sense), so the two versions of the theory are exactly equivalent in their treatment of stopper-sided games (which is the main practical value of this theory).

5. Idempotents

If G is loopfree, then of course

$$G + \overline{G} = 0,$$

since second player can win by following the mirror-image strategy. But if G is loopy, then in general

$$G + \overline{G} = D \mathbin{\&} \overline{D}$$

for some stopper D, the **degree** of G. Often the degree will be an **idempotent**: $D + D = D$. In such cases the games of degree D are closed under addition, and in this section we'll see that they admit a surprisingly rich structure theory.

Definition 5.1. The **degree of loopiness** (or just **degree**) of G, denoted by G°, is defined by

$$G^\circ = G \mathbin{⅄} \overline{G}.$$

If G° is an idempotent, then we say that G° is **stable** and that G has **stable degree**.

Examples. We've seen that

$$\mathbf{on} + \mathbf{off} = \mathbf{on} \mathbin{\&} \mathbf{off},$$

so that $\mathbf{on}^\circ = \mathbf{off}^\circ = \mathbf{on}$. It's easily checked that $\{\mathbf{on} \mid \mathbf{off}\}^\circ = \mathbf{on}$ as well.

Likewise, we have $\mathbf{over}^\circ = \mathbf{over}$ and $\boldsymbol{+}_{\mathbf{on}}^\circ = \boldsymbol{+}_{\mathbf{on}}$. However

$$\uparrow^{[\mathbf{on}]} + \downarrow_{[\mathbf{on}]} = \uparrow^{\mathbf{on}} \mathbin{\&} \downarrow_{\mathbf{on}},$$

so that $(\uparrow^{[\mathbf{on}]})^\circ = \uparrow^{\mathbf{on}}$, which is rather smaller than $\uparrow^{[\mathbf{on}]}$.

Example. The game

$$G = \{\mathbf{on} \mid 0\}$$

presents an interesting case. We have $\mathbf{on} + \overline{G} = \{\mathbf{on} \mid \mathbf{dud}\}$, so that

$$G + \overline{G} = \big\{\{\mathbf{on} \mid 0\}, \{\mathbf{on} \mid \mathbf{dud}\} \;\big\|\; \{0 \mid \mathbf{off}\}, \{\mathbf{dud} \mid \mathbf{off}\}\big\}.$$

This is a plumtree, so by picking plums as in Section 4, we find that $G + \overline{G}$ has onside

$$\big\{\{\mathbf{on} \mid 0\}, \{\mathbf{on} \mid \mathbf{on}\} \;\big\|\; \{0 \mid \mathbf{off}\}, \{\mathbf{on} \mid \mathbf{off}\}\big\},$$

obtained by replacing each **dud** with **on**. This simplifies to

$$\{\mathbf{on} \,\|\, 0 \mid \mathbf{off}\}$$

after removing dominated options.

Therefore $G^\circ = \{\mathbf{on} \,\|\, 0 \mid \mathbf{off}\}$; but $G + G = \mathbf{on}$, so that $(G+G)^\circ = \mathbf{on}$. This shows that the games of degree G° are *not* closed under sums.

The above example shows that G° is not necessarily stable. However, in the case where D *is* stable, there is a rich structure theory governing the games of degree D. For G and H of stable degree D, write

$$\begin{aligned} G \equiv H \pmod{D} \quad &\text{if and only if} \quad G \mathbin{\curlywedge} D = H \mathbin{\curlywedge} D \\ &\text{if and only if} \quad G \mathbin{\curlyvee} \overline{D} = H \mathbin{\curlyvee} \overline{D}. \end{aligned}$$

(The latter implication is a consequence of the fact that D is an idempotent; we'll prove it shortly.) Then the **class** of G is its equivalence class modulo D, denoted by $[G]$. It turns out that under suitable assumptions on D, we can associate to every game G of degree D a unique **variety** $V(G) \in [D]$. Then for G and H of degree D,

$$G = H \quad \text{if and only if} \quad [G] = [H] \text{ and } V(G) = V(H),$$

so that every game of degree D is uniquely characterized by its class and variety. Moreover, we'll see that for each class $[G]$, the variety mapping

$$V : [G] \to [D]$$

is an upsum- and downsum-preserving bijection, so that all of the classes of degree D have identical structure.

Example: Varieties of $\uparrow^{\mathbf{on}}$

We've seen that $\uparrow^{[\mathbf{on}]}$ is a game of atomic weight 1 with degree $\uparrow^{\mathbf{on}}$. Its multiples form a natural sequence of games with atomic weight n:

$$\uparrow^{[\mathbf{on}]},\ 2 \cdot \uparrow^{[\mathbf{on}]},\ 3 \cdot \uparrow^{[\mathbf{on}]},\ \ldots,$$

whose canonical forms (for $n \geq 1$) are given by

$$(n+1) \cdot \uparrow^{[\mathbf{on}]} = \big\{0 \;\big|\; n \cdot \uparrow^{[\mathbf{on}]} + *\big\}.$$

(This identity can be easily verified by showing that the difference of the two forms is a $\mathscr{D}$-position.) For comparison, note the parallel with the loopfree analogues $\Uparrow n$:

$$(n+1)\cdot\uparrow = \{0 \mid n\cdot\uparrow + *\}.$$

Now consider the upsum

$$U = \uparrow^{[\mathbf{on}]} \,\hat{+}\, \uparrow^{\mathbf{on}}.$$

U has degree $\uparrow^{\mathbf{on}}$ and atomic weight 1 (since $\uparrow^{[\mathbf{on}]}$ and $\uparrow^{\mathbf{on}}$ have atomic weights 1 and 0, respectively, and both have degree $\uparrow^{\mathbf{on}}$); but it is distinct from $\uparrow^{[\mathbf{on}]}$, with canonical form given by

$$U = \{0 \mid \uparrow^{\mathbf{on}}*\}.$$

Observe the close resemblance to the canonical forms of $\uparrow^{[n]} + \uparrow^{n}$ (cf. Section II.4):

$$\uparrow^{[n]} + \uparrow^{n} = \{0 \mid \uparrow^{n}*\},$$

so that U behaves like the limit of the uptimal sequence 0.2, 0.12, 0.112, ... Since $\uparrow^{\mathbf{on}}$ is an idempotent, the multiples $n\cdot U$ have the form $n\cdot\uparrow^{[\mathbf{on}]} + \uparrow^{\mathbf{on}}$, and we have a new sequence

$$U,\ 2\cdot U,\ 3\cdot U,\ \ldots.$$

Putting all these facts together, we see two complementary sequences:

$$\begin{array}{ccccccccc}
\cdots & -3\cdot U & -2\cdot U & -U & \downarrow_{\mathbf{on}} & \uparrow^{[\mathbf{on}]} & 2\cdot\uparrow^{[\mathbf{on}]} & 3\cdot\uparrow^{[\mathbf{on}]} & \cdots \\
\cdots & -3\cdot\uparrow^{[\mathbf{on}]} & -2\cdot\uparrow^{[\mathbf{on}]} & -\uparrow^{[\mathbf{on}]} & \uparrow^{\mathbf{on}} & U & 2\cdot U & 3\cdot U & \cdots
\end{array}$$

Every column in this diagram represents a class, and each class is represented by two distinct varieties, $\uparrow^{\mathbf{on}}$ and $\downarrow_{\mathbf{on}}$. These are certainly not the only classes of $\uparrow^{\mathbf{on}}$: for example, we can translate any of the above games by a loopfree game such as $*$. Every such class exhibits games of both varieties.

There are other varieties as well. For example, let

$$V = \pm(U*) = \{0 \mid \uparrow^{\mathbf{on}} \,\|\, \downarrow_{\mathbf{on}} \mid 0\}.$$

Then we have $V \,\hat{+}\, \uparrow^{\mathbf{on}} = \uparrow^{\mathbf{on}}$, so that $V \equiv 0 \pmod{\uparrow^{\mathbf{on}}}$ and hence is a new variety. Every class of degree $\uparrow^{\mathbf{on}}$ contains a unique element of variety V.

Properties of Degrees and Classes

Proposition 5.2 (Downsum Absorbancy Rule). *If $G^{\circ} \geq X \geq 0$, then $G \,\check{+}\, X = G$.*

Proof. Certainly $G \curlyvee X \geq G$, since $X \geq 0$. But also

$$G \curlywedge \overline{G} \geq X,$$

so by the A, B, C Property (Proposition 4.17 on page 320),

$$G \geq G \curlyvee X. \qquad \square$$

Proposition 5.3. *For all loopy games G and H:*

(a) $(-G)^\circ = G^\circ$.
(b) $(G^\circ)^\circ = G^\circ$.
(c) $G^\circ \curlyvee G^\circ = G^\circ$.
(d) $(G \curlywedge H)^\circ \leq G^\circ \curlywedge H^\circ$.
(e) $G^\circ \curlyvee H^\circ \leq (G \curlyvee H)^\circ$.

Proof. (a) is immediate from the definitions.

(b) Let $D = G^\circ$. Taking $X = D$ in the Downsum Absorbancy Rule, we have $G \curlyvee D = G$. Therefore

$$\overline{G} \curlyvee G \curlyvee D = \overline{G} \curlyvee G,$$

so that $\overline{D} \curlyvee D = \overline{D}$. Negating this gives $D \curlywedge \overline{D} = D$, i.e., $D^\circ = D$.
(c) This follows directly from the Downsum Absorbancy Rule, since

$$(G^\circ)^\circ \geq G^\circ \geq 0.$$

(d) We have $(G \curlywedge H)^\circ = (G \curlywedge H) \curlywedge (\overline{G} \curlyvee \overline{H})$, whereas

$$G^\circ \curlywedge H^\circ = (G \curlywedge \overline{G}) \curlywedge (H \curlywedge \overline{H}) = (G \curlywedge H) \curlywedge (G \curlywedge \overline{H}).$$

(e) By the Downsum Absorbancy Rule, $G^\circ \curlyvee G = G$ and $H^\circ \curlyvee H = H$, so

$$(G^\circ \curlyvee H^\circ) \curlyvee (G \curlyvee H) = G \curlyvee H.$$

Therefore by the A, B, C Property,

$$G^\circ \curlyvee H^\circ \leq (G \curlyvee H) \curlywedge \overline{(G \curlyvee H)} = (G \curlyvee H)^\circ. \qquad \square$$

Proposition 5.4. *If G and H have stable degree D, then $G \equiv H \pmod{D}$ if and only if*

$$G \curlywedge D \geq H \geq G \curlyvee \overline{D}.$$

Proof. By the A, B, C Property, the latter condition is equivalent to

$$G \curlywedge D \geq H \quad \text{and} \quad H \curlywedge D \geq G.$$

The forward implication follows immediately, since if $G \equiv H \pmod{D}$, then certainly $G \curlywedge D \geq H \curlywedge D \geq H$, and likewise $H \curlywedge D \geq G$.

For the reverse implication, note that since D is an idempotent,

$$G \curlywedge D = G \curlywedge D \curlywedge D \geq H \curlywedge D,$$

and likewise $H \curlywedge D \geq G \curlywedge D$. □

Varieties

Definition 5.5. Let $D \geq 0$ be an idempotent. A **variety** of degree D is an element of the class $[D]$.

Note that if V is a variety of degree D, then $D \geq V \geq \overline{D}$, so that

$$D \curlywedge D \geq V \curlywedge D \geq \overline{D} \curlywedge D$$

and we have $V \curlywedge D = D$. Now if $V \geq 0$, then the Downsum Absorbancy Rule implies $V \curlyvee D = D$, so necessarily $V = D$. Similarly, $V \leq 0$ implies $V = \overline{D}$, and it follows that *all varieties except for D and $\overline{D}$ are confused with* 0. The following lemma further refines this picture.

Lemma 5.6. *If V is a variety of stable degree D, then*

$$V \curlyvee D = V = V \curlywedge \overline{D}.$$

Proof. Since $D \geq 0$, monotonicity of upsum and downsum gives

$$V \curlyvee D \geq V \geq V \curlywedge \overline{D}.$$

It remains to show that

$$V \curlywedge \overline{D} \geq V \curlyvee D.$$

By the A, B, C Property, this is equivalent to

$$V \curlywedge \overline{V} \geq D \curlyvee D,$$

which follows from the assumption $V^\circ = D$ (using the fact that D is an idempotent). □

Consider the identity established in Lemma 5.6:

$$V \curlyvee D = V = V \curlywedge \overline{D}.$$

This identity generalizes to arbitrary G in a natural manner:

$$G \curlyvee (\overline{G} \curlywedge D) = V(G) = G \curlywedge (\overline{G} \curlyvee \overline{D}).$$

Here $\overline{G} \curlywedge D$ and $\overline{G} \curlyvee \overline{D}$ are the top and bottom elements of $[\overline{G}]$, just as D and $\overline{D}$ are the top and bottom elements of $[\overline{V}]$ (which is the same class as $[V]$, since $V \curlywedge D = \overline{V} \curlywedge D = D$). This generalized identity, in fact, will be taken as the *definition* of $V(G)$.

Unfortunately, it is an open problem whether the generalized identity holds for arbitrary G (and, consequently, whether every G necessarily has a well-defined variety). The best we can do is to prove that $V(G)$ is defined for every G that satisfies a certain **stability** condition.

Definition 5.7. For all G of stable degree D, we define

$$\hat{V}(G) = G \curlywedge (\overline{G} \curlyvee \overline{D}) \quad \text{and} \quad \check{V}(G) = G \curlyvee (\overline{G} \curlywedge D).$$

When $\hat{V}(G) = \check{V}(G)$, then their common value is the **variety** of G, denoted by $V(G)$.

In the absence of a stability condition, the following lemma is the most that we can prove:

Lemma 5.8. *Suppose that G has stable degree D. Then $\hat{V}(G)$ and $\check{V}(G)$ are varieties of degree D, and $\hat{V}(G) \geq \check{V}(G)$.*

So the issue is that $\hat{V}(G)$ might be *strictly* greater than $\check{V}(G)$, and it is here that we will need an additional hypothesis.

Proof of Lemma 5.8. Let $V = \hat{V}(G)$. Then certainly

$$G \curlywedge \overline{G} \curlywedge \overline{D} \geq V \geq G \curlyvee \overline{G} \curlyvee \overline{D},$$

and since $G^\circ = D$ it follows that $D \geq V \geq \overline{D}$.

We next show that $V^\circ = D$. From the identity $G \curlywedge D \curlywedge \overline{D} = G \curlywedge D$ and the A, B, C Property, we obtain

$$G \curlywedge D \geq (G \curlywedge D) \curlyvee D.$$

Therefore

$$\overline{G} \curlyvee (G \curlywedge D) \geq \overline{G} \curlyvee (G \curlywedge D) \curlyvee D,$$

that is,

$$\overline{V} \geq \overline{V} \curlyvee D.$$

By the A, B, C Property,

$$V \curlywedge \overline{V} \geq D$$

so that $V^\circ \geq D$. This shows that $\hat{V}(G)$ is a variety of degree D, and the proof for $V(G)$ is similar.

Finally, the claim that $\hat{V}(G) \geq \check{V}(G)$ is equivalent (by the A, B, C Property) to the inequality

$$G \curlywedge \overline{G} \geq (G \curlywedge D) \curlyvee (\overline{G} \curlywedge D).$$

But $G \curlywedge \overline{G} = D$, and

$$\begin{aligned}(G \curlywedge D) \curlyvee (\overline{G} \curlywedge D) &\leq G \curlywedge D \curlywedge \overline{G} \curlywedge D \\ &= G^\circ \curlywedge D \\ &= D \curlywedge D \\ &= D.\end{aligned}$$

∎

The Stability Conjecture

We can now state the definition of stability.

Definition 5.9. Suppose that G has stable degree D. Then G is **stable** if

$$(G \mathbin{\curlywedge} D) \mathbin{\curlyvee} (\overline{G} \mathbin{\curlywedge} D) \geq 0.$$

Conjecture (Stability Conjecture)**.** Every game of stable degree is stable.

Note that for all G, we have $G \mathbin{\curlywedge} D \geq G \mathbin{\curlyvee} \overline{D}$ (trivially), so that on

$$(G \mathbin{\curlywedge} D) + (\overline{G} \mathbin{\curlywedge} D)$$

Left can certainly find a survival strategy. If G is stable, then she can in fact find a *winning* strategy, and this is quite a strong hypothesis. Its power derives from the following (much more general) lemma.

Lemma 5.10. *If $A \mathbin{\curlyvee} B \geq 0$, then $G \mathbin{\curlyvee} B \geq G \mathbin{\curlywedge} \overline{A}$, for all G.*

Proof. Let $T = G \mathbin{\curlyvee} B$. Then Right, playing second, can survive $G+B+\overline{T}$.

Consider the following swivel chair diagram:

$$\begin{matrix} G + B + \overline{T} \\ \overline{G} + A + T \end{matrix}$$

Left, playing second, uses the following strategy on $\overline{G} + A + T$. On the columns $G+\overline{G}$ and $\overline{T}+T$, she plays the mirror image strategy; on $B+A$ she plays Left's known winning strategy. Thus all moves for Right on $\overline{G}+A+T$ either receive responses on $\overline{G}+A+T$ or are swiveled to moves for *Left* on $G+B+\overline{T}$. In the latter case, Left responds using the known *Right* survival strategy on $G+B+\overline{T}$, possibly transferring the response back to $\overline{G}+A+T$.

There are guaranteed to be just finitely many moves on $B+A$, since Left is following a winning strategy there. This means that play will never get "trapped" on $G+B+\overline{T}$, since plays on G or $\overline{T}$ are automatically transferred back. This shows that Left survives on $\overline{G}+A+T$. Moreover, since G is a stopper and play on A ultimately terminates, infinite play cannot be confined to $\overline{G}+A$. So we've shown that

$$T \mathrel{\hat{\geq}} G + \overline{A},$$

which implies

$$T \geq G \mathbin{\curlywedge} \overline{A}.$$

□

Theorem 5.11. *If G is stable, then $\hat{V}(G) \leq \check{V}(G)$.*

Proof. Let $A = G \curlywedge D$ and $B = \overline{G} \curlywedge D$. Since G is stable, we have $A \curlyvee B \geq 0$, so by Lemma 5.10

$$G \curlyvee B \geq G \curlywedge \overline{A},$$

which is precisely the conclusion of the theorem. □

Together with Lemma 5.8 on page 327, Theorem 5.11 shows that $V(G)$ is defined whenever G is stable.

Example. Let's classify all the varieties of **on**. We know that **on** and **off** are varieties. Any other variety V must satisfy

$$\mathbf{on} > V > \mathbf{off}.$$

Now from Theorem 5.11 we have that $\mathbf{on} \curlyvee V = V$, so Right, playing second, can survive

$$\mathbf{on} + V - V = (\mathbf{on} + V) + (\mathbf{on} - V).$$

But Left can win $\mathbf{on} - V$ playing first (since $\mathbf{on} > V$). If Left could also win $\mathbf{on} + V$ playing second, then she could win overall; so Right must be able to survive $\mathbf{on} + V$ playing *first*. Since **on** has no Right options, this shows $\mathbf{on} \leq -V^R$ for some V^R, i.e., $V^R = \mathbf{off}$. A symmetric argument shows that some $V^L = \mathbf{on}$, so that $V = \{\mathbf{on} \mid \mathbf{off}\}$.

We conclude that **on** has exactly three varieties: **on**, **off**, and $\{\mathbf{on} \mid \mathbf{off}\}$.

Example. Although **on** has just three varieties, most idempotents have many. For example we've already remarked that $\uparrow^{\mathbf{on}}$ has the variety

$$\{0 \mid \uparrow^{\mathbf{on}} \,\|\, \downarrow_{\mathbf{on}} \mid 0\}$$

and we can construct many more such varieties, for example

$$\{x\downarrow_{\mathbf{on}} \mid \uparrow^{\mathbf{on}} \,\|\, \downarrow_{\mathbf{on}} \mid -x\uparrow^{\mathbf{on}}\}$$

for any number x. Exercise 5.1 explores this theme in more detail.

Algebra of Varieties

We can now show that the value of every stable game G is uniquely characterized by its class and variety.

Theorem 5.12. *If G is stable of degree D, then*

$$(G \curlywedge D) \curlyvee V(G) = G.$$

Proof. Certainly

$$\begin{aligned}(G \curlywedge D) \curlyvee V(G) &= G \curlyvee (G \curlywedge D) \curlyvee (\overline{G} \curlywedge D) \\ &\geq G,\end{aligned}$$

since G is stable. The claim that it is $\leq G$ is equivalent (by the A, B, C Property) to the inequality

$$(G \curlywedge D) \curlyvee (\overline{G} \curlywedge D) \leq G^\circ,$$

which follows from the fact that $G^\circ = D$ (as in the proof of Lemma 5.8). $\square$

Theorem 5.13. *If $G^\circ = H^\circ$ and both G and H are stable, then*

$$V(G \curlywedge H) = V(G) \curlywedge V(H) \quad \text{and} \quad V(G \curlyvee H) = V(G) \curlyvee V(H).$$

Proof. We'll show that $V(G \curlywedge H) = V(G) \curlywedge V(H)$; the proof for downsums is identical. First of all,

$$\begin{aligned} V(G \curlywedge H) &= G \curlywedge H \curlywedge (\overline{G} \curlyvee \overline{H} \curlyvee \overline{D}) \\ &= G \curlywedge H \curlywedge ((\overline{G} \curlyvee \overline{D}) \curlyvee (\overline{H} \curlyvee \overline{D})) \\ &\leq G \curlywedge H \curlywedge ((\overline{G} \curlyvee \overline{D}) \curlywedge (\overline{H} \curlyvee \overline{D})) \\ &= (G \curlywedge (\overline{G} \curlyvee \overline{D})) \curlywedge (H \curlywedge (\overline{H} \curlyvee \overline{D})) \\ &= V(G) \curlywedge V(H). \end{aligned}$$

This shows that $V(G \curlywedge H) \leq V(G) \curlywedge V(H)$. For the other direction, note that

$$\begin{aligned} V(G) \curlywedge V(H) &= (G \curlywedge (\overline{G} \curlyvee \overline{D})) \curlywedge (H \curlywedge (\overline{H} \curlyvee \overline{D})) \\ &= (G \curlywedge H) \curlywedge (\overline{G} \curlyvee \overline{D}) \curlywedge (\overline{H} \curlyvee \overline{D}) \\ &\leq (G \curlywedge H) \curlyvee (\overline{G} \curlywedge D) \curlyvee (\overline{H} \curlywedge D); \end{aligned}$$

this last inequality involves two applications of Lemma 5.10, using the fact that G and H are both stable. It therefore suffices to show that

$$(G \curlywedge H) \curlyvee (\overline{G} \curlywedge D) \curlyvee (\overline{H} \curlywedge D) \leq (G \curlywedge H) \curlywedge (\overline{G} \curlyvee \overline{H} \curlyvee \overline{D})$$

(here the right-hand side is just the definition of $V(G \curlywedge H)$), which by the A, B, C Property is equivalent to showing that

$$(\overline{G} \curlywedge D) \curlyvee (\overline{H} \curlywedge D) \curlyvee (G \curlywedge H \curlywedge D) \leq (G \curlywedge H)^\circ.$$

But $(G \curlywedge H)^\circ = D$, and certainly

$$\begin{aligned} &(\overline{G} \curlywedge D) \curlyvee (\overline{H} \curlywedge D) \curlyvee (G \curlywedge H \curlywedge D) \\ &\leq (\overline{G} \curlywedge D) \curlywedge (\overline{H} \curlywedge D) \curlywedge (G \curlywedge H \curlywedge D) \\ &= G^\circ \curlywedge H^\circ \curlywedge D \\ &= D. \end{aligned}$$

$\square$

Theorem 5.14. *Suppose that D is stable and let V be a variety of D. Then every stable class of D contains exactly one member of variety V.*

Proof. Uniqueness is clear, since Theorem 5.12 shows that the value of a stable G depends only on its class and variety.

For existence, fix a stable class $[G]$ and put

$$H = (G \curlywedge D) \curlyvee V.$$

Then certainly $[H] = [G]$, since $G \curlywedge D \geq H \geq \overline{G} \curlyvee D$. Moreover, from Theorem 5.13 we have

$$V(H) = (V(G) \curlywedge D) \curlyvee V = D \curlyvee V = V,$$

so that H has class $[G]$ and variety V. $\square$

Exercises

5.1 *More varieties of* $\uparrow^{\mathbf{on}}$.

(a) Using *cgsuite*, investigate all sums of the four varieties
$$\{1 + G \mid \uparrow^{\mathbf{on}} \,\|\, \downarrow_{\mathbf{on}} \mid -1 + H\}$$
as G and H range over $\uparrow^{\mathbf{on}}$ and $\downarrow_{\mathbf{on}}$.

(b) Let $G = \{0 \mid \uparrow^{\mathbf{on}} \,\|\, \downarrow_{\mathbf{on}}\}$. Determine its variety V. Show that
$$G,\ G \curlywedge G,\ G \curlywedge G \curlywedge G,\ \ldots$$
exhibits a sequence of strictly increasing varieties of $\uparrow^{\mathbf{on}}$,
$$V < V \curlywedge V < V \curlywedge V \curlywedge V < \cdots.$$

5.2 Classify all *dicotic* varieties of **over**.

5.3 If D is an idempotent and $D < \mathbf{on}$, then $D \leq \mathbf{over}$.

5.4 If $G^{\circ} = 0$, then G is necessarily equal to a loopfree game. (Apply Exercise 1.4(c) on page 289.)

5.5 (a) $G \curlyvee \mathbf{+}_{\mathbf{on}} = G$ for all G, unless G is equal to a loopfree game. (Use Exercise 5.4.)

(b) $G \curlyvee \mathbf{+}_{\mathbf{over}} = G$ for all dicotic G, unless G is equal to a loopfree game.

5.6 Every game of degree $\mathbf{+}_{\mathbf{on}}$ has the form $G + \mathbf{+}_{\mathbf{on}}$ or $G + \mathbf{-}_{\mathbf{on}}$, with G loopfree.

5.7 For each $m \geq 2$, let $\mathbf{star}_m = \{0 \,\|\, 0, *m \mid 0, \mathbf{pass}\}$. Show that:

(a) $\mathbf{star}_m$ is an idempotent that absorbs $*m$ and $\uparrow^2$, but it is confused with $*a$ for every $a > 1$ with $a \neq m$.

(b) If G is an idempotent and $G \geq \mathbf{star}_m$ for every $m \geq 2$, then $G \geq \mathbf{over}$.

5.8 Let $D \geq 0$ be an idempotent and suppose that $G^{\circ} \leq D$ (but is not necessarily equal to D). Then G is D-**stable** if $(G \curlywedge D) \curlyvee (\overline{G} \curlywedge D) \geq 0$. If G is D-stable, then we say $\hat{V}_D(G)$ and $\check{V}_D(G)$ are the **upsum** and **downsum varieties** of G, respectively (here we write the subscript D for clarity, to emphasize the dependence on D).

(a) $\hat{V}_D(G)$ and $\check{V}_D(G)$ are necessarily varieties of degree D.

(b) If G is D-stable, then $\hat{V}_D(G) \leq \check{V}_D(G)$, but the inequality might be strict.

(c) If G and H are D-stable and $H^\circ = D$, then

$$\hat{V}(G \mathbin{+\!\!\!\!\!\wedge} H) = \hat{V}_D(G) \mathbin{+\!\!\!\!\!\wedge} \hat{V}(H) \quad \text{and} \quad \check{V}(G \mathbin{+\!\!\!\!\!\vee} H) = \check{V}_D(G) \mathbin{+\!\!\!\!\!\vee} \check{V}(H).$$

So G behaves like it has variety $\hat{V}_D(G)$ in upsums with games of degree D, or variety $\check{V}_D(G)$ in downsums.

Notes

Almost all of the material in this section is based upon *Winning Ways*, where it was first introduced. Almost none of it was actually proved there, and I know of no other sources that exist at the time of this writing.

It's a fascinating theory—but, like the theory of sides in Section 4, it is incomplete. For one thing, it is unknown whether the sum of finite stoppers is necessarily stopper-sided—so that all of the applications of upsum and downsum in this section are, in the most general case, suspect. On top of this, there is of course the Stability Conjecture itself. Until these questions are resolved, the best that we can offer is a rich structure theory for stable games whose degree is known to be a stopper.

Within this context, there are nonetheless many interesting questions worth pursuing. The specific class and variety structure of various idempotents deserves more attention. For some idempotents, such as **on** and $+_{\mathbf{on}}$, there are simple answers (see Exercise 5.6, for example); but most appear to have a more intricate structure.

Open Problem. Investigate the class and variety structure of various idempotents such as $\uparrow^{\mathbf{on}}$, $\mathbf{star}_m$, and ♠.

Here $\mathbf{star}_m$ is defined in Exercise 5.7, and ♠ is given by

$$♠ = \{0^2 \parallel \{\mathbf{on} \mid 0^4\}\}.$$

A partial analysis of the structure of ♠ appears in Chapter 11 of *Winning Ways*, but there is doubtless more to be discovered. This is a theory that's ripe for further exploration.

Chapter VII

Temperature Theory

The canonical form of a short game G contains sufficient information to determine $o(G + X)$, for any choice of X. The resulting *canonical theory*, as described in Chapter II, gives rise to elegant algebraic representations of game values that are a successful tool in many contexts.

Unfortunately there are many situations in which canonical forms retain "too much" information: structure that *in theory* is intrinsic to the value of G but *in practice* is rarely relevant. In Section II.5, we encountered one such example, the following 2×7 AMAZONS position (it might be worthwhile to review the discussion on pages 118–121):

$G =$ 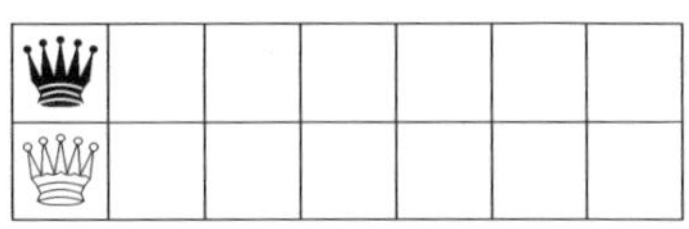 .

In Section II.5 we showed that G has thermograph identical to ± 9, with Left's move to $+9$ given by the "obvious" option G^{L_1}:

$G^{L_1} =$ 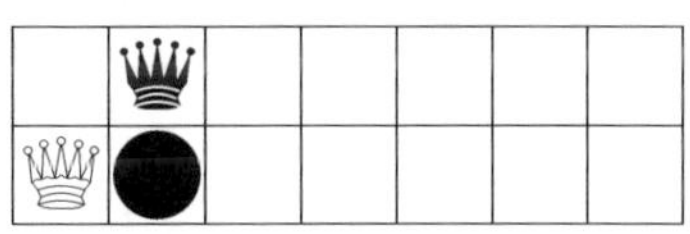 .

The source of trouble in G is the "risky" option G^{L_2}:

$G^{L_2} =$  .

We previously showed that G^{L_2} contributes to the canonical form of G, but not to its thermograph. From the perspective of the canonical theory, G^{L_2} is unavoidably relevant: if Right (say) moves from $G - G$ to $G - G^{L_2}$, Left's only winning response is to $G^{L_2} - G^{L_2}$. So if $X = -G$, then on $G + X$ we must know all about G^{L_2} in order to play well. But for most other choices of X, it's sufficient to know just the "obvious" move G^{L_1}. In such cases the mountain of information carried by G^{L_2} is extraneous noise.

The *orthodox theory* cuts through this noise with a different approach. Rather than seek to preserve $o(G+X)$ for *all* games X, the orthodox theory focuses on a single, "most typical" value of X. Centering on best play in a *typical* environment dramatically simplifies the theory.

Orthodoxy is closely tied to thermography: a Left option G^L is orthodox if and only if it contributes to the thermograph of G. In the case of the 2×7 AMAZONS position given above, G has exactly the same thermograph as ± 9, so its thermograph is entirely determined by G^{L_1} and $-G^{L_1}$. The troubling option G^{L_2}, though canonical, is **unorthodox** and can be disregarded in most circumstances.

In Section 1, we formalize the definition of enriched environment and prove that it aligns with the temperature theory of Section II.5. Section 2 introduces the orthodox theory and some of its consequences, and Sections 3 through 5 explore extensions of this theory to loopy games.

1. Enriched Environments

Recall the following example from Section II.5 (page 114):

$$G = \int^{100} *3 + \int^{99} *5 + \int^{98} *8.$$

Left, playing first on G, should decline to move on the hottest component, preferring instead the option

$$G^L = \int^{100} *3 + \int^{99} *5 + \left(98 + \int^{98} *6\right).$$

From the study of NIM, we know that moving to G^L guarantees her the last move overall. Each time she declines to move on the hottest component, she gives up at most two points—but getting the last move is worth at least 98 points.

The dominant feature of G is the steep drop in temperature below $t = 98$ (all the way down to $t = -1$). This drop places a huge premium on getting the last move at temperature 98. As a result, the NIM-like combinatorics of play above $t = 98$ are critically important.

Similarly consider a slight modification of G:

$$H = G \pm 3 \pm 2 \pm 1.$$

A well-played game of H will begin with a fierce battle to get the last move on G. The loser of this battle will be entitled to play first on $\pm 3 \pm 2 \pm 1$, an advantage that is worth $3 - 2 + 1 = 2$ points, but this "consolation prize" is in the noise. Here, the dominant feature is the temperature drop from $t = 98$ to $t = 3$, which puts a premium on winning the combinatorial battle at high temperatures.

Now consider a third variant:

$$J = G \pm 97 \pm 96 \pm 95 \pm \cdots \pm 1.$$

On J it is no longer crucial to get the last move at temperature 98. For example, if Left plays NIM at high temperatures, sacrificing a few points in order to move last on G, Right will simply respond by taking the 97-point switch. The value of an extra move at temperature 98 is diluted, because there are compensating moves available at temperature 97. The reader might wish to check that on J, the optimal first move is simply to play on the hottest component (of temperature 100).

The compound

$$\pm 97 \pm 96 \pm 95 \pm \cdots \pm 1$$

is known as a **coupon stack**, and the individual switches $\pm x$ are its **coupons**. The coupon stack serves as an **enriched environment** for G. On the sum J, the players can safely ignore the fine combinatorial structure of G, because the environment is rich enough to ensure an adequate supply of strong moves at all temperatures. Neither player can gain more than one point by getting the last move at any temperature t, because her opponent will always have a compensating move available at temperature $t - 1$. Since the cost of moving on a cooler component is at least one point, this implies that it's always correct to play on the hottest component.

Suppose we modify G so that it's cheaper to move on one of the cooler components:

$$G' = \int^{98+2\delta} *3 + \int^{98+\delta} *5 + \int^{98} *8,$$

where $\delta = 1/16$, say. Then we simply introduce a more granular environment:

$$J' = G' \pm (98 - \delta) \pm (98 - 2\delta) \pm (98 - 3\delta) \pm \cdots \pm \delta.$$

On J' it will always be correct to play on the hottest component.

These sorts of enriched environments are artificial constructions, but they serve a useful theoretical purpose. The Left and Right stops of $G + \mathscr{E}_t$ (where $\mathscr{E}_t$ denotes a sufficiently rich environment) are closely related to the thermographic scores $L_t(G)$ and $R_t(G)$ from Section II.5. This provides an alternative formulation of the temperature theory with important practical implications. Positions that arise in practice (in games such as GO) often have components that span a broad range of temperatures. Such positions naturally approximate a rich environment. Therefore, playing on the hottest component will *usually* be correct—it will fail only when the environment is badly impoverished.

The resulting *orthodox theory* of combinatorial games is concerned with best play in a rich environment. A move on G is **orthodox** if it achieves the optimal score on $G + \mathscr{E}_t$, for a sufficiently rich environment $\mathscr{E}_t$. In each of the above examples, the orthodox options are on the component of temperature 100.

The orthodox theory should be contrasted with the *canonical theory* of Chapter II. We might say that

the *canonical theory* informs optimal play in *every* environment,

whereas

the *orthodox theory* informs optimal play in a *typical* environment.

Throughout this chapter, the variable δ will denote a rational number of the form 2^{-n}, with $n \geq 0$.

Definition 1.1. Fix numbers $\delta > 0$ and $t \geq 0$, with t an integer multiple of δ. The **coupon stack of temperature** t **and granularity** δ, denoted by $\mathscr{E}_t^\delta$, is defined by

$$\mathscr{E}_t^\delta = \pm\delta \pm 2\delta \pm 3\delta \pm \cdots \pm (t-\delta) \pm t.$$

Lemma 1.2. *For all $\delta > 0$ and $t \geq 0$ with t an integer multiple of δ,*

$$\mathscr{E}_t^\delta = \left\{ \mathscr{E}_{t-\delta}^\delta + t \;\middle|\; \mathscr{E}_{t-\delta}^\delta - t \right\}.$$

(The only undominated moves on $\mathscr{E}_t^\delta$ are those on the largest coupon.)

Proof. Clearly $\mathscr{E}_t^\delta = \mathscr{E}_{t-\delta}^\delta \pm t$. All of the incentives of $\mathscr{E}_{t-\delta}^\delta$ are dominated by those of $\pm t$, so the result follows from the Dominated Incentives Theorem. □

The following proposition characterizes the stops of $\mathscr{E}_t^\delta$.

Proposition 1.3. *For all $\delta > 0$ and $t \geq 0$ with t an integer multiple of δ,*

$$L(\mathscr{E}_t^\delta) = \begin{cases} t/2 & \text{if } t/\delta \text{ is even;} \\ t/2 + \delta/2 & \text{if } t/\delta \text{ is odd,} \end{cases}$$

and $R(\mathscr{E}_t^\delta) = -L(\mathscr{E}_t^\delta)$.

Proof. For fixed value of δ, the proof proceeds by a straightforward induction on t/δ, using Lemma 1.2. □

Definition 1.4. Let G be a short game and let $\delta > 0$ and $t \geq 0$, with t an integer multiple of δ. The **enriched scores** $L_t^\delta(G)$ and $R_t^\delta(G)$ are given by

$$L_t^\delta(G) = L\big(G + \mathscr{E}_t^\delta\big) - \tfrac{t}{2} \quad \text{and} \quad R_t^\delta(G) = R\big(G + \mathscr{E}_t^\delta\big) + \tfrac{t}{2}.$$

The adjustment of $t/2$ in Definition 1.4 reflects the intrinsic value of moving first on the environment. By Proposition 1.3, it is equal to $L(\mathscr{E}_t^\delta)$ when t/δ is even. When t/δ is odd, it is slightly more convenient (for technical reasons) to use an adjustment of $t/2$, rather than $t/2 + \delta/2$.

This adjustment ensures that $L_t^\delta(0) = 0$ whenever t is an integer multiple of 2δ.

$L_t^\delta(G)$ and $R_t^\delta(G)$ can be written recursively:

Proposition 1.5. *Let G be a short game and let $\delta > 0$ and $t > 0$, with t an integer multiple of δ. Then*

$$L_t^\delta(G) = \max_{G^L}\big(R_t^\delta(G^L) - t,\ R_{t-\delta}^\delta(G)\big),$$
$$R_t^\delta(G) = \min_{G^R}\big(L_t^\delta(G^R) + t,\ L_{t-\delta}^\delta(G)\big).$$

Proof. The Left options of $G + \mathscr{E}_t^\delta$ have the form

$$G^L + \mathscr{E}_t^\delta \quad \text{and} \quad G + \mathscr{E}_{t-\delta}^\delta + t$$

(using Lemma 1.2). Therefore

$$\begin{aligned} L_t^\delta(G) &= L(G + \mathscr{E}_t^\delta) - \tfrac{t}{2} \\ &= \max_{G^L}\big(R(G^L + \mathscr{E}_t^\delta) - \tfrac{t}{2},\ R(G + \mathscr{E}_{t-\delta}^\delta) + \tfrac{t}{2}\big) \\ &= \max_{G^L}\big(R_t^\delta(G^L) - t,\ R_{t-\delta}^\delta(G)\big). \end{aligned}$$ □

The recursion in Proposition 1.5 provides another conceptual model for the temperature theory. We can think of $L_t^\delta(G)$ as describing an auction with a minimum decrement of δ. To start the auction, Left is given the option to play at temperature t. If she declines, then Right is offered the same opportunity at temperature $t - \delta$, and so on. Eventually one of the players will choose to play on G (or, if G is numberish, the auction will continue all the way down to temperature 0).

Theorem 1.6. *Let G be a short game and let $\delta > 0$ and $t \geq 0$, with t an integer multiple of δ. Then*

$$\lim_{\delta \to 0} L_t^\delta(G) = L_t(G) \quad \textit{and} \quad \lim_{\delta \to 0} R_t^\delta(G) = R_t(G).$$

Theorem 1.6 is an immediate corollary of the following lemma. Let $\lambda_t(G)$, $\rho_t(G)$, $\tilde{\lambda}_t(G)$, and $\tilde{\rho}_t(G)$ be as defined in Section II.5.

Lemma 1.7. *Let G be a short game and let $\delta > 0$ and $t \geq 0$, with t an integer multiple of δ. Then*

$$|L_t^\delta(G) - \lambda_t(G)| \leq n\delta \quad \textit{and} \quad |R_t^\delta(G) - \rho_t(G)| \leq n\delta$$

where $n = \mathrm{b}(G)$.

Proof. Let δ be fixed for the remainder of this proof. We proceed by induction on t/δ. The $t = 0$ case is immediate, since

$$L_0^\delta(G) = L(G) = \lambda_0(G).$$

For $t > 0$, we proceed by induction on $n = \mathrm{b}(G)$. We'll show that:

(i) $L_t^\delta(G) \leq \lambda_t(G) + n\delta$;

(ii) $L_t^\delta(G) \geq \lambda_t(G) - n\delta$.

The corresponding inequalities for $R_t^\delta(G)$ follow by symmetry.

(i) $L_t^\delta(G) \leq \lambda_t(G) + n\delta$. From Proposition 1.5 we have

$$L_t^\delta(G) = \max_{G^L} \left(R_t^\delta(G^L) - t,\ R_{t-\delta}^\delta(G) \right).$$

There are two cases.

Case 1: $L_t^\delta(G) = R_t^\delta(G^L) - t$ for some G^L. Then by induction on n,

$$\begin{aligned} R_t^\delta(G^L) - t &\leq \rho_t(G^L) - t + (n-1)\delta && \text{((ii), inductively)} \\ &\leq \lambda_t(G) + (n-1)\delta. \end{aligned}$$

Case 2: $L_t^\delta(G) = R_{t-\delta}^\delta(G)$. Then by induction on t/δ,

$$\begin{aligned} R_{t-\delta}^\delta(G) &\leq \rho_{t-\delta}(G) + n\delta && \text{((ii), inductively)} \\ &\leq \rho_t(G) + n\delta && (\rho \text{ is nondecreasing in } t) \\ &\leq \lambda_t(G) + n\delta. \end{aligned}$$

(ii) $L_t^\delta(G) \geq \lambda_t(G) - n\delta$. Again there are two cases.

Case 1: $t - \delta \geq t(G)$. Then by induction on t/δ,

$$\begin{aligned} L_t^\delta(G) &\geq R_{t-\delta}^\delta(G) && \text{(Proposition 1.5)} \\ &\geq \rho_{t-\delta}(G) - n\delta && \text{((i), inductively)} \\ &= \rho_t(G) - n\delta && (\text{since } t - \delta \geq t(G)) \\ &= \lambda_t(G) - n\delta && (\text{since } t \geq t(G)). \end{aligned}$$

Case 2: $t - \delta < t(G)$. If $t \leq t(G)$, then $\lambda_t(G) = \tilde{\lambda}_t(G)$; otherwise, we have $|t - t(G)| \leq \delta$. In either event, it follows that

$$|\lambda_t(G) - \tilde{\lambda}_t(G)| \leq \delta.$$

Now choose G^L with $\tilde{\lambda}_t(G) = \rho_t(G^L) - t$. Then

$$\begin{aligned} L_t^\delta(G) &\geq R_t^\delta(G^L) - t && \text{(Proposition 1.5)} \\ &\geq \rho_t(G^L) - t - (n-1)\delta && \text{((i), inductively)} \\ &= \tilde{\lambda}_t(G) - (n-1)\delta \\ &\geq \lambda_t(G) - n\delta && \text{(since } |\lambda_t(G) - \tilde{\lambda}_t(G)| \leq \delta). \qquad \square \end{aligned}$$

Negative Temperatures

If x is a number, then $t(x) = -1/2^n$, where $x = m/2^n$ in lowest terms. The definition of $L_t^\delta(G)$ can be extended down to $t = -1$ in a way that's consistent with this observation. Negative-temperature coupons aren't suitable for this purpose, because we have $\{x \mid -x\} = 0$ when $x < 0$. The solution is to use the recursive form of Proposition 1.5 directly in the *definition* of $L_t^\delta(G)$ and $R_t^\delta(G)$. Since the ordinary stops $L(G)$ and $R(G)$ are computed at $t = 0$, we need a new basis for the recursion at $t = -1$.

To determine the **fullstops** of G, we force play to continue even when G is a number. Either player may elect to pass at any time, and play ends after two consecutive passes. Formally:

Definition 1.8. The **Left** (resp. **Right**) **fullstop** of G, denoted $L_{\text{full}}(G)$ (resp. $R_{\text{full}}(G)$), is given by

$$\begin{aligned} L_{\text{full}}(G) &= \max_{G^L}\big(R_{\text{full}}(G^L) + 1,\ R^*_{\text{full}}(G)\big), \\ R_{\text{full}}(G) &= \min_{G^R}\big(L_{\text{full}}(G^R) - 1,\ L^*_{\text{full}}(G)\big), \\ L^*_{\text{full}}(G) &= \max_{G^L}\big(R_{\text{full}}(G^L) + 1,\ 0\big), \\ R^*_{\text{full}}(G) &= \min_{G^R}\big(L_{\text{full}}(G^R) - 1,\ 0\big). \end{aligned}$$

The fullstop is always an integer, and we'll shortly see that it's equivalent to the score of G at temperature -1.

We can now state the revised definitions of L_t^δ and R_t^δ. This should be viewed as a replacement for Definition 1.4. We'll soon show that the two definitions converge in the limiting case, and in particular Theorem 1.6 remains valid with the revised definition.

Definition 1.9. Let G be a short game and let $\delta > 0$ and $t \geq -1$, with t an integer multiple of δ. The **enriched scores** $L_t^\delta(G)$ and $R_t^\delta(G)$ are given by

$$L_t^\delta(G) = \begin{cases} L_{\text{full}}(G) & \text{if } t = -1; \\ \max_{G^L}\big(R_t^\delta(G^L) - t,\ R_{t-\delta}^\delta(G)\big) & \text{otherwise;} \end{cases}$$

$$R_t^\delta(G) = \begin{cases} R_{\text{full}}(G) & \text{if } t = -1; \\ \min_{G^R}\big(L_t^\delta(G^R) + t,\ L_{t-\delta}^\delta(G)\big) & \text{otherwise.} \end{cases}$$

The revised version of Theorem 1.6 is identical, but with $t = -1$ as the base case in place of $t = 0$.

Theorem 1.10. *Let G be a short game and let $\delta > 0$ and $t \geq -1$, with t an integer multiple of δ. Then*

$$\lim_{\delta \to 0} L_t^\delta(G) = L_t(G) \quad \textit{and} \quad \lim_{\delta \to 0} R_t^\delta(G) = R_t(G)$$

using the revised Definition 1.9.

The $t = -1$ case is treated by the following proposition.

Proposition 1.11. $L_{\text{full}}(G) = L_{-1}(G)$ *and* $R_{\text{full}}(G) = R_{-1}(G)$ *for all* G.

Proof. We have

$$\tilde{L}_{-1}(G) = \max_{G^L}\big(R_{-1}(G^L) + 1\big) \quad \text{and} \quad \tilde{R}_{-1}(G) = \min_{G^R}\big(L_{-1}(G^R) - 1\big).$$

By induction, we can assume that $R_{\text{full}}(G^L) = R_{-1}(G^L)$, etc. Therefore

$$\begin{aligned} L_{\text{full}}(G) &= \max\big(\tilde{L}_{-1}(G),\ R_{\text{full}}^*(G)\big) \\ &= \max\big(\tilde{L}_{-1}(G),\ \min(\tilde{R}_{-1}(G),\ 0)\big). \end{aligned}$$

If $\tilde{L}_{-1}(G) \geq \tilde{R}_{-1}(G)$, then

$$\min(\tilde{R}_{-1}(G),\ 0) \leq \tilde{R}_{-1}(G) \leq \tilde{L}_{-1}(G)$$

and it follows that $L_{\text{full}}(G) = \tilde{L}_{-1}(G)$. Since also $L_{-1}(G) = \tilde{L}_{-1}(G)$ in this case (Proposition II.5.12 on page 108), we have $L_{\text{full}}(G) = L_{-1}(G)$.

If $\tilde{L}_{-1}(G) < \tilde{R}_{-1}(G)$, then $L_{-1}(G) = R_{-1}(G) = n$, the simplest integer with

$$\tilde{L}_{-1}(G) \leq n \leq \tilde{R}_{-1}(G).$$

If $n \geq 0$, then $\tilde{L}_{-1}(G) = n$ and $\min(\tilde{R}_{-1}(G),\ 0) = 0$, so $L_{\text{full}}(G) = n$. If $n < 0$, then $\tilde{R}_{-1}(G) = n$, and again $L_{\text{full}}(G) = n$.

In all cases, $L_{\text{full}}(G) = L_{-1}(G)$, and the same argument works for $R_{\text{full}}(G) = R_{-1}(G)$. $\square$

Proof of Theorem 1.10. When $t = -1$, this reduces to Proposition 1.11. For $t > -1$ the proof is identical to the $t > 0$ case from Lemma 1.7. $\square$

Exercises

1.1 If x is a number, then $L_{\text{full}}(x) = \lceil x \rceil$ and $R_{\text{full}}(x) = \lfloor x \rfloor$.

1.2 Fix G and $\delta > 0$. Suppose $t(H)$ is an integer multiple of 2δ, for every subposition H of G. Prove that $L_t^\delta(G) = L_t(G)$ and $R_t^\delta(G) = R_t(G)$ whenever t is an integer multiple of 2δ. (In particular, the limits in Theorem 1.6 converge to a constant for all sufficiently small δ.)

Notes

The orthodox theory was introduced by Berlekamp in a 1996 paper, *The economist's view of combinatorial games* [**Ber96**]. Berlekamp sought a playable GO variant in which orthodox play can be enforced. The classical temperature theory, defined in terms of cooling rates set by auction, proved to be unsuitable for this purpose—professional GO players balked at the idea of paying a "tax" in order to move. Berlekamp's insight was that paying a tax of t points is essentially equivalent to declining a coupon of temperature t, leading to the equivalent formulation of the temperature theory in terms of coupon stacks.

A subsequent paper on AMAZONS [**Ber00b**] presented a more formal exposition of the orthodox theory, including a proof of Theorem 1.6. The proof given here is a modification of a simpler proof due to Bewersdorff [**Bew04**].

The term *fullstop* is new to this book.

2. Orthodoxy

Theorem 1.6 (and its generalization, Theorem 1.10) serves as the essential basis of the orthodox theory: optimal play of G in the "typical" environment $\mathscr{E}_t^\delta$ coincides exactly with the classical thermography of Section II.5, with an error of at most δ. The central ideas of the temperature theory can therefore be used to define the main ingredients of orthodoxy.

It will sometimes be tedious to keep track of the δ terms explicitly, so we'll use the shorthand $\mathscr{E}_t$ for the limiting value as $\delta \to 0$. For example,

$$L(G + \mathscr{E}_t) \quad \text{is shorthand for} \quad \lim_{\delta \to 0} L(G + \mathscr{E}_t^\delta).$$

Definition 2.1. Let G be a short game. A Left option G^L is said to be **orthodox at temperature** t (or simply t-**orthodox**) if

$$R_t(G^L) - t = L_t(G).$$

Likewise, a Right option G^R is **orthodox at temperature** t if

$$L_t(G^R) + t = R_t(G).$$

An option is **orthodox** if it is orthodox at temperature $t(G)$.

Definition 2.2. A short game G is **dormant** at temperature t if $t > t(G)$; **active** at t otherwise.

If G is active at t, then $L_t(G) = \tilde{L}_t(G)$ and $R_t(G) = \tilde{R}_t(G)$, so there is guaranteed to be at least one orthodox move for each player. If G is dormant at t, then there might or might not be orthodox moves available.

Example. Let

$$G = \{6 \mid 1, \pm 2\}.$$

Since $m(\pm 2) < m(1)$, Right's option ± 2 dominates at high temperatures. We can see this by cooling G: we have $G_2 = \{4 \mid 3, 2*\}$, and Right's move to 3 is dominated by $2*$. This also shows that

$$m(G) = t(G) = 3.$$

Which Right option is t-orthodox, and for which t? Note that

$$L_t(\pm 2) = \begin{cases} 2 - t & \text{if } t \le 2; \\ 0 & \text{otherwise.} \end{cases}$$

Therefore when $1 < t \le 3$, we have $L_t(\pm 2) < L_t(1)$, so ± 2 is Right's only t-orthodox move. Conversely, when $t < 1$, then $L_t(\pm 2) > L_t(1)$ and 1 is his only t-orthodox option. *Both* options are orthodox at exactly $t = 1$; *neither* is orthodox at $t > 3$.

It's useful to think about this example in terms of the enriched sum $G + \mathscr{E}_t^\delta$. For $1 < t < 3$ Right must play from G to ± 2; if Left responds with a move on ± 2, then Right can take a coupon. However when $t < 1$, Right's move to ± 2 is inferior: Left will revert to 2, and the coupon of size t won't provide adequate compensation. So Right's orthodox move is to cut his losses and play to 1. For $t > 3$, Right will prefer not to play on G at all, and to play in the environment instead.

Example. Let

$$H = \{9 \mid 1 \,\|\, 0\}.$$

Then $m(H) = t(H) = 1$. Both players have just one option, which is necessarily orthodox at all $t \le 1$. However Left's option $H^L = \{9 \mid 1\}$ is *also* t-orthodox when $1 < t \le 4$. This is because

$$R_t(H^L) = \begin{cases} 1 + t & \text{if } t \le 4; \\ 5 & \text{otherwise,} \end{cases}$$

so that when $1 < t \le 4$ we have

$$R_t(H^L) - t = 1.$$

Therefore Left can safely play her move to H^L at all temperatures $t \le 4$. Right will be forced to respond locally, even though there are attractive moves in the environment.

This example shows that some moves can be orthodox at temperatures well above $t(H)$. The key property of H is that Left's move to H^L is **unstable**: it raises the local temperature. The distinction between stable and unstable moves is a key ingredient of the orthodox theory.

Theorem 2.3 (Orthodox Forecast Theorem). *Let G be a short game and suppose $t \ge t(G)$. Then*

$$L(G + \mathscr{E}_t) = m(G) + \frac{t}{2} \quad \textit{and} \quad R(G + \mathscr{E}_t) = m(G) - \frac{t}{2}.$$

Proof. $L_t(G) = L(G + \mathscr{E}_t) - \frac{t}{2}$ for all t, and $L_t(G) = m(G)$ for $t \ge t(G)$. The argument for $R(G + \mathscr{E}_t)$ is identical. □

Consider a sum such as

$$G_1 + G_2 + \cdots + G_k.$$

The Orthodox Forecast Theorem suggests that

$$m(G_1) + m(G_2) + \cdots + m(G_k) + \frac{t}{2}$$

is a good approximation to

$$L(G_1 + G_2 + \cdots + G_k),$$

provided that the sum is suitably rich. In particular, if any component G_i is equal to $\mathscr{E}_t^{\delta}$, then the forecast is accurate to within a bounded multiple of δ.

If the sum is impoverished, then the forecast might be off by as much as $t/2$, and the size of the error depends on the fine combinatorial structure of G. For example, on the badly impoverished games

$$G = \pm t \quad \text{and} \quad H = \pm t \pm (t - \epsilon)$$

we have $L(G) = t$ and $L(H) = \epsilon$, although the forecast in both cases is $t/2$.

Definition 2.4. The **Left orthodox forecast** of a sum $G_1 + \cdots + G_k$ is equal to

$$m(G_1) + \cdots + m(G_k) + \frac{t}{2},$$

where $t = \max_i t(G_i)$. Likewise, the **Right orthodox forecast** is

$$m(G_1) + \cdots + m(G_k) - \frac{t}{2}.$$

Ambient Temperature

Let $G = G_1 + G_2 + \cdots + G_k$, and consider the enriched sum

$$G + \mathscr{E}_t^\delta,$$

with $t \geq t(G_i)$ for all i. Orthodox play of $G+\mathscr{E}_t^\delta$ tends to follow a predictable course. The opening consists of alternating exchanges of coupons, until the value of the top coupon drops below the temperature of some component G_i. Then there occurs a sequence of moves on G; this might temporarily raise the temperature of some of the components, possibly even above the initial temperature t. Eventually the temperatures of all components settle below the temperature of the coupon stack, and there ensues another sequence of coupon exchanges.

This pattern—long sequences of coupon exchanges, punctuated by high-temperature battles on G—continues until all components are numbers. The sum of these numbers determines the final **score** of the run.

At any point during such an orthodox run, the **ambient temperature** t is the value of the hottest remaining coupon. A move that heats the temperature of some G_i above t is said to be **unstable**. In orthodox play, unstable moves entail a response in the *same* component; **stable** moves permit a response *elsewhere*.

We've already seen that the orthodox forecasts are exactly correct in the presence of a rich environment. We'll soon prove that either player can achieve her orthodox forecast by responding locally to unstable moves, and otherwise moving in the hottest remaining component.

Many of these concepts still apply when G is played in the absence of a rich environment. In that case, the ambient temperature is defined to be the *lowest* temperature yet observed during the current run. The definitions of stable and unstable are the same. Now, however, stable moves might cause arbitrarily large temperature drops. Such moves are highly valuable to the player who makes them. As we saw in the NIM-like example from Section 1, exploiting these **thermal shocks** often demands attention to the fine combinatorial structure of G.

Definition 2.5. Let $G = G_1 + \cdots + G_k$. The **board temperature** of G is given by

$$t_{\mathrm{b}}(G) = \max\{0, t(G_1), \ldots, t(G_k)\}.$$

Note that $t_{\mathrm{b}}(G)$ is *not* the same as $t(G)$. For example, if

$$G = (\pm t) + (\pm t),$$

then $t_{\mathrm{b}}(G) = t$, whereas $t(G) = -1$. In particular, $t_{\mathrm{b}}(G)$ is sensitive to the specific componentwise representation of G. In practice we'll avoid mentioning this dependency explicitly, since it will usually be clear from the context.

Definition 2.6. Let $G = G_1 + \cdots + G_k$ and let $\vec{G}$ be an alternating run of G. Let $H \in \vec{G}$ be a subposition of G that occurs during the run:

$$H = H_1 + \cdots + H_k.$$

The **ambient temperature** $t_{\mathrm{a}}(H)$ is the minimum value of $t_{\mathrm{b}}(X)$, as X ranges over all ancestors of H in $\vec{G}$ (including H itself).

Observe that:

- $t_{\mathrm{a}}(H)$ is sensitive *both* to the componentwise representation of G *and* to the structure of the run $\vec{G}$. These dependencies will usually be clear from the context.
- Ambient temperature decreases monotonically throughout the run and never falls below 0, even when all of the components are numbers.

Definition 2.7. Suppose H and J are consecutive elements of $\vec{G}$ (so that J is an option of H). The (**signed**) **temperature drop at** J is given by

$$\Delta t_{\mathrm{a}}(J) = \begin{cases} t_{\mathrm{a}}(H) - t_{\mathrm{a}}(J) & \text{if } J \text{ is a Left option of } H; \\ -\big(t_{\mathrm{a}}(H) - t_{\mathrm{a}}(J)\big) & \text{if } J \text{ is a Right option of } H. \end{cases}$$

Since ambient temperature decreases monotonically throughout a run, $\Delta t_{\mathrm{a}}(J)$ is always nonnegative when J is a Left option of H, and nonpositive when it's a Right option.

Definition 2.8. Let $\vec{G}$ be an alternating run of G. The **score** of $\vec{G}$ is the first element of $\vec{G}$ in which all components are equal to numbers.

A Naïve Strategy: hotstrat

The Orthodox Forecast Theorem shows that either player can achieve her orthodox forecast in a suitably rich environment. We'll now discuss efficient strategies for orthodox play in practice.

The most basic strategy is hotstrat:

Always play an orthodox move on the hottest component.

Unfortunately hotstrat sometimes gives bad advice even in a rich environment. The reason is that the opponent might play an unstable move, heating a component up above the ambient temperature. When this happens, the environment becomes temporarily impoverished. At the new, higher temperature, a clever opponent can exploit the poverty of the environment to defeat hotstrat by arbitrarily large amounts. The following example illustrates this phenomenon.

Example. Let

$$G = \left\{ \int^{10} \uparrow \,\middle|\, -4 \right\} \quad \text{and} \quad H = \left\{ 10 \,\middle|\, \int^{9} * \right\}.$$

Then $t(G) = 2$ and $t(H) = 1$, so $\mathscr{E}_2$ is a rich environment for $G+H$. Consider

$$G + H + \mathscr{E}_2.$$

The Left orthodox forecast is

$$m(G) + m(H) + \frac{2}{2} = -2 + 9 + 1 = 8.$$

Left's orthodox move (and the one selected by hotstrat) is to play first on G to

$$\int^{10} \uparrow + H + \mathscr{E}_2.$$

Now Right's orthodox move is to play on $\int^{10} \uparrow$. But suppose Right plays instead on H, to

$$\int^{10} \uparrow + \int^{9} * + \mathscr{E}_2$$

(perhaps realizing his game is lost and hoping to stir things up).

Here hotstrat goes astray, selecting the inferior move on $\int^{10} \uparrow$, and after Right's response on $\int^{9} *$ the position is

$$10 + -9 + \mathscr{E}_2.$$

After the exchange of coupons on $\mathscr{E}_2$, the final score is $10 - 9 + 1 = 2$, far short of the Left orthodox forecast.

Starting at a higher ambient temperature doesn't improve the situation; for example, on

$$G + H + \mathscr{E}_{10}$$

hotstrat will advise an exchange of coupons until the ambient temperature falls to 2, after which play might proceed just as before.

A Refined Strategy: sentestrat

In the previous example, Right created a situation involving multiple infinitesimals, each heated above the ambient temperature:

$$\int^{10} \uparrow + \int^{9} * + \mathscr{E}_2.$$

hotstrat failed because the locally orthodox move on the hottest component, $\int^{10} \uparrow$, is not orthodox overall. Indeed we know from the study of infinitesimals that on $\uparrow + *$, the correct move is to $\uparrow + 0$, not $0 + *$. The temperature drop from 9 to 2 ensures that it's more important to get the last move at temperature 9 than to play on the hottest component, so the situation resembles the NIM-like position from Section 1.

Fortunately, a slight refinement of hotstrat is sufficient to overcome this problem. This modification is sentestrat (pronounced *SEN-tay-strat*) and we can state it like so:

> If your opponent has just moved on a component whose temperature is now $\geq$ the ambient temperature t, respond *locally* with a t-orthodox move on that component.
> Otherwise, play a t-orthodox move on the hottest component.

sentestrat is an orthodox strategy: it guarantees the orthodox forecast in a sufficiently rich environment. In impoverished environments sentestrat still suggests reasonably good moves, guaranteeing that Left achieves the orthodox forecast to within an error of $t/2$.

More precisely, suppose Left moves first on the compound

$$G = G_1 + \cdots + G_k$$

and plays according to sentestrat. We'll prove that:

- The score is at least $m(G)$.
- Moreover, if one of the components is an $\mathscr{E}_t^\delta$, then the score is at least $m(G) + t/2$ (the orthodox forecast), to within a bounded multiple of δ.

If Right moves first and Left plays according to sentestrat, then the guaranteed scores are $m(G) - t$ and $m(G) - t/2$ instead of $m(G)$ and $m(G) + t/2$.

In particular, "unbounded failures" (such as are possible with hotstrat) cannot happen, regardless of the environment.

All of these results are consequences of a more general result, the Orthodox Accounting Theorem, which we'll now prove. The theorem gives a precise accounting of deviations from the orthodox forecast, assuming Left is playing sentestrat.

Orthodox Accounting

Theorem 2.9 (Orthodox Accounting Theorem). *Fix $G_1, \ldots, G_k$. Suppose Left plays first (second) on the compound*

$$G = G_1 + \cdots + G_k$$

and follows sentestrat. *Then she is guaranteed a score of at least*

$$z + \frac{1}{2}\sum_{i=1}^{n} \Delta t_i,$$

where z is the Left (Right) orthodox forecast and Δt_i are the signed temperature drops of the corresponding run.

Proof. Put $t = t_{\mathrm{b}}(G) = \max_i t(G_i)$. Let x and y be the Left and Right orthodox forecasts of G, respectively, so that $x - y = t$. We'll show that each subposition

$$Y = Y_1 + \cdots + Y_k,$$

with Right to move, satisfies

$$(\dagger) \qquad R_t(Y_1) + R_t(Y_2) + \cdots + R_t(Y_k) - \frac{t}{2} \geq y,$$

so long as the ambient temperature remains at t. Likewise, each subposition

$$X = X_1 + \cdots + X_k,$$

with Left to move, satisfies

$$(\ddagger) \qquad L_t(X_1) + R_t(X_2) + \cdots + R_t(X_k) + \frac{t}{2} \geq x,$$

so long as the ambient temperature remains at t and assuming that Right's previous move was played on X_1.

The initial position G satisfies both ($\dagger$) and ($\ddagger$) automatically, since $L_t(G_i) = R_t(G_i) = m(G_i)$ for each component G_i (by definition of t).

Now consider a position Y satisfying ($\dagger$). Assume without loss of generality that Right moves on Y_1; then certainly $L_t(Y_1^R) - t \geq R_t(Y_1)$, so the new position X satisfies ($\ddagger$).

Next consider X satisfying ($\ddagger$). If $t(X_1) \geq t$, then since Right has just moved on X_1, sentestrat recommends a move that is orthodox at temperature t, say X_1^L. Therefore $R_t(X_1^L) = L_t(X_1) + t$, and the new position satisfies ($\dagger$).

Finally, suppose instead that $t(X_1) < t$, but some other component has temperature $\geq t$. Then sentestrat recommends an orthodox move on the hottest such component, say from X_2 to X_2^L. We have $R_t(X_2^L) = L_t(X_2) + t$, and again the new position satisfies ($\dagger$).

This shows that the recurrences (†) and (‡) are satisfied, so long as the ambient temperature remains at t. Eventually the ambient temperature will drop, say to $t' < t$. If it happens after a move by Left, then we have

$$\Delta t_1 = t - t'$$

and the first position Y reached after the temperature drop satisfies

$$R_t(Y_1) + R_t(Y_2) + \cdots + R_t(Y_k) - \frac{t}{2} \geq y.$$

Since $t(Y_i) < t$ for all Y_i, this implies

$$m(Y_1) + m(Y_2) + \cdots + m(Y_k) - \frac{t}{2} \geq y.$$

By induction on G, Left is guaranteed a score on Y of at least

$$\begin{aligned} & m(Y_1) + \cdots + m(Y_k) - \frac{t'}{2} + \frac{1}{2}\sum_{i=2}^{n} \Delta t_i \\ &= m(Y_1) + \cdots + m(Y_k) - \frac{t}{2} + \frac{1}{2}\Delta t_1 + \frac{1}{2}\sum_{i=2}^{n} \Delta t_i \\ &\geq y + \frac{1}{2}\sum_{i=1}^{n} \Delta t_i. \end{aligned}$$

The same argument works if the temperature drop happens after a move by Right, this time using the fact that $\Delta t_1 = -(t - t')$. □

Corollary 2.10. *Suppose Left follows* sentestrat *on the compound*

$$G = G_1 + \cdots + G_k.$$

Then she is guaranteed a score of at least $m(G)$ playing first, or $m(G) - t$ playing second, where $t = \max_i t(G_i)$.

Proof. By the Orthodox Accouting Theorem, Left's guaranteed score is

$$z + \frac{1}{2}\sum_{i=1}^{n} \Delta t_i,$$

where z is the appropriate orthodox forecast. Certainly

$$\sum_{i=1}^{n} \Delta t_i \geq \sum_{i=1}^{n} -|\Delta t_i| = -t.$$

(In the worst-case scenario, all of the signed temperature drops are negative.) Therefore the score is at least

$$z - \frac{t}{2},$$

which is $m(G)$ if z is the Left orthodox forecast, and $m(G) - t$ if z is the Right orthodox forecast. □

Theorem 2.11. *Fix $G_1, \ldots, G_k$ and let $t \geq \max_i t(G_i)$. Suppose Left plays first (second) on the compound*

$$G = G_1 + \cdots + G_k + \mathscr{E}_t^\delta$$

and follows sentestrat. *Then she is guaranteed at least the Left (Right) orthodox forecast, to within a bounded multiple of δ.*

Proof. For each temperature drop Δt_i, let m_i be the number of moves played on $G_1 + \cdots + G_k$ since the preceding temperature drop. Since Left is following sentestrat, she won't play on $\mathscr{E}_t^\delta$ if a hotter component is present. It follows that at most $m_i + 1$ moves were played on $\mathscr{E}_t^\delta$ since the preceding drop, so that

$$|\Delta t_i| \leq (m_i + 1)\delta.$$

In particular, we have

$$\sum_{m_i \geq 1} |\Delta t_i| \leq 2n\delta,$$

where n is the formal birthday of $G_1 + \cdots + G_k$. But successive temperature drops with $m_i = 0$ alternate between Left and Right, so that

$$\sum_{m_i = 0} \Delta t_i \geq -n\delta.$$

(Here the error term $n\delta$ arises because moves on $G_1 + \cdots + G_k$ might break the parity of moves on $\mathscr{E}_t^\delta$; but there can be at most n of the former.)

Therefore we obtain a bound

$$\sum \Delta t_i \geq -3n\delta,$$

and since n is a constant independent of δ, the result follows from the Orthodox Accounting Theorem. □

Suppose Left moves first on the compound

$$G = G_1 + \cdots + G_k$$

and both players follow sentestrat. Then they can be certain the final score will fall between $m(G)$ and $m(G) + t$. If Left (say) wishes to influence the score more precisely, then she must exploit the combinatorial structure of G in a way that increases her total share of the resulting temperature drops (often by deviating from sentestrat and making unorthodox moves). The situation can become quite complicated, since the effort of getting the last move at temperature t might (as a side effect) perturb the combinatorics at some lower temperature $t' < t$.

The presence of a coupon stack puts a bound on the steepness of any particular temperature drop. Consequently, in a sufficiently rich environment such combinatorial considerations can be disregarded: either player is guaranteed the orthodox forecast following sentestrat; and by symmetry neither can hope to do better.

Exercises

2.1 (a) Give an example of a game G with an orthodox option G^L that is canonically senseless (cf. Exercise II.2.6 on page 68).
(b) Show that for every t there is a Left option of G that is both sensible *and* t-orthodox.
(c) True or false: There exists a G with an orthodox option G^L that is Inf-senseless (cf. Exercise II.6.11 on page 135).

Notes

Most of the material in this section is due to Berlekamp [**Ber96, Ber02**]. Details of its origin are described in the notes to Section 1; a summary of its extensions to loopy games can be found in Sections 3 through 5.

3. Generalized Temperature

PUSH is played with black and white tokens on a $1 \times n$ strip. Left may slide any black token one space to the left. If the space is occupied by another token, then it is pushed to the left along with any tokens immediately behind it. Tokens pushed off the board are removed from the game. In the example below, Left can push the white tokens off the board in three moves:

Likewise, Right may slide any white token one space to the right, pushing any tokens in its immediate path.

PUSH exhibits some familiar values. A single black token at the far end of a $1 \times n$ strip obviously has value n, and it's easy to construct a ± 1:

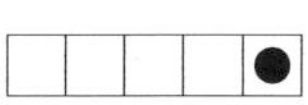 $= 5,$ 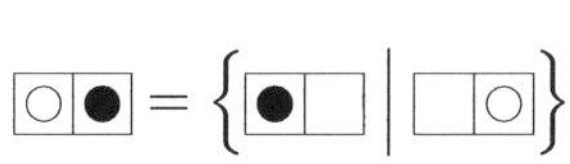 $= \pm 1.$

There are also loopy values. Consider the following position:

$K =$ 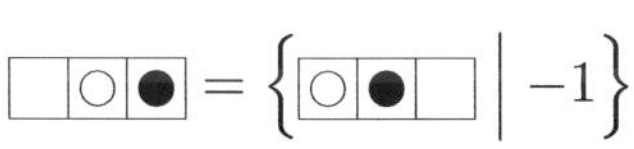 $= \{1 \mid K \,\|\, -1\}.$

Here K has an option K^L from which Right can return to K. This situation is called a **ko**. We'll draw the game graph of K using an arc to illustrate the ko:

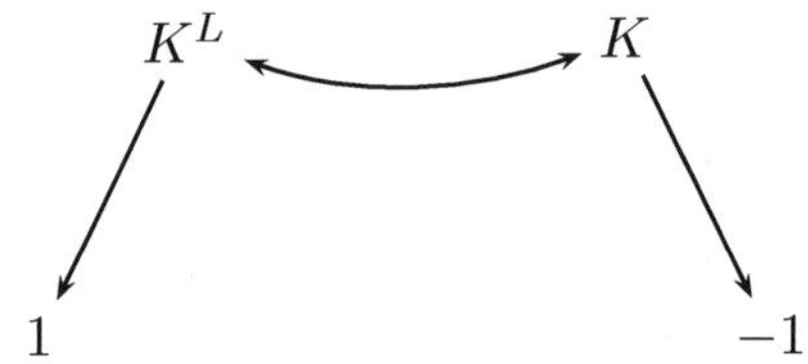

Loopy positions such as K obviously introduce the possibility of infinite runs. For example, in the sum $K \pm \frac{1}{2}$ neither player will be eager to allow her opponent to win the ko. If Left moves first on $K \pm \frac{1}{2}$, then both players will prefer to move indefinitely from K to K^L and back, since the alternative is to concede the ko and lose the game.

In Chapter VI, we addressed this situation by designating $K \pm \frac{1}{2}$ to be a draw if Left moves first. This accurately models many games with loopy positions, including CHECKERS and CHESS.[1] However other games, notably GO, instead introduce various prohibitions on returning to a prior position. It turns out that this type of restriction—a **koban** rule—leads to a more refined temperature theory. This theory is necessary in order to model games with actual koban rules (such as GO), and it also provides useful insights into games for which infinite plays are drawn.

Our version of the koban rule is stated as follows:[2]

No position may be repeated during an individual play of G.

If $K \pm \frac{1}{2}$ is played under the koban rule, then after Left's move to $K^L \pm \frac{1}{2}$, Right is prohibited from moving back to $K \pm \frac{1}{2}$ and is forced to take the coupon. However the cycle in K remains combinatorially relevant, since in a play of

$$K + X$$

there might be *local* repetition in the component K, even though every subposition of $K + X$ is *globally* unique.

This phenomenon is illustrated by the example

$$K + K + K.$$

[1] CHESS uses a "three-repetition rule" in which play is declared a draw if the same position recurs for the third time. It's easy to see that with perfect play, this is equivalent to the definition of draw in Chapter VI.

[2] The situation for GO itself is not so straightforward. While our version of the koban rule agrees with tournament GO rules in most situations, there are certain exotic positions for which the GO rules are more complicated. Moreover, different regional dialects of GO sometimes give conflicting interpretations of such positions. This is one of the reasons for avoiding a more detailed study of GO in this book.

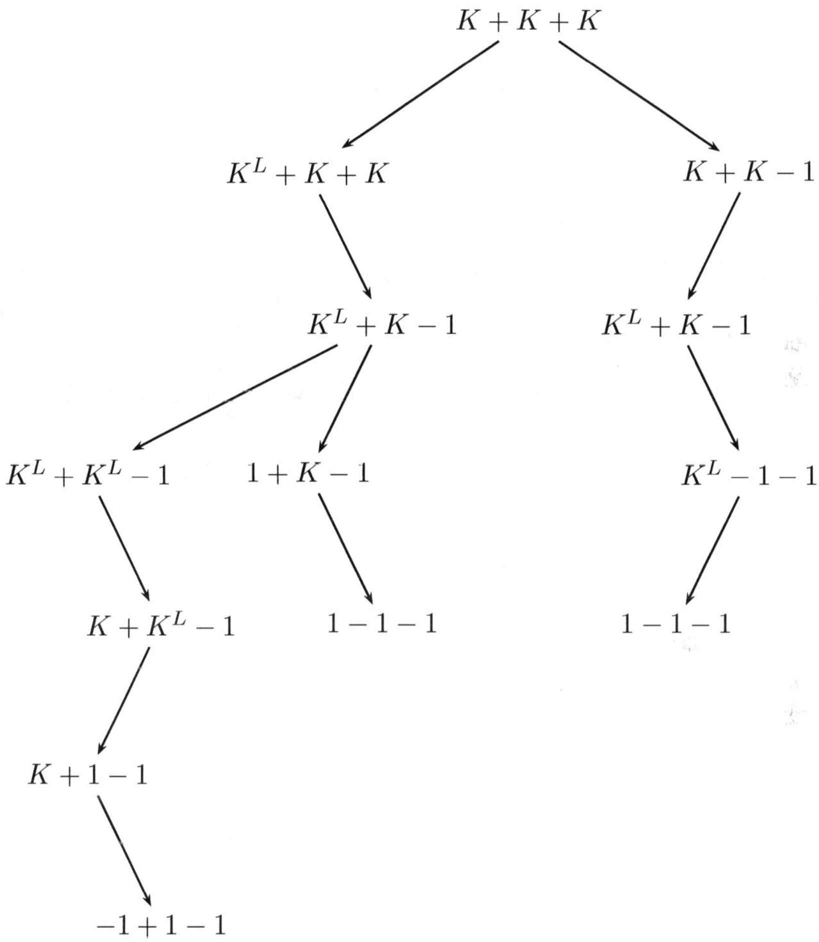

Figure 3.1. The Left and Right stops of $K + K + K$ are -1.

Typical plays are illustrated in Figure 3.1 on the preceding page. Left, playing first, can only move to $K^L + K + K$. Now Right is banned from playing on K^L and so must move to $K^L + K - 1$. If Left now moves to $1 + K - 1$, Right will play to $1 - 1 - 1$, yielding a final score of -1. If instead Left moves to $K^L + K^L - 1$, Right will revert to $K + K^L - 1$ (*local* repetition on the first component), after which play proceeds just as before. In all cases the Left stop is -1. A similar analysis shows that the Right stop is also -1.

No loopfree game G can satisfy $L(3\cdot G) = R(3\cdot G) = -1$, since this would imply $m(G) = -\frac{1}{3}$, which is plainly impossible. We'll soon see, however, that in fact $m(K) = -\frac{1}{3}$ in a precise sense.

Toward a Temperature Theory

Here's an outline of our plan for introducing a generalized temperature theory. The first step is to define analogues of $L_t(G)$ and $R_t(G)$ by modeling the play of G in an enriched environment:

- Define $L_t^\delta(G)$ and $R_t^\delta(G)$ in terms of the sum $G + \mathscr{E}_t^\delta$, played with a koban restriction.
- Define $L_t(G)$ and $R_t(G)$ as limits of $L_t^\delta(G)$ and $R_t^\delta(G)$. By Theorem 1.10 (page 340), this is consistent with the loopfree definition.

The next goal (which we'll tackle in Section 4) will be to prove that these limits exist. It turns out to be quite difficult to do this for all loopy games G, so our focus will be on a particular class of **simple** loopy games, which have an especially tractable structure. The strategy here is much the same as for loopfree games:

- Develop an explicit **thermographic calculus** for computing trajectories $\lambda_t(G)$ and $\rho_t(G)$ recursively, for all simple games G.
- Show that $\lambda_t(G)$ and $\rho_t(G)$ agree with $L_t^\delta(G)$ and $R_t^\delta(G)$, respectively, to within a bounded multiple of δ.

Finally, in Section 5 we'll prove a version of the Orthodox Accounting Theorem for sums of simple loopy games.

$L_t(G)$ and $R_t(G)$

We proceed with the definitions of $L_t^\delta(G)$ and $R_t^\delta(G)$. For technical reasons that will shortly become clear, it's difficult to define loopy analogues of the stops $L(G)$ and $R(G)$, and the solution is to generalize the fullstops $L_{\text{full}}(G)$ and $R_{\text{full}}(G)$ instead. The following definitions are modeled on Definitions 1.8 and 1.9.

Recall that if $\mathscr{A}$ is a set of games, then $\mathscr{A}[G]$ is shorthand for $\mathscr{A} \cup \{G\}$.

Definition 3.1. Let G be loopy and let $\mathscr{A}$ be a set of games. The **Left** (resp. **Right**) **fullstop** of G, with $\mathscr{A}$ **kobanned**, is defined by

$$\begin{aligned}
L_{\text{full}}(G \setminus \mathscr{A}) &= \max_{G^L \notin \mathscr{A}} \big(R_{\text{full}}(G^L \setminus \mathscr{A}[G]) + 1,\ R^*_{\text{full}}(G \setminus \mathscr{A})\big),\\
R_{\text{full}}(G \setminus \mathscr{A}) &= \min_{G^R \notin \mathscr{A}} \big(L_{\text{full}}(G^R \setminus \mathscr{A}[G]) - 1,\ L^*_{\text{full}}(G \setminus \mathscr{A})\big),\\
L^*_{\text{full}}(G \setminus \mathscr{A}) &= \max_{G^L \notin \mathscr{A}} \big(R_{\text{full}}(G^L \setminus \mathscr{A}[G]) + 1,\ 0\big),\\
R^*_{\text{full}}(G \setminus \mathscr{A}) &= \min_{G^R \notin \mathscr{A}} \big(L_{\text{full}}(G^R \setminus \mathscr{A}[G]) - 1,\ 0\big).
\end{aligned}$$

$L_{\text{full}}(G)$ and $R_{\text{full}}(G)$ are shorthand for $L_{\text{full}}(G \setminus \emptyset)$ and $R_{\text{full}}(G \setminus \emptyset)$.

Example. If G is loopfree, then it is necessarily not a proper subposition of itself. Therefore

$$R_{\text{full}}(G^L \setminus \mathscr{A}[G]) = R_{\text{full}}(G^L \setminus \mathscr{A}),$$

since the inclusion of G in $\mathscr{A}[G]$ has no effect. In particular,

$$L_{\text{full}}(G) = \max_{G^L} \big(R_{\text{full}}(G^L) + 1,\ R^*_{\text{full}}(G)\big),$$

agreeing with the earlier definition of $L_{\text{full}}(G)$ for loopfree G.

Example. For the one-point ko K introduced above, we have $R_{\text{full}}(K) = -2$ (Right moves to -1 immediately), and $L_{\text{full}}(K) = 3$ (Left moves to $+1$ in two moves, and Right can only pass).

We can now define $L^\delta_t(G)$ and $R^\delta_t(G)$ for loopy games G. Although we're relying on the recursive formulation from Definition 1.9 (that works for negative temperatures), the underlying conceptual model is an environment consisting of a stack of coupons.

Definition 3.2. Let G be loopy and let $\mathscr{A}$ be a set of games. Fix $\delta > 0$ and $t > -1$, with t an integer multiple of δ. The **enriched scores** with $\mathscr{A}$ **kobanned**, $L^\delta_t(G \setminus \mathscr{A})$ and $R^\delta_t(G \setminus \mathscr{A})$, are defined by

$$\begin{aligned}
L^\delta_t(G \setminus \mathscr{A}) &= \max_{G^L \notin \mathscr{A}} \big(R^\delta_t(G^L \setminus \mathscr{A}[G]) - t,\ R^\delta_{t-\delta}(G)\big),\\
R^\delta_t(G \setminus \mathscr{A}) &= \min_{G^R \notin \mathscr{A}} \big(L^\delta_t(G^R \setminus \mathscr{A}[G]) + t,\ L^\delta_{t-\delta}(G)\big).
\end{aligned}$$

We write $L^\delta_t(G)$ and $R^\delta_t(G)$ as shorthand for $L^\delta_t(G \setminus \emptyset)$ and $R^\delta_t(G \setminus \emptyset)$. For the base case $t = -1$, we define

$$L^\delta_{-1}(G) = L_{\text{full}}(G) \quad \text{and} \quad R^\delta_{-1}(G) = R_{\text{full}}(G)$$

for all δ, just as in the loopfree case.

The recursive expression for $L_t^\delta(G \setminus \mathscr{A})$ in Definition 3.2 models Left's various options on the sum $G + \mathscr{E}_t^\delta$, with $\mathscr{A}$ kobanned. She is permitted a move to any G^L not in the kobanned set $\mathscr{A}$; then in the recursive expression for G^L, the position G is incorporated into $\mathscr{A}$, signifying that both players are now kobanned from returning to G. Alternatively, she might choose to take a coupon, lowering the temperature to $t - \delta$. In that case, the entire set $\mathscr{A}$ is replaced by $\emptyset$: the move in the (loopfree) environment $\mathscr{E}_t^\delta$ permanently alters the global position and therefore "resets" the local koban on G.

Definition 3.3. Let G be a loopy game. For $t \geq -1$ we define the **Left** and **Right scores** of G by

$$L_t(G) = \lim_{\delta \to 0} L_t^\delta(G) \quad \text{and} \quad R_t(G) = \lim_{\delta \to 0} R_t^\delta(G)$$

whenever these limits exist.

Some Examples

The following examples will help motivate many of the succeeding ideas. Each example is coupled with an informal discussion about the putative limits $L_t(G)$ and $R_t(G)$. These discussions are intended to provide motivating intuitions about such examples, rather than formal arguments. The formal proofs will emerge in Section 4, as a byproduct of the thermographic calculus.

Example. We begin with the example K defined on page 351:

$$K = \{1 \mid K \parallel -1\}.$$

Its game graph and thermograph are shown in Figure 3.2. Suppose Right plays first on $K + \mathscr{E}_t^\delta$, to $-1 + \mathscr{E}_t^\delta$. Then Left will respond on the coupon stack, yielding an enriched score of

$$-1 + L(\mathscr{E}_t^\delta) + \frac{t}{2} = -1 + t.$$

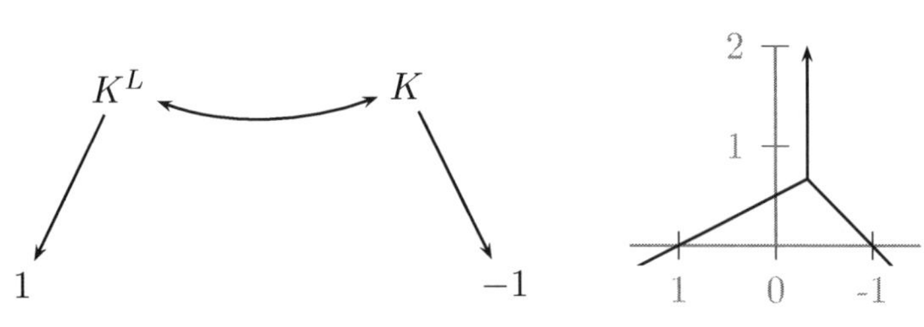

Figure 3.2. The thermograph of K $\left(m(K) = -\frac{1}{3},\ t(K) = \frac{2}{3}\right)$.

(The $t/2$ term is from Definition 1.4.) Likewise, if Left plays first on $K + \mathscr{E}_t^\delta$, Right will take a coupon, leaving the position $K^L - t + \mathscr{E}_{t-\delta}^\delta$. Assuming Left continues with a move from K^L to 1, we obtain a score of

$$1 - t + R(\mathscr{E}_{t-\delta}^\delta) - \frac{t}{2} = 1 - 2t.$$

When $t \le \frac{2}{3}$, then $1 - 2t \ge -1 + t$, so both players will be eager to play on K. When $t > \frac{2}{3}$, neither will be eager to play on K and they will simply exchange coupons until the temperature falls below $\frac{2}{3}$. This suggests the thermograph shown in Figure 3.2:

$$L_t(K) = \begin{cases} 1 - 2t & \text{if } t \le \frac{2}{3}; \\ -\frac{1}{3} & \text{otherwise;} \end{cases} \qquad R_t(K) = \begin{cases} -1 + t & \text{if } t \le \frac{2}{3}; \\ -\frac{1}{3} & \text{otherwise.} \end{cases}$$

There is one loose end in this argument. What if Left, after playing to $K^L + \mathscr{E}_t^\delta$, responds to Right's move on $\mathscr{E}_t^\delta$ by taking another coupon? Then Right is free to return to K (since the *global* position has changed due to the moves on $\mathscr{E}_t^\delta$). The two coupons cancel each other out, so it is now Left's move on $K + \mathscr{E}_{t-2\delta}^\delta$—and, since Right has just played on K, Left's only option is to take a third coupon. Therefore Left gains nothing by playing this way: she might as well have simply played on $\mathscr{E}_t^\delta$ in the first place.

This is a central argument. After Left's initial move to K^L, Right's response on $\mathscr{E}_t^\delta$ is known as a **forcing pass**. We can safely assume that Left *must* play again on K^L, because failure to do so permits Right to return to K, leaving Left no better off than had she simply played in the environment in the first place. Once she makes the initial decision to move from K to K^L, Left has committed herself to playing out the ko. (Note that we haven't *proved* this yet; we've just asserted it based on general principles. The proof will follow in Section 4.)

In the above example, we explicitly accounted for the δ terms arising in the play of $\mathscr{E}_t^\delta$. In the succeeding examples, we'll suppress the δ terms in order to avoid excessive bookkeeping. The symbol $\mathscr{E}_t$ will represent the limiting case of a coupon stack with a "sufficiently small" δ.

Example. Consider the following generalization of K to a $1 \times n$ strip:

$K_n =$ 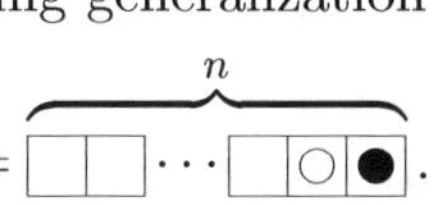

(Note that $K = K_3$.) Right can move directly to -1, for a score of $-1 + t$. Left can move to $+1$ in $n - 1$ steps—and at each step, Right makes a forcing pass, so that the final score is $1 - (n-1)t$. Just as in the previous example, we have $t(K_n) = 2/n$ and $m(K_n) = -1 + 2/n$.

The game graph and thermograph of K_5 are shown in Figure 3.3.

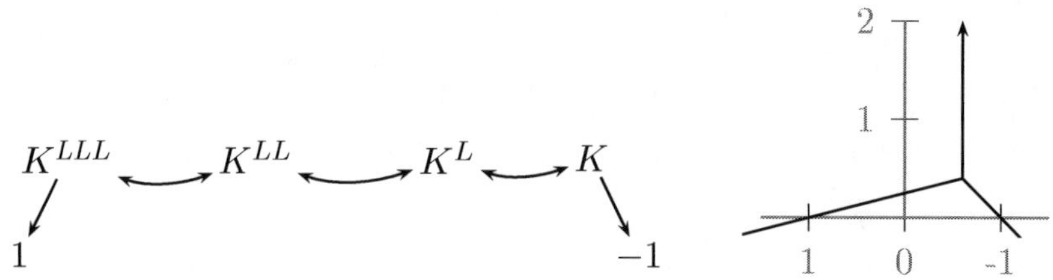

Figure 3.3. The thermograph of K_5 $\left(m(K_5) = -\frac{3}{5},\ t(K_5) = \frac{2}{5}\right)$.

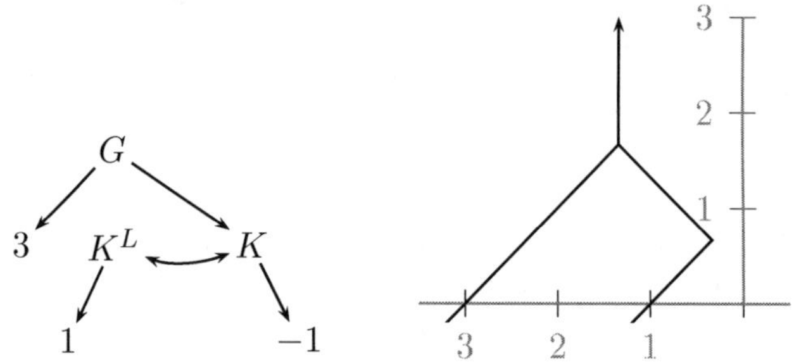

Figure 3.4. Thermograph of $G = \{3 \mid K\}$ $\left(m(G) = \frac{4}{3},\ t(G) = \frac{5}{3}\right)$.

Example. With K as before, consider $G = \{3 \mid K\}$. The game graph and thermograph of G are pictured in Figure 3.4. If Right moves first on G and $t \geq t(K)$, then the score is $m(K)+t = -\frac{1}{3}+t$. If Left moves first on G, then of course the score is $3-t$. This accounts for the shape of the thermograph above $t = \frac{2}{3}$. But something startling happens at lower temperatures. If $t < t(K)$ and Right moves first on G, then Left will follow with a move on K. Right will take a coupon, and Left will move from K^L to 1. Therefore *Right* ends up with a coupon, for a final score of $1-t$, even though he moved first! Among temperatures below $t(K)$, Right would prefer to play at the *highest* temperature possible.

This is a situation that cannot happen in a loopfree context. If Right moves first on a loopfree game H, then there are only two possibilities: either Left collects a coupon when H stabilizes or neither player does. These correspond to a slope of $+1$ or 0 for $R_t(H)$.

Here we see a third possibility: the slope of $R_t(G)$ below $t(K)$ is -1. It's not hard to see that by replacing K with a suitable K_n, we can construct thermographs exhibiting any desired negative Right slope.

Example. $G = \{1 \mid K\}$ illustrates an even more remarkable phenomenon. Its thermograph is shown in Figure 3.5 on the next page. Here the Left and Right walls of the thermograph coincide: at all temperatures t, the final score is the same no matter who moves first. When $t \leq \frac{2}{3}$, Right will end up with a coupon regardless of who moves first, and the score will be $1-t$. It's

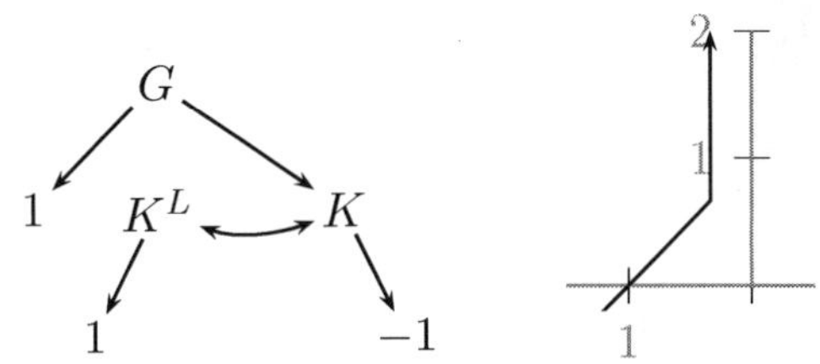

Figure 3.5. The thermograph of $G = \{1 \mid K\}$ $\big(m(G) = \frac{1}{3},\ t(G) = \frac{2}{3}\big)$.

an odd situation: Left has no reason to play on G at all, and she'll happily take a coupon instead, trying to drive down the ambient temperature and decrease the value of Right's inevitable coupon. Conversely, Right will be eager to play on G at the earliest opportunity (below $t = \frac{2}{3}$).

Example. Figure 3.6 gives an example with multiple active temperature regions. At temperatures $t < 2$, neither player will be eager to play into the ko: her opponent would win the ko and collect two points, so the top coupon does not provide adequate compensation. Therefore at such temperatures the game behaves exactly like ± 1: active when $t \leq 1$ and dormant when $1 < t < 2$.

When $2 < t < 4$, however, the game is more exciting. If Left plays into the ko, then Right is presented with an unpalatable choice: fight the ko and give Left a coupon, leading to a score of $t - 2$; or allow Left nine points in exchange for two coupons, leading to a score of $7 - 2t$. These are both strictly positive, so at these temperatures Left will be eager to play on G (and, by symmetry, so will Right).

The region $2 < t < 3$ illustrates a phenomenon unique to loopy positions: at such temperatures, play on G becomes strictly *more* valuable as t increases. Playing into the ko forces the opponent to fight it, giving up a

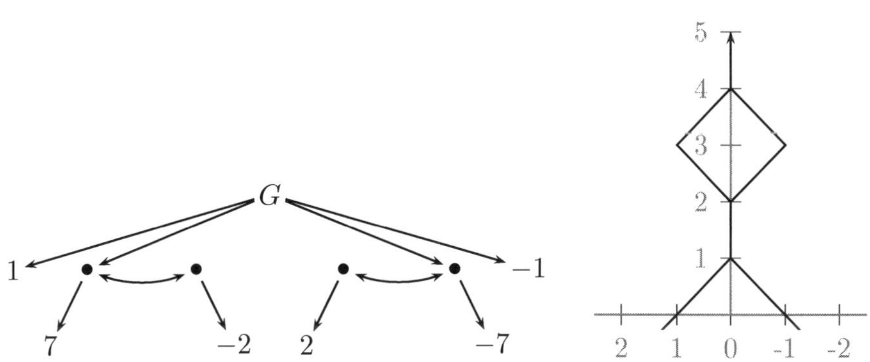

Figure 3.6. The thermograph of G has multiple active temperature regions.

coupon in the process; so whoever plays first on G would like that coupon to be as *large* as possible.

When $t \geq 3$ we are above the mast temperature of G^L, so the game behaves like ± 4, becoming dormant again when $t > 4$. (Here 4 is the mast value of G^L.)

A Refined Orthodoxy

These examples demonstrate the need for a refinement of the orthodox methods from Section 2. The main complication involves examples such as Figure 3.6, in which G is dormant at various temperatures $t < t(G)$. Contrast the following with Definition 2.2 on page 342.

Definition 3.4. A loopy game G is **dormant** at temperature t if its thermograph is a vertical mast just below t; **active** otherwise.

If G is dormant at t, then the **activation temperature** of G below t is the greatest $t' < t$ such that G is active at t'.

(By "a vertical mast just below t," we mean that for some $\epsilon > 0$ and constant value x, we have $L_u(G) = R_u(G) = x$ for all $u \in [t - \epsilon, t]$.)

Definition 3.5. The **temperature** $t(G)$ is the greatest t such that G is active at t (equivalently, it's the activation temperature of G below ∞). The **mast value** $m(G)$ is the value of $L_t(G)$ at $t = t(G)$ (which is necessarily equal to $R_t(G)$).

In particular, if $t > t(G)$, then G is necessarily dormant at t, and $t(G)$ is the activation temperature of G below t. The example in Figure 3.6 is also dormant at temperatures on the interval $]1, 2]$, with activation temperature 1.

Note that all of these definitions are contingent on proving that such a "greatest" value of t necessarily exists; this will have to wait until Section 4.

A major goal of the refined orthodoxy is to prove a generalized Orthodox Accounting Theorem that works for loopy games. The starting point is the following direct generalization of `sentestrat` (cf. the loopfree `sentestrat` on page 347):

If your opponent has just moved on a component that is now active at the ambient temperature t, respond *locally* with a t-orthodox move on that component. Otherwise, play a t-orthodox move on the component with hottest activation temperature below t. If the only suggested move is kobanned, play a t-orthodox move on the *available* component with hottest activation temperature below t.

Unfortunately this generalization doesn't quite work, since the failure to play a kobanned move might lead to an arbitrarily large loss. In particular, if G has a much hotter (unstable) subposition H and at some point Left (say) is kobanned from playing on H, then her resulting loss might be substantially greater than the ambient temperature set by G. Resolving this obstacle will require new techniques that we'll explore in Section 5.

Cold Kos

If G is loopfree, then $t(G) < 0$ if and only if G is equal to a number. This is one of the fundamental results of the loopfree theory. Among its consequences:

- Every loopfree G has **stops** $L(G)$ and $R(G)$. These are necessarily equal to $L_0(G)$ and $R_0(G)$, the t-intercepts of its thermograph.
- If $t(G) < 0$, then $t(G)$ necessarily has the form $-1/2^n$ for some n.
- The entire subzero thermograph of G (that is, the values of $L_t(G)$ and $R_t(G)$ for $-1 \le t < 0$) is completely determined by its values and slopes just below $t = 0$.

If G is loopy, then the situation is more complicated: there exist games with arbitrary negative temperatures that are not equal to numbers. For example, the game G in Figure 3.7 has mean $\frac{2}{3}$ and temperature $-\frac{1}{3}$. Similarly, H (also in Figure 3.7) has mean $\frac{1}{4}$ and temperature $-\frac{1}{4}$ but is *not* equal to the number $\frac{1}{4}$. The differences between H and $\frac{1}{4}$ are made clear by comparing their thermographs, as in Figure 3.8. Games with any desired negative

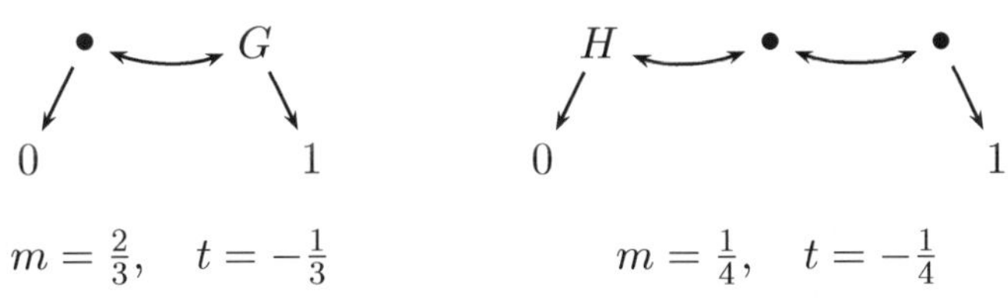

Figure 3.7. Some examples of cold kos.

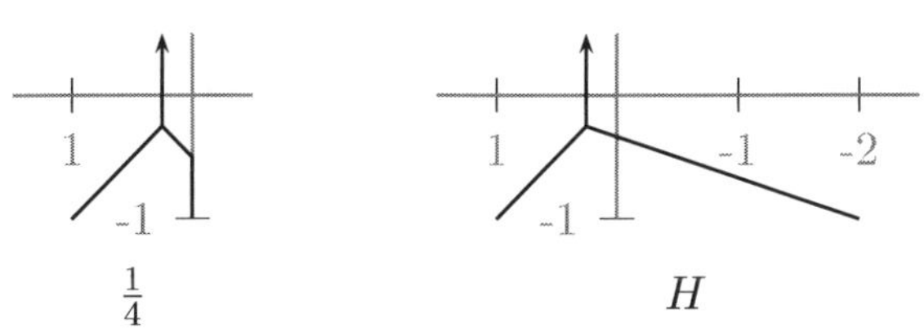

Figure 3.8. Thermographs compared: the number $\frac{1}{4}$ and the loopy game H from Figure 3.7.

temperature can be similarly constructed (Exercise 4.6). Likewise, subzero thermographs for loopy games can have arbitrarily complicated structure.

In the loopfree theory, negative temperatures can be "short-circuited" by replacing stopping positions with equivalent numbers, so subzero thermography isn't strictly necessary. In the generalized theory, a proper accounting of negative temperatures is an essential ingredient. This is the reason why the definitions of $L_t^\delta(G)$ and $R_t^\delta(G)$ in Definition 3.2 necessarily involve the use of $L_{\text{full}}(G)$ and $R_{\text{full}}(G)$ as the base case.

Simple Loopy Games

The definitions of $L_t(G)$ and $R_t(G)$ (and the related orthodoxy) apply to all loopy games. However, the proof that they exist (and the Orthodox Accounting Theorem) will involve a thermographic calculus for a specialized class of **simple** loopy games. Extending this calculus to more complicated games is difficult for technical reasons and is beyond the scope of this book.

Definition 3.6. A **ko** in a loopy game is an alternating 2-cycle.

Definition 3.7. Suppose H is a Left option of G. If G is also a Right option of H (so that G and H form a ko), then we say H is a **Left ko option** of G (and correspondingly G is a **Right ko option** of H).

The game K defined on page 351 has a Left ko option K^L; and K^L in turn has K as a Right ko option.

Definition 3.8. Let G be a loopy game. We say that G is **simple** if:

(i) the only loops in G are kos; and
(ii) every subposition of G has at most one Left ko option and at most one Right ko option.

In particular, all of the examples discussed so far in this chapter are simple. Figure 3.9 on the next page summarizes several of them, along with a few further examples. A simple game G necessarily consists of a ko, or an iterated sequence of kos, together with various **exits** (moves to subpositions from which G is unreachable). These exits provide key recursion points that will help to simplify the thermographic calculus in Section 4.

Although simple games form a small subset of all loopy games, they nonetheless are an interesting class of positions that demonstrate widely varied behavior.[3] This class is not closed under sums, but we'll nonetheless be able to prove some nontrivial theorems about *sums* of simple games (including a version of the Orthodox Accounting Theorem) by componentwise application of the thermographic calculus.

[3]In particular, almost all Go positions that arise in practice are sums of simple components.

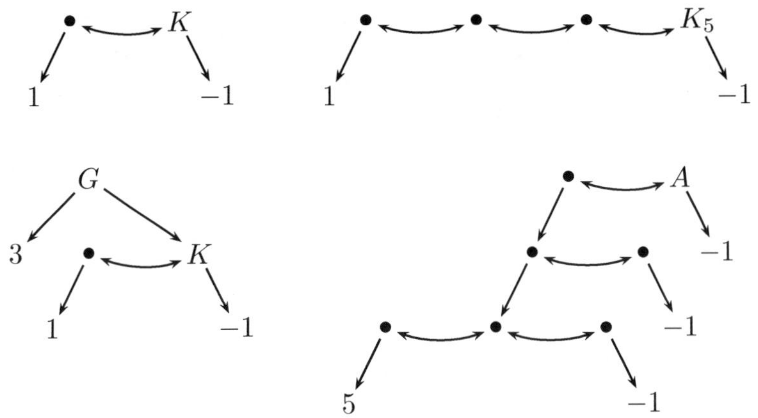

Figure 3.9. Several examples of simple games.

Note that when G is simple, then $L_t(G \setminus \mathscr{A})$ depends on at most one element of $\mathscr{A}$, since G is a subposition of at most one of its Left options—namely, its (unique) Left ko option. Therefore, in runs involving G, we can safely assume that $\mathscr{A}$ is always a singleton and write $L_t(G \setminus H)$ as shorthand for $L_t(G \setminus \{H\})$.

Complex Loops

There are two ways that a game G might fail to be simple:

- some subposition of G has multiple Left (or Right) ko options, as in Figure 3.10; or
- G has long cycles—for example, $K + K$ has the 4-cycle pictured in Figure 3.11 on the following page (in addition to many 2-cycles).

Complex loops like these can be hard to analyze systematically, though it's sometimes possible to determine $L_t(G)$ and $R_t(G)$ for specific positions by a direct calculation.

Example. Consider the game G in Figure 3.10. On $G + \mathscr{E}_t$ Left can defer playing from G to $+1$ until the environment reaches temperature -1. Any

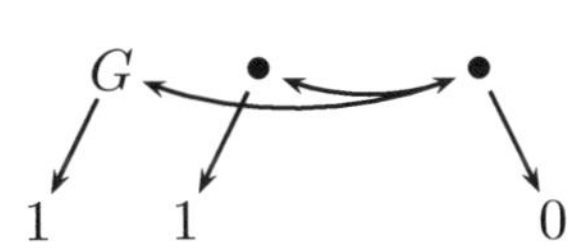

Figure 3.10. A game G whose Right option G^R has two Left ko options.

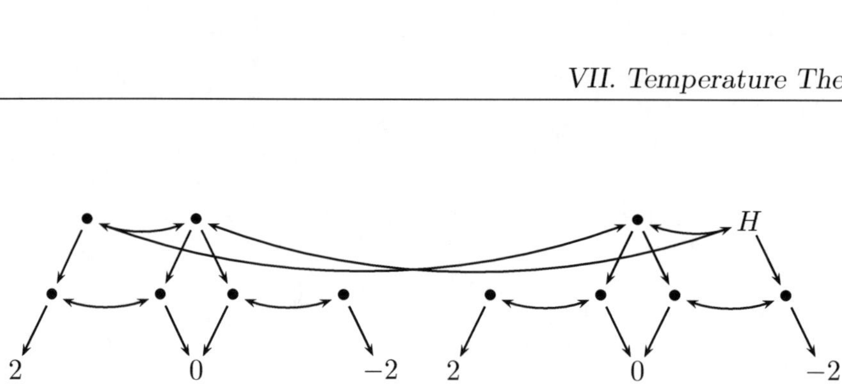

Figure 3.11. Full game graph of the double ko $H = K + K$.

time Right moves from G to G^R, Left simply reverts to the isomorphic position G', forcing an exchange of coupons (since Right is now kobanned from playing to G^R again). If Right later moves from G' to G^R, Left reverts back to G, forcing another exchange of coupons.

As the ambient temperature decreases, there might be many transitions from G to G' and back, but Left will continue to play on the environment rather than exit the ko. Eventually the temperature of the environment will fall to -1, and Left will exit the ko, for a score of $+2$ (one point at temperature -1). This shows that $L_t(G) = 2$ for all $t \geq -1$. We also have $R_t(G) = 2$, since if Right makes an opening move on G, Left will simply revert to G'.

Effectively, Right's option to G^R has no value: G in isolation behaves like the integer $2 = \{1 \mid \}$. However, the behavior of G in sums can be more subtle. There are circumstances in which Right's option is useful; see the succeeding analysis on page 393.

A similar argument shows that G^R has exactly the same thermograph as $\{2 \mid 0\}$. At temperatures $t \leq 1$, Right will exit the ko, moving to 0; and Left will move to G (or G') and thereafter play as before.

4. Generalized Thermography

We can now develop a thermographic calculus for simple loopy games, showing in particular that $L_t(G)$ and $R_t(G)$ exist for all such games G. Recall the recursive construction of loopfree thermographs from Section II.5:

- First, thermographic **scaffolds** $\tilde{\lambda}_t(G)$ and $\tilde{\rho}_t(G)$ are defined recursively by
$$\tilde{\lambda}_t(G) = \max_{G^L} \big(\rho_t(G^L) - t\big) \quad \text{and} \quad \tilde{\rho}_t(G) = \min_{G^R} \big(\lambda_t(G^R) + t\big).$$
- Then the **walls** $\lambda_t(G)$ and $\rho_t(G)$ are obtained by "intersecting" the scaffolds. At active temperatures $t \leq t(G)$, the walls and scaffolds coincide. At dormant temperatures $t > t(G)$, the walls $\lambda_t(G)$ and $\rho_t(G)$ are both equal to $m(G)$, forming an infinite vertical mast.

- Finally, we show recursively that the thermographic walls $\lambda_t(G)$ and $\rho_t(G)$ are identical to the Left and Right scores $L_t(G)$ and $R_t(G)$.

The technique for simple loopy games G is much the same, but with new complications arising at each step:

- The recursive definition of scaffolds is more complicated, since it must account for kos.
- The scaffolds need not be monotonic, and they might cross each other several times, creating multiple active and dormant temperature regions. (Figure 3.6 on page 359 is an example.) The definition of thermographic intersection must be generalized to take this into account.
- The proof that "walls = scores" also must account for kos.

Thermographic Intersection

We begin with the definition of generalized thermographic intersection. This will be a key ingredient of all thermographic techniques in the sequel. First we give some preliminary definitions. Recall that a thermographic **trajectory** is a continuous piecewise linear function $f : [-1, \infty] \to \mathbb{Q}$ (Definition II.5.5 on page 105). Here $[-1, \infty]$ denotes an interval of rational numbers (that is, a convex subset of $\mathbb{Q}$), just as in Section II.5.

Definition 4.1. A **scaffold** is a pair of trajectories $S = (l, r)$. We say l is the **Left scaffold** and r is the **Right scaffold** of S.

Definition 4.2. Let $S = (l, r)$ be a scaffold and let $t \geq -1$. We say t is a **hill temperature** of S if $l_t > r_t$, and a **cave temperature** if $l_t \leq r_t$.

A temperature t is a **junction point** of S if either l or r changes slope at t.

Definition 4.3. A **cave** of a scaffold S is a maximal interval $[b, c]$, with $-1 \leq b < c \leq \infty$, such that every $t \in [b, c]$ is a cave temperature. We say b is the **base** and c the **chimney** of the cave.

A point (t, x) is said to be **inside the cave** if $t \in [b, c]$ and $x \in [l_t, r_t]$.

Examples. A loopfree scaffold $(\tilde{l}, \tilde{r})$ always has exactly one cave, $[t(G), \infty]$. Here $t(G)$ is the base and ∞ is the chimney.

In the general case, there might be multiple hills and multiple caves. For example, the game G from Figure 3.6 (page 359) has the scaffolds pictured in Figure 4.1(a) on the next page. There are two caves, $[1, 2]$ and $[4, \infty]$, separated by the hill region $[2, 4]$. In the thermographic intersection (Figure 4.1(b)), there are two separate masts, one for each cave.

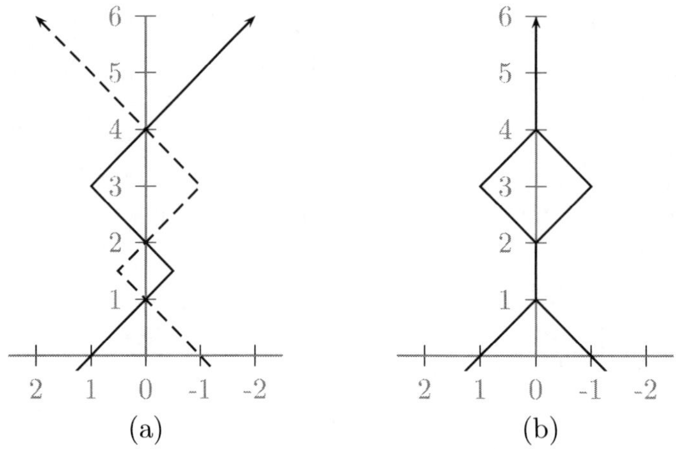

Figure 4.1. (a) Scaffolds of the game G from Figure 3.6. For clarity, the Left and Right scaffolds are pictured with a soLid and bRoken line, respectively. (b) The thermographic intersection of these scaffolds.

In the thermography of loopfree games, it is always true that:

(i) When t is a hill temperature, the walls coincide with their scaffolds: $l_t = \tilde{l}_t$ and $r_t = \tilde{r}_t$.

(ii) When t is a cave temperature, then $l_t = r_t$, and the point (t, l_t) lies inside the cave. (That is, the walls form a mast through the cave.)

(iii) There is exactly one cave: $[t(G), \infty]$.

(iv) The mast through this cave is always a straight vertical line of constant value $m(G)$.

In the generalized thermographic calculus, (i) and (ii) remain true, but not (iii) nor (iv). We've already seen an example with multiple masts (Figure 4.1), and in general these masts aren't necessarily straight vertical lines. For example, in Figure 4.2(a) on the facing page, it is not possible to draw an infinite vertical mast extending upwards from the base of the cave. In such cases the Left and Right walls still coincide, forming a mast through the cave, but the mast is "crooked" as in Figure 4.2(b). To complete the definition of thermographic intersection, we must describe the shape of this mast.

It is helpful to imagine an ideal balloon released into the cave, starting at the point of the base. The balloon travels freely upwards until it is blocked by one of the cave walls. In that event, the balloon travels upwards at a nonzero slope, hugging the wall until it is again free to rise vertically. Eventually the balloon reaches the chimney of the cave (or forms an infinite mast in the case $c = \infty$).

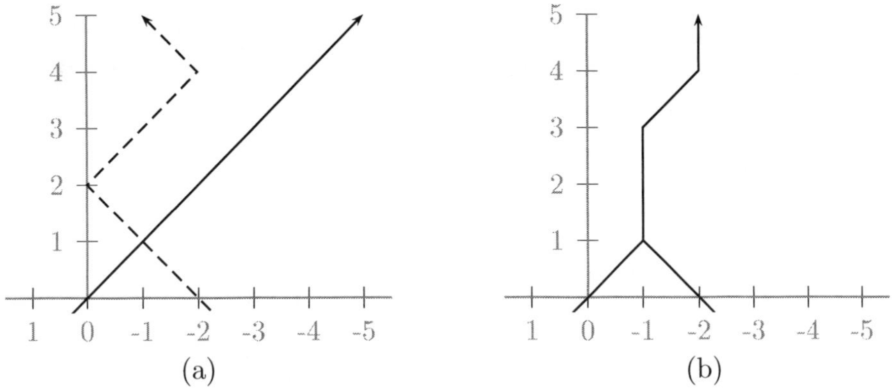

Figure 4.2. (a) A cave that cannot be traversed by a straight vertical mast. (b) The thermographic intersection; observe the distinctive crooked mast.

When $b > -1$, it must be the case that $\tilde{l}_b = \tilde{r}_b$ (since there is a hill just below b), so this "balloon trajectory" uniquely defines a mast through the cave. When $b = -1$, it is possible that $\tilde{l}_b < \tilde{r}_b$. In that case, the base of the cave is a broad interval rather than a point, and there is a range of possible initial values for the balloon. It turns out that the correct choice is the *simplest* number in the closed interval $[\tilde{l}_{-1}, \tilde{r}_{-1}]$ (we'll prove this later). This corresponds to the case where one (or both) of the players would prefer to pass in computing the fullstop.

We'll now give formal definitions of the balloon trajectory and thermographic intersection.

Definition 4.4. Let $[b, c]$ be a cave of $S = (l, r)$. The **balloon trajectory** $f : [b, c] \to \mathbb{Q}$ is defined as follows. Let $f(b)$ be the simplest number in the closed interval $[l_b, r_b]$. Supposing $x = f(t_0)$ is known for some t_0, we define $f(t)$ over an interval $[t_0, t_2]$ according to the following rules:

- Let $t_1 \geq t_0$ be the *largest* value such that (t, x) is in the cave for all $t \in [t_0, t_1]$. Put $f(t) = x$ for all $t \in [t_0, t_1]$. (We include the case $t_1 = \infty$.)
- If $t_1 = c$, then we are done.
- Let $t_2 > t_1$ be the least junction point of S strictly greater than t_1. (Again, we allow $t_2 = \infty$, in the case where no such junction point exists.) If $l_{t_2} > x$, then we let $f(t) = l_t$ for $t \in [t_1, t_2]$. If $r_{t_2} < x$, then $f(t) = r_t$ for $t \in [t_1, t_2]$. (If neither is true, then this implies (t, x) is in the cave for all $t \in [t_1, t_2]$, contradicting the choice of t_1.)

The choices of t_2 in Definition 4.4 form a strictly increasing sequence of junction points of S, so the recursion must stop after a finite number of steps.

Definition 4.5. Let $S = (\tilde{l}, \tilde{r})$. The **thermographic intersection** (l, r) is given as follows.

- For each hill temperature t of S, we put $l_t = \tilde{l}_t$ and $r_t = \tilde{r}_t$.
- On each cave $[b, c]$ of S, we put $l_t = r_t = f(t)$, where f is the balloon trajectory on $[b, c]$.

When a cave mast has positive slope, it is necessarily in contact with the Left scaffold; when it has negative slope, it must be in contact with the Right scaffold. This is a key feature—in fact, as we'll see shortly, it's why the balloon trajectory "works." The following lemma is an important consequence.

Lemma 4.6. *Let $\tilde{S} = (\tilde{l}, \tilde{r})$ be a scaffold with thermographic intersection $S = (l, r)$. If $l_t > l_u$ for some $t > u$, then necessarily $l_v = \tilde{l}_v$ for some $v \in [u, t]$. Likewise, if $r_t < r_u$ for $t > u$, then $r_v = \tilde{r}_v$ for some $v \in [u, t]$.*

Proof. If some $v \in [u, t]$ is a hill temperature, then we are done. Otherwise, $[u, t]$ lies entirely within a cave, and so l_v is equal to the corresponding balloon trajectory for all $v \in [u, t]$. But the balloon trajectory is defined so that:

- If a segment has positive slope, then it must coincide with $\tilde{l}$.
- If a segment has negative slope, then it must coincide with $\tilde{r}$.

The lemma follows immediately. □

Thermal Intensity

When G is loopfree, then every segment of its thermograph has slope 0, 1, or -1. This isn't true for generalized thermographs, which can have arbitrary slopes. However any *particular* thermograph certainly has bounded slope, since it has just finitely many segments. This is one of its central properties.

Definition 4.7. Let $S = (l, r)$ be a scaffold. The **thermal intensity** (or just **intensity**) of S is the maximum value of $|k|$, as k ranges over the slopes of all segments occuring in l or r.

In particular, if $S = (l, r)$ has intensity k, then $|l_t - l_{t-\delta}| \le k\delta$ and $|r_t - r_{t-\delta}| \le k\delta$, for all t and δ. This gives the following analytic characterization of thermographic intersection.

Proposition 4.8. *Let $\tilde{S} = (\tilde{l}, \tilde{r})$ be a scaffold with thermographic intersection $S = (l, r)$. Then:*

(a) $l_t \geq r_t$ *for all* t.

(b) *If* t *is a hill temperature, then* $l_t = \tilde{l}_t$ *and* $r_t = \tilde{r}_t$.

(c) *If* t *is a cave temperature, then* $l_t = r_t$.

(d) *For all* $\delta > 0$ *and all* t *with* $l_t > l_{t-\delta}$, *we have*

$$|l_t - \tilde{l}_t| \leq 2k\delta$$

where k *is the thermal intensity of* $\tilde{S}$. *Likewise, if* $r_t < r_{t-\delta}$, *then* $|r_t - \tilde{r}_t| \leq 2k\delta$.

Proof. (a)–(c) are straight from the definition of S and are restated here for convenience. It remains to prove (d).

From Lemma 4.6, we know that $l_v = \tilde{l}_v$ for some $v \in [t - \delta, t]$. But this implies

$$|l_t - \tilde{l}_t| \leq |l_t - l_v| + |\tilde{l}_v - \tilde{l}_t| = k\delta + k\delta$$

by definition of thermal intensity. □

In fact, conditions (a)–(d) in Proposition 4.8 uniquely characterize the thermographic intersection of S (Exercise 4.5), though we won't need this fact in the succeeding proofs.

The Thermographic Calculus

We can now describe the thermographic calculus for simple loopy games.

Definition 4.9. Let G be simple. We define trajectories $\lambda_t(G)$ and $\rho_t(G)$ (the **walls** of G), together with $\tilde{\lambda}_t(G)$ and $\tilde{\rho}_t(G)$ (the **scaffolds** of G), as follows.

If G has no Left ko option, then

$$\tilde{\lambda}_t(G) = \max_{G^L} \left(\rho_t(G^L) - t\right)$$

exactly as in the loopfree case. If G has a Left ko option H, then

$$\tilde{\lambda}_t(G) = \max\left(\rho_t(G^L) - t,\ \min\left(\lambda_t(H^R),\ \tilde{\lambda}_t(H) - t\right)\right)$$

where G^L ranges over all Left options except H, and H^R ranges over all Right options except G.

Likewise, if G has no Right ko option, then

$$\tilde{\rho}_t(G) = \min_{G^R} \left(\lambda_t(G^R) + t\right)$$

and if G has a Right ko option J, then

$$\tilde{\rho}_t(G) = \min\left(\lambda_t(G^R) + t,\ \max\left(\rho_t(J^L),\ \tilde{\rho}_t(J) + t\right)\right)$$

where G^R ranges over all Right options except J, and J^L ranges over all Left options except G.

Then $(\lambda_t(G), \rho_t(G))$ is the thermographic intersection of $(\tilde{\lambda}_t(G), \tilde{\rho}_t(G))$.

We give some remarks on Definition 4.9:

- The assumption that G is simple guarantees that the recursion is well-founded. If G^L is a Left ko option of G, then the definition of $\tilde{\lambda}_t(G)$ explicitly does not involve $\tilde{\rho}_t(G^L)$. Conversely, if G^L is not a Left ko option of G, then G cannot be a subposition of G^L (this would contradict the assumption that every cycle in G is an alternating 2-cycle).
- The clause $\tilde{\lambda}_t(H) - t$ in the expression
$$\min\left(\lambda_t(H^R),\ \tilde{\lambda}_t(H) - t\right)$$
reflects Right's forcing pass. The idea is that if Left moves from G to H, then Right may either move to some H^R or take a coupon (of value t) and compel Left to play again from H. The use of the scaffold $\tilde{\lambda}_t(H)$ rather than the wall $\lambda_t(H)$ reflects the compulsory nature of Left's second move.

Example. Let K be as defined on page 351. Certainly $\tilde{\rho}_t(K) = -1 + t$. Now K^L is a Left ko option of K, so
$$\tilde{\lambda}_t(K) = \tilde{\lambda}_t(K^L) - t = 1 - 2t.$$
The scaffold is a hill when $t < \frac{2}{3}$, and a cave when $t \geq \frac{2}{3}$, agreeing with the thermograph in Figure 3.2.

Example. Now consider K_n. We have
$$\tilde{\lambda}_t(K) = \tilde{\lambda}_t(K^L) - t = \tilde{\lambda}_t(K^{LL}) - 2t = \cdots = 1 - (n-1)t.$$
This agrees with our calculation in Figure 3.3.

It will be convenient to write
$$\lambda_t(G \setminus H) = \begin{cases} \max\limits_{G^L \neq H} \left(\rho_t(G^L) - t,\ \tilde{\rho}_t(G)\right) & \text{if } H \text{ is a Left option of } G; \\ \lambda_t(G) & \text{otherwise;} \end{cases}$$
and likewise,
$$\rho_t(G \setminus H) = \begin{cases} \min\limits_{G^R \neq H} \left(\lambda_t(G^R) + t,\ \tilde{\lambda}_t(G)\right) & \text{if } H \text{ is a Right option of } G; \\ \rho_t(G) & \text{otherwise.} \end{cases}$$

With this definition, we can write $\tilde{\lambda}_t(G)$ and $\tilde{\rho}_t(G)$ succinctly:
$$\tilde{\lambda}_t(G) = \max_{G^L} \left(\rho_t(G^L \setminus G) - t\right) \quad \text{and} \quad \tilde{\rho}_t(G) = \min_{G^R} \left(\lambda_t(G^R \setminus G) + t\right).$$

If G has a Left ko option H and $\rho_t(H \setminus G) = \tilde{\lambda}_t(H)$, so that Right's best option (on H with G kobanned) is to issue a forcing pass, then we have in particular

$$\tilde{\lambda}_t(H) - \lambda_t(G) = \rho_t(H \setminus G) - \lambda_t(G) \leq t.$$

The following key lemma shows that $\tilde{\lambda}_t(H)$ can be replaced by $\lambda_t(H)$ in this inequality. (This can be interpreted as showing that Left can gain at most t by moving from G to H.)

Lemma 4.10 (Forcing Pass Lemma). *Suppose G is simple and has a Left ko option H. Assume $\rho_t(H \setminus G) = \tilde{\lambda}_t(H)$. Then*

$$\lambda_t(H) - \lambda_t(G) \leq t.$$

Proof. There are three cases.

Case 1: $\lambda_t(H) = \tilde{\lambda}_t(H)$. Then

$$\begin{aligned}\lambda_t(H) - \lambda_t(G) &= \tilde{\lambda}_t(H) - \lambda_t(G)\\ &= \rho_t(H \setminus G) - \lambda_t(G)\\ &\leq t.\end{aligned}$$

Case 2: $\lambda_t(H) \neq \tilde{\lambda}_t(H)$, but $\rho_t(G) = \tilde{\rho}_t(G)$. Then t is a cave temperature of H, so $\lambda_t(H) = \rho_t(H)$ and

$$\begin{aligned}\lambda_t(H) - \lambda_t(G) &= \rho_t(H) - \lambda_t(G)\\ &\leq \tilde{\rho}_t(H) - \lambda_t(G)\\ &\leq \lambda_t(G \setminus H) + t - \lambda_t(G).\end{aligned}$$

So we must show that $\lambda_t(G \setminus H) - \lambda_t(G) \leq 0$. Now

$$\lambda_t(G \setminus H) = \max_{G^L \neq H} \left(\rho_t(G^L) - t,\ \tilde{\rho}_t(G)\right)$$

since H is a Left ko option of G. Certainly $\left(\rho_t(G^L) - t\right) - \lambda_t(G) \leq 0$ for every $G^L \neq H$; to complete the case, we must verify that $\tilde{\rho}_t(G) - \lambda_t(G) \leq 0$. But this follows immediately from the hypothesis $\rho_t(G) = \tilde{\rho}_t(G)$.

Case 3: $\lambda_t(H) \neq \tilde{\lambda}_t(H)$ and $\rho_t(G) \neq \tilde{\rho}_t(G)$. Then fix the largest $u < t$ such that $\lambda_u(H) = \tilde{\lambda}_u(H)$ or $\rho_u(G) = \tilde{\rho}_u(G)$. By Cases 1 and 2, we have

$$\lambda_u(H) - \lambda_u(G) \leq u.$$

But by Lemma 4.6, we have $\lambda_t(H) \leq \lambda_u(H)$ and $\lambda_t(G) \geq \lambda_u(G)$. □

Proof of the Thermographic Calculus

The following theorem generalizes Theorem 1.6 to simple loopy games.

Theorem 4.11. *Let G be simple and let $t \geq -1$. Then*

$$L_t(G) = \lambda_t(G) \quad \textit{and} \quad R_t(G) = \rho_t(G),$$

and in particular $L_t(G)$ and $R_t(G)$ exist.

The proof will be by induction, with the **subposition count**, defined as follows, used as a surrogate for birthday.

Definition 4.12. The **subposition count** $\mathrm{s}(G)$ is equal to the number of distinct subpositions of G. Likewise, $\mathrm{s}(G \setminus H)$ is the number of distinct subpositions of G that are reachable without passing through H.

The base case $t = -1$ in Theorem 1.6 is given by the following proposition (cf. Proposition 1.11 on page 340).

Proposition 4.13. $L_{\mathrm{full}}(G) = \lambda_{-1}(G)$ *and* $R_{\mathrm{full}}(G) = \rho_{-1}(G)$ *for all simple games G.*

Proof. We'll show that, in fact,

$$L_{\mathrm{full}}(G \setminus H) = \lambda_{-1}(G \setminus H) \quad \text{and} \quad R_{\mathrm{full}}(G \setminus H) = \rho_{-1}(G \setminus H)$$

for all G and H. The proof is by induction on $\mathrm{s}(G \setminus H)$. Now we know that

$$\tilde{\lambda}_{-1}(G) = \max_{G^L} \bigl(\rho_{-1}(G^L \setminus G) + 1\bigr) \quad \text{and} \quad \tilde{\rho}_{-1}(G) = \min_{G^R} \bigl(\lambda_t(G^R \setminus G) - 1\bigr).$$

There are two cases.

Case 1: H is *not* a Left ko option of G. Then $\mathrm{s}(G \setminus H) = \mathrm{s}(G)$ (since G is simple), so by induction we can assume that $R_{\mathrm{full}}(G^L \setminus G) = \rho_{-1}(G^L \setminus G)$, etc. Therefore

$$\begin{aligned} L_{\mathrm{full}}(G) &= \max\bigl(\tilde{\lambda}_{-1}(G),\ R^*_{\mathrm{full}}(G)\bigr) \\ &= \max\bigl(\tilde{\lambda}_{-1}(G),\ \min(\tilde{\rho}_{-1}(G),\ 0)\bigr). \end{aligned}$$

The rest of Case 1 is just like the proof of Proposition 1.11. In the case where $\tilde{\lambda}_{-1}(G) < \tilde{\rho}_{-1}(G)$, we have that -1 is a cave temperature of G, so the common value of $\lambda_{-1}(G) = \rho_{-1}(G)$ is *definitionally* chosen to be the simplest integer that fits between them (see Definition 4.4 on page 367).

Case 2: H *is* a Left ko option of G. Then

$$\lambda_{-1}(G \setminus H) = \max_{G^L \neq H} \bigl(\rho_{-1}(G^L) + 1,\ \tilde{\rho}_{-1}(G)\bigr).$$

Likewise,

$$L_{\mathrm{full}}(G \setminus H) = \max_{G^L \neq H} \bigl(R_{\mathrm{full}}(G^L \setminus G) + 1,\ R^*_{\mathrm{full}}(G \setminus H)\bigr).$$

Since G is simple, G cannot be a subposition of any $G^L \neq H$ (this would imply that H has multiple Left ko options); nor can H be a subposition of any G^R (this would imply that G has a long cycle). Therefore

$$L_{\text{full}}(G \setminus H) = \max_{G^L \neq H} \left(R_{\text{full}}(G^L) + 1,\ R^*_{\text{full}}(G)\right).$$

By induction, we can assume that $\rho_{-1}(G^L) = R_{\text{full}}(G^L)$ for all $G^L \neq H$ and that $\lambda_{-1}(G^R \setminus G) = L_{\text{full}}(G^R \setminus G)$ for all G^R. Therefore just as in Case 1,

$$L_{\text{full}}(G) = \max\left(\tilde{\lambda}_{-1}(G),\ \min(\tilde{\rho}_{-1}(G),\ 0)\right)$$

and the rest of the proof is the same. □

For the inductive step we'll need the following notion of thermal intensity of a *game* (previously, thermal intensity was defined only for scaffolds):

Definition 4.14. Let G be simple. The **thermal intensity** (or just **intensity**) of G is the maximum value of k, as k ranges over the thermal intensity of the scaffolds of all subpositions of G.

Now the remainder of Theorem 4.11 is an immediate corollary of the following approximation lemma (cf. Lemma 1.7 on page 338).

Lemma 4.15. *Let G be simple. Fix $\delta > 0$ and $t \geq -1$, with t an integer multiple of δ. Then*

$$|L^\delta_t(G) - \lambda_t(G)| \leq 2kn\delta \quad \textit{and} \quad |R^\delta_t(G) - \rho_t(G)| \leq 2kn\delta,$$

where $n = \mathrm{s}(G)$ and k is the thermal intensity of G.

Proof. The granularity δ will remain fixed for the duration of the proof. We will show that

$$(\dagger) \qquad |L^\delta_t(G) - \lambda_t(G)| \leq 2k\mathrm{s}(G)\delta$$

for all t and G. The corresponding relation for $\rho_t(G)$ follows by symmetry. The value of k will remain fixed as the maximum thermal intensity of all games under consideration.

The proof is by induction on t/δ. The base case, $t = -1$, is given by Proposition 4.13. For a fixed value of $t > -1$, we proceed by induction on an index variable n. We'll prove the following two inequalities, which together imply (†):

(i) $L^\delta_t(G) \leq \lambda_t(G) + 2kn\delta$ whenever $n = \mathrm{s}(G)$;

(ii) $L^\delta_t(G) \geq \lambda_t(G) - 2kn\delta$ whenever $n = \mathrm{s}(G)$.

We'll actually prove a stronger hypothesis in place of (i); this generalization will be an important part of the induction:

(i′) $L^\delta_t(G \setminus H) \leq \lambda_t(G \setminus H) + 2kn\delta$ whenever $n = \mathrm{s}(G \setminus H)$.

In addition, we'll prove a third inequality in tandem with (i′) and (ii). Define the **Left subposition number** of G as follows:

$$\tilde{\mathrm{s}}^L(G) = \max_{G^L} \mathrm{s}(G^L \setminus G).$$

It is the number of subpositions that can be reached if Left moves first from G and play is never permitted to pass through G again. (G itself is specifically *excluded* from the count.) Then:

(iii) $L_t^\delta(G) \geq \tilde{\lambda}_t(G) - 2kn\delta$ whenever $n = \tilde{\mathrm{s}}^L(G)$.

Note the similarity between (ii) and (iii); however (iii) does not follow from (ii), because (iii) is bounded by a tighter value of n. This tighter bound will be central to the induction. In particular, we always have

$$\tilde{\mathrm{s}}^L(G) < \mathrm{s}(G)$$

since G is counted in $\mathrm{s}(G)$ but not $\tilde{\mathrm{s}}^L(G)$.

We'll now prove (i′), (ii), and (iii).

(i′) $L_t^\delta(G \setminus H) \leq \lambda_t(G \setminus H) + 2kn\delta$ whenever $n = \mathrm{s}(G \setminus H)$. There are three cases.

Case 1: $L_t^\delta(G \setminus H) = R_t^\delta(G^L) - t$ for some G^L, where G^L is *not* a Left ko option of G. Then

$$\begin{aligned} L_t^\delta(G \setminus H) &= R_t^\delta(G^L) - t \\ &\leq \rho_t(G^L) - t + 2k(n-1)\delta && \text{((ii), inductively)} \\ &\leq \lambda_t(G) + 2k(n-1)\delta. \end{aligned}$$

Case 2: $L_t^\delta(G \setminus H) = R_t^\delta(G^L \setminus G) - t$, where G^L is a Left ko option of G. If $\rho_t(G^L \setminus G) = \lambda_t(G^{LR} \setminus G^L) + t$ for some G^{LR}, then G^{LR} cannot be a Right ko option of G^L (since G is the unique such Right ko option), so G^L cannot be a subposition of G^{LR} (since G is simple). Therefore

$$\begin{aligned} R_t^\delta(G^L \setminus G) &\leq L_t^\delta(G^{LR}) + t \\ &\leq \lambda_t(G^{LR}) + t + 2k(n-1)\delta \\ &= \rho_t(G^L \setminus G) + 2k(n-1)\delta \\ &\leq \lambda_t(G) + 2k(n-1)\delta. \end{aligned}$$

If instead $\rho_t(G^L \setminus G) = \tilde{\lambda}_t(G^L)$, then

$$\begin{aligned} R_t^\delta(G^L \setminus G) &\le L_{t-\delta}^\delta(G^L) \\ &\le \lambda_{t-\delta}(G^L) + 2k(n-1)\delta && \text{((ii), inductively)} \\ &\le \lambda_t(G^L) + 2k(n-1)\delta + k\delta && \text{(definition of intensity)} \\ &\le \lambda_t(G) + t + 2kn\delta && \text{(Forcing Pass Lemma).} \end{aligned}$$

Case 3: $L_t^\delta(G \setminus H) = R_{t-\delta}^\delta(G)$. (Left's best move is to play in the environment.) If H is a Left ko option of G, then

$$\begin{aligned} R_{t-\delta}^\delta(G) &\le \tilde{\rho}_{t-\delta}(G) + 2k(n-1)\delta && \text{(iii), inductively)} \\ &\le \tilde{\rho}_t(G) + 2k(n-1)\delta + k\delta && \text{(definition of intensity)} \\ &\le \lambda_t(G \setminus H) + 2kn\delta. \end{aligned}$$

If H is not a Left ko option of G, but $\rho_{t-\delta}(G) \le \rho_t(G)$, then

$$\begin{aligned} R_{t-\delta}^\delta(G) &\le \rho_{t-\delta}(G) + 2kn\delta && \text{((ii), inductively on } t) \\ &\le \rho_t(G) + 2kn\delta \\ &\le \lambda_t(G) + 2kn\delta && \text{(Proposition 4.8(a)).} \end{aligned}$$

Otherwise, by Proposition 4.8(d) we have

$$|\rho_t(G) - \tilde{\rho}_t(G)| \le k\delta.$$

Therefore

$$\begin{aligned} R_{t-\delta}^\delta(G) &\le \tilde{\rho}_{t-\delta}(G) + 2k(n-1)\delta && \text{((iv), inductively)} \\ &\le \tilde{\rho}_t(G) + 2k(n-1)\delta + k\delta && \text{(definition of intensity)} \\ &\le \rho_t(G) + 2k(n-1)\delta + 2k\delta && \text{(Proposition 4.8(d))} \\ &\le \lambda_t(G) + 2kn\delta && \text{(Proposition 4.8(a)).} \end{aligned}$$

(ii) $L_t^\delta(G) \ge \lambda_t(G) - 2kn\delta$ whenever $n = \mathrm{s}(G)$. There are two cases.

Case 1: $t - \delta$ is a cave temperature of G and $\lambda_t(G) \le L_{t-\delta}(G)$. Then

$$\begin{aligned} L_t^\delta(G) &\ge R_{t-\delta}^\delta(G) \\ &\ge \rho_{t-\delta}(G) - 2kn\delta && \text{(induction on } t) \\ &= \lambda_{t-\delta}(G) - 2kn\delta && \text{(since } t-\delta \text{ is a cave temperature)} \\ &\ge \lambda_t(G) - 2kn\delta. \end{aligned}$$

Case 2: Otherwise, it must be the case that

$$|\lambda_t(G) - \tilde{\lambda}_t(G)| \le 2k\delta.$$

(Either $t - \delta$ is a hill temperature, so that $L_{t-\delta}(G) = \tilde{\lambda}_{t-\delta}(G)$, or else $\lambda_t(G) > L_{t-\delta}(G)$, in which case we can apply Proposition 4.8(d).) Therefore

$$\begin{aligned} L_t^\delta(G) &\geq \tilde{\lambda}_t(G) - 2k(n-1)\delta \\ &\geq \lambda_t(G) - 2kn\delta. \end{aligned}$$

(iii) $L_t^\delta(G) \geq \tilde{\lambda}_t(G) - 2kn\delta$ whenever $n = \tilde{\mathrm{s}}^L(G)$. Fix a Left option G^L such that $\tilde{\lambda}_t(G) = \rho_t(G^L \setminus G) - t$; then

$$\begin{aligned} L_t^\delta(G) &\geq R_t^\delta(G^L \setminus G) - t \\ &\geq \rho_t(G^L \setminus G) - t + 2kn\delta && \text{((i), inductively)} \\ &= \tilde{\lambda}_t(G). \end{aligned}$$

□

In light of Theorem 4.11, we'll drop the notation $\lambda_t(G)$ and $\rho_t(G)$ in favor of $L_t(G)$ and $R_t(G)$. We'll also write $\tilde{L}_t(G)$ and $\tilde{R}_t(G)$ in place of $\tilde{\lambda}_t(G)$ and $\tilde{\rho}_t(G)$. The main consequences of the thermographic calculus are summarized as follows:

Theorem 4.16 (cf. Theorem II.5.11 on page 108)**.** *For all simple G:*

(a) *$L_t(G)$, $R_t(G)$, $m(G)$, and $t(G)$ are well-defined.*

(b) *$L_t(G)$ and $R_t(G)$ are piecewise linear in t. Every segment has integer slope.*

(c) *$L_t(G) \geq R_t(G)$ for all t.*

Proof. These are all immediate consequences of Theorem 4.11. □

Exercises

4.1 For each example on pages 356–360, use the thermographic calculus to prove that its thermograph matches the asserted diagram.

4.2 Let $a, b \in \mathbb{D}$ with $a > b > 0$. Determine mast values and temperatures for the games G and H shown below, in terms of a and b:

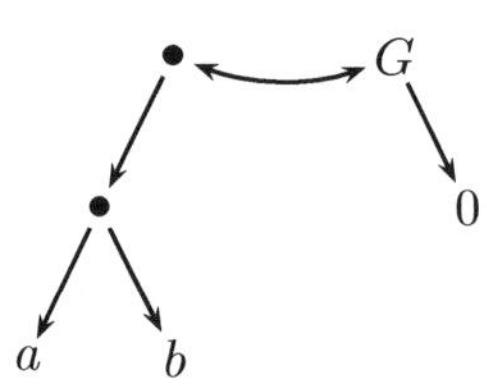

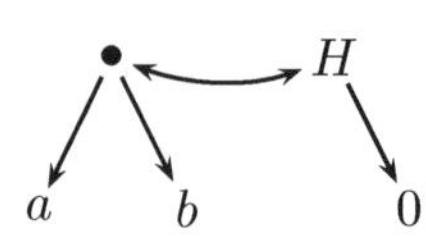

4.3 Exhibit a game G whose scaffolds match Figure 4.2(a) on page 367.

4.4 *Multiple ko from the same position.*

(a) Compute thermographs of the games G, G', and H shown at right (cf. Figure 3.10).

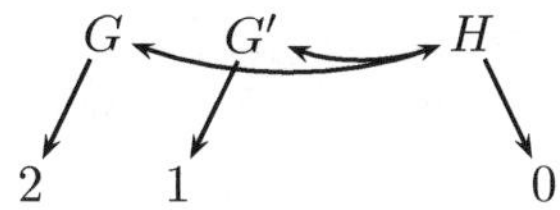

(b) What if the exits to 2 and 1 are replaced by arbitrary simple games X and Y? Extend to a general analysis of such positions.

4.5 Prove that the conditions in Proposition 4.8(a)–(d) uniquely characterize the thermographic intersection.

4.6 (a) For every $x \in \mathbb{Q}$ and every $t > -1$, there is a simple game G with $m(G) = x$ and $t(G) = t$.

(b) However, if G is any game with $t(G) = -1$, then $m(G)$ is an integer.

4.7 Fix a cave temperature t of the scaffold $S = (l, r)$. For $x \in [l_t, r_t]$, define the **quenching temperature** at x, denoted $q(x)$, as the *smallest* temperature such that (t', x) is still in the cave, for all $t' \in [q(x), t]$. (Envision a ball dropped into the cave from (t, x). The ball travels straight down through the cave, and $q(x)$ is the temperature at which it finally collides with the cave wall.)

Prove that in the thermographic intersection of S, the mast value at t is equal to the unique x of lowest quenching temperature, except possibly when some $q(x) = -1$.

Notes

Generalized thermography was introduced by Berlekamp [**Ber96**] and has been explored by Berlekamp, Müller, and Spight [**BMS96**], Spight [**Spi99, Spi02**], and others, mainly in the context of specific applications to Go.

There is obvious room for improvement in the scope of the theory. The calculus presented here applies only to simple games; but $L_t(G)$ and $R_t(G)$ are *defined* for all G, so it is a well-defined problem to extend the calculus to encompass more complicated cycles. William Fraser made partial progress in his doctoral thesis [**Fras02**], but no fully general method is known.

Open Problem. Develop a thermographic calculus that works for all loopy games.

The generalized temperature theory was invented in order to better understand Go. Other examples of games with koban—Push (introduced in Section 3) and the related game Woodpush (described in *Lessons in Play*)—seem rather contrived. While advances in Go are major achievements, it would be valuable to have some other examples to which the temperature theory can be meaningfully applied.

Open Problem. Invent some new loopy games that play well under a koban restriction, as a source of further examples for the generalized temperature theory.

5. Komaster Thermography

The thermographic calculus presented in Section 4 correctly determines $L_t(G)$ and $R_t(G)$, and therefore $m(G)$ and $t(G)$, for all simple games G.

However, these invariants behave poorly in sums, even in a rich environment, presenting major obstacles to an extension of the loopfree orthodoxy. Here are some of the more troubling observations:

Mast values don't add. If G is simple and H is loopfree, then $G + H$ is also simple, and in particular $m(G + H)$ is defined by the thermographic calculus. In general, however,

$$m(G + H) \neq m(G) + m(H).$$

This is why we've carefully called $m(G)$ the "mast value" of G, rather than its "mean value."

Temperatures don't bound. Likewise, in general it might be true that

$$t(G + H) > \max\{t(G), t(H)\},$$

violating a key property of loopfree temperatures.

Thermographs don't compose. $R_t(G^L)$ and $L_t(G^R)$ are defined for all G^L and G^R (assuming G is simple), but in general we have

$$\tilde{L}_t(G) \neq \max_{G^L} \big(R_t(G^L) - t\big) \quad \text{and} \quad \tilde{R}_t(G) \neq \min_{G^R} \big(L_t(G^R) + t\big).$$

Because of the koban restriction, the computation of $L_t(G)$ and $R_t(G)$ is highly sensitive to the fact that we "started at" G. For this reason, the thermograph of G can't be computed from the thermographs of its subpositions, a serious obstacle to generalizing the Orthodox Accounting Theorem.

The following examples illustrate these phenomena.

Example. Let $K = \{1 \mid K \parallel -1\}$, as in Section 3, and recall (page 351 and page 370) that

$$\tilde{L}_t(K) = L_t(K) = 1 - 2t \quad \text{for } t \le \tfrac{2}{3}.$$

Likewise (by symmetry)

$$R_t(K^L) = -1 + 2t \quad \text{for } t \le \tfrac{2}{3}.$$

This shows that thermographs don't compose, since $\tilde{L}_0(K) = 1$, while $R_0(K^L) = -1$. Whoever moves first (Left on K, or Right on K^L) gains a tactical advantage from the koban that is not reflected in the converse thermograph.

Example. Now consider the **threat** of size 5 given by

$$\theta_5 = \big\{10 \mid 0 \parallel \big\}.$$

θ_5 is a game of mean 0 and temperature -1. In fact $\theta_5 = 0$ in the canonical sense, since it's a second-player win, so it can safely be disregarded whenever it appears in a sum of loopfree games. However, θ_5 can have a dramatic

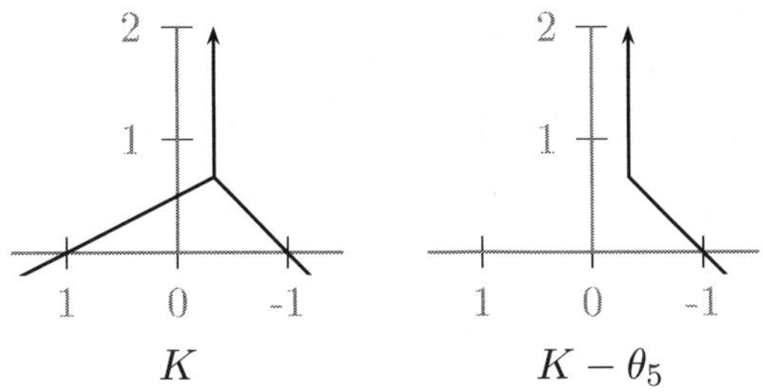

Figure 5.1. The thermographs of K and $K - \theta_5$.

effect in sums involving loopy games, because it can be used to counteract the koban. For example, consider

$$K - \theta_5$$

where as before, $K = \{1 \mid K \parallel -1\}$. Suppose Left moves first at temperature $t \le \frac{2}{3}$, to $K^L - \theta_5$. If Left had moved from K to K^L in isolation, Right would be forced to take a coupon. But from $K^L - \theta_5$ Right can instead play on $-\theta_5$, to

$$K^L + \{0 \mid -10\}.$$

Left clearly cannot afford to ignore the threat, so she must revert to $K^L + 0$. Right, no longer kobanned, returns to K, after which *Left* is forced to take a coupon. Finally Right moves from K to -1. This shows that

$$L_t(K - \theta_5) = -1 + t.$$

(A formal proof can be obtained using the calculus of Section 4.)

Figure 5.1 compares the thermographs of K and $K - \theta_5$. On $K - \theta_5$, the thermograph is a mast at all temperatures.

Example. A more disturbing example is the game G pictured in Figure 5.2. We have $\tilde{L}_t(G) = 0$ whenever $0 \le t \le 1$: Left can move to H in two moves, giving up a coupon; then Right moves to 0 in two moves, also giving up a

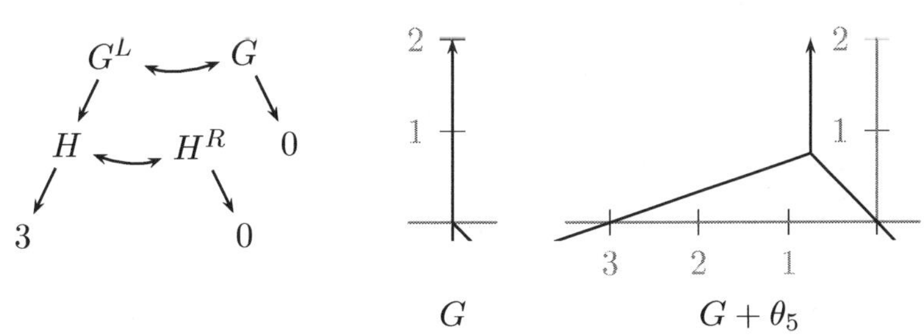

Figure 5.2. G and $G + \theta_5$ have different mast values.

coupon, for a net score of 0. Since $\tilde{R}_t(G) = t$, this shows that

$$L_t(G) = R_t(G) = 0$$

for all $t \geq 0$. Therefore G has mast value 0. But consider $G + \theta_5$. Now when Right moves from H to H^R, Left can issue her threat, moving to

$$H^R + \{10 \mid 0\}.$$

Right must revert to $H^R + 0$. Now Left, free of the koban, can move to 3 in two moves, giving up two more coupons. The final score is $3 - 3t$. So we have

$$\tilde{L}_t(G) = 3 - 3t \quad \text{and} \quad \tilde{R}_t(G) = t.$$

The scaffolds meet at $t = \frac{3}{4}$, and we see that $m(G + \theta_5) = \frac{3}{4}$. Since

$$m(G) = m(\theta_5) = 0,$$

this shows that

$$m(G + \theta_5) \neq m(G) + m(\theta_5).$$

Moreover,

$$t(G + \theta_5) > \max\{t(G), t(\theta_5)\},$$

since $t(G) = 0$ and $t(\theta_5) = -1$, but $t(G + \theta_5) = \frac{3}{4}$.

Threat Environments

As these examples show, large threats provide opportunities to disrupt the koban, which can affect the outcome of a sum even though the threats have mean 0. Small threats can have an even more subtle effect. For example, consider

$$\theta_{1/2} = \{1 \mid 0 \parallel \}.$$

Play on $K - \theta_{1/2}$ begins just as for $K - \theta_5$, but now when Right issues his threat, Left will ignore it and complete the ko anyway. The threat is too small to enable Right to win the ko outright, but it provides him with some compensation: we see that $L_t(K - \theta_{1/2}) = 0$ whenever $t \leq \frac{1}{2}$. (For $t > \frac{1}{2}$ Right won't bother to issue the threat at all, and we have $L_t(K - \theta_{1/2}) = 1 - 2t$, just as for K in isolation.)

The thermograph is shown in Figure 5.3 on the next page; observe how it is intermediate between K and $K - \theta_5$. Thermographs of some related games are also included for comparison.

Other types of threats are still more subtle:

- "Complex threats" of the form $\{X \mid 0 \parallel \}$, where X is some more complicated game.

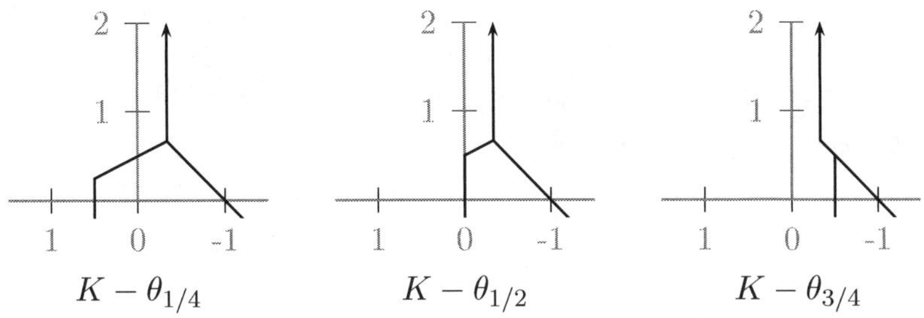

Figure 5.3. The thermographs of $K - \theta_t$ for various t.

- Games like $G = \{10 \mid 0 \parallel -1\}$, which is like θ_5 but with a one-point "bonus." In a loopfree setting G behaves just like $\{2 \mid 0 \parallel -1\}$ (in the orthodox sense), but in sums with loopy games the size of the threat matters. Even in a rich environment, it might sometimes be to Right's advantage to remove the threat rather than play on a hotter component.

The combinatorics of these types of threats can become quite difficult. In close battles where the size and texture of threats matters, combinatorial interactions between components can be significant, even in a rich environment.

To address these difficulties, we now introduce a new type of **threat environment**, composed of a large number of very large threats. In the presence of a threat environment, the combinatorics of threat structure are not relevant, because complex threats are dominated by the abundance of larger threats in the environment.

Definition 5.1. The **standard threat of size** x, denoted by θ_x, is

$$\theta_x = \big\{2x \mid 0 \parallel \,\big\}.$$

Definition 5.2. The nth **threat environment**, denoted by Θ_n, is

$$\Theta_n = n \cdot \theta_n.$$

We'll shortly prove that if G is simple, then the thermograph of $G + \Theta_n$ converges to a constant for all sufficiently large values of n. For such n, we say that Left is **komaster** on the sum $G + \Theta_n$. Likewise, for sufficiently large n we say that Right is komaster on the sum $G - \Theta_n$. It's clear that these environments are biased, so that $G + \Theta_n$ is at least as favorable to Left as G, and $G - \Theta_n$ is at least as favorable to Right. Either player would prefer to be komaster, if possible.

The following definitions apply to all loopy games G (not necessarily simple).

Definition 5.3. For $t \geq -1$ we define the **Left komaster scores** $L_t^\sharp(G)$ and $R_t^\sharp(G)$, and the **Right komaster scores** $L_t^\flat(G)$ and $R_t^\flat(G)$, by

$$L_t^\sharp(G) = \lim_{n\to\infty} L_t(G + \Theta_n), \qquad R_t^\sharp(G) = \lim_{n\to\infty} R_t(G + \Theta_n),$$

$$L_t^\flat(G) = \lim_{n\to\infty} L_t(G - \Theta_n), \qquad R_t^\flat(G) = \lim_{n\to\infty} R_t(G - \Theta_n),$$

whenever these limits exist. The **Left komaster thermograph** of G is the pair $(L_t^\sharp(G), R_t^\sharp(G))$, and the **Right komaster thermograph** is defined analogously.

We'll prove below that all of these limits converge when G is simple. In the meantime, the following propositions apply whenever all of the referenced quantities are defined.

Proposition 5.4. *$L_t^\sharp(G) \geq L_t(G) \geq L_t^\flat(G)$ and $R_t^\sharp(G) \geq R_t(G) \geq R_t^\flat(G)$, with equality when G is loopfree.*

Proof. $G + \Theta_n$ introduces no new Right options to G, so necessarily

$$L_t^\delta(G + \Theta_n) \geq L_t^\delta(G)$$

for all t and δ. This shows that $L_t^\sharp(G) \geq L_t(G)$, and the same argument works in all cases.

When G is loopfree, we have equality because $\Theta_n = 0$ in the canonical sense. □

Definition 5.5. $m^\sharp(G)$ and $t^\sharp(G)$ are defined to be the mast value and temperature of the thermograph $\big(L_t^\sharp(G), R_t^\sharp(G)\big)$, whenever they exist. $m^\flat(G)$ and $t^\flat(G)$ are similarly defined in terms of $\big(L_t^\flat(G), R_t^\flat(G)\big)$.

Proposition 5.6. *$m^\sharp(G) \geq m(G) \geq m^\flat(G)$, with equality if G is loopfree.*

Proof. Immediate from Proposition 5.4. □

The point of this theory is the following definition:

Definition 5.7. G is **placid** if $m^\sharp(G) = m^\flat(G)$; **hyperactive** otherwise.

Hyperactivity arises when a subposition of G contains a ko that is hotter than G itself. In orthodox play of G, the environment (no matter how rich) will be depleted before the kofight begins, so it will not provide adequate compensation to the loser of the kofight. Figure 5.2 on page 379 gives an example of a hyperactive position; most of the other examples we've seen so far are placid.

Recall that the game G in Figure 5.2 is sensitive to threats in the environment: $m(G) \neq m(G+\theta_5)$, as we showed above. This is precisely because the subposition H is hotter than G itself. Orthodox play of $G + \mathscr{E}_t^\delta$ depletes the environment down to temperature 0 before any moves are made on G, and the environment is therefore badly impoverished relative to H. This poverty creates the opportunity for Left to profit from her threat.

Hyperactive positions are rife with technical complications that make orthodox accounting difficult. Conversely, if all the components of a sum are placid, then the orthodox forecast is independent of the threat environment, and orthodox accounting is more straightforward. The essential idea is that either player can achieve the orthodox forecast (in a suitably rich environment) by *assuming* her opponent is komaster and playing accordingly—simply declining to fight any ko. The presence of threats in other components cannot disrupt this strategy, because the assumption about the threat environment is already maximally pessimistic.

A slight extension of this idea can be used to handle sums involving at most one hyperactive component with known komaster (see Exercise 5.4). Sums of multiple hyperactive components require more advanced ("dogmatic") techniques that are beyond the scope of this book; these are briefly addressed in the notes on page 396.

The Komaster Calculus

We'll now develop a calculus for computing komaster thermographs. For clarity, we'll give the calculus for the Left komaster scores, $L_t^\sharp(G)$ and $R_t^\sharp(G)$; their duals $L_t^\flat(G)$ and $R_t^\flat(G)$ can be obtained symmetrically.

Definition 5.8. Let G be simple. We define $\tilde{\lambda}_t^\sharp(G)$, $\tilde{\rho}_t^\sharp(G)$, $\lambda_t^\sharp(G)$, and $\rho_t^\sharp(G)$ recursively as follows. If G has no Left ko, then

$$\tilde{\lambda}_t^\sharp(G) = \max_{G^L} \left(\rho_t^\sharp(G^L) - t\right)$$

just as in Definition 4.9. If G has a Left ko H, then

$$\tilde{\lambda}_t^\sharp(G) = \max\left(\rho_t^\sharp(G^L) - t,\ \min\left(\lambda_t^\sharp(H^R),\ \tilde{\lambda}_t^\sharp(H) - t\right)\right)$$

where G^L ranges over all Left options except H, and H^R ranges over all Right options except G, again just as in Definition 4.9.

For $\tilde{\rho}_t^\sharp(G)$, however, we put

$$\tilde{\rho}_t^\sharp(G) = \min_{G^R} \left(\lambda_t^\sharp(G^R) + t\right)$$

whether or not G has a Right ko. Then $(\lambda_t^\sharp(G), \rho_t^\sharp(G))$ is the thermographic intersection of $(\tilde{\lambda}_t^\sharp(G), \tilde{\rho}_t^\sharp(G))$.

The definition of $\tilde{\rho}_t^\sharp(G)$ imposes no restrictions on Left's responses after a move to G^R, even when G^R is a Right ko of G. This reflects the fact that Left can use a threat to return to G. However if Left chooses to do so, then Right may issue a forcing pass. This is because Left has a finite number of threats; continually using a threat to return to the same position will eventually exhaust her threat stack.

We have the following recurrences:

$$\tilde{\lambda}_t^\sharp(G) = \max_{G^L}\bigl(\rho_t^\sharp(G^L \setminus G) - t\bigr),$$
$$\tilde{\rho}_t^\sharp(G) = \min_{G^R}\bigl(\lambda_t^\sharp(G^R) + t\bigr),$$

with $\rho_t^\sharp(G^L \setminus G)$ defined just as in Section 4:

$$\rho_t^\sharp(G \setminus H) = \begin{cases} \min\limits_{G^R \neq H} \bigl(\lambda_t^\sharp(G^R) + t,\ \tilde{\lambda}_t^\sharp(G)\bigr) & \text{if } H \text{ is a Right option of } G; \\ \rho_t(G) & \text{otherwise.} \end{cases}$$

The following lemma shows that Left can never gain by "frivolously" using one of her threats (that is, playing a threat for any reason other than to return to a kobanned position).

Lemma 5.9. *For all G and all $k, n \geq 0$,*

$$L_t(G + k\theta_n) = \max_{G^L}\bigl(R_t(G^L + k\theta_n \setminus G + k\theta_n) - t\bigr).$$

(That is, Left's moves on $k\theta_n$ don't contribute to the calculation.)

Proof. Let $H \cong G + (k-1)\theta_n + \{2n \mid 0\}$. Right has a move from H to $G + (k-1)\theta_n$, so

$$R_t(H) - t \leq L_t(G + (k-1)\theta_n).$$

By induction on k,

$$L_t(G + (k-1)\theta_n) = \max_{G^L}\bigl(R_t(G^L + (k-1)\theta_n \setminus G + (k-1)\theta_n) - t\bigr).$$

Therefore for some choice of G^L, we have

$$\begin{aligned} R_t(H) &\leq R_t(G^L + (k-1)\theta_n \setminus G + (k-1)\theta_n) \\ &\leq R_t(G^L + k\theta_n \setminus G + k\theta_n), \end{aligned}$$

the last inequality due to the fact that θ_n has no Right option. $\square$

We need one more lemma, showing that a sufficiently large threat must be answered:

Lemma 5.10. *If $n \geq \mathrm{s}(G)$, then for all $m \geq 0$ and $t \geq -1$,*

$$R_t(G + \Theta_m + \{2n \mid 0\}) = L_t(G + \Theta_m) + t.$$

(Right is forced to play from $\{2n \mid 0\}$ to 0.)

Proof. Every other option for Right has the form $G^R + \Theta_m + \{2n \mid 0\}$, but

$$\begin{aligned} L_t(G^R + \Theta_m + \{2n \mid 0\}) + t &\geq R_t(G^R + \Theta_m + 2n) \\ &\geq R_t(G^R + 2n) \\ &\geq n \qquad\qquad (\text{since } \mathrm{s}(G^R) \leq n). \end{aligned}$$

Likewise, $L_t(G+\Theta_m)+t \leq n$, so necessarily dominates all other options. □

Now we can prove the main theorem:

Theorem 5.11. *Let $n \geq \mathrm{s}(G)$. Then*

$$L_t(G + \Theta_n) = \lambda_t^\sharp(G) \quad \textit{and} \quad R_t(G + \Theta_n) = \rho_t^\sharp(G)$$

for all $t \geq -1$.

Proof. The proof is by induction on $\mathrm{s}(G)$. For a fixed choice of $\mathrm{s}(G)$, we'll prove the following, for all G and all $n \geq \mathrm{s}(G)$:

(i) $\tilde{\lambda}_t^\sharp(G) = \tilde{\lambda}_t(G + \Theta_{n-1})$.

(ii) $\tilde{\rho}_t^\sharp(G) = \tilde{R}_t(G + \Theta_n)$.

These clearly imply the theorem. We'll also prove:

(iii) $\lambda_t^\sharp(G) = L_t(G + \Theta_n \setminus H + \Theta_n)$, for H a Left ko option of G.

(iv) $\rho_t^\sharp(G \setminus J) = \rho_t(G + \Theta_{n-1} \setminus J + \Theta_{n-1})$, for J a Right ko option of G.

With $\mathrm{s}(G)$ fixed, we first prove (i) and (iv) by induction on $\tilde{\mathrm{s}}^L(G)$, followed by (ii) and (iii) by induction on $\tilde{\mathrm{s}}^R(G)$. (The **subposition numbers** $\tilde{\mathrm{s}}^L(G)$ and $\tilde{\mathrm{s}}^R(G)$ are as defined on page 374.)

(i) is straightforward:

$$\begin{aligned} \tilde{\lambda}_t^\sharp(G) &= \max_{G^L} \left(\rho_t^\sharp(G^L \setminus G) - t\right) \\ &= \max_{G^L} \left(\rho_t(G^L + \Theta_{n-1} \setminus G + \Theta_{n-1}) - t\right) \\ &= \tilde{\lambda}_t(G + \Theta_{n-1}). \end{aligned}$$

The middle identity follows by induction: when G^L is a Left ko option of G, then $\tilde{\mathrm{s}}^L(G^L) < \tilde{\mathrm{s}}^L(G)$, so we can apply (iv) inductively; otherwise, G is not a subposition of G^L (since G is simple), so $\mathrm{s}(G^L) < \mathrm{s}(G)$ and the theorem applies inductively. The latter inequality is Lemma 5.9.

(iv) is similar:

$$\begin{aligned}\rho_t^\sharp(G \setminus J) &= \min_{G^R \neq J} \left(\lambda_t^\sharp(G^R) + t,\ \tilde{\lambda}_t^\sharp(G)\right) \\ &= \min_{G^R \neq J} \left(\lambda_t(G^R + \Theta_{n-1}) + t,\ \tilde{\lambda}_t(G + \Theta_{n-1})\right) \\ &= \rho_t(G + \Theta_{n-1} \setminus J + \Theta_{n-1}).\end{aligned}$$

As before, the middle identity follows by induction: $\mathrm{s}(G^R) < \mathrm{s}(G)$ for all G^R distinct from J; and $\tilde{\lambda}_t^\sharp(G) = \tilde{\lambda}_t(G + \Theta_{n-1})$ since we have already proved (i) for G.

For (ii) we have

$$\begin{aligned}\tilde{\rho}_t^\sharp(G) &= \min_{G^R} \left(\lambda_t^\sharp(G^R) + t\right) \\ &= \min_{G^R} \left(L_t(G^R + \Theta_n \setminus G + \Theta_n) + t\right) \\ &= \tilde{R}_t(G + \Theta_n).\end{aligned}$$

Here the middle inequality follows by applying (iii) when G^R is a Right ko of G (in which case $\tilde{\mathrm{s}}^R(G^R) < \tilde{\mathrm{s}}^R(G)$) and applying the theorem inductively otherwise.

(iii) We have already proved (i) for G, so that $\tilde{\lambda}_t(G) = \tilde{\lambda}_t(G + \Theta_{n-1})$. It therefore suffices to show that

$$L_t(G + \Theta_n \setminus H + \Theta_n) = \tilde{\lambda}_t(G + \Theta_{n-1}).$$

But $n \geq 1$ (as G must have at least one subposition, namely G), so certainly

$$\begin{aligned}L_t(G + \Theta_n \setminus H + \Theta_n) &\geq R_t(G + \Theta_{n-1} + \{2n \mid 0\}) && (\Theta_n \geq \Theta_{n-1} + \theta_n) \\ &= L_t(G + \Theta_{n-1}) && \text{(Lemma 5.10)},\end{aligned}$$

and the converse inequality is easy, since

$$L_t(G + \Theta_{n-1}) = L_t(G + \Theta_n) \geq L_t(G + \Theta_n \setminus H + \Theta_n). \qquad \square$$

We'll now drop the use of λ and ρ, and we'll write simply $L_t^\sharp(G)$, $R_t^\sharp(G)$, $\tilde{L}_t^\sharp(G)$, and $\tilde{R}_t^\sharp(G)$ for the appropriate quantities. Theorem 5.11 (and its dual) give the following recurrences:

$$\begin{aligned}\tilde{L}_t^\sharp(G) &= \max_{G^L} \left(R_t^\sharp(G^L \setminus G) - t\right), & \tilde{R}_t^\sharp(G) &= \min_{G^R} \left(L_t^\sharp(G^R) + t\right), \\ \tilde{L}_t^\flat(G) &= \max_{G^L} \left(R_t^\flat(G^L) - t\right), & \tilde{R}_t^\flat(G) &= \min_{G^R} \left(L_t^\flat(G^R \setminus G) + t\right),\end{aligned}$$

where

$$L_t^\flat(G \setminus H) = \max_{G^L \neq H} \big(R_t^\flat(G^L) - t,\ \tilde{R}_t^\flat(G)\big),$$

$$R_t^\sharp(G \setminus H) = \min_{G^R \neq H} \big(L_t^\sharp(G^R) + t,\ \tilde{L}_t^\sharp(G)\big).$$

Generalized Orthodox Accounting

We can now state the Orthodox Accounting Theorem for placid simple games. The **Left orthodox forecast** of a sum $G_1 + \cdots + G_k$ is defined just as in the loopfree case (Definition 2.4 on page 343):

$$m(G_1) + \cdots + m(G_k) + \frac{t}{2}.$$

Likewise, the **Right orthodox forecast** is given by

$$m(G_1) + \cdots + m(G_k) - \frac{t}{2}.$$

Definition 5.12. Let $G = G_1 + \cdots + G_k$. The **Left-biased board temperature** of G is given by

$$t_{\mathrm{b}}^\sharp(G) = \max\{0, t^\sharp(G_1), \ldots, t^\sharp(G_k)\}.$$

Likewise, the **Left-biased board activation temperature** of G below t is the largest $t' \le t$ such that at least one $G_i + \Theta$ is active at t'.

Definition 5.13. Let $G = G_1 + \cdots + G_k$ and fix any alternating run $\vec{G}$ of G. We define the **Left-biased ambient temperature** $t_{\mathrm{a}}^\sharp(H)$ of each $H \in \vec{G}$ as follows.

- $t_{\mathrm{a}}^\sharp(G) = t_{\mathrm{b}}^\sharp(G)$, the board temperature of G.
- Suppose J is the successor of H in $\vec{G}$. Then $t_{\mathrm{a}}^\sharp(J)$ is the board activation temperature of $J + \Theta$ below $t_{\mathrm{a}}^\sharp(H)$.

Right-biased board temperature, board activation temperature, and ambient temperature are defined analogously, with $^\flat$ in place of $^\sharp$. We also have the following analogue of the **score** of a run. Here arbitrary games of negative temperature take the place of numbers, so that the score is defined as soon as the ambient temperature becomes negative:

Definition 5.14 (cf. Definition 2.8 on page 345). Let $\vec{G}$ be an alternating run of G. The **Left-biased score** of $\vec{G}$ is given by

$$m^\sharp(X_1) + \cdots + m^\sharp(X_k),$$

where $X = X_1 + \cdots + X_k$ is the first element of $\vec{G}$ with $t_{\mathrm{b}}^\sharp(X) < 0$.

We can now give the generalized version of sentestrat.

> Assume that your opponent is komaster, so that all biases are in his favor.
>
> If your opponent has just moved on a component that is now active at the ambient temperature t, respond *locally* with a t-orthodox move on that component. Otherwise, play a t-orthodox move on the component with hottest activation temperature below t.
>
> If the only suggested move is kobanned, play a t-orthodox move on the *available* component with hottest activation temperature below t.

Signed temperature drops are defined just as in the loopfree case (Definition 2.7 on page 345), and they play exactly the same role in the Orthodox Accounting Theorem. However, the koban clause in the new sentestrat entails another type of adjustment in our accounting:

Definition 5.15. Let $\vec{G}$ be a run played according to sentestrat, and suppose H and J are successive elements of G, with J a Left option of H. Let $t = t_{\mathrm{a}}^{\flat}(H)$ and let t' be the Right-biased activation temperature (below t) of the component that Left actually played on from H to J. The corresponding **ko adjustment** is given by

$$\Delta u(J) = -(t - t').$$

The ko adjustment is 0 except when Left plays on a component whose activation temperature is strictly less than the ambient temperature. Under sentestrat, this can only happen when she was kobanned from playing at the ambient temperature; this is why it is called a "ko adjustment." If G is loopfree, then the ko adjustments are necessarily identically 0.

Theorem 5.16 (Orthodox Accounting Theorem)**.** *Let $G_1, \ldots, G_k$ be simple and placid. Suppose Left plays first (second) on the compound*

$$G = G_1 + \cdots + G_k$$

and follows sentestrat*. Then she is guaranteed a score of at least*

$$z + \frac{1}{2}\sum_{i=1}^{m} \Delta t_i + \sum_{j=1}^{n} \Delta u_j$$

where z is the Left (Right) orthodox forecast, Δt_i are the signed temperature drops of the corresponding run, and Δu_j are the signed ko adjustments.

Proof. Let x and y be the Left and Right orthodox forecasts of G. Since each component G_i is placid, we have $m(G_i) = m^\flat(G_i)$, so in fact

$$x = m^\flat(G_1)+\cdots+m^\flat(G_k)+\frac{t_\mathrm{b}^\flat(G)}{2} \quad \text{and} \quad y = m^\flat(G_1)+\cdots+m^\flat(G_k)-\frac{t_\mathrm{b}^\flat(G)}{2}.$$

We'll prove the following statement by induction: if t is *any* Right-biased board activation temperature of G, then Left (following sentestrat with the ambient temperature "starting" at t) is guaranteed a score of at least

$$z_t + \frac{1}{2}\sum_{i=1}^{m} \Delta t_i + \sum_{j=1}^{n} \Delta u_j,$$

where z_t is one of x_t or y_t, depending on who moves first:

$$x_t = m_t^\flat(G_1) + \cdots + m_t^\flat(G_k) + \frac{t}{2} \quad \text{and} \quad y_t = m_t^\flat(G_1) + \cdots + m_t^\flat(G_k) - \frac{t}{2}.$$

Here $m_t^\flat(G_i)$ denotes the common value of $L_t^\flat(G_i) = R_t^\flat(G_i)$ (which must be the same, since t is an activation temperature). Note that this stronger statement reduces to the original theorem in the case $t = t_\mathrm{b}^\flat(G)$, which is the Right-biased board activation temperature of G below ∞.

The proof is much the same as the proof of Theorem 2.9 (page 348). We'll show that each subposition

$$Y = Y_1 + \cdots + Y_k,$$

with Right to move, satisfies

$$(\dagger) \qquad R_t^\flat(Y_1) + R_t^\flat(Y_2) + \cdots + R_t^\flat(Y_k) - \frac{t}{2} + \sum_{j=1}^{n'} \Delta u_j \geq y_t,$$

so long as at least one Y_i remains active at t. Likewise, each subposition

$$X = X_1 + \cdots + X_k,$$

with Left to move, satisfies

$$(\ddagger) \qquad L_t^\flat(X_1 \setminus Z) + R_t^\flat(X_2) + \cdots + R_t^\flat(X_k) + \frac{t}{2} + \sum_{j=1}^{n'} \Delta u_j \geq x_t,$$

so long as at least one X_i remains active at t and assuming that if Right has just played a move, then it was from Z to X_1. In both cases, n' represents the number of ko adjustments applied so far during the run of G.

By the inductive hypothesis, both ($\dagger$) and ($\ddagger$) are satisfied automatically for the initial position G. (Here, obviously, $n' = 0$ and the ko adjustment term contributes nothing to the forecast.)

Now consider a position Y satisfying (†). Assume without loss of generality that Right moves on Y_1; then certainly $L_t^\flat(Y_1^R \setminus X_1) + t \geq R_t^\flat(Y_1)$, so the new position X satisfies (‡), with $Z = X_1$.

Next consider X satisfying (‡). There are two cases.

Case 1: X_1 is active at t. If there is some t-orthodox move X_1^L distinct from Z, then sentestrat recommends it, and we have $L_t^\flat(X_1 \setminus Z) = R_t^\flat(X_1^L) - t$. The recurrence (†) follows.

If Z is the only t-orthodox Left option of X_1, then necessarily $L_t^\flat(X_1 \setminus Z) = R_t^\flat(X_1)$. In this case sentestrat recommends a move elsewhere, say (without loss of generality) from X_2 to X_2^L. This entails a ko adjustment equal to

$$\Delta u = L_t^\flat(X_2) - R_t^\flat(X_2^L) + t.$$

In particular,

$$\begin{aligned} R_t^\flat(X_2^L) + \Delta u &= L_t^\flat(X_2) + t \\ &\geq R_t^\flat(X_2) + t. \end{aligned}$$

Therefore

$$R_t^\flat(X_1) + R_t^\flat(X_2^L) + \Delta u \geq L_t^\flat(X_1 \setminus Z) + R_t^\flat(X_2) + t,$$

from which the recurrence (†) follows.

Case 2: X_1 is dormant at t. Then $L_t^\flat(X_1 \setminus Z) = R_t^\flat(X_1)$. In this case sentestrat recommends a move on some other component, say from X_2 to X_2^L. We are assuming that at least one component is active at t, so either X_2^L is t-orthodox or else all t-orthodox options are kobanned.

If X_2^L is t-orthodox, then

$$R_t^\flat(X_2^L) = L_t^\flat(X_2) - t \geq R_t^\flat(X_2) - t$$

and (†) follows. If all t-orthodox moves are kobanned, then there is a ko adjustment just as in Case 1.

This shows that the recurrences (†) and (‡) are satisfied, so long as at least one component is active at t. Eventually all components will become dormant at t. Let t' be the (Right-biased) board activation temperature below t, so that the ambient temperature drops from t to t'. If this happens after a move by Left, then we have

$$\Delta t_1 = t - t'$$

and the first position Y reached after the temperature drop satisfies

$$R_t^\flat(Y_1) + R_t^\flat(Y_2) + \cdots + R_t^\flat(Y_k) - \frac{t}{2} \geq y_t.$$

Since each Y_i is dormant at t, we can write

$$m_t^\flat(Y_1) + m_t^\flat(Y_2) + \cdots + m_t^\flat(Y_k) - \frac{t}{2} \geq y_t,$$

and since t' is the activation temperature below t, we have $m_{t'}^\flat(Y_i) = m_t^\flat(Y_i)$ for each i, so that

$$m_{t'}^\flat(Y_1) + m_{t'}^\flat(Y_2) + \cdots + m_{t'}^\flat(Y_k) - \frac{t}{2} \geq y_t.$$

By induction on G, Left is guaranteed a score on Y of at least

$$\begin{aligned} &m_{t'}^\flat(Y_1) + \cdots + m_{t'}^\flat(Y_k) - \frac{t'}{2} + \frac{1}{2}\sum_{i=2}^{n} \Delta t_i \\ &= m_{t'}^\flat(Y_1) + \cdots + m_{t'}^\flat(Y_k) - \frac{t}{2} + \frac{1}{2}\Delta t_1 + \frac{1}{2}\sum_{i=2}^{n} \Delta t_i \\ &\geq y_t + \frac{1}{2}\sum_{i=1}^{n} \Delta t_i. \end{aligned}$$

The same argument works if the temperature drop happens after a move by Right, this time using the fact that $\Delta t_1 = -(t - t')$. □

We arrive at the main result of the generalized orthodox theory:

Theorem 5.17. *Let $G_1, \ldots, G_k$ be simple and placid. Suppose Left plays first (second) on the compound*

$$G = G_1 + \cdots + G_k + \mathscr{E}_t^\delta$$

and follows sentestrat. *Then she is guaranteed at least the Left (Right) orthodox forecast, to within a bounded multiple of δ.*

Proof. The argument is just the same as for Theorem 2.11 on page 350. The total number of moves on G between successive temperature drops is bounded by $\mathrm{s}(G)$; likewise, there can be at most $\mathrm{s}(G)$ ko adjustments, each of magnitude at most $\mathrm{s}(G)\delta$. Since $\mathrm{s}(G)$ is a constant independent of δ, these terms are negligible in the limiting case. □

The following corollary of the Orthodox Accounting Theorem gives significant insight into arbitrary sums of placid simple games. (Note that the generalized temperature theory of Section 4 cannot be applied directly to such sums, since they are not necessarily simple.)

Theorem 5.18. *If $G = G_1 + \cdots + G_k$ is a sum of placid simple games, then $m(G)$ is defined and*

$$m(G) = m(G_1) + \cdots + m(G_k).$$

Moreover, G is necessarily dormant at all $t > t_b(G)$ (so that if $t(G)$ exists, then $t(G) \le \max\{t(G_1), \ldots, t(G_k)\}$).

Proof. Let $x = m(G_1) + \cdots + m(G_k)$. Theorem 5.17 shows that for all $t \ge t_b(G)$, we have

$$|L_t^\delta(G) - x| \le k\delta \quad \text{and} \quad |R_t^\delta(G) - x| \le k\delta$$

for some constant k. Therefore $L_t(G) = R_t(G) = x$ for all such t, and the theorem follows. □

Komonster Thermography

If we write out all the limits in the definition of $L_t^\sharp(G)$, we have

$$L_t^\sharp(G) = \lim_{n\to\infty} \lim_{\delta\to 0} L_t^\delta(G + \Theta_n).$$

This construction models an environment with a large number n of threats, and a fine granularity δ for the coupon stack. The order in which the limits appear is important. The number of threats n is large enough to win any kofight on G—but not so large that Left can drive down the ambient temperature in the process, because n is small in relation to $1/\delta$ (in the limiting case).

Consider again the game $K = \{1 \mid K \parallel -1\}$. On

$$K - \Theta_n + \mathscr{E}_t^\delta$$

Right's supply of threats ensures that he will always win the kofight. In order to do so he must give up a coupon to Left, so he would prefer to win the ko at the lowest temperature possible. However at any $t \le 2/3$, Left will start the kofight, playing from K to K^L; Right will use up a kothreat and return to K; and Left will take a coupon. At this point Right might choose to take another coupon—but if he does so, Left will simply repeat the whole sequence again. After n iterations Right will run out of threats and be forced to complete (or else lose) the ko. If Right chooses to play all n iterations, there will be a total of $2n$ plays on $\mathscr{E}_t^\delta$, lowering the temperature by $2n\delta$, which is neglible as $\delta \to 0$. Therefore we can assume Right is forced to complete the ko at temperature t. This is why Left's play on $\mathscr{E}_t^\delta$ acts as a forcing pass, *even if Right is komaster.*

A different model is obtained by interchanging the limits in $L_t^\sharp(G)$:

$$L_t^{\sharp\sharp}(G) = \lim_{\delta\to 0} \lim_{n\to\infty} L_t^\delta(G + \Theta_n).$$

Here Left has a supply of threats that is large *in relation to* $1/\delta$. Left's advantage on $L_t^{\sharp\sharp}(G)$ is much greater than on $L_t^\sharp(G)$. Now, when faced with a kofight, she can afford to win it at any temperature she chooses. Consider

$$K - \Theta_n + \mathscr{E}_t^\delta$$

again, this time with $n \gg 1/\delta$. Now when Left plays from K to K^L, Right can continue to issue kothreats even as the environment becomes depleted—all the way down to temperature -1. It's not hard to verify (and we'll prove in a moment) that, in fact,

$$L_t^{\flat\flat}(K) = -2$$

for all $t \geq -1$. Moreover

$$R_t^{\flat\flat}(K) = -2$$

since *Right* won't bother to play on K until temperature -1. It's almost as if Left's move from K to K^L doesn't exist at all: Right has so many threats that he can defeat all attempts to utilize it.

The thermograph $\left(L_t^{\sharp\sharp}(G), R_t^{\sharp\sharp}(G)\right)$ is the **Left komonster** thermograph of G, and the **Right komonster** thermograph is defined analogously in terms of $\left(L_t^{\flat\flat}(G), R_t^{\flat\flat}(G)\right)$. The komonster theory has a remarkably easy solution for simple games.

Theorem 5.19. *Let G be simple, and let H be the loopfree game obtained by deleting all Right ko options from all subpositions of G. Then the Left komonster thermograph of G is identical to the (ordinary) thermograph of H.*

Proof Sketch. Consider a run of $G + \mathscr{E}_t^\delta$. Each time Right moves from some subposition $X + \mathscr{E}_t^\delta$ to a Right ko option $X^R + \mathscr{E}_t^\delta$, Left can use one of her threats to return to $X + \mathscr{E}_t^\delta$. This forces Right to make a different move on X or else take a coupon, both of which can be mirrored in the play of H. □

It follows that komonster thermographs have all the same properties as ordinary loopfree thermographs (cf. Theorem II.5.11 on page 108).

This shows that for simple games, komonster assumptions coincide with the onside and offside from Chapter VI. We have the following pleasing spectrum of temperature theories:

$$\text{onside} = \text{Left komonster} > \text{Left komaster} > \text{ordinary} >$$
$$> \text{Right komaster} > \text{Right komonster} = \text{offside}$$

in which the inequalities indicate decreasing favorability to Left.

But beware! If G is not simple, then the situation can be much more subtle. Consider the game G shown in Figure 3.10 on page 363. In the analysis on page 363 we showed that $L_t(G) = R_t(G) = 2$ for all $t \geq -1$. This remains true *even if Right is komonster.* Consider a run of the sum

$$G - \Theta_n.$$

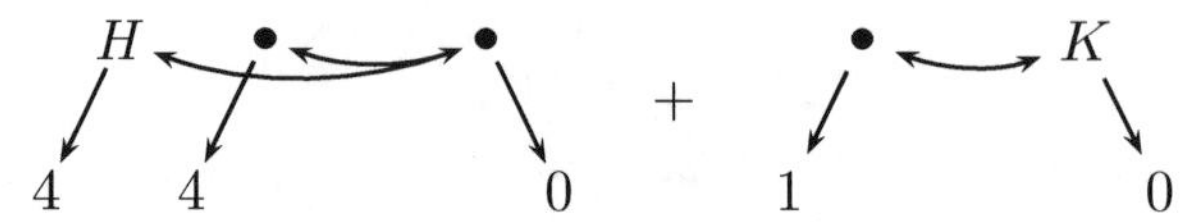

Figure 5.4. The subtle effects of a complex loop.

Each time Right uses a threat to return to G^R, Left can counter by moving to the other Left ko option. Eventually Right will run out of threats, no matter how large n is. Therefore $L_t^{\flat\flat}(G) = R_t^{\flat\flat}(G) = 2$ for all $t \geq -1$.

However, the offside of G is $\{1 \mid -1\}$, obtained by deleting both Left ko options of G^R. The same simplification isn't possible with a koban in effect, even if Right is komonster. So for general G (not necessarily simple), the identity "onside = Left komonster" does not hold.

This example illustrates another subtle point. Although G behaves just like the number 2 when played *in isolation*, its effect on sums can be very different. For example, consider the compound shown in Figure 5.4, in which a hotter variant of G is played next to an ordinary one-point ko.

Let's calculate the (ordinary) thermographs for H, K, and $H+K$. Certainly

$$L_t(H) = R_t(H) = 5$$

for all t (by the preceding arguments), and K has the familiar thermograph of temperature $\frac{1}{3}$ shown in Figure 3.2 on page 356. But something startling happens when the two of them are put together. Whenever Left moves from K to K^L on the sum

$$H + K,$$

Right can play from H to H^R, forcing Left to return to H'. This sequence frees Right to play back to K. Therefore Right can defer moving from K to 0 all the way down to temperature -1, and we have

$$L_t(H + K) = R_t(H + K) = 4$$

for all t. The component H is ostensibly worth 5 points to Left, but in the sum $H + K$ it has a surprising side-effect: it promotes Right to komonster on the component K. Right's option to move from H (or H') to H^R, while useless in isolation, essentially functions as an infinitely large supply of threats. Left has the power to remove this threat supply (by playing from H or H' to 4), but doing so erodes her own advantage on H.

Amazingly, Right's "effective komonster" status is preserved even if Left is explicitly made komonster. For on the sum

$$H + K + \Theta_n$$

the key battle occurs on K—and no matter how large the value of n, Left will have just finitely many threats with which to fight this battle, whereas Right effectively has infinitely many. We see that

$$L_t^{\sharp\sharp}(H+K) = R_t^{\sharp\sharp}(H+K) = 4$$

so the thermograph of $H+K$ is always the same, regardless of the threat environment! The presence of H gives Right a sort of "super-komonster" status that is resilient even against actual komonster assumptions.

However, this status only works at certain temperatures, for suppose we replace K by a hotter variant and consider the sum

$$H + K', \quad \text{where} \quad K' = \{8 \mid K' \parallel 0\}.$$

In this case, if Left opens by moving to $H + (K')^L$ and Right then plays to $H^R + (K')^L$, Left will simply ignore the threat on H and complete the ko on K'. The infinitely many threats granted by H all have bounded temperature, and in this respect they are *weaker* than the threats in a true komonster environment. This type of situation is still not fully understood.

Exercises

5.1 Determine the biased thermographs of each example on pages 356–360.

5.2 Convert the proof sketch of Theorem 5.19 into a formal induction.

5.3 If G is simple, then $t(G) \le \min\{t^\sharp(G), t^\flat(G)\}$.

5.4 *Orthodox accounting with one hyperactive component.*

(a) Let $G_1, \ldots, G_k$ be simple. Suppose that G_1 is hyperactive and each of $G_2, \ldots, G_k$ is placid. Suppose Left plays first (second) on the compound $G = G_1 + \cdots + G_k + \Theta$. Show that Left can guarantee a score of at least

$$z + \frac{1}{2}\sum_{i=1}^{m} \Delta t_i + \sum_{j=1}^{n} \Delta u_j,$$

where Δt_i are the signed temperature drops, Δu_j are the signed ko adjustments, and

$$z = m^\sharp(G_1) + m(G_2) + \cdots + m(G_k) \pm t/2,$$

with the sign of $t/2$ depending on who moves first. (Use the following modification of sentestrat: Left assumes that Right will win any ko on $G_2 + \cdots + G_k$, but she fights any ko on G_1 using threats from Θ. Use the fact that $m(G_i) = m^\flat(G_i)$ for all $i \ne 1$, just as in the proof of Theorem 5.16.)

(b) Conclude that on $G_1 + \cdots + G_k + \mathscr{E}_t^\delta + \Theta$, either player can achieve the biased orthodox forecast z, to within a bounded multiple of δ.

(c) Why doesn't this work if more than one component is hyperactive?

Notes

Like most of the material in this chapter, the ideas in this section originated with Berlekamp [**Ber96**]. Fraser's extension of the thermographic calculus to certain complex games [**Fras02**], described in the notes to Section 4 on page 377, applies to the komaster calculus as well.

The komaster calculus resolves the orthodox theory for sums of placid simple games (Theorem 5.17 on page 391). It also works for sums with just one hyperactive component and an explicit komaster (Exercise 5.4). For sums involving several hyperactive components, the situation is less clear, and the components can interact in subtle ways.

A recent innovation due to Berlekamp and Spight shows promise in simplifying the theory of hyperactive positions. They define a **neutral threat environment** containing a balanced supply of threats for both players. The environment contains threats at all temperatures up to t with granularity δ:

$$\Theta_t^\delta = \big(\theta_t + \theta_{t-\delta} + \cdots + \theta_0\big) + \big(-\theta_t - \theta_{t-\delta} - \cdots - \theta_0\big).$$

In the *dogmatic theory*, the environment consists of a coupon stack together with a neutral threat stack. We define

$$L_t^\natural(G) = \lim_{\delta\to 0} L_t(G + \Theta_t^\delta) \quad \text{and} \quad R_t^\natural(G) = \lim_{\delta\to 0} R_t(G + \Theta_t^\delta)$$

and $m^\natural(G)$ and $t^\natural(G)$ are the corresponding mast value and temperature. Then if $G_1, \ldots, G_k$ are simple, the orthodox forecast

$$m^\natural(G_1) + \cdots + m^\natural(G_k) \pm \frac{t^\natural(G)}{2}$$

can be achieved by either player, on the compound $G_1 + \cdots + G_k + \Theta_u^\delta + \mathscr{E}_t^\delta$ for sufficiently large u, to within a bounded multiple of δ.

The basic ideas of the dogmatic theory are sketched in [**Spi03**], but much work remains in order to fill in the details. In addition, the existing theory applies only to simple games; extending it to complex loopy games remains an open problem.

Open Problem. Develop a formal treatment of the dogmatic theory.

Open Problem. Extend the dogmatic theory to (any reasonable subclass of) complex loopy games.

Chapter VIII

Transfinite Games

The preceding chapters focused primarily on games with finitely many positions. Such a finiteness assumption is an essential ingredient of certain key results, including the existence of canonical forms and mean values. Other aspects of the theory, however, generalize almost verbatim when the finiteness requirement is dropped, resulting in an expanded universe of **long games** with an extraordinarily rich structure.

Just as short game values form a countable group $\mathbb{G}$, the values of long games form a Group[1] **PG** whose domain is a proper Class. Perhaps the most surprising and mysterious aspect of **PG** is that it contains a transfinite number system, the **surreal numbers**, that is defined in exactly the same way as $\mathbb{D}$ for short games. The surreals form a totally ordered, real-closed Field that subsumes both the ordinals and the reals.

Equally remarkable is the transfinite generalization of the impartial theory. The nim arithmetic of Section IV.5 extends verbatim to long games: just as short impartial games form a field isomorphic to $(\mathbb{N}, \oplus, \otimes)$, long impartial games form a Field of characteristic 2 isomorphic to $(\mathbf{ON}, \oplus, \otimes)$. This Field, denoted $\mathbf{ON}_2$, has a unique and captivating inductive structure.

All of the games mentioned in this chapter (aside from a few exercises in Section 1) are loopfree and are considered under the normal play convention. The study of loopy and misère-play transfinite games has not progressed beyond infancy. The many mysteries in this chapter suggest that even among loopfree games, the existing body of knowledge barely scratches the surface.

[1]It is a now-standard convention, introduced by Conway in ONAG, to capitalize the names of structures whose domain is a proper Class.

1. The Group PG

Short games were defined hierarchically in Chapter II:

$$\tilde{\mathbb{G}}_{n+1} = \Big\{\{\mathscr{G}^L \mid \mathscr{G}^R\} : \mathscr{G}^L, \mathscr{G}^R \subset \tilde{\mathbb{G}}_n\Big\}$$

and

$$\tilde{\mathbb{G}} = \bigcup_{n \geq 0} \tilde{\mathbb{G}}_n.$$

But why stop there? Continuing into the transfinite, let

$$\tilde{\mathbb{G}}_\omega = \Big\{\{\mathscr{G}^L \mid \mathscr{G}^R\} : \mathscr{G}^L, \mathscr{G}^R \subset \tilde{\mathbb{G}}\Big\}.$$

In this definition $\mathscr{G}^L$ and $\mathscr{G}^R$ are permitted to be infinite sets of options, so that we have games such as

$$\omega = \{0, 1, 2, 3, \ldots \mid \} \qquad \text{(“omega”)}$$

and

$$*\omega = \{0, *, *2, *3, \ldots \mid 0, *, *2, *3, \ldots\} \qquad \text{(“star omega”)}.$$

More generally,

Definition 1.1. For all ordinals $\alpha \in \mathbf{ON}$ we define

$$\tilde{\mathbb{G}}_\alpha = \Big\{\{\mathscr{G}^L \mid \mathscr{G}^R\} : \mathscr{G}^L, \mathscr{G}^R \subset \bigcup_{\beta < \alpha} \tilde{\mathbb{G}}_\beta\Big\}.$$

Then a **long game** is an element of

$$\tilde{\mathbf{PG}} = \bigcup_{\alpha \in \mathbf{ON}} \tilde{\mathbb{G}}_\alpha.$$

Here $\tilde{\mathbf{PG}}$ denotes the Class of all Partizan Games. As in Chapter II, we write $\tilde{\mathrm{b}}(G)$ for **formal birthday** of G: here it is the least *ordinal* α such that $G \in \tilde{\mathbb{G}}_\alpha$.

The Fundamental Theorem generalizes practically verbatim, via slight modification of the proof of Theorem I.1.5. For clarity, here's the full argument:

Theorem 1.2 (Fundamental Theorem of Combinatorial Game Theory, Long Form)**.** *Let G be a long combinatorial game, and assume normal play. Either Left can force a win playing first on G or else Right can force a win playing second, but not both.*

Proof. Consider a typical Left option of G, say G^L. $\tilde{\mathrm{b}}(G^L) < \tilde{\mathrm{b}}(G)$, so we may assume by transfinite induction (and symmetry) that either Right can force a win playing first on G^L or else Left can force a win playing second.

If Right can win *all* such G^L playing first, then certainly he can win G playing second regardless of Left's opening move. Conversely, if Left can win *any* such G^L playing second, then he can win G by moving to it. Exactly one of these two possibilities must hold. □

Therefore to each long game G, we can assign one of the four traditional outcome classes $\mathscr{L}$, $\mathscr{R}$, $\mathscr{N}$, $\mathscr{P}$. As always we denote by $o(G)$ the outcome class of G. The definitions of $G+H$ and $-G$ are exactly the same as for short games, and we have the usual equivalence relation:

Definition 1.3. Let G and H be long games. Then we write

$$G = H \quad \text{if} \quad o(G+X) = o(H+X) \text{ for every long game } X.$$

Likewise,

$$G \geq H \quad \text{if} \quad o(G+X) \geq o(H+X) \text{ for every long game } X.$$

Definition 1.4. The **game value** of G is its equivalence class modulo $=$. The Class of game values is denoted by **PG**.[2]

A surprising amount of theory from Chapter II translates verbatim to **PG**. For example:

Proposition 1.5. *Let G, H, and J be long games. Then:*

(a) *If $G \geq H$, then $G+J \geq H+J$.*

(b) $0+G \cong G$.

(c) $G \geq 0$ *if and only if* $o(G) \geq \mathscr{P}$.

(d) $G-G=0$.

(e) $G \geq H$ *if and only if* $o(G-H) \geq \mathscr{P}$.

Proof. All of the proofs are identical to the corresponding proofs from Chapter II. □

Theorem 1.6. PG *is a partially ordered Abelian Group.*

Proof. Addition is commutative and associative on **PG**. Proposition 1.5 shows that 0 is an identity, that every element of **PG** has an inverse, and that $\geq$ respects the additive structure of **PG**. □

[2]There is a slight complication with this definition. Since $\tilde{\mathbf{PG}}$ is a proper Class, the equivalence class of any particular G is, in general, a proper Class also—so that we cannot legally form the collection of all such Classes. This is a familiar problem in set theory, which can be conveniently resolved by restricting each equivalence class to those elements of minimal set-theoretic rank. It's our intention to avoid getting terribly bogged down in these kinds of metamathematical details, so we'll avoid these types of comments as much as possible throughout this chapter. There is a cursory discussion in Section B.3; for a more detailed treatment, see [**Jec02**] or [**Kun83**].

It follows from Proposition 1.5 that when G and H are both short, then the relation $G = H$ agrees with our earlier definition from Chapter II. We can therefore identify $\mathbb{G}$ with the corresponding subgroup of $\mathbf{PG}$, so that

$$\mathbb{G} \subset \mathbf{PG}.$$

As in Chapter II, we write $\mathbb{G}_\alpha$ for the set of values of games in $\tilde{\mathbb{G}}_\alpha$. The **birthday** of G, denoted by $\mathrm{b}(G)$, is the least α with $G \in \mathbb{G}_\alpha$. Clearly

$$\mathbb{G}_\alpha \subset \mathbb{G}_\beta \quad \text{whenever} \quad \alpha \leq \beta$$

and

$$\mathbf{PG} = \bigcup_{\alpha \in \mathbf{ON}} \mathbb{G}_\alpha.$$

We also write **IG** for the Subclass of impartial values, $\mathbf{PG}^0$ for the Subclass of dicotic values, and $\mathbb{G}^0_\alpha = \mathbf{PG}^0 \cap \mathbb{G}_\alpha$.

The Reals ...

Some remarkable values are born on day ω. We've already defined ω and $*\omega$, and of course there's also

$$-\omega = \{\ \mid 0, -1, -2, -3, \ldots\}.$$

But we also have, say,

$$G = \left\{0, \tfrac{1}{4}, \tfrac{5}{16}, \ldots \;\middle|\; \tfrac{1}{2}, \tfrac{3}{8}, \tfrac{11}{32}, \ldots\right\},$$

whose options are successive rational approximations to $\frac{1}{3}$. In fact it's easily checked that

$$G + G + G = 1$$

by showing that $G + G + G - 1$ is a second-player win.

The definition of G is really just a Dedekind cut, and indeed Dedekind's argument translates directly into the language of combinatorial games.

Definition 1.7. For $x \in \mathbb{R}$ we denote by $[\![x]\!]$ the game $\{\mathcal{X}^L \mid \mathcal{X}^R\}$, where

$$\mathcal{X}^L = \{q \in \mathbb{D} : q < x\} \quad \text{and} \quad \mathcal{X}^R = \{q \in \mathbb{D} : q > x\}.$$

Theorem 1.8. *Let $x, y \in \mathbb{R}$. Then $x \geq y$ if and only if $[\![x]\!] \geq [\![y]\!]$.*

Proof. The $\Rightarrow$ direction is clear from the definition, since the Left (resp. Right) options of $[\![x]\!]$ form a superset (resp. subset) of those of $[\![y]\!]$.

For the $\Leftarrow$ direction, suppose $x < y$ and fix any $q \in \mathbb{D}$ with $x < q < y$. Then on

$$[\![x]\!] - [\![y]\!]$$

Right has a move to $\llbracket x \rrbracket - q$. Now every $\llbracket x \rrbracket^L < q$, by choice of q. Moreover, from $\llbracket x \rrbracket - q^R$ Right can revert to $q - q^R < 0$. This shows that

$$\llbracket x \rrbracket - q \leq 0$$

and therefore $\llbracket x \rrbracket - \llbracket y \rrbracket \lhd\!\vert\ 0$. □

Therefore the mapping $x \mapsto \llbracket x \rrbracket$ induces an isomorphism of $\mathbb{R}$ onto a subgroup of $\mathbb{G}_\omega$. Henceforth we'll identify $\mathbb{R}$ with this subgroup and write simply x in place of $\llbracket x \rrbracket$.

... and the Surreals

This shows that on day ω, we obtain the reals via Dedekind cuts of dyadic rationals. In hindsight, the genetic construction of dyadic rationals from Chapter II feels much like a Dedekind construction too: in both cases, new numbers are formed in the gaps between "cuts" of simpler ones.

Conway noticed this phenomenon soon after discovering the axiomatic theory of combinatorial games. He also observed that Cantor's construction of the ordinals also fits perfectly into the game-theoretic framework, since for all $\alpha \in \mathbf{ON}$ we can write

$$\alpha = \{\mathscr{G}^L \mid \}, \quad \text{where} \quad \mathscr{G}^L = \{\beta \in \mathbf{ON} : \beta < \alpha\}.$$

Conway's supreme insight was that the two constructions can be merged into a single genetic definition of number. Once Dedekind cuts are seen to be step ω of a natural inductive process, there is nothing stopping us from extending it "all the way up" into Cantor's transfinite.

Definition 1.9. A long game x is a **surreal number** if $y^L < y^R$ for every subposition y of x and every y^L and y^R. We denote the Class of (values of) surreal numbers by **SN**.

We've already seen that $\mathbb{R} \subset \mathbf{SN}$ and $\mathbf{ON} \subset \mathbf{SN}$. One might say that

Dedekind focused on just one *step* of the construction (step ω),

whereas

Cantor focused on just one *side* of the construction.

By relaxing these restrictions, Conway arrived at a number system that subsumes both the reals and the ordinals. But there is much more to **SN**, even on day ω: for example the number

$$\iota = \{0 \mid 1, \tfrac{1}{2}, \tfrac{1}{4}, \tfrac{1}{8}, \ldots\}.$$

Certainly $\iota < x$ for every positive $x \in \mathbb{D}$ (and thus every positive real), but also $\iota > 0$, since $\iota^L = 0$. So ι is an infinitesimal number.

$$
\begin{array}{rcl}
\omega - 1 = & \{0,1,2,3,\dots \mid \omega\} & (\omega+1) \\
\omega - 2 = & \{0,1,2,3,\dots \mid \omega, \omega-1\} & (\omega+2) \\
\frac{1}{2}\omega = & \{0,1,2,3,\dots \mid \omega, \omega-1, \omega-2, \omega-3, \dots\} & (\omega\times 2) \\
\frac{1}{2}\omega - 1 = & \{0,1,2,3,\dots \mid \frac{1}{2}\omega\} & (\omega\times 2+1) \\
\frac{1}{4}\omega = & \{0,1,2,3,\dots \mid \frac{1}{2}\omega, \frac{1}{2}\omega-1, \frac{1}{2}\omega-2, \dots\} & (\omega\times 3) \\
\frac{1}{8}\omega = & \{0,1,2,3,\dots \mid \frac{1}{4}\omega, \frac{1}{4}\omega-1, \frac{1}{4}\omega-2, \dots\} & (\omega\times 4) \\
\sqrt{\omega} = & \{0,1,2,3,\dots \mid \omega, \frac{1}{2}\omega, \frac{1}{4}\omega, \dots\} & (\omega^2) \\
\omega^{1/4} = & \{0,1,2,3,\dots \mid \sqrt{\omega}, \frac{1}{2}\sqrt{\omega}, \frac{1}{4}\sqrt{\omega}, \dots\} & (\omega^3) \\
\omega^{1/\omega} = & \{0,1,2,3,\dots \mid \sqrt{\omega}, \omega^{1/4}, \omega^{1/8}, \dots\} & (\omega^\omega) \\
\frac{1}{3}\omega = & \{0, \frac{1}{4}\omega, \frac{5}{16}\omega, \dots \mid \frac{1}{2}\omega, \frac{3}{8}\omega, \frac{11}{32}\omega, \dots\} & (\omega^2) \\
\pi\omega = & \{3\omega, \frac{25}{8}\omega, \frac{201}{64}\omega, \dots \mid \frac{7}{2}\omega, \frac{13}{4}\omega, \frac{51}{16}\omega, \frac{101}{32}\omega, \dots\} & (\omega^2) \\
\frac{1}{\omega} = & \{0 \mid 1, \frac{1}{2}, \frac{1}{4}, \dots\} & (\omega) \\
\frac{2}{\omega} = & \{\frac{1}{\omega} \mid 1, \frac{1}{2}, \frac{1}{4}, \dots\} & (\omega+1) \\
\frac{1}{2\omega} = & \{0 \mid \frac{1}{\omega}\} & (\omega+1) \\
\omega^{-2} = & \{0 \mid \frac{1}{\omega}, \frac{1}{2\omega}, \frac{1}{4\omega}, \dots\} & (\omega\times 2) \\
\frac{1}{2}\omega^{-2} = & \{0 \mid \omega^{-2}\} & (\omega\times 2+1) \\
\omega^{-\omega} = & \{0 \mid \omega^{-1}, \omega^{-2}, \omega^{-4}, \dots\} & (\omega^2)
\end{array}
$$

Figure 1.1. Some simple surreals and their ordinal birthdays.

Venturing past day ω we encounter a whole host of fascinating characters, such as those in Figure 1.1 on the preceding page. Some of their names are easily justified; for example, if G is the game called "$\omega - 1$," then it's easy to check that $G - \omega + 1$ is a second-player win. In Section 2 we'll show that the surreals have a multiplicative structure that motivates names like $\sqrt{\omega}$. Note that $\frac{1}{\omega}$ is the game we called ι above.

The following observations are straightforward generalizations of the theory of $\mathbb{D}$ (from Section II.3).

Proposition 1.10. *If x is a number, then $x^L < x < x^R$ for each x^L and x^R. If x and y are numbers, then so is $x + y$.*

Proof. Identical to the proofs of Propositions II.3.2 and II.3.3 (page 69). $\square$

Theorem 1.11. $\mathbf{SN}$ *is totally ordered: either $x \geq y$ or $y \geq x$, for all $x, y \in \mathbf{SN}$.*

Proof. If $x \lhd\!\!| \; y$, then either $x^R \leq y$ or $x \leq y^L$. But $x < x^R$ and $y^L < y$, so in both cases $x < y$. $\square$

Theorem 1.12 (Archimedean Principle). *For any $G \in \mathbf{PG}$, there is some ordinal α such that $\alpha > G > -\alpha$.*

Proof. Let $\alpha = \mathrm{b}(G)$. Then on $\alpha - G$, Right's options are all of the form $\alpha - G^L$, which Left can revert to $\beta - G^L$, with $\beta = \mathrm{b}(G^L)$. This shows, by (transfinite) induction, that $\mathrm{b}(G) \geq G$ for all G. $\square$

Theorem 1.13 (Simplicity Theorem). *Let G be a long game and put*

$$\mathcal{I} = \{x \in \mathbf{SN} : G^L \lhd\!\!| \; x \lhd\!\!| \; G^R \text{ for every } G^L \text{ and } G^R\}.$$

If $\mathcal{I} \neq \emptyset$, then G is equal to the unique $x \in \mathcal{I}$ of smallest birthday.

Proof. Identical to the proofs of Proposition II.3.9 and Theorem II.3.10. $\square$

Theorem 1.14. *If G has no Right options, then G is equal to an ordinal.*

Proof. The interval $\mathcal{I} = \{x \in \mathbf{SN} : x \; |\!\!\rhd G^L \text{ for every } G^L\}$ is unbounded, so it must contain an ordinal. By the Simplicity Theorem, G is equal to the least ordinal in $\mathcal{I}$. $\square$

Correspondingly, if G has no Left options, then G is equal to $-\alpha$ for some ordinal α. Some useful related terminology:

Definition 1.15. The **absolute value** of $x \in \mathbf{SN}$, denoted $|x|$, is given by

$$|x| = \begin{cases} x & \text{if } x \geq 0; \\ -x & \text{if } x < 0. \end{cases}$$

Definition 1.16. $x \in \mathbf{SN}$ is a **signed ordinal** if $|x|$ is an ordinal.

Not all theorems about finite numbers generalize cleanly. The Number Avoidance and Number Translation Theorems are badly false, as the following example demonstrates.

Example. Consider the game $\pm\mathbb{N}$ defined by

$$\pm\mathbb{N} = \pm(0, 1, 2, 3, \ldots) = \{0, 1, 2, 3, \ldots \mid 0, -1, -2, -3, \ldots\}.$$

It's easy to see that Number Translation can't hold for $\pm\mathbb{N}$. For if n is any integer, then

$$\pm\mathbb{N} = \{n,\ n+1,\ n+2,\ n+3,\ \ldots \mid n,\ n-1,\ n-2,\ n-3,\ \ldots\},$$

whereas $\pm\mathbb{N}$ is certainly *not* equal to $\pm\mathbb{N} + n$. Indeed, on the sum

$$\pm\mathbb{N} \pm \mathbb{N} + 1$$

Left's *only* winning move is to play from 1 to 0. This shows quite dramatically that Number Avoidance fails in the general case.

From Infinitesimal to Small

The game that we called $\frac{1}{\omega}$,

$$\tfrac{1}{\omega} = \{0 \mid 1, \tfrac{1}{2}, \tfrac{1}{4}, \tfrac{1}{8}, \ldots\},$$

is infinitesimal in the sense that $\frac{1}{\omega} \ll x$ for every real number x. However it's easily seen that

$$\uparrow \ll \tfrac{1}{\omega}$$

by a slight generalization of the Lawnmower Theorem, and this remains true if $\frac{1}{\omega}$ is replaced by any positive surreal.

Theorem 1.17. *If G is dicotic and $x > 0$ is a surreal number, then $G < x$.*

Proof. The proof is just like the Lawnmower Theorem (Theorem II.4.2 on page 83). If Right plays first on $x - G$, then it must be to some $x' - G'$ with $x' > 0$ and G' dicotic, so by induction Left has a winning move. If Left plays first, then she can move to some $x - G^R$ (also winning by induction) unless $G \cong 0$; but in that case, the assumption $x > 0$ guarantees the win. □

The term **small** is reserved for games that are bounded by every positive surreal, distinguishing them from merely infinitesimal games (that are bounded by the positive reals).

Definition 1.18. A long game G is **infinitesimal** if $-x < G < x$ for every real number $x > 0$.

Definition 1.19. A long game G is **small** if $-x < G < x$ for every surreal number $x > 0$. We say G is **all-small** if every subposition of G is small.

Theorem 1.20. *$G \in \mathbf{PG}^0$ if and only if G is all-small.*

Proof. The $\Rightarrow$ direction is Theorem 1.17. For $\Leftarrow$, we can assume that every subposition H of G with $H = 0$ also satisfies $H \cong 0$ (since putting G into such a form does not change its value). But if H is a subposition of G with $H \neq 0$, then H must have at least one Right option: otherwise Theorem 1.14 would imply that H is an ordinal, contradicting the assumption that H is small. Likewise, H must have at least one Left option. This shows that G is dicotic. □

Large Numbers and $\mathbf{PG}^\infty$

SN includes a host of numbers that exceed all the integers. Certainly $\omega > n$ for any n, but there are other surreals that are much smaller than ω, yet still greater than any integer. Figure 1.1 gives the examples

$$\sqrt{\omega},\ \omega^{1/4},\ \ldots,\ \omega^{1/\omega}$$

and the same ideas can be used to form numbers like ω^{1/ω^2}, etc.

The gap between $\mathbb{N}$ and the larger surreals turns out to be an important threshold in the surreal hierarchy, and this motivates the following definitions.

Definition 1.21. A surreal number x is **large** if $|x| > n$ for all integers n.

Definition 1.22. A game G is **moderate** if $-x < G < x$ for every large surreal $x > 0$. We say G is **all-moderate** if every subposition of G is moderate.

This is just like the definition of small, except we require only that G be bounded by *large* surreals. We denote the Class of all moderate game values by $\mathbf{PG}^\infty$, and we write $\mathbb{G}_\alpha^\infty = \mathbb{G}_\alpha \cap \mathbf{PG}^\infty$, paralleling the definitions of $\mathbf{PG}^0$ and $\mathbb{G}_\alpha^0$. It is clear that $\mathbf{PG}^\infty$ is a Group.

Certainly $\mathbb{G} \subset \mathbf{PG}^\infty$, and also $\mathbb{R} \subset \mathbf{PG}^\infty$ and $\mathbf{PG}^0 \subset \mathbf{PG}^\infty$. So $\mathbf{PG}^\infty$ is quite a rich Class, and to a remarkable degree it behaves much like a macroscopic analogue of $\mathbf{PG}^0$. We'll now investigate some of its surprising properties.

Recall the game $\pm\mathbb{N}$ defined above, given by

$$\pm\mathbb{N} = \pm(0, 1, 2, 3, \ldots) = \{0, 1, 2, 3, \ldots \mid 0, -1, -2, -3, \ldots\}.$$

If n is an integer, then certainly $\pm\mathbb{N} \ngtr n$. However if $x > 0$ is any *large* surreal, then $\pm\mathbb{N} < x$, since on $x \pm \mathbb{N}$ Left can win simply by declining to move

	$\mathbf{PG}^0$	$\mathbf{PG}^\infty$
elements	all-small	all-moderate
bounded by	positive surreals	large surreals
signed ordinal subpositions	must be 0	must be integers
standard unit	$\uparrow$	1
commensurate with	infinitesimals	moderate games

Figure 1.2. Some of the parallels between $\mathbf{PG}^0$ and $\mathbf{PG}^\infty$.

on x until the $\pm\mathbb{N}$ component is resolved. So we have $\pm\mathbb{N} \in \mathbf{PG}^\infty$, though it's confused with all the integers. Even more surprising is the following game:

$$\infty = \{\mathbb{N} \mid \pm\mathbb{N}\} = \{0, 1, 2, 3, \ldots \,\|\, 0, 1, 2, 3, \ldots \mid 0, -1, -2, -3, \ldots\}.$$

It's easily checked that $\infty > n$ for all integers n; but also $\infty < x$ for all large surreals x, by the same argument used for $\pm\mathbb{N}$. So $\infty \in \mathbf{PG}^\infty$, but it exceeds all the moderate numbers! We can think of ∞ as falling within the "gap" between $\mathbb{N}$ and the large surreals, much as $\uparrow$ lies between 0 and the positive surreals.

We have the following easy characterization of all-moderate games.

Theorem 1.23. *The following are equivalent, for a long game G:*

(i) *G is all-moderate.*

(ii) *No subposition of G is equal to an* infinite *signed ordinal.*

Note that Theorem 1.23 is just a macroscopic Lawnmower Theorem. Indeed, if we replace "moderate" by "small" and "infinite" by "nonzero," then the result is exactly the classical Lawnmower Theorem. Figure 1.2 summarizes the parallels between $\mathbf{PG}^0$ and $\mathbf{PG}^\infty$.

Proof. (i) $\Rightarrow$ (ii) is immediate, since an infinite signed ordinal is *a priori* large. For the converse, suppose x is large. Then on $x - G$, Left can win by playing arbitrarily on $-G$ until she is no longer able to do so. At that point the position must have value $x' - \alpha$, where $x' > x$ and α is equal to a signed ordinal. By the assumption (ii), α must be equal to an integer, and since x is large, this implies $x' - \alpha > 0$. $\square$

What are the largest elements of $\mathbf{PG}^\infty$? Certainly we can form $\infty + \infty$, $3 \cdot \infty$, etc., but we can also iterate the definition of ∞: writing

$$G = \Big\{0,\ \infty,\ 2 \cdot \infty,\ \ldots \,\Big\|\, 0,\ \infty,\ 2 \cdot \infty,\ \ldots \,\Big|\, 0,\ -\infty,\ -2 \cdot \infty,\ \ldots\Big\},$$

we have $G > n \cdot \infty$ for all n, just as $\infty > n \cdot 1$.

Iterating this construction, define

$$\infty^0 = 1$$

and for all $\alpha \in \mathbf{ON}$ with $\alpha \geq 1$,

$$\infty^\alpha = \left\{ n \cdot \infty^\beta \,\middle\|\, n \cdot \infty^\beta \mid -n \cdot \infty^\beta \right\}_{\beta<\alpha,\, n\geq 0}.$$

Here the subscripts indicate that β ranges over all ordinals strictly less than α, and n ranges over all non-negative integers. In particular, note that $\infty^1 = \infty$ and $\infty^2 = G$.

Clearly $\infty^\beta \ll \infty^\alpha$ whenever $\beta < \alpha$; and also $\infty^\alpha \in \mathbf{PG}^\infty$, by Theorem 1.23. This construction spans the whole of $\mathbf{PG}^\infty$, illuminating the vast hierarchy from moderate to large.

Theorem 1.24. *If $G \in \mathbf{PG}^\infty$, then $-\infty^\alpha < G < \infty^\alpha$ for some $\alpha \in \mathbf{ON}$.*

Proof. If G is equal to an integer, then the conclusion is trivial, so assume that it is not. Since $G \in \mathbf{PG}^\infty$, this implies that G is not equal to a signed ordinal, so it must have options for both players.

Now for each subposition H of G, we can assume by induction that

$$-\infty^{\alpha(H)} < H < \infty^{\alpha(H)}$$

for some $\alpha(H) \in \mathbf{ON}$. Let

$$\alpha = \sup_H \big(\alpha(H)\big)$$

and consider

$$3 \cdot \infty^\alpha - G.$$

Right's moves on $-G$ obviously lose, so suppose Right moves on ∞^α to

$$2 \cdot \infty^\alpha + (\infty^\alpha)^R - G.$$

Left can respond with an arbitrary move on $-G$, to some $-G^R$. Now Right's moves are to

$$\infty^\alpha + 2 \cdot (\infty^\alpha)^R - G^R \qquad \text{or} \qquad 2 \cdot \infty^\alpha - n \cdot \infty^\beta - G^R.$$

But both of these moves are losing: the former since $2 \cdot (\infty^\alpha)^R = 0$; and the latter because $\infty^\alpha - n \cdot \infty^\beta$ and $\infty^\alpha - G^R$ are both positive. $\square$

The Small

Many constructions from $\mathbb{G}$ have natural generalizations to **PG**. We have the **transfinite nimbers**:

$$*\alpha = \{*\beta \mid *\beta\}_{\beta<\alpha}.$$

Transfinite tinies can be defined just as in the finite case,

$$+_\alpha = \{0 \parallel 0 \mid -\alpha\}.$$

The uptimals generalize too:

$$\uparrow^{[\alpha]} = (* : \alpha) - * \quad \text{and} \quad \uparrow^{\alpha} = \{0 \mid \downarrow_{[\beta]}*\}_{\beta<\alpha}.$$

For example,

$$\uparrow^{[\omega]} = \{0,\ \uparrow,\ \uparrow^{[2]},\ \uparrow^{[3]},\ \ldots \mid *\}, \qquad \uparrow^{[\omega+1]} = \{\uparrow^{[\omega]} \mid *\},$$
$$\uparrow^{\omega} = \{0 \mid *,\ \downarrow*,\ \downarrow_{[2]}*,\ \downarrow_{[3]}*,\ \ldots\}, \qquad \uparrow^{\omega+1} = \{0 \mid \downarrow_{[\omega]}*\}.$$

For the most part these generalizations enjoy many of the properties one would expect. Some of these properties are explored in the exercises.

Now it follows from the Archimedean Principle that the games $+_\alpha$ are the smallest positive long games. However, the transfinite analogue of the sequence

$$0,\ \uparrow,\ \Uparrow,\ \Uparrow\!\!\!\uparrow,\ \ldots$$

is less clear. There doesn't seem to be a natural way to define "$\omega \cdot \uparrow$," and in fact we'll shortly see compelling evidence that no such generalization exists.

The correct continuation is suggested by the parallel with $\mathbf{PG}^\infty$. Norton product (Definition II.7.6 on page 141) generalizes verbatim to all-moderate games:

Definition 1.25. For all-moderate G, the **Norton product** of G by $\uparrow$ is given by

$$G \cdot \uparrow = \begin{cases} n \cdot \uparrow & \text{if } G \text{ is equal to an integer } n; \\ \{G^L \cdot \uparrow + \Uparrow* \mid G^R \cdot \uparrow + \Downarrow*\} & \text{otherwise.} \end{cases}$$

Now consider

$$\infty \cdot \uparrow = \Big\{0, \uparrow, \Uparrow, \ldots \Big\| 0, \uparrow, \Uparrow, \ldots \Big| 0, \downarrow, \Downarrow, \ldots\Big\}$$

and more generally,

$$\infty^\alpha \cdot \uparrow = \Big\{n \cdot (\infty^\beta \cdot \uparrow) \Big\| n \cdot (\infty^\beta \cdot \uparrow) \Big| -n \cdot (\infty^\beta \cdot \uparrow)\Big\}_{\beta<\alpha,\, n\geq 0}.$$

We arrive at the following remarkable theorem.

Theorem 1.26. *If $G \in \mathbf{PG}^0$, then $-\infty^\alpha \cdot \uparrow < G < \infty^\alpha \cdot \uparrow$ for some α.*

Proof. Identical to the proof of Theorem 1.24. The case where G is an integer cannot arise, since it is presumed that $G \in \mathbf{PG}^0$. $\square$

Theorem 1.26 is the clearest possible indication that there is no meaningful definition of $\omega \cdot \uparrow$. We have $\infty^\alpha < \omega$ for all α, but $G < \infty^\alpha \cdot \uparrow$ for some α, whenever G is dicotic.

The remote star relation can also be generalized verbatim:

$$G \sim H \quad \text{if} \quad o(G + X + \text{☆}) = o(H + X + \text{☆}) \text{ for every } X,$$

in which ☆ denotes a sufficiently remote *transfinite* nimber. When $G \sim H \cdot \uparrow$ for some $H \in \mathbf{PG}^\infty$, then we say H is the **atomic weight** of G. This raises the following tantalizing question.

Question. Can the atomic weight calculus be generalized to $\mathbf{PG}^0$?

Such a calculus would establish a surjective atomic weight homomorphism $\mathrm{aw} : \mathbf{PG}^0 \to \mathbf{PG}^\infty$, extending the usual mapping $\mathrm{aw} : \mathbb{G}^0 \to \mathbb{G}$ and cementing the relationship between $\mathbf{PG}^0$ and $\mathbf{PG}^\infty$.

Group Structure of PG

In Chapter III we showed that $\mathbb{G}$ has no elements of odd order. By contrast, **PG** contains elements of *every* finite order, as the following example demonstrates.

Example. Let

$$G = \oint^{1*} \tfrac{1}{3}.$$

Here the definition of $\oint^{1*}$ is exactly the same as for short games (Definition II.5.23 on page 114), and Theorem II.5.24 generalizes verbatim. This shows that

$$G + G + G = 1.$$

However $G \neq \frac{1}{3}$, since from G Left can move to $1*$. (In the definition of G, we can assume that $\frac{1}{3}$ has 0 as a Left option, since $\oint^{1*} \frac{1}{3}$ is independent of the form of $\frac{1}{3}$.)

This shows that $G - \frac{1}{3}$ has order 3 in **PG**. We can obviously construct elements of any finite order in the same way, replacing $\frac{1}{3}$ by a suitable $\frac{1}{n}$.

Lattice Structure of PG

The lattice structure of $\mathbb{G}_\alpha$ generalizes the structure of finite $\mathbb{G}_n$ in a straightforward way (Exercise 1.9). Conversely, the structure of $\mathbb{G}^0_\alpha$ is less clear. The preceding discussion of infinitesimals can be generalized to show that if $\alpha \geq \omega$, then $\mathbb{G}^0_\alpha$ has no maximal element (Exercise 1.10). Since for $1 \leq n < \omega$, $\mathbb{G}^0_n$ has exactly two maximal elements ($\uparrow(n{-}1)$ and $\uparrow(n{-}1)*$), this shows that the $\alpha \geq \omega$ case is fundamentally different.

Open Problem. Describe the partial-order structure of $\mathbb{G}^0_\alpha$, for $\alpha \geq \omega$.

Exercises

1.1 For every long game G, there is a K such that $K = G$ and K has no reversible options.

1.2 If $\alpha > \beta$, then $+_\alpha \ll +_\beta$.

1.3 Determine the smallest positive game born by day α, for all ordinals α. (There are three separate cases, depending on whether α is a limit ordinal; a limit ordinal + 1; or an ordinal of the form $\beta + n$ for $n \geq 2$.)

1.4 (a) $\uparrow^{[\alpha]} + \uparrow^\alpha + * = \{0 \mid \uparrow^\alpha\}$ for all $\alpha \in \mathbf{ON}$.
(b) If $\alpha > \beta$, then $\uparrow^\alpha \ll \uparrow^\beta$, for all $\alpha, \beta \in \mathbf{ON}$.
(c) Write out the options for $\uparrow^{[\omega]}*$ and $\uparrow^\omega *$.
(d) $\uparrow^{[\alpha+1]} - \uparrow^{\alpha+1} = \uparrow^{[\alpha]}$ for all $\alpha \in \mathbf{ON}$. What can one say about $\uparrow^{[\omega]} - \uparrow^\omega$?

1.5 If α is an ordinal of the form ω^β, then $\bigcup_{\xi<\alpha} \mathbb{G}_\xi$ is a group.

1.6 Show that $\infty + \infty = \{\mathbb{N} \parallel \mathbb{N} \mid 0\}$.

1.7 A **Dedekind section** of **SN** is a pair of Classes $(\mathbf{X}, \mathbf{Y})$ with $\mathbf{X} \cup \mathbf{Y} = \mathbf{SN}$, such that $x > y$ for all $x \in \mathbf{X}$, $y \in \mathbf{Y}$. For $G \in \mathbf{PG}$, the **Left** and **Right sections** of G are defined by

$$\mathbf{L}(G) = \big(\{x \in \mathbf{SN} : x \geq G\}, \{x \in \mathbf{SN} : x \lhd\!\!| \; G\}\big),$$
$$\mathbf{R}(G) = \big(\{x \in \mathbf{SN} : x \;|\!\!\rhd G\}, \{x \in \mathbf{SN} : x \leq G\}\big).$$

A section $(\mathbf{X}, \mathbf{Y})$ is **numeric** if it has the form $\mathbf{L}(x)$ or $\mathbf{R}(x)$, for some surreal number x.
(a) If $G \notin \mathbf{SN}$, then $\mathbf{L}(G) = \sup\big(\mathbf{R}(G^L)\big)$ and $\mathbf{R}(G) = \inf\big(\mathbf{L}(G^R)\big)$, where $\sup(\mathbf{X}_i, \mathbf{Y}_i) = (\bigcup \mathbf{X}_i, \bigcap \mathbf{Y}_i)$ and $\inf(\mathbf{X}_i, \mathbf{Y}_i) = (\bigcap \mathbf{X}_i, \bigcup \mathbf{Y}_i)$.
(b) The Left and Right sections of every short game are numeric.
(c) If x is equal to a number and G is not, *and* if the Left and Right sections of G are numeric, then $G + x = \big\{G^L + x \bigm| G^R + x\big\}$.
(d) For all G, there is a *set* $\mathcal{A} \subset \mathbf{SN}$ such that $\mathbf{L}(G)$ can be expressed as either the sup or the inf of the sections of elements of $\mathcal{A}$.

1.8 Define

$$\int^{\mathbb{N}} G = \begin{cases} G & \text{if } G \text{ is equal to a number;} \\ \left\{ n + \int^{\mathbb{N}} G^L \,\middle|\, -n + \int^{\mathbb{N}} G^R \right\}_{n \in \mathbb{N}} & \text{otherwise.} \end{cases}$$

(a) True or false: If $G > 0$, then $\int^{\mathbb{N}} G > n$ for all $n \in \mathbb{N}$.
(b) Let $G = \int^{\mathbb{N}} \uparrow$. Investigate the relative size of $\infty - G$.

1.9 (a) $\mathbb{G}_\alpha$ is a complete distributive lattice, for all α.
(b) Describe the join-irreducibles of $\mathbb{G}_\alpha$ in terms of $\bigcup_{\beta<\alpha} \mathbb{G}_\beta$.

1.10 (a) Every element of $\mathbb{G}^0_\omega$ is $\leq \uparrow n$ for some n.
(b) Find an increasing sequence of games $U_n \in \mathbb{G}^0_{\omega+1}$ such that every element of $\mathbb{G}^0_{\omega+1}$ is $\leq U_n$ for some n.
(c) Generalize to higher $\mathbb{G}^0_\alpha$.

1.11 *Universal embedding property for Abelian groups.*
(a) The generalized Norton product (cf. Exercise II.7.15 on page 149) extends to long games, with exactly the same definition. Show that for all $r, s \in \mathbb{R}$ and all $U > 0$, we have $(r + s) \cdot U = r \cdot U + s \cdot U$, and $r \geq s$ if and only if $r \cdot U \geq s \cdot U$.
(b) Let $G = \frac{2}{n} \cdot \{2x \mid x\} - \frac{3x}{2}$, with $n \geq 2$ and $x \in \mathbf{SN}$. Show that $n \cdot G = 0$ and that $k \cdot G \mathrel{|\!\rhd} y$ for $1 \leq k < n$ and all $y < \frac{x}{n}$.
(c) Prove Theorem 1.27, cited below. (By Zorn's Lemma, it suffices to assume that $\mathcal{B}$ is generated by $\mathcal{A} \cup \{a\}$, for some a. If $na \notin \mathcal{A}$ for any $n > 0$, then define $g(na)$ to be a sufficiently large ordinal. Otherwise, let $n \in \mathbb{N}$ be least with $na \in \mathcal{A}$, fix an H with $n \cdot H = f(na)$, and define $g(na)$ to be $H + J$, where J is given by (b) for some sufficiently large α.)

1.12 *Surreal pseudonumbers.* A **transfinite loopy game** is a loopy game whose graph might have infinitely many vertices. A transfinite stopper G is a **pseudonumber** if for every subposition H of G and every H^L and H^R, we have $H^L < H^R$ (cf. Exercise VI.2.8 on page 300). Show that:
(a) If G is a pseudonumber, then $G^L \leq G \leq G^R$ for each G^L and G^R.
(b) The pseudonumbers are totally ordered by $\geq$.
(c) If $\mathcal{A}$ is a set of pseudonumbers, then $\hat{\mathcal{A}} = \{\mathcal{A} \mid \mathbf{pass}\}$ is a pseudonumber and is a least upper bound for $\mathcal{A}$.
(d) The ENTREPRENEURIAL CHESS position in Figure I.4.4(b) (on page 32) has value $\mathbf{on}$ & $\hat{\mathbb{Z}}$.

Notes

Most of this section essentially paraphrases ONAG. There has also been substantial further progress in characterizing the abstract structure of **PG**. First David Moews proved that **PG** has the universal embedding property for Abelian groups [**Moe02**]:

Theorem 1.27 (Moews). *Suppose that $\mathcal{B}$ is an Abelian group (whose domain is a set) and let $\mathcal{A} < \mathcal{B}$ be a subgroup. Then for any embedding $f : \mathcal{A} \to$ **PG** of Abelian groups, there is an embedding $g : \mathcal{B} \to$ **PG** extending f.*

It is straightforward to show that this characterizes the structure of **PG** as an Abelian group, providing a sort of high-level analogue to Theorem III.3.19

(page 176). Moews's result was substantially improved by Jacob Lurie [**Lur02**], who showed that "Abelian group" can be replaced by "partially ordered Abelian group" in Theorem 1.27, confirming a longstanding conjecture of Conway. Lurie's result characterizes the structure of **PG** as a partially ordered Abelian group. A similar (but much easier) result can be proved for **SN** (see Exercise 3.3 on page 436).

The remaining material in this section is largely neglected in the literature (aside from the surreal numbers, which are discussed in the notes to Section 2 on page 423). The terms "moderate" and "all-moderate" are new, though the concepts are Conway's.

2. Surreal Numbers

In Section 1 we saw that **SN** relates to the structure of **PG** in much the same way that $\mathbb{D}$ relates to $\mathbb{G}$. We'll now show that under a suitable definition of multiplication, **SN** is a totally ordered Field with its own spectacularly rich structure theory. Section 3 discusses the algebraic and analytic structure of **SN** in more detail, culminating in a proof that **SN** is real-closed: every odd-degree polynomial with coefficients in **SN** has a root in **SN**.

We begin with the definition of product.

Definition 2.1. The **Conway product** of long games x and y, denoted by $x \times y$ (or just xy), is given by

$$xy = \{x^Ly + xy^L - x^Ly^L,\ x^Ry + xy^R - x^Ry^R \mid \\ \mid x^Ly + xy^R - x^Ly^R,\ x^Ry + xy^L - x^Ry^L\}.$$

Here the expression $x^Ly+xy^L-x^Ly^L$ is understood to range over all *pairs* of options (x^L, y^L), and likewise for the remaining terms in Definition 2.1. Although Conway product is defined for all long games, it is primarily of interest when x and y are surreal numbers.

The definition is motivated by the following observation. Certainly each $x^L < x$ and each $y^L < y$. So assuming that multiplication has the properties of a Field, it must be the case that

$$(x - x^L)(y - y^L) > 0,$$

which implies

$$xy - x^Ly - xy^L + x^Ly^L > 0,$$

yielding a lower bound of $x^Ly+xy^L-x^Ly^L$ for xy. This justifies the inclusion of $x^Ly + xy^L - x^Ly^L$ as a Left option of xy, and the remaining options in Definition 2.1 are similarly motivated. The fact that this definition "works" (in the sense that no *other* options are needed in order for **SN** to enjoy a Field structure) is a theorem that we shall prove shortly.

Proposition 2.2. *For all long games x, y, and z:*

(a) $x0 \cong 0$.

(b) $xy \cong yx$.

(c) $x1 \cong x$.

(d) $(xy)z = x(yz)$.

(e) $(x+y)z = xz + yz$.

Proof. (a) and (b) are immediate from the definition. For (c),

$$x1 \cong \{x^L 1 + x0 - x^L 0 \mid x^R 1 + x0 - x^R 0\} \cong \{x^L 1 \mid x^R 1\}$$

(since 1 has no Right options), and the result follows by induction.

For (d) a typical Left option of $(xy)z$ is

$$(x^L y)z + (xy^L)z + (xy)z^L - (x^L y^L)z - (x^L y)z^L - (xy^L)z^L + (x^L y^L)z^L$$

which by induction is equal to

$$x^L(yz) + x(y^L z) + x(yz^L) - x^L(y^L z) - x^L(yz^L) - x(y^L z^L) + x^L(y^L z^L),$$

a typical Left option of $x(yz)$. The other cases are all the same, with various x^L, y^L, z^L replaced by x^R, y^R, z^R.

For (e) a typical Left option of $(x+y)z$ is

$$\begin{aligned}
&(x+y)^L z + (x+y)z^L - (x+y)^L z^L \\
&= (x^L + y)z + (x+y)z^L - (x^L + y)z^L \\
&= x^L z + yz + xz^L + yz^L - x^L z^L - yz^L \\
&= x^L z + xz^L - x^L z^L + yz \\
&= (xz)^L + yz.
\end{aligned}$$

For the first step in this chain of equalities, we might have chosen $(x + y^L)$ as a Left option of $x+y$, obtaining instead a number of the form $xz + (yz)^L$. The remaining cases are identical. □

Proposition 2.3. *For all surreal numbers x, x_1, x_2, and y:*

(a) *xy is a surreal number.*

(b) *If $x_1 = x_2$, then $x_1 y = x_2 y$.*

(c) *If $x_1 \le x_2$, then $x_1 y - x_1 y^L \le x_2 y - x_2 y^L$ and $x_1 y^R - x_1 y \le x_2 y^R - x_2 y$, for all y^L and y^R. Moreover, if $x_1 < x_2$, then both succeeding inequalities are also strict.*

Proof. The proof will be by induction on an integer parameter n. At stage n, we'll first prove (a) for all x and y with $\tilde{\mathrm{b}}(x) + \tilde{\mathrm{b}}(y) = n$, and then we'll prove (b) and (c) for all x_1, x_2, and y with

$$\max\left\{\tilde{\mathrm{b}}(x_1), \tilde{\mathrm{b}}(x_2)\right\} + \tilde{\mathrm{b}}(y) = n.$$

Moreover, for fixed value of n, the proofs of (b) and (c) will proceed by induction on $\tilde{\mathrm{b}}(x_1) + \tilde{\mathrm{b}}(x_2)$.

(a) By induction (and the fact that **SN** is a Group), we can assume that every option of xy is a surreal number, so we must prove that $(xy)^L < (xy)^R$ for every $(xy)^L$ and $(xy)^R$. We'll show that

$$x^{L_1}y + xy^L - x^{L_1}y^L < x^{L_2}y + xy^R - x^{L_2}y^R$$

for all x^{L_1}, x^{L_2}, y^L, and y^R; the remaining cases are all proved the same way.

First assume $x^{L_1} \le x^{L_2}$. Then by inductive application of (c), we have

$$x^{L_1}y - x^{L_1}y^L \le x^{L_2}y - x^{L_2}y^L.$$

Likewise, a separate inductive application of (c) gives

$$y^L x - y^L x^{L_2} < y^R x - y^R x^{L_2}$$

via the substitution $(x_1, x_2, y) \leftarrow (y^L, y^R, x)$, using the fact that $y^L < y^R$. Applying these inequalities in succession to $(xy)^L$:

$$\begin{aligned} x^{L_1}y + xy^L - x^{L_1}y^L &\le x^{L_2}y + xy^L - x^{L_2}y^L \\ &< x^{L_2}y + xy^R - x^{L_2}y^R. \end{aligned}$$

If instead $x^{L_1} \ge x^{L_2}$, then by a similar argument

$$\begin{aligned} x^{L_1}y + xy^L - x^{L_1}y^L &< x^{L_1}y + xy^R - x^{L_1}y^R \\ &\le x^{L_2}y + xy^R - x^{L_2}y^R. \end{aligned}$$

(b) Here we must show that $(x_1y)^L < x_2y < (x_1y)^R$ for all $(x_1y)^L$ and $(x_1y)^R$, and likewise with x_1 and x_2 interchanged. We'll show that

$$x_1^L y + x_1 y^L - x_1^L y^L < x_2 y$$

for all x_1^L and y^L; the other cases have similar proofs. By induction on n, we can apply (b) to obtain

$$x_1 y^L = x_2 y^L,$$

and by induction on $\tilde{b}(x_1) + \tilde{b}(x_2)$, we obtain from (c)

$$x_1^L y - x_1^L y^L < x_2 y - x_2 y^L,$$

using the fact that $x_1^L < x_2$. Combining these observations yields the desired conclusion.

(c) This follows directly from (b) when $x_1 = x_2$, so assume $x_1 < x_2$. Then either $x_1 < x_1^R \le x_2$ for some x_1^R or else $x_1 \le x_2^L < x_2$ for some x_2^L. The argument in both cases is identical, so assume the former.

Now by inductive application of (a), $x_1 y$ is a number, so that

$$x_1 y < x_1^R y + x_1 y^L - x_1^R y^L \quad \text{and} \quad x_1^R y + x_1 y^R - x_1^R y^R < x_1 y$$

for all y^L and y^R. We can rewrite these inequalities as

$$x_1 y - x_1 y^L < x_1^R y - x_1^R y^L \quad \text{and} \quad x_1 y^R - x_1 y < x_1^R y^R - x_1^R y.$$

Also, by inductive application of (c), we have

$$x_1^R y - x_1^R y^L \le x_2 y - x_2 y^L \quad \text{and} \quad x_1^R y^R - x_1^R y \le x_2 y^R - x_2 y.$$

It follows immediately that

$$x_1 y - x_1 y^L < x_2 y - x_2 y^L \quad \text{and} \quad x_1 y^R - x_1 y < x_2 y^R - x_2 y. \qquad \square$$

Proposition 2.4. *If $x_1 \le x_2$ and $y \ge 0$, then $x_1 y \le x_2 y$.*

Proof. If $y = 0$, then $x_1 y = x_2 y = 0$ by Proposition 2.3(b), so assume $y > 0$. Then some $y^L \ge 0$. Now Proposition 2.3(b) shows that the conclusion is independent of the form of y, so we can assume that every $y^L \ge 0$.

We must show that every $(x_1 y)^L < x_2 y$ and $x_1 y <$ every $(x_2 y)^R$. We'll show that

$$x_1^L y + x_1 y^L - x_1^L y^L < x_2 y$$

for all x_1^L and y^L; the remaining cases are similar. First of all, since $x_1^L < x_2$, Proposition 2.3(c) implies that

$$x_1^L y + x_2 y^L - x_1^L y^L < x_2 y.$$

But $y^L \ge 0$, so by induction on y we have $x_1 y^L \le x_2 y^L$, and the conclusion follows. $\square$

Theorem 2.5. **SN** *is a totally ordered commutative Ring with identity* 1.

Proof. Proposition 2.3(a) shows that Conway product induces a binary operation over **SN**. By Proposition 2.2, this operation is commutative, associative, and distributive over + and has identity 1. Proposition 2.4 shows that it respects the total ordering $\le$. $\square$

Field Structure

Theorem 2.5 shows that the surreal numbers have a Ring structure under Conway multiplication. We'll now show, by explicit construction, that every surreal number has a multiplicative inverse in this Ring.

Definition 2.6. Let x be a positive surreal number. We define the **inverse** y of x by

$$y = \left\{0, \frac{1+(x^R-x)y^L}{x^R}, \frac{1+(x^L-x)y^R}{x^L} \;\middle|\; \frac{1+(x^L-x)y^L}{x^L}, \frac{1+(x^R-x)y^R}{x^R}\right\}.$$

Here $1/x^R$ and $1/x^L$ denote the inverses of x^L and x^R, defined recursively, and x^L is understood to range over all *strictly positive* Left options of x.

Note that y^L and y^R appear in the definition of y. This indicates that the definition of y is itself recursive: it's the closure formed by iteratively feeding newly constructed options back into the definition. (One could instead define successive approximations $\mathscr{Y}_n^L$ and $\mathscr{Y}_n^R$ to the sets of Left and Right options for $y = \{\mathscr{Y}^L \mid \mathscr{Y}^R\}$, taking $\mathscr{Y}^L$ and $\mathscr{Y}^R$ to be the unions of these approximations. However the form in Definition 2.6, given by Conway, is considerably more transparent and is easily translated into more formal language.)

Example. Let $x = 3 = \{2 \mid \}$. Then there is no x^R, and the unique x^L gives $x^L - x = -1$. So the definition of $y = 1/x$ reduces to

$$y = \{0, \tfrac{1}{2}(1-y^R) \mid \tfrac{1}{2}(1-y^L)\}.$$

We have the initial $y^L = 0$. Feeding this into the definition gives the next $y^R = \frac{1}{2}(1-0) = \frac{1}{2}$. Then we obtain a new $y^{L'} = \frac{1}{2}\left(1-\frac{1}{2}\right) = \frac{1}{4}$, followed by a new $y^{R'} = \frac{1}{2}\left(1-\frac{1}{4}\right) = \frac{3}{8}$, and so on. Iterating gives

$$y = \{0, \tfrac{1}{4}, \tfrac{5}{16}, \ldots \mid \tfrac{1}{2}, \tfrac{3}{8}, \tfrac{11}{32}, \ldots\},$$

which is exactly the form of $\frac{1}{3}$ given in Section 1.

Example. Let $x = \omega + 1 = \{\omega \mid \}$. Just as in the previous example, there is no x^R, and we have $x^L - x = -1$. The same iteration gives the inverse

$$y = \{0,\ \tfrac{1}{\omega} - \tfrac{1}{\omega^2},\ \ldots \mid \tfrac{1}{\omega},\ \tfrac{1}{\omega} - \tfrac{1}{\omega^2} + \tfrac{1}{\omega^3},\ \ldots\},$$

in which the Left and Right options are successive Taylor series approximations to $1/(\omega+1)$.

Proposition 2.7. *With x and y as in Definition 2.6:*

(a) *$xy^L < 1 < xy^R$ for all y^L and y^R.*

(b) *y is a surreal number.*

(c) *$xy = 1$.*

Proof. (a) Certainly $x0 < 1$. Among the remaining cases, a typical Left option has the form

$$y^L = \frac{1+(x^R-x)y^{L'}}{x^R}$$

for some "earlier" $y^{L'}$. Presuming inductively that $xy^{L'} < 1$, we now have

$$xy^L = \frac{x + (x^R - x)xy^{L'}}{x^R} < \frac{x + (x^R - x)}{x^R} = 1.$$

The remaining cases are similar.

(b) $y^L \geq y^R$ would imply $xy^L \geq xy^R$, contradicting (a).

(c) Let $z = xy$. Consider a typical Left option of z of the form

$$z^L = x^L y + xy^L - x^L y^L.$$

If $x^L = 0$, then $z^L = xy^L$, which is < 1 by (a). Otherwise,

$$\begin{aligned} z^L &= x^L y + (x - x^L)y^L \\ &= 1 + x^L y - \big(1 + (x^L - x)y^L\big) \\ &= 1 + x^L y - x^L y^R \\ &= 1 + x^L(y - y^R) \\ &< 1. \end{aligned}$$

The remaining Left options are handled similarly, so that every $z^L < 1$, and therefore $z \leq 1$. Likewise, a similar argument shows that every $z^R > 1$, and we also have $z > 1^L$ since 0 is a Left option of z (taking $x^L = y^L = 0$ in the definition of z^L). This shows that $z \geq 1$ also. □

Theorem 2.8. **SN** *is a totally ordered Field.*

Proof. Proposition 2.7 shows that every $x \in \mathbf{SN}$ has a multiplicative inverse, which together with Theorem 2.5 establishes the result. □

Sign Sequences

For each $\alpha \in \mathbf{ON}$, write

$$\mathcal{M}_\alpha = \{x \in \mathbf{SN} : \mathrm{b}(x) \leq \alpha\},$$
$$\mathcal{O}_\alpha = \{x \in \mathbf{SN} : \mathrm{b}(x) < \alpha\}.$$

Now if $x \in \mathbf{SN}$ with birthday α, then x partitions the elements of $\mathcal{O}_\alpha$ into two disjoint intervals,

$$\mathcal{X}^L = \{y \in \mathcal{O}_\alpha : y < x\} \quad \text{and} \quad \mathcal{X}^R = \{y \in \mathcal{O}_\alpha : y > x\}.$$

The Simplicity Theorem shows that $x = \{\mathcal{X}^L \mid \mathcal{X}^R\}$. This suggests approximating x by considering similar partitions of $\mathcal{O}_\beta$ for $\beta < \alpha$.

Definition 2.9. Let $x \in \mathbf{SN}$ of birthday α. For $\beta \in \mathbf{ON}$ we define the βth **approximation to** x, denoted by x_β, by

$$x_\beta = \{\mathcal{X}^L_\beta \mid \mathcal{X}^R_\beta\},$$

where

$$\mathcal{X}_\beta^L = \{y \in \mathcal{O}_\beta : y < x\} \quad \text{and} \quad \mathcal{X}_\beta^R = \{y \in \mathcal{O}_\beta : y > x\}.$$

Clearly $x_0 = 0$ for all x, and since $\mathrm{b}(x_\beta) \le \beta$, we necessarily have $x_\beta \ne x$ whenever $\beta < \mathrm{b}(x)$. Moreover if $\beta \ge \mathrm{b}(x)$, then the Simplicity Theorem gives $x_\beta = x$, so these approximations eventually converge to x.

Examples. Let $x = \frac{5}{8}$. Then

$$\begin{aligned} x_0 &= \{\,\mid\,\} &&= 0,\\ x_1 &= \{0 \mid\,\} &&= 1,\\ x_2 &= \{-1, 0 \mid 1\} &&= \tfrac{1}{2},\\ x_3 &= \left\{-2, -1, -\tfrac{1}{2}, 0, \tfrac{1}{2} \;\middle|\; 1, 2\right\} &&= \tfrac{3}{4},\\ x_4 &= \left\{\ldots, \tfrac{1}{4}, \tfrac{1}{2} \;\middle|\; \tfrac{3}{4}, 1, \ldots\right\} &&= \tfrac{5}{8} = x. \end{aligned}$$

Similarly, if we let $y = \frac{1}{3}$, then we obtain

$$0,\ 1,\ \tfrac{1}{2},\ \tfrac{1}{4},\ \tfrac{3}{8},\ \tfrac{5}{16},\ \ldots,\ \tfrac{1}{3}$$

for

$$y_0,\ y_1,\ y_2,\ y_3,\ y_4,\ y_5,\ \ldots,\ y_\omega.$$

For some values of β, we have $x_\beta < x$, while for others $x_\beta > x$. The sequence of *signs* of the differences $x - x_\beta$ is a fundamental invariant of x.

Definition 2.10. A **sign sequence** is a mapping $\sigma : \alpha \to \{+1, -1\}$, for some ordinal $\alpha \in \mathbf{ON}$. We say α is the **length** (or **domain**) of the sequence and write $\alpha = \mathrm{dom}(\sigma)$.

Sign sequences are ordered lexicographically: we write $\sigma < \tau$ to mean $\sigma(\beta) < \tau(\beta)$, where β is the least ordinal at which they differ. (Here we pretend that $\sigma(\beta) = 0$ for all $\beta \ge \mathrm{dom}(\sigma)$, and likewise for τ.) With this ordering, the sign sequences form a totally ordered Class.

If $\alpha = \mathrm{dom}(\sigma)$ and $\beta \le \alpha$, then $\sigma \upharpoonright \beta$ denotes the **restriction** of σ to β (i.e., the sign sequence of length β that is identically equal to σ on its domain).

Definition 2.11. Let $x \in \mathbf{SN}$ of birthday α. The **sign expansion** of x, denoted by σ_x, is the sign sequence of length α given by

$$\sigma_x(\beta) = \begin{cases} 1 & \text{if } x > x_\beta; \\ -1 & \text{if } x < x_\beta, \end{cases}$$

for all $\beta < \alpha$.

Examples. For $x = \frac{5}{8}$, we have $x_0 < x$ and $x_2 < x$, while $x_1 > x$ and $x_3 > x$. So

$$\sigma_x = (1, -1, 1, -1),$$

which we can abbreviate $+-+-$. Likewise, $y = \frac{1}{3}$ has the sign expansion

$$\sigma_y = +--+-+-+-\cdots$$

of length ω.

Proposition 2.12. *If $\beta \leq \alpha = \mathrm{b}(x)$ and $y = x_\beta$, then $\sigma_y = \sigma_x \upharpoonright \beta$.*

Proof. This is clear, since $(x_\beta)_\gamma = x_\gamma$ for all β and γ. □

The lexicographic ordering on sign expansions coincides with the usual ordering on **SN**:

Theorem 2.13. *If $x > y$, then $\sigma_x > \sigma_y$.*

Proof. The proof is by induction on x and y. Since the theorem does not depend on the form of x, we can assume that every option of x is equal to x_β, for some $\beta < \mathrm{b}(x)$. Then by Proposition 2.12, the options of x all have the form $\sigma \upharpoonright \beta$, and in particular $\sigma_x > \sigma_{x^L}$ for all x^L, and $\sigma_x < \sigma_{x^R}$ for all x^R. We can make the same assumption regarding y.

Now if $x > y$, then either some $x^L \geq y$ or $x \geq$ some y^R. The proof is the same in both cases, so assume the former. Then by induction, $\sigma_{x^L} \geq \sigma_y$. But we also have $\sigma_x > \sigma_{x^L}$, by our choice of the form of x, and it follows that $\sigma_x > \sigma_y$. □

Theorem 2.14. *The mapping $x \mapsto \sigma_x$ is an order-preserving bijection from* **SN** *onto the Class of sign sequences.*

Proof. Theorem 2.13 shows that the map is order-preserving and one-to-one. To see that it's onto, fix a sign sequence σ of length α. By induction, we can assume that for each $\beta < \alpha$, there is some x_β with $\sigma \upharpoonright \beta = \sigma_{x_\beta}$. Put

$$x = \{\mathcal{X}^L \mid \mathcal{X}^R\}$$

where

$$\mathcal{X}^L = \{x_\beta : (\sigma \upharpoonright \beta) < \sigma\} \quad \text{and} \quad \mathcal{X}^R = \{x_\beta : (\sigma \upharpoonright \beta) > \sigma\}.$$

By Theorem 2.13, x is a number, and clearly $\mathrm{b}(x) \leq \alpha$ and each x_β is the βth approximation to x. Since $x - x_\beta = \sigma(\beta)$, we conclude that $\sigma_x = \sigma$. □

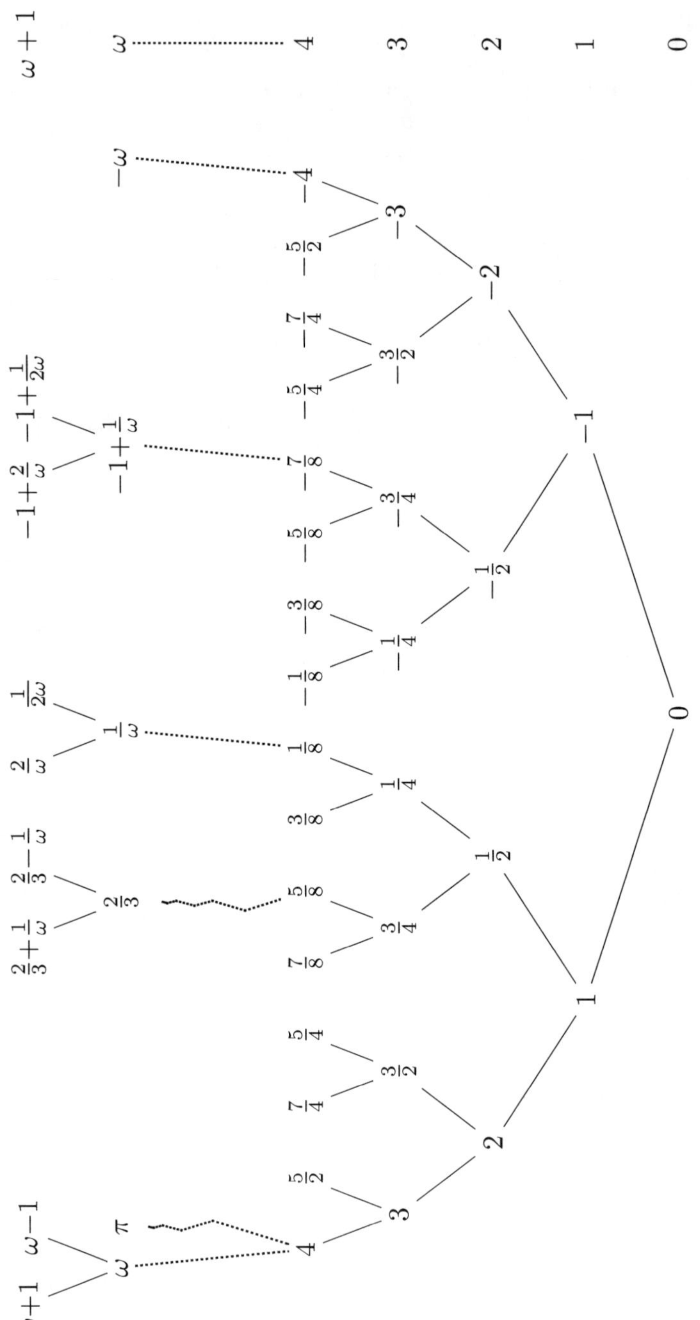

Figure 2.1. The transfinite number tree (with birthdays labeled on the right).

The Number Tree

In Section II.3 we saw that the dyadic rationals can be arranged into a binary tree of height ω (cf. Figure II.3.1 on page 74). Similarly, the surreal numbers can be arranged into a binary tree of height **ON**. A fragment of this tree is shown in Figure 2.1 on the preceding page.

Each number with birthday α has exactly two successors on day $\alpha + 1$. If α is a limit ordinal, then the numbers born on day α correspond one-to-one with branches through the tree below α.

The sign expansion σ_x serves as a "map" for locating a specific number x in this tree: a $+$ in the sign expansion corresponds to a left fork of the tree; a $-$ corresponds to a right fork. This gives a particularly elegant way of visualizing the relationship between sign expansions and the hereditary structure of **SN**.

Berlekamp's Sign-Expansion Rule

Berlekamp's Sign-Expansion Rule is a method for translating between the binary expansion of a real number and its sign expansion.

Let $x > 0$ be a positive real number. If x is an integer, then its sign expansion consists entirely of a corresponding number of $+$ symbols:

$$\sigma_5 = {+}{+}{+}{+}{+}.$$

Suppose that x is a noninteger. First write out the (finite or infinite) binary expansion of x, for example:

$$5\tfrac{13}{32} = 101.01101 \qquad \text{or} \qquad \pi = 11.0010010000111\cdots.$$

This expansion has three components: the integer part; the "binary point"; and the fractional part. The sign expansion σ_x is obtained by concatenating translations for each of these parts.

- The integer part translates to a corresponding number of $+$ signs.
- The binary point translates to the exact string $+-$.
- The fractional part translates directly in binary: $0 \to -$ and $1 \to +$, except that *if* the binary expansion is finite, then the terminal "1" is omitted.

Here is a typical example:

$$\begin{array}{rcccc} x = 5\tfrac{13}{32} = & 101 & . & 0\,1\,1\,0\,1 & \\ \sigma_x = & \overbrace{{+}{+}{+}{+}{+}} & \overbrace{{+}{-}} & {-}{+}{+}{-} & \quad \text{(length 11)}. \end{array}$$

Here's one with an infinite binary expansion:

$$\begin{aligned}\pi = \;& 11 \;\; . \;\; 0\,0\,1\,0\,0\,1\,0\,0\,0\,0\,1\,1\,1\,\cdots \\ \sigma_\pi = \;& \overbrace{+++}\;\overbrace{+-}\;\;--+--+----+++\cdots \quad \text{(length } \omega).\end{aligned}$$

The sign expansion for a negative real $x < 0$ can obviously be computed by negating the sequence for $-x$. We leave as an exercise the easy proof that Berlekamp's Sign-Expansion Rule "works."

Elements of $\mathcal{M}_\omega$

Sign sequences are a powerful tool for penetrating the structure of **SN**. For example, one can quickly determine all the surreals born by day ω by reasoning in terms of their sign sequences:

- If σ_x is finite, then $x \in \mathbb{D}$, with its value given by Berlekamp's Rule.
- If σ_x has infinitely many $+$'s and infinitely many $-$'s, then $x \in \mathbb{R}$, again by Berlekamp's Rule.
- If σ_x has finitely many $-$'s (but at least one) and infinitely many $+$'s, then necessarily
$$\sigma_x = \tau{-}{+}{+}{+}{+}{+}\cdots$$
for some finite sign sequence τ. Here $x = y - \frac{1}{\omega}$, where $y \in \mathbb{D}$ is such that $\sigma_y = \tau$.
- Likewise, if σ_x has just finitely many $+$'s (but at least one) and infinitely many $-$'s, then $x = y + \frac{1}{\omega}$ for some $y \in \mathbb{D}$.
- Finally, if σ_x consists *entirely* of $+$'s or $-$'s, then $x = \omega$ or $-\omega$.

All of these assertions are easily verified by writing x in terms of its approximations and checking that the appropriate differences are second-player wins. Since the above list exhausts all sign sequences of length $\le \omega$, it necessarily exhausts all of $\mathcal{M}_\omega$. Therefore

$$\mathcal{M}_\omega = \mathbb{R} \cup \left(\mathbb{D} + \tfrac{1}{\omega}\right) \cup \left(\mathbb{D} - \tfrac{1}{\omega}\right) \cup \{-\omega, \omega\}.$$

It's instructive to go a step further and classify $\mathcal{M}_{\omega+1}$. This case is actually somewhat simpler, since if $\mathrm{b}(x) = \omega + 1$, then σ_x is obtained by appending a single $+$ or $-$ to the sign sequence σ_y of some y with $\mathrm{b}(y) = \omega$.

- If $y \in \mathbb{R}$, then $x = y + \frac{1}{\omega}$ or $y - \frac{1}{\omega}$ (depending on the sign of $\sigma_x(\omega)$).
- If $y = z + \frac{1}{\omega}$, for $z \in \mathbb{D}$, then $x = z + \frac{2}{\omega}$ or $x = z + \frac{1}{2\omega}$ (again, depending on the sign of $\sigma_x(\omega)$).
- Likewise, if $y = z - \frac{1}{\omega}$, then $x = z - \frac{2}{\omega}$ or $x = z - \frac{1}{2\omega}$.
- If $y = \omega$, then $x = \omega + 1$ or $\omega - 1$; likewise for $y = -\omega$.

So

$$\begin{aligned}\mathcal{M}_{\omega+1} = \quad & \left(\mathbb{R} + \left\{-\tfrac{1}{\omega}, 0, \tfrac{1}{\omega}\right\}\right) \\ & \cup \left(\mathbb{D} + \left\{-\tfrac{2}{\omega}, -\tfrac{1}{2\omega}, \tfrac{1}{2\omega}, \tfrac{2}{\omega}\right\}\right) \\ & \cup \left(\{-\omega, \omega\} + \{-1, 0, 1\}\right).\end{aligned}$$

Exercises

2.1 If n is an integer in canonical form and G is any long game, then

$$G \times n = \overbrace{G + G + \cdots + G}^{n \text{ times}}.$$

2.2 $\mathcal{O}_\alpha$ is a group if and only if $\alpha = \omega^\beta$ for some β.

2.3 Determine all elements of $\mathcal{M}_{\omega+2}$.

2.4 There exist long games G, H, and J such that $G = H$, but $G \times J \neq H \times J$.

2.5 *Explicit square roots.* Let $x \geq 0$, and define the **square root** y of x by

$$y = \left\{ \sqrt{x^L},\ \frac{x + y^L y^R}{y^L + y^R} \;\middle|\; \sqrt{x^R},\ \frac{x + y^L y^{L'}}{y^L + y^{L'}},\ \frac{x + y^R y^{R'}}{y^R + y^{R'}} \right\},$$

where $\sqrt{x^L}$ and $\sqrt{x^R}$ are defined recursively and $(y^L, y^{L'})$ and $(y^R, y^{R'})$ range over pairs chosen so that the relevant denominator is nonzero. Prove that $y^2 = x$.

Notes

Conway arrived at the surreal numbers in the early 1970s, almost by accident; the fascinating story of their discovery is told in detail in Appendix C. They quickly captured the imagination of mathematicians around the world. In an interesting twist, they first appeared in print in a 1974 novella by Donald Knuth [**Knu74**], who also coined the term "surreal number." This is a rare instance of a substantial new mathematical theory first introduced in print by a work of fiction!

In 1976 Conway finally published ONAG, which includes much of the essential material on surreals and remains a standard reference. A more formal treatment was later given by Harry Gonshor [**Gon86**]. Norman Alling's book [**All87**] employs a different approach, with an emphasis on classical analytic and topological properties of **SN**.

The proof of Proposition 2.3 given here is due to Schleicher and Stoll [**SS06**]; the remainder of our treatment mostly follows ONAG. More recent advances are discussed in the notes to Section 3 on page 437.

Exercise 2.5: Clive Bach.

3. The Structure of Surreal Numbers

Having established that **SN** has a Field structure, we can proceed to investigate its algebraic and analytic properties. The key result is that every surreal number can be expressed in **normal form** as a transfinite sum of terms $\omega^y \times r$, with $r \in \mathbb{R}$. Here ω^y denotes the surreal ω-**power** of y, a well-behaved generalization of ordinal exponentiation.

It turns out that such normal forms can be combined and manipulated in just the same way as ordinary power series, and as a consequence many algebraic and analytic properties of $\mathbb{R}$ can be lifted to **SN**. In particular, this technique provides a proof that **SN** is real-closed, which is the main result of this section.

Surreal ω-Powers

We begin with a description of the ω-power and its properties.

Definition 3.1. For each $x \in \mathbf{SN}$, we define the ω-**power** of x by

$$\omega^x = \left\{0, r\omega^{x^L} \;\middle|\; r\omega^{x^R}\right\}_{r\in\mathbb{R}^+}.$$

Proposition 3.2. *If $x = y$, then $\omega^x = \omega^y$; if $x < y$, then $\omega^x \ll \omega^y$.*

Proof. First suppose $x < y$. Then either $x \le$ some y^L or some $x^R \le y$. The proof is the same in both cases, so assume the former. Then by induction, $\omega^x \le \omega^{y^L}$. But for all $r \in \mathbb{R}^+$ we have $r\omega^{y^L} < \omega^y$, so it follows that $\omega^x \ll \omega^y$.

Next suppose $x = y$. Then for each x^L, we have $x^L < y$, so by induction $\omega^{x^L} \ll \omega^y$. In particular, $r\omega^{x^L} < \omega^y$ for each $r \in \mathbb{R}^+$. Since $0 < \omega^y$, we have demonstrated that every $(\omega^x)^L < \omega^y$. The other cases are all the same. □

Crucially, every positive surreal number is **commensurate** with exactly one ω-power, in the following sense.

Definition 3.3. Positive surreal numbers x and y are **commensurate** if there exist real numbers $r, s \in \mathbb{R}$ such that $x < ry$ and $y < sx$.

Note that for all $x, y > 0$, either $x \ll y$ or $y \ll x$ or else x and y are commensurate.

Proposition 3.4. *Every $x > 0$ is commensurate with ω^y for exactly one choice of y, and ω^y is the simplest number commensurate with x.*

Proof. Uniqueness is clear from Proposition 3.2, so it suffices to prove existence.

Fix a surreal number $x > 0$. Then we can write x in the form

$$x = \{0, x^L \mid x^R\},$$

with all $x^L, x^R > 0$ and all $\mathrm{b}(x^L), \mathrm{b}(x^R) < \mathrm{b}(x)$. If x is commensurate with any of its options, then we are done (by induction).

Otherwise, it must be the case that $x^L \ll x$ for all x^L and $x \ll x^R$ for all x^R. By induction, each x^L is commensurate with some ω^{y^L}, and each x^R with some ω^{y^R}. Then for every x^L there is some $r \in \mathbb{R}$ such that $x^L < r\omega^{y^L} < x$, and likewise for every x^R, so that writing $y = \{y^L \mid y^R\}$,

$$x = \left\{0, r\omega^{y^L} \;\middle|\; r\omega^{y^R}\right\} = \omega^y.$$

We've shown that every number x is either commensurate with a number of smaller birthday or else of the form ω^y. The proposition follows. □

Proposition 3.5. $\omega^{x+y} = \omega^x\omega^y$ *for all* $x, y \in \mathbf{SN}$.

Proof. Consider a typical Left option of $\omega^x\omega^y$ of the form

$$t = (\omega^x)^L\omega^y + \omega^x(\omega^y)^L - (\omega^x)^L(\omega^y)^L.$$

When $(\omega^y)^L = 0$, then $t = (\omega^x)^L\omega^y$. This is 0 when $(\omega^x)^L = 0$ and otherwise ranges over $r\omega^{x^L}\omega^y$, which by induction is equal to $r\omega^{x^L+y}$. Likewise, when $(\omega^x)^L = 0$, then t ranges over $r\omega^{x+y^L}$. This shows that every Left option of ω^{x+y} is also a Left option of $\omega^x\omega^y$.

When neither term in the expression for t is 0, then

$$\begin{aligned} t &= r\omega^{x^L}\omega^y + \omega^x s\omega^{y^L} - r\omega^{x^L}s\omega^{y^L} \\ &= r\omega^{x^L+y} + s\omega^{x+y^L} - rs\omega^{x^L+y^L} \end{aligned}$$

for some $r, s \in \mathbb{R}^+$, the latter identity following by induction. Put $z = \max\{x^L + y, x + y^L\}$; then certainly $z > x^L + y^L$. So t is positive and commensurate with ω^z and so is dominated by some Left option of ω^{x+y}.

The remaining Left options of $\omega^x\omega^y$ have the form

$$r\omega^{x^R+y} + s\omega^{x+y^R} - rs\omega^{x^R+y^R},$$

which is always negative and therefore dominated by 0. So we've shown that every Left option of ω^{x+y} is a Left option of $\omega^x\omega^y$ and that every Left option of $\omega^x\omega^y$ is dominated by a Left option of ω^{x+y}. A similar argument shows that the same is true for Right options. □

Normal Form

A well-known result in set theory (the *Cantor Normal Form Theorem*, Theorem B.3.14 on page 468) shows that every ordinal α can be written uniquely in the form

$$\alpha = \sum_{i<k} \left(\omega^{\gamma_i} \times n_i\right)$$

where $\langle \gamma_i : i < k \rangle$ is a decreasing sequence of ordinals and each $n_i \in \mathbb{N}$.

The *Conway Normal Form Theorem* neatly generalizes this result to the surreal numbers, with one key difference. In the Cantor normal form, the sequence $\langle \gamma_i : i < k \rangle$ is necessarily finite, since there can be no infinite decreasing sequence of ordinals. Conversely, there *do* exist infinite decreasing sequences of surreals (of arbitrary ordinal length), and surreal normal forms must take this into account. Here's the statement of the theorem:

Theorem 3.6 (Conway Normal Form Theorem). *Every $x \in \mathbf{SN}$ can be written uniquely in the form*

$$x = \sum_{\xi < \alpha} \left(\omega^{y_\xi} \times r_\xi \right),$$

where $\langle y_\xi : \xi < \alpha \rangle$ is a decreasing sequence of surreals and each $r_\xi \in \mathbb{R}$.

This theorem, of course, depends on an appropriate definition of transfinite series. We'll give this definition now. The idea is that at limit ordinals, the value of the (partial) sum is the *simplest* surreal number that fits all prior partial sums:

Definition 3.7. The value of the series

$$\sum_{\xi < \alpha} \left(\omega^{y_\xi} \times r_\xi \right),$$

with $\langle y_\xi : \xi < \alpha \rangle$ a decreasing sequence of surreals and $r_\xi \in \mathbb{R}$, is defined as follows. Let Σ_β denote the βth partial sum; we may presume by induction that Σ_β is defined for all $\beta < \alpha$. If α is a successor ordinal, say $\alpha = \beta + 1$, then simply put

$$\Sigma_\alpha = \Sigma_\beta + \omega^{y_\beta} r_\beta.$$

If α is a limit ordinal, then

$$\Sigma_\alpha = \left\{ \Sigma_\beta + \omega^{y_\beta}(r_\beta - s) \;\middle|\; \Sigma_\beta + \omega^{y_\beta}(r_\beta + s) \right\}_{\beta < \alpha,\, s \in \mathbb{R}^+}.$$

The following lemma encapsulates the successor stages in the proof of the Conway Normal Form Theorem.

Lemma 3.8. *For each $x \neq 0$, there exist $y \in \mathbf{SN}$ and $r \in \mathbb{R}$ such that*

$$x = \omega^y r + x', \quad \textit{where } |x'| \ll |x|.$$

Proof. First assume that $x > 0$. Let $y \in \mathbf{SN}$ be such that x is commensurate with ω^y. Put

$$r = \sup\{s \in \mathbb{R} : \omega^y s \le x\}$$

(where the supremum is taken over $\mathbb{R}$ as in classical analysis) and write

$$x' = x - \omega^y r.$$

Now for all $s \in \mathbb{R}^+$, we have

$$\omega^y(r-s) < x < \omega^y(r+s)$$

so that $-\omega^y < x'/s < \omega^y$. Therefore $|x'| \ll \omega^y$.

If $x < 0$, then we can apply this argument to $-x$ and negate the resulting values of r and x'. □

Proof of Theorem 3.6. We'll define $x_\alpha, y_\alpha \in \mathbf{SN}$ and $r_\alpha \in \mathbb{R}$ in such a way that

$$x = x_\alpha + \sum_{\xi<\alpha} \omega^{y_\xi} r_\xi$$

for all ordinals α, and $|x_\alpha| \ll |x_\beta|$ whenever $\alpha > \beta$ (unless $x_\beta = 0$). As before, we denote by Σ_α the αth partial sum, so that $x = x_\alpha + \Sigma_\alpha$.

(Note that in Section 2, we used the symbol x_β to represent the βth sign-approximation of x, and here we are using it for the βth normal-form remainder. These are *not* the same thing, but the collision of notation should cause no confusion; it will always be clear from the context which is being discussed.)

For $\alpha = 0$ we take $x_0 = x$. At a successor ordinal $\alpha = \beta + 1$, write

$$x_\beta = \omega^{y_\beta} r_\beta + x_\alpha$$

as in Lemma 3.8; this defines y_β, r_β, and x_α.

Suppose instead that α is a limit ordinal. Then y_ξ and r_ξ are defined for all $\xi < \alpha$, and this determines x_α. Fix $\beta < \alpha$; we must show that $|x_\alpha| \ll |x_\beta|$. Now

$$\Sigma_{\beta+1} - \omega^{y_\beta} s < \Sigma_\alpha < \Sigma_{\beta+1} + \omega^{y_\beta} s$$

for all $s \in \mathbb{R}^+$ (directly from Definition 3.7). Therefore

$$|x_\alpha - x_{\beta+1}| = |\Sigma_\alpha - \Sigma_{\beta+1}| \ll \omega^{y_\beta}.$$

But $|x_{\beta+1}| \ll \omega^{y_\beta}$, so this implies $|x_\alpha| \ll \omega^{y_\beta}$, which suffices since ω^{y_β} is commensurate with x_β.

This completes the definition of x_α. To finish the existence proof, we must show that some $x_\alpha = 0$, so that the least such α gives a normal form.

But for each α, we see that Σ_α is the *simplest* number that satisfies

$$\Sigma_\beta + \omega^{y_\beta}(r_\beta - s) < \Sigma_\alpha < \Sigma_\beta + \omega^{y_\beta}(r_\beta + s) \quad \text{for all } \beta < \alpha \text{ and } s \in \mathbb{R}.$$

(This is the definition of Σ_α when α is a limit ordinal and is easily seen to be true when α is a successor ordinal as well.) But x also satisfies these inequalities, so necessarily $\mathrm{b}(\Sigma_\alpha) \leq \mathrm{b}(x)$. This is true for all $\alpha \in \mathbf{ON}$. Since $\mathbb{G}_{\mathrm{b}(x)}$ is a set and $\mathbf{ON}$ is a proper Class, there must exist some α for which $\Sigma_\alpha = \Sigma_{\alpha+1}$, that is, $x_\alpha = 0$.

For uniqueness, suppose that

$$x = \sum_{\xi<\alpha} \omega^{y_\xi} r_\xi \quad \text{and} \quad x' = \sum_{\xi<\alpha'} \omega^{y'_\xi} r'_\xi$$

are surreal numbers in normal form, with partial sums Σ_β and Σ'_β and remainders x_β and x'_β. Suppose there is some β with $\Sigma_\beta \neq \Sigma'_\beta$, and fix the least such β; we must show that $x \neq x'$. Without loss of generality, assume $\Sigma_\beta > \Sigma'_\beta$. Then by minimality of β,

$$\omega^{y_\beta} r_\beta > \omega^{y'_\beta} r'_\beta$$

so that

$$\Sigma_\beta - \Sigma'_\beta \geq s\omega^{y_\beta}$$

for some $s \in \mathbb{R}^+$. But $|x_\beta| \ll \omega^{y_\beta}$ and $|x'_\beta| \ll \omega^{y'_\beta} \leq \omega^{y_\beta}$, so in fact

$$x - x' \geq s\omega^{y_\beta} + (x_\beta - x'_\beta) > 0. \qquad \square$$

Algebra of Normal Forms

Although normal forms are transfinite expressions, their sums and products behave just like ordinary power series: sums are obtained by adding up the coefficients of each term ω^y; products are obtained via the expected multiplication rule. This miraculous fact is the foundation of surreal analysis. The following definitions will be convenient.

Definition 3.9. Let $x, y \in \mathbf{SN}$. We denote by $\pi_y(x)$ the coefficient of ω^y in the normal form of x. (If ω^y does not appear as a term, then $\pi_y(x) = 0$.) Then the **support** of x is given by

$$\operatorname{supp}(x) = \{y \in \mathbf{ON} : \pi_y(x) \neq 0\}.$$

Clearly $\operatorname{supp}(x)$ is always a set and is anti-well-ordered (i.e., is the range of a decreasing ordinal sequence of surreals). The main theorems are the *sum rule* and *product rule* for normal forms:

Theorem 3.10. *Let* $w, x, y \in \mathbf{SN}$. *Then* $\pi_y(w + x) = \pi_y(w) + \pi_y(x)$.

Theorem 3.11. *Let* $w, x, y \in \mathbf{SN}$. *Then*

$$\pi_y(wx) = \sum_{u+v=y} \pi_u(w)\pi_v(x).$$

Note that the sum in Theorem 3.11 appears to range over an infinite number of possible terms. However, we'll shortly prove a combinatorial lemma (Lemma 3.18, below) showing that there can be just finitely many pairs (u, v) with $u \in \operatorname{supp}(w)$, $v \in \operatorname{supp}(x)$, and $u + v = y$, so the expression in Theorem 3.11 makes sense.

We begin by proving Theorem 3.10. The basic idea is to dissect the normal forms of w and x into a finite number of terms whose behavior can be easily characterized.

Definition 3.12. $x \in \mathbf{SN}$ is ***y*-truncated** if $z > y$ for every $z \in \operatorname{supp}(x)$.

Theorem 3.13. *The following are equivalent, for $x, y \in \mathbf{SN}$ and $r \in \mathbb{R}$:*

(i) $\pi_y(x) = r$.

(ii) $x = a + \omega^y r + b$, *for some a, b such that a is y-truncated and $|b| \ll \omega^y$.*

Proof. Let

$$x = \sum_{\xi<\alpha} \omega^{y_\xi} r_\xi$$

be the normal form of x. Put $\beta = \sup\{\xi < \alpha : y_\xi > y\}$. Then certainly the βth partial sum Σ_β is y-truncated. If $y_\beta = y$, then

$$x = \Sigma_\beta + \omega^y r + x_{\beta+1}$$

with $r = r_\beta$. (Here $x_{\beta+1}$ is the remainder term, as in the proof of Theorem 3.6.) Otherwise $y_\beta < y$, so that

$$x = \Sigma_\beta + \omega^y r + x_\beta$$

with $r = 0$. In both cases $\pi_y(x) = r$, and the expression for x has the form (ii). □

Lemma 3.14. *The following are equivalent, for $x, y \in \mathbf{SN}$:*

(i) *x is y-truncated.*

(ii) $x = \{x - \omega^{y+\epsilon} \mid x + \omega^{y+\epsilon}\}$ *for some $\epsilon > 0$.*

Proof. For (i) ⇒ (ii), let $\epsilon = \{y \mid \operatorname{supp}(x)\} - y$. Then $\epsilon > 0$, but $y + \epsilon < z$ for all $z \in \operatorname{supp}(x)$. The conclusion follows. (ii) ⇒ (i) follows from the definition of normal form. □

Lemma 3.15. *If w and x are y-truncated, then so is $w + x$.*

Proof. For some $\epsilon > 0$, we can write

$$w = \left\{w - \omega^{y+\epsilon} \;\middle|\; w + \omega^{y+\epsilon}\right\} \quad \text{and} \quad x = \left\{x - \omega^{y+\epsilon} \;\middle|\; x + \omega^{y+\epsilon}\right\}$$

and it follows immediately that

$$w + x = \left\{w + x - \omega^{y+\epsilon} \;\middle|\; w + x + \omega^{y+\epsilon}\right\}.$$ □

We can now prove that normal forms obey the sum rule:

Proof of Theorem 3.10. By Theorem 3.13, we have

$$w = a + \omega^y r + c \quad \text{and} \quad x = b + \omega^y s + d$$

where a and b are y-truncated, $r = \pi_y(w)$, $s = \pi_y(x)$, and $|c|, |d| \ll \omega^y$. Then

$$w + x = (a + b) + \omega^y(r + s) + (c + d).$$

Now $a + b$ is y-truncated by Lemma 3.15, and certainly $|c + d| \ll \omega^y$, so it follows from Theorem 3.13 that $\pi_y(w + x) = r + s$. □

We now prove the product rule. The first step is to consider the case where one of the multiplicands is a primitive term of the form ω^z.

Lemma 3.16. *If x is y-truncated, then $x\omega^z$ is $(y + z)$-truncated.*

Proof. We have

$$x = \left\{x - \omega^{y+\epsilon} \;\middle|\; x + \omega^{y+\epsilon}\right\}$$

for all sufficiently small $\epsilon > 0$. Then a typical Left option of $x\omega^z$ has the form

$$\begin{aligned} &(x - \omega^{y+\epsilon})\omega^z + x(\omega^z)^L - (x - \omega^{y+\epsilon})(\omega^z)^L \\ &= x\omega^z - \omega^{y+z+\epsilon} + \omega^{y+\epsilon}(\omega^z)^L. \end{aligned}$$

A similar calculation applies to the remaining Left and Right options of $x\omega^z$. But $(\omega^z)^L \ll \omega^z$, so that $\omega^{y+\epsilon}(\omega^z)^L \ll \omega^{y+z+\epsilon}$, and it follows that

$$x\omega^z = \left\{x\omega^z - \omega^{y+z+\epsilon} \;\middle|\; x\omega^z + \omega^{y+z+\epsilon}\right\}.$$ □

Theorem 3.17. *Let $x, y, z \in$ **SN**. Then $\pi_{y+z}(x\omega^z) = \pi_y(x)$.*

Proof. Using Theorem 3.13, write

$$x = a + r\omega^y + b,$$

with a y-truncated, $r = \pi_y(x)$, and $|b| \ll \omega^y$. Then

$$\omega^z x = \omega^z a + r\omega^{y+z} + \omega^z b.$$

Here $\omega^z a$ is $(y+z)$-truncated, and by Lemma 3.16, $|\omega^z b| \ll \omega^{y+z}$. Therefore Lemma 3.14 gives $\pi_{y+z}(x) = r$. □

We'll now prove the full product rule, Theorem 3.11. The following lemma shows that the sum in the statement of the theorem is well-defined.

Lemma 3.18. *Suppose $\mathcal{A}$ and $\mathcal{B}$ are well-ordered sets of surreal numbers. Then $\mathcal{A} + \mathcal{B}$ is also well-ordered. Moreover, for any $x \in$ **SN**, there are at most finitely many pairs (a, b) with $a \in \mathcal{A}$, $b \in \mathcal{B}$, and $a + b = x$.*

Proof. If the lemma is false (for either reason), then there must exist an infinite nonincreasing sequence

$$a_1 + b_1 \geq a_2 + b_2 \geq a_3 + b_3 \geq \cdots$$

such that all pairs (a_i, b_i) are distinct. Suppose (for contradiction) that such a sequence exists. By passing to an appropriate subsequence, we can assume that $a_1 \leq a_2 \leq a_3 \leq \cdots$ (otherwise there would be a descending terminal segment of the a_i, contradicting the assumption on $\mathcal{A}$). But this implies $b_1 \geq b_2 \geq b_3 \geq \cdots$. Since $\mathcal{B}$ is well-ordered, the b_i must be eventually constant; hence the a_i must also be eventually constant, contradicting the uniqueness assumption. □

Proof of Theorem 3.11. Temporarily write $w \circledast x$ for the formal product of w and x, given by

$$\pi_y(w \circledast x) = \sum_{u+v=y} \pi_u(w)\pi_v(x)$$

for all y. (This is well-defined, since $\mathrm{supp}(w) + \mathrm{supp}(x)$ is anti-well-ordered, by Lemma 3.18.) Our goal is to prove that $wx = w \circledast x$ for all w and x.

Now a typical Left option of wx has the form

$$(wx)^L = w^L x + wx^L - w^L x^L.$$

By induction on the length of the normal forms of w and x, we may suppose that

$$(wx)^L = w^L \circledast x + w \circledast x^L - w^L \circledast x^L,$$

which we can rewrite (using the ordinary rules of algebra) as

$$(wx)^L = w \circledast x - (w - w^L) \circledast (x - x^L).$$

By Theorem 3.10, we have

$$w - w^L = \omega^u r + a \quad \text{and} \quad x - x^L = \omega^v s + b$$

for some $u \in \mathrm{supp}(w)$, $v \in \mathrm{supp}(x)$, $r, s \in \mathbb{R}$, and $|a| \ll \omega^u$, $|b| \ll \omega^v$. Therefore

$$(w - w^L) \circledast (x - x^L) = \omega^{u+v} rs + c$$

with $|c| \ll \omega^{u+v}$. The same calculation works for the remaining options of wx, so that

$$wx = \{w \circledast x - \omega^{u+v} t \mid w \circledast x + \omega^{u+v} t\}_{u \in \mathrm{supp}(w),\, v \in \mathrm{supp}(x),\, t \in \mathbb{R}} = w \circledast x. \quad \square$$

Power Series

We can now define convergence of sequences and series of surreals. This will be a key ingredient in the proof that **SN** is real-closed. Here a series will be an ordinary *countable* series such as those encountered in classical analysis:

$$x = \sum_{n<\omega} x_n = \sum_{n=0}^{\infty} x_n,$$

of length exactly ω.

Definition 3.19. Let $x_0, x_1, x_2, \ldots$ be a sequence of surreals (of length exactly ω) and let $x \in \mathbf{SN}$. We write

$$x = \lim_{n\to\infty} x_n$$

to mean that

$$\pi_y(x) = \lim_{n\to\infty} \pi_y(x_n) \quad \text{for all } y \in \mathbf{SN}.$$

In this case we say that x_n **converges to** x.

Definition 3.20. Let $x_0, x_1, x_2, \ldots$ be a sequence of surreals (of length exactly ω) and let $x \in \mathbf{SN}$. We write

$$x = \sum_{n=0}^{\infty} x_n$$

to mean that the partial sums of the series converge to x.

The main application is the following crucial theorem.

Theorem 3.21. *Let f be a formal power series with real coefficients,*

$$f(x) = \sum_{n=0}^{\infty} x^n c_n, \qquad \textit{each } c_n \in \mathbb{R}.$$

Then $f(\zeta)$ converges absolutely for every infinitesimal $\zeta \in \mathbf{SN}$.

The proof of this theorem involves expanding the normal forms of the terms ζ^n. This requires the following generalization of Lemma 3.18.

Lemma 3.22. *Suppose $\mathcal{A} \subset \mathbf{SN}^+$ is a well-ordered set of positive surreal numbers. Then the set $\mathcal{S}$ of finite sums of elements of $\mathcal{A}$ is also well-ordered. Moreover, each element of $\mathcal{S}$ can be written as a sum of elements of $\mathcal{A}$ in just finitely many ways.*

Proof. If the lemma is false (for either reason), there must exist an infinite nonincreasing sequence from $\mathcal{S}$,

$$s_1 \geq s_2 \geq s_3 \geq \cdots$$

such that each s_i is expressible as the sum of k_i elements of $\mathcal{A}$,

$$s_i = a_i^1 + a_i^2 + \cdots + a_i^{k_i} \qquad (a_i^1 \geq a_i^2 \geq \cdots \geq a_i^{k_i})$$

and each tuple of summands $(a^1, a^2, \ldots, a^k)$ occurs at most once. (Note that the superscripts here are *indices*, not exponents.) Assume (for contradiction) that such a sequence exists.

Now by repeated application of Lemma 3.18, we see that $k\mathcal{A}$ is well-ordered for each k, and each element of $k\mathcal{A}$ can be written as a sum of k elements of $\mathcal{A}$ in just finitely many ways. This shows that the values k_i cannot concentrate on any particular k and so must be unbounded. By passing to an appropriate subsequence, we can assume that each $k_i > i$.

Now consider the sequence obtained by listing just the *first* summand of each s_i:

$$a_1^1, a_2^1, a_3^1, \ldots.$$

Since $\mathcal{A}$ is well-ordered, this sequence must have a nondecreasing subsequence. Let $\mathcal{I}_1$ index such a subsequence, so that if $i, i' \in \mathcal{I}_1$ and $i < i'$, then $a_i^1 \leq a_{i'}^1$.

Next list the *second* summand of each s_i, this time starting with $i = 2$ (the assumption $k_i > i$ implies that such a second summand always exists):

$$a_2^2, a_3^2, a_4^2, \ldots.$$

Again since $\mathcal{A}$ is well-ordered, there must exist an index set $\mathcal{I}_2 \subset \mathcal{I}_1$ such that if $i, i' \in \mathcal{I}_2$ and $i < i'$, then $a_i^2 \leq a_{i'}^2$.

Continuing like this, we obtain nested sets of indices

$$\mathcal{I}_1 \supset \mathcal{I}_2 \supset \mathcal{I}_3 \supset \cdots$$

such that for each j, the subsequence of $a_j^j, a_{j+1}^j, a_{j+2}^j, \ldots$ indexed by $\mathcal{I}_j$ is nondecreasing. Now diagonalize: choose any

$$\mathcal{I} = \{i_1, i_2, i_3, \ldots\} \qquad (i_1 < i_2 < i_3 < \cdots)$$

such that $i_j \in \mathcal{I}_j$. By passing to the subsequence indexed by $\mathcal{I}$, we can assume that

$$(\dagger) \qquad a_i^j \leq a_{i'}^j \quad \text{whenever} \quad j \leq i \leq i'.$$

Consider an element s_i, and compare it to the subsequent element s_{k_i}:

$$\begin{aligned} s_i &= a_i^1 + a_i^2 + \cdots + a_i^i + \cdots + a_i^{k_i}, \\ s_{k_i} &= a_{k_i}^1 + a_{k_i}^2 + \cdots + a_{k_i}^i + \cdots + a_{k_i}^{k_i} + \cdots. \end{aligned}$$

Certainly $s_i \geq s_{k_i}$. But s_{k_i} has strictly more terms than s_i, and the additional terms are all strictly positive (by a key hypothesis of the lemma).

Thus there must be some j for which $a_i^j > a_{k_i}^j$. Now the assumption (†) implies $j > i$, so that

$$a_i^i \geq a_i^j > a_{k_i}^j \geq a_{k_i}^{k_i}.$$

This is true for all i, so if we let $d_1 = 1$ and $d_{i+1} = k_{d_i}$, we have

$$a_{d_1}^{d_1} > a_{d_2}^{d_2} > a_{d_3}^{d_3} > \cdots$$

contradicting the assumption that $\mathcal{A}$ is well-ordered. □

Proof of Theorem 3.21. Fix an infinitesimal $\zeta \in \mathbf{SN}$. We must show that the sequence

$$\sum_{n=0}^{\infty} \zeta^n c_n$$

converges, i.e., that

$$\sum_{n=0}^{\infty} \pi_y(\zeta^n) c_n$$

converges for all $y \in \mathbf{SN}$.

Let $\mathcal{A} = \operatorname{supp}(\zeta)$ and let $\mathcal{S}$ be the set of all finite sums of elements of $\mathcal{A}$. Then $\operatorname{supp}(\zeta^n) \subset \mathcal{S}$ for all n, by Theorem 3.11.

Now $\mathcal{A}$ is anti-well-ordered and (since ζ is infinitesimal) consists entirely of negative numbers. By Lemma 3.22, $\mathcal{S}$ is also anti-well-ordered, so we can write it as a decreasing sequence

$$\mathcal{S} = \{y_\xi : \xi < \alpha\}.$$

Moreover, Lemma 3.22 shows that each element of $\mathcal{S}$ can be expressed as a sum from $\mathcal{A}$ in just finitely many ways. So for fixed y, there are at most finitely many n with $\pi_y(\zeta^n) \neq 0$, so each sequence $\sum \pi_{y_\xi}(\zeta^n) c_n$ is trivially convergent. Let r_ξ be its limit. Then

$$f(\zeta) = \sum_{\xi < \alpha} \omega^{y_\xi} r_\xi.$$

□

Algebraic Properties of SN

We'll now show that **SN** is a real-closed Field: every odd-degree polynomial over **SN** has a root in **SN**. This remarkable result provides a complete characterization of the algebraic structure of **SN**. The following is an easy partial result.

Theorem 3.23. *Every $x > 0$ has a positive n^{th} root in* **SN**, *for all $n \geq 1$.*

Proof. By factoring out the highest-order term in the normal form of x, we can write

$$x = \omega^y r(1 + \zeta),$$

with $r \in \mathbb{R}$ and ζ an infinitesimal. Now let $f(t)$ be the Taylor series expansion of $(1+t)^{1/n}$. By Theorem 3.21, $f(\zeta)$ converges, so that

$$\omega^{y/n} r^{1/n} f(\zeta)$$

is an nth root of x. □

The generalization to arbitrary polynomials requires a multivariate version of Theorem 3.21:

Theorem 3.24. *Let f be a formal power series with real coefficients,*

$$f(x_1, \ldots, x_k) \in \mathbb{R}[[x_1, \ldots, x_k]].$$

Then $f(\zeta_1, \ldots, \zeta_k)$ converges absolutely for every choice of infinitesimals $\zeta_1, \ldots, \zeta_k$.

Proof. The proof is similar to Theorem 3.21. Let

$$\mathcal{A} = \bigcup_{i=1}^{k} \operatorname{supp}(\zeta_i).$$

If $\pi_y(\zeta_1^{n_1} \zeta_2^{n_2} \cdots \zeta_k^{n_k}) \neq 0$ for some $n_1, \ldots, n_k$, then y can be expressed as a finite sum of elements of $\mathcal{A}$. So by Lemma 3.22, at most finitely many monomials in $\zeta_1, \ldots, \zeta_k$ contain y in their support. The rest of the proof is just like Theorem 3.21. □

Theorem 3.25. *Every odd-degree polynomial with coefficients in* **SN** *has a root in* **SN**.

Proof. Write

$$f(x) = x^n + \sum_{i-0}^{n-1} x^i c_i, \quad n \text{ odd}.$$

By a suitable translation $x \mapsto x + t$, we may assume that $c_{n-1} = 0$. If all the remaining coefficients are also 0, then $f(x)$ has the trivial root $x = 0$. Otherwise, by a further transformation of the form $x \mapsto ux$, we can assume that $\max_i |c_i| = 1$. Then f has the form

$$f(x) = x^n + \sum_{i=0}^{n-2} x^i (r_i + \zeta_i),$$

with each ζ_i infinitesimal, each $r_i \in \mathbb{R}$, and at least one $r_i \neq 0$.

Now let

$$g(x) = x^n + \sum_{i=0}^{n-2} x^i r_i.$$

Certainly $g(x)$ has a root $a \in \mathbb{R}$ (since $\mathbb{R}$ is real-closed), so we can write

$$(\dagger) \qquad g(x) = g_1(x)g_2(x),$$

with $g_1(x) = (x-a)^m$ and a not a root of $g_2(x)$. Since $c_{n-1} = 0$, it cannot be the case that $g(x) = (x-a)^m$, so $\deg(g_2) \geq 1$.

Next consider the ring $\mathcal{R} = \mathbb{R}[\zeta_0, \ldots, \zeta_{n-2}]$. The completion of $\mathcal{R}$ over the ideal $(\zeta_0, \ldots, \zeta_{n-2})$ is the ring

$$\hat{\mathcal{R}} = \mathbb{R}[[\zeta_0, \ldots, \zeta_{n-2}]]$$

consisting of all formal power series in the variables $\zeta_0, \ldots, \zeta_{n-2}$ over $\mathbb{R}$. By classical results in commutative algebra (cf. Hensel's Lemma [**Eis95**]), the factorization $(\dagger)$ of g lifts to a factorization of f,

$$f(x) = f_1(x)f_2(x),$$

such that $\deg(f_i) = \deg(g_i)$, and the coefficients of f_i are in $\hat{\mathcal{R}}$. But $\zeta_0, \ldots, \zeta_{n-2}$ are infinitesimals, so by Theorem 3.24, all their power series converge. Therefore the coefficients of f_i are in **SN**.

Since f has odd degree, one of f_1 or f_2, say f_1, must have odd degree. But $1 \leq \deg(f_1) < n$, so by induction f_1 has a root in **SN**. Therefore f has a root in **SN**. □

Exercises

3.1 Express $\sqrt[3]{\omega+1}$ and $1/\sqrt[3]{\omega+1}$ in normal form.

3.2 **SN**$[i]$ is (up to isomorphism) the unique algebraically closed Field whose domain is a proper Class. (Here $\mathbf{SN}[i] \cong \mathbf{SN}[x]/(x^2+1)$.)

3.3 *Universal embedding property for totally ordered fields.* Suppose $\mathcal{F}$ is a totally ordered field (whose domain is a set) and let $\mathcal{E} < \mathcal{F}$ be a subfield. Then for any embedding $f : \mathcal{E} \to \mathbf{SN}$ of totally ordered fields, there is an embedding $g : \mathcal{F} \to \mathbf{SN}$ extending f.

3.4 *Universal embedding property for fields.* Suppose $\mathcal{F}$ is a field (whose domain is a set) and let $\mathcal{E} < \mathcal{F}$ be a subfield. Then for any embedding $f : \mathcal{E} \to \mathbf{SN}[i]$ of fields, there is an embedding $g : \mathcal{F} \to \mathbf{SN}[i]$ extending f.

3.5 *Surreal number theory.* x is an **omnific integer** if $x = \{x-1 \mid x+1\}$. The Class of omnific integers is denoted by **OZ**.

(a) The following are equivalent, for all $x \in \mathbf{SN}$: (i) $x \in \mathbf{OZ}$; (ii) $\pi_0(x)$ is an integer, and $\pi_y(x) = 0$ for all $y < 0$; (iii) $\sigma_x(\alpha) = \sigma_x(\alpha+1)$ for all $\alpha \in \mathbf{ON}$.

(b) **OZ** is a Subring of **SN**, and every $x \in \mathbf{SN}$ is the quotient of two omnific integers.

(c) *Euclidean algorithm.* For any $a, b \in \mathbf{OZ}$ with $b > 0$, there are unique $q, r \in \mathbf{OZ}$ with $a = bq + r$ and $0 \le r < b$.
(d) $(\omega, \omega\sqrt{2})$ is a nonprincipal ideal of $\mathbf{OZ}$.

3.6 A surreal number x is a **generalized epsilon number** if $\omega^x = x$ (cf. Definition B.3.16 on page 468). x is **reducible** if all the exponents in its normal form have strictly smaller birthday than x, and **irreducible** otherwise.
(a) The following are equivalent: (i) x is irreducible; (ii) x is a generalized epsilon number; (iii) $\mathrm{b}(x)$ is an epsilon number.
(b) For $y \in \mathbf{SN}$, define $\epsilon(y)$ recusively by
$$\epsilon(y) = \Big\{\alpha,\ \epsilon(y^L)+1,\ \omega^{\epsilon(y^L)+1},\ \omega^{\omega^{\epsilon(y^L)+1}},\ \dots \ \Big| \ \epsilon(y^R)-1,\ \omega^{\epsilon(y^R)-1},\ \omega^{\omega^{\epsilon(y^R)-1}},\ \dots\Big\}$$
where α ranges over all ordinals $< \epsilon_0$. Prove that x is a generalized epsilon number iff $x = \epsilon(y)$ for some y, and show that $\mathrm{b}(x) = \epsilon(\mathrm{b}(y))$.
(c) x is a **quasi-epsilon number** if $\omega^{-x} = x$. Show that every quasi-epsilon number is infinitesimal, and find a recursive enumeration $\epsilon^{-1}(y)$ of the quasi-epsilon numbers (analogous to $\epsilon(y)$ for ordinary epsilon numbers). What can we say about $\mathrm{b}(x)$?

Notes

Most of the material in this section is due to Conway, and our treatment largely recapitulates ONAG, with substantial elaboration. The proofs of Theorem 3.11 and Lemmas 3.18 and 3.22 (mostly glossed over by Conway) are due to Gonshor [**Gon86**].

Further progress has been sporadic. Notably, Martin Kruskal found a suitable definition for $\exp(x)$ that has the expected analytic properties, which was further explored by Gonshor (see [**Gon86**] for discussion). Writing $\log(x)$ for its inverse, one can then make the usual definition $x^y = \exp(y \times \log(x))$. Note that this is *not* the same notion of exponential as the ω-power used in the normal form.

Gonshor has also made the following conjecture:

Conjecture. $\mathrm{b}(x \times y) \le \mathrm{b}(x) \times \mathrm{b}(y)$ for all $x, y \in \mathbf{SN}$.

Although the analogue for addition ($\mathrm{b}(x + y) \le \mathrm{b}(x) + \mathrm{b}(y)$) is completely trivial, Gonshor's conjecture has proved to be surprisingly resistant to attack, and the best-known bounds for $\mathrm{b}(xy)$ are quite poor. Ehrlich and van den Dries have made partial progress, by proving Gonshor's conjecture in the case where x and y are both of the form $\omega^z \times r$ [**vdDE01**].

In the same paper, Ehrlich and van den Dries showed that $\mathcal{O}_\alpha$ is a ring if and only if $\alpha = \omega^{\omega^\beta}$ for some β; and a field if and only if α is an epsilon number. Moreover, each such field is closed under Kruskal–Gonshor exponentiation. The proofs of these results are surprisingly intricate, despite their deceptive similarity to Exercise 2.2 (which is elementary).

Ehrlich has also made notable progress in understanding the universality properties of surreals; his recent comprehensive survey [**Ehr12**] is a good entrypoint into the subject. [**Ehr94**] is another substantial (though older) reference.

We conclude with perhaps the most important open problem in the theory of surreal numbers:

Question. Does there exist a reasonable general definition of surreal integration?

Exercise 3.6(c): Gonshor.

4. Transfinite Nim Arithmetic

Recall (from Section 1) that **IG** is the Class of (transfinite) impartial games, and in particular the **transfinite nimbers** are defined recursively by

$$*\alpha = \{*\beta \mid *\beta\}_{\beta<\alpha} \quad \text{for all } \alpha \in \mathbf{ON}.$$

The Sprague–Grundy Theorem and its proof generalize verbatim: if G is a long impartial game, then $G = *\alpha$ for some ordinal α. We write $\mathscr{G}(G) = \alpha$, the (transfinite) **nim value** of G, which can be described recursively:

$$\mathscr{G}(G) = \operatorname*{mex}_{G' \in G} \mathscr{G}(G').$$

The transfinite **nim-sum** is given by

$$\alpha \oplus \beta = \operatorname*{mex}_{\substack{\alpha'<\alpha \\ \beta'<\beta}} \left(\alpha' \oplus \beta,\ \alpha \oplus \beta'\right)$$

(cf. Proposition IV.1.4 on page 181), so that

$$\mathscr{G}(G+H) = \mathscr{G}(G) \oplus \mathscr{G}(H).$$

Then just as in the finite case, we have the *nim-addition rule*,

$$*\alpha + *\beta = *\gamma, \quad \text{where} \quad \gamma = \alpha \oplus \beta.$$

The nim-sum of ordinal 2-powers behaves just like the nim-sum of integer 2-powers:

$$2^\alpha \oplus 2^\beta = \begin{cases} 0 & \text{if } \alpha = \beta; \\ 2^\alpha + 2^\beta & \text{if } \alpha \neq \beta. \end{cases}$$

This can be proved directly by generalizing Bouton's Theorem, and it will also emerge as a byproduct of the succeeding analysis. (Here 2^α denotes *ordinal* exponentiation, as described in Section B.3; thus, for example, $2^\omega = \omega$.)

Now any ordinal can be written as the sum of ordinal 2-powers (cf. Theorem B.3.14 on page 468), so this observation completely describes the behavior of the transfinite nim-sum operation. To determine the sum $\alpha \oplus \beta$, we can express α and β as sums of 2-powers and "add without carrying"—canceling out the common exponents and adding up the remaining ones using standard ordinal arithmetic. This shows that as a Group, **IG** has the structure of $\mathbb{Z}_2^{\mathbf{ON}}$, the direct sum of a proper Class of copies of $\mathbb{Z}_2$.

The proofs of all these assertions are identical to the corresponding finite proofs (Theorem I.1.1, Theorem IV.1.3, etc.), making the additive theory of **IG** a rather trivial extension of the finite case. What makes **IG** special is that nim-*product* generalizes too, and the resulting Field structure is utterly unlike anything else in the theory of combinatorial games.

Nim Product and the Field $\mathbf{ON}_2$

If G and H are impartial, then Definition 2.1 (page 412) reduces to

$$G \times H = \{G' \times H + G \times H' + G' \times H' : G' \in G \text{ and } H' \in H\},$$

where just as in Chapter IV we identify G with the set of its options, and likewise for H and $G \times H$. The *nim-multiplication rule* follows:

$$*\alpha \times *\beta = *\gamma, \quad \text{where} \quad \gamma = \alpha \otimes \beta,$$

where the product operation $\otimes$ is defined like so:

Definition 4.1. The (transfinite) **nim-product** $\alpha \otimes \beta$ is given by

$$\alpha \otimes \beta = \operatorname*{mex}_{\substack{\alpha' < \alpha \\ \beta' < \beta}} \big((\alpha' \otimes \beta) \oplus (\alpha \otimes \beta') \oplus (\alpha' \otimes \beta')\big).$$

Remarkably, this definition is identical to the definition of nim-product over $\mathbb{N}$ from Section IV.5. There we showed that $(\mathbb{N}, \oplus, \otimes)$ is a quadratically closed field of characteristic 2. The expanded structure $(\mathbf{ON}, \oplus, \otimes)$ is denoted $\mathbf{ON}_2$, and we'll now show that it's an algebraically closed Field. Just as in the finite case, the proof that $\mathbf{ON}_2$ is a commutative Ring is a routine induction.

Lemma 4.2. $(\mathbf{ON}, \oplus, \otimes)$ *is a commutative Ring with identity* 1. *Moreover, if* $\alpha \otimes \beta = 0$, *then* $\alpha = 0$ *or* $\beta = 0$.

Proof. Identical to Lemmas IV.5.4 and IV.5.5 on page 216. □

Proceeding as in Section IV.5, denote by $\mathcal{P}_\gamma$ the set of all predecessors of γ, for each $\gamma \in \mathbf{ON}$:[3]

$$\mathcal{P}_\gamma = \{\alpha \in \mathbf{ON} : \alpha < \gamma\}.$$

We'll show that each ordinal γ is a solution to the *algebraically simplest* equation over $\mathcal{P}_\gamma$ that is not satisfied by any element of $\mathcal{P}_\gamma$. By algebraically

[3]In set theory, it is customary to *define* each ordinal γ as the set of its predecessors. Therefore $\gamma = \mathcal{P}_\gamma$ and, strictly speaking, there is no need for the extra bit of notation. To reduce confusion, however, we'll stick to the notation "γ" when we intend to treat γ as an individual, and "$\mathcal{P}_\gamma$" when we intend to discuss the set of its predecessors.

simplest, we mean the earliest of the following operations: sum; product; inverse; root of polynomial. Stated precisely, we have the following extraordinary theorem of Conway.

Theorem 4.3 (The Simplest Extension Theorem). *Let* $\gamma \in \mathbf{ON}$.

(a) *If* $\mathcal{P}_\gamma$ *is not a group (under* $\oplus$*), then* $\gamma = \alpha \oplus \beta$*, where* (α, β) *is the lexicographically least pair of ordinals with* $\alpha \oplus \beta \notin \mathcal{P}_\gamma$.

(b) *If* $\mathcal{P}_\gamma$ *is a group but not a ring, then* $\gamma = \alpha \otimes \beta$*, where* (α, β) *is the lexicographically least pair of ordinals with* $\alpha \otimes \beta \notin \mathcal{P}_\gamma$.

(c) *If* $\mathcal{P}_\gamma$ *is a ring but not a field, then* $\gamma \otimes \alpha = 1$*, where* α *is the least nonzero ordinal with no nim-inverse in* $\mathcal{P}_\gamma$.

(d) *If* $\mathcal{P}_\gamma$ *is a field but not algebraically closed, then* γ *is a nim-root of the lexicographically least polynomial over* $\mathcal{P}_\gamma$ *with no nim-root in* $\mathcal{P}_\gamma$. *(Higher-degree coefficients are higher-order bits in the lexicographic ordering of polynomials.)*

(e) *If* $\mathcal{P}_\gamma$ *is an algebraically closed field, then* γ *is transcendental over* $\mathcal{P}_\gamma$.

We'll prove Theorem 4.3 in tandem with the following lemma. As in Section IV.5, we denote by $\alpha \oslash \beta$ the unique ordinal γ such that $\beta \otimes \gamma = \alpha$. Likewise, we denote by $\alpha^{\textcircled{n}}$ the nth power of α in $\mathbf{ON}_2$:

$$\alpha^{\textcircled{n}} = \overbrace{\alpha \otimes \alpha \otimes \cdots \otimes \alpha}^{n \text{ times}}.$$

Lemma 4.4. *Let* $\gamma \in \mathbf{ON}$.

(a′) *Suppose that* $\mathcal{P}_\gamma$ *is a group. Then* $(\gamma \times \beta) \oplus \alpha = (\gamma \times \beta) + \alpha$ *for all* $\beta \in \mathbf{ON}$ *and all* $\alpha < \gamma$.

(b′) *Suppose that* $\mathcal{P}_\gamma$ *is a ring and that for some* $\delta \leq \gamma$, $\mathcal{P}_\delta$ *is a group and every nonzero* $\alpha \in \mathcal{P}_\delta$ *has a nim-inverse in* $\mathcal{P}_\gamma$. *Then* $\gamma \otimes \alpha = \gamma \times \alpha$ *for all* $\alpha < \delta$.

(c′) *Suppose that* $\mathcal{P}_\gamma$ *is a field and that every polynomial of degree* $\leq n$ *has a nim-root in* $\mathcal{P}_\gamma$. *Then*

$$\bigoplus_{i=0}^{n} (\gamma^{\textcircled{i}} \otimes \alpha_i) = \sum_{i=0}^{n} (\gamma^i \times \alpha_i)$$

for all $\alpha_0, \ldots, \alpha_n < \gamma$.

Proof of Theorem 4.3 and Lemma 4.4.

(a) For every $\alpha' < \alpha$ and $\beta' < \beta$, we have $\alpha' \oplus \beta < \gamma$ and $\alpha \oplus \beta' < \gamma$, by choice of (α, β). This shows that $\alpha \oplus \beta \leq \gamma$, and since $\alpha \oplus \beta \notin \mathcal{P}_\gamma$, we have equality.

(a′) Each ordinal less than $\gamma \times \beta$ has the form $(\gamma \times \beta') + \gamma'$, with $\beta' < \beta$ and $\gamma' < \gamma$. By induction on β, this is equal to $(\gamma \times \beta') \oplus \gamma'$, so that

$$(\gamma \times \beta) \oplus \alpha = \operatorname*{mex}_{\alpha',\beta',\gamma'} \big((\gamma \times \beta') \oplus \gamma' \oplus \alpha,\ (\gamma \times \beta) \oplus \alpha'\big).$$

But $\mathcal{P}_\gamma$ is a group, so as γ' ranges over $\mathcal{P}_\gamma$, the nim-sum $\gamma' \oplus \alpha$ also ranges over $\mathcal{P}_\gamma$. Therefore

$$(\gamma \times \beta) \oplus \alpha = \operatorname*{mex}_{\alpha',\beta',\gamma'} \big((\gamma \times \beta') \oplus \gamma',\ (\gamma \times \beta) \oplus \alpha'\big).$$

By induction, we can rewrite the excludents as $(\gamma \times \beta') + \gamma'$ and $(\gamma \times \beta) + \alpha'$, which together range precisely over the ordinals less than $(\gamma \times \beta) + \alpha$.

(b) For every $\alpha' < \alpha$ and $\beta' < \beta$, we have

$$\alpha' \otimes \beta \in \mathcal{P}_\gamma \quad \text{and} \quad \alpha \otimes \beta' \in \mathcal{P}_\gamma \quad \text{and} \quad \alpha' \otimes \beta' \in \mathcal{P}_\gamma,$$

by choice of (α, β). Therefore since $\mathcal{P}_\gamma$ is a group, it follows that

$$(\alpha' \otimes \beta) \oplus (\alpha \otimes \beta') \oplus (\alpha' \otimes \beta') \in \mathcal{P}_\gamma.$$

This shows that $\alpha \otimes \beta \leq \gamma$, and since $\alpha \otimes \beta \notin \mathcal{P}_\gamma$, we have equality.

(b′) Here

$$\begin{aligned} \gamma \otimes \alpha &= \operatorname*{mex}_{\substack{\gamma'<\gamma \\ \alpha'<\alpha}} (\gamma \otimes \alpha') \oplus (\gamma' \otimes \alpha) \oplus (\gamma' \otimes \alpha') \\ &= \operatorname*{mex}_{\substack{\gamma'<\gamma \\ \alpha'<\alpha}} (\gamma \otimes \alpha') \oplus \big(\gamma' \otimes (\alpha \oplus \alpha')\big). \end{aligned}$$

Now since $\alpha' < \alpha < \delta$ and $\mathcal{P}_\delta$ is a group, we have $\alpha \oplus \alpha' \in \mathcal{P}_\delta$. By hypothesis, this implies that $\alpha \oplus \alpha'$ has a nim-inverse in $\mathcal{P}_\gamma$. So as γ' ranges over $\mathcal{P}_\gamma$, the expression

$$\gamma' \otimes (\alpha \oplus \alpha')$$

also ranges over $\mathcal{P}_\gamma$ (for any fixed value of α'). Therefore

$$\gamma \otimes \alpha = \operatorname*{mex}_{\substack{\gamma'<\gamma \\ \alpha'<\alpha}} (\gamma \otimes \alpha') \oplus \gamma'.$$

By induction and (a′), we can rewrite each excludent as $(\gamma \times \alpha') + \gamma'$, which is the typical ordinal less than $\gamma \times \alpha$.

(c) Fix the largest $\delta \leq \alpha$ such that $\mathcal{P}_\delta$ is a group. (Such a "largest" value of δ must exist, since the union of groups is also a group.) Since $\mathcal{P}_{\delta \times 2}$ is a group, we then have $\delta \leq \alpha < \delta \times 2$, so that by the nim-addition rule $\alpha \oplus \delta < \delta$. Put $\beta = \alpha \oplus \delta$.

We'll show that

$$\gamma \otimes \delta = (\gamma \otimes \beta) \oplus 1$$

which certainly implies that $\gamma \otimes \alpha = 1$. Now from (a′) and (b′), we have $(\gamma \otimes \beta) \oplus 1 = (\gamma \times \beta) + 1$, so it suffices to show that

$$\begin{aligned}(\gamma \times \beta) + 1 &= \gamma \otimes \delta \\ &= \operatorname*{mex}_{\substack{\gamma'<\gamma \\ \delta'<\delta}} (\gamma \otimes \delta') \oplus (\gamma' \otimes \delta) \oplus (\gamma' \otimes \delta') \\ (\dagger) \qquad &= \operatorname*{mex}_{\substack{\gamma'<\gamma \\ \delta'<\delta}} (\gamma \otimes \delta') \oplus \big(\gamma' \otimes (\delta \oplus \delta')\big).\end{aligned}$$

We first show that all ordinals less than $\gamma \times \beta$ are excludents in the expression (†). Such an ordinal has the form $(\gamma \times \beta') + \gamma'$, for some $\beta' < \beta$ and $\gamma' < \gamma$. Since $\beta' < \beta < \delta$, we have

$$\delta \oplus \beta' = \delta + \beta' < \delta + \beta = \alpha.$$

Therefore $\delta \oplus \beta'$ is invertible in $\mathcal{P}_\gamma$, and we can write

$$\gamma^* = \gamma' \oslash (\delta \oplus \beta') \quad \text{for some} \quad \gamma^* \in \mathcal{P}_\gamma.$$

But this shows that

$$\begin{aligned}(\gamma \times \beta') + \gamma' &= (\gamma \otimes \beta') \oplus \gamma' && \text{(applying (a′) and (b′))} \\ &= (\gamma \otimes \beta') \oplus \big(\gamma^* \otimes (\delta \oplus \beta')\big),\end{aligned}$$

which has the form of an excludent for (†).

Likewise, $\gamma \times \beta = \gamma \otimes \beta$ is an excludent, since we can take $\gamma' = 0$ and $\delta' = \beta$ in the expression (†).

Finally, $(\gamma \times \beta) + 1 = (\gamma \otimes \beta) \oplus 1$ itself is *not* an excludent: this would imply $\delta' = \beta$, and therefore $\gamma' \otimes (\delta \oplus \beta) = 1$, a contradiction since $\delta \oplus \beta = \alpha$.

(c′) We first show that $\gamma^{\textcircled{n}} = \gamma^n$. If $\alpha < \gamma^n$, then by considering the ordinal expansion of α in base γ, we can write

$$\alpha = \sum_{i=0}^{n-1} (\gamma^i \times \alpha_i), \quad \text{with each } \alpha_i \in \mathcal{P}_\gamma.$$

By induction on n, we have

$$\alpha = \bigoplus_{i=0}^{n-1} (\gamma^{\textcircled{i}} \otimes \alpha_i).$$

Therefore

$$\alpha = \gamma^{\textcircled{n}} \oplus f(\gamma),$$

where f is a monic polynomial of degree n with coefficients in $\mathcal{P}_\gamma$. By hypothesis, every such polynomial splits in $\mathcal{P}_\gamma$, so we can write

$$\alpha = \gamma^{\textcircled{n}} \oplus \bigotimes_{j=1}^{n} (\gamma \oplus \beta_j),$$

in which $\beta_1, \ldots, \beta_n$ are the nim-roots of f in $\mathcal{P}_\gamma$. But this is exactly the form of a typical excludent for $\gamma^{\boxed{n}}$ (this can be seen by inductively expanding the definition of $\otimes$). This shows that every ordinal $\alpha < \gamma^n$ is an excludent for $\gamma^{\boxed{n}}$. The same chain of reasoning in reverse shows that γ^n cannot itself be an excludent, establishing that $\gamma^{\boxed{n}} = \gamma^n$.

To complete the proof of (c′), it suffices to show that $\gamma^{\boxed{n}} \otimes \alpha = \gamma^n \times \alpha$ for any $\alpha < \gamma$ (since we can then apply (a′) to obtain the full expression). The argument is similar to (b′). Write

$$\gamma^{\boxed{n}} \otimes \alpha = \gamma^n \otimes \alpha = \operatorname*{mex}_{\substack{\gamma' < \gamma^n \\ \alpha' < \alpha}} (\gamma^n \otimes \alpha') \oplus \big(\gamma' \otimes (\alpha \oplus \alpha')\big).$$

Since $\alpha' < \alpha < \gamma$ and $\mathcal{P}_\gamma$ is a field, we know that $\alpha \oplus \alpha'$ has a nim-inverse in $\mathcal{P}_\gamma$. Moreover, $\mathcal{P}_{\gamma^n}$ is closed under nim-multiplication by ordinals $< \gamma$ (by induction on n). It follows that as γ' ranges over $\mathcal{P}_{\gamma^n}$, the expression

$$\gamma' \otimes (\alpha \oplus \alpha')$$

also ranges over $\mathcal{P}_{\gamma^n}$ (for any fixed value of α'). Therefore

$$\gamma^n \otimes \alpha = \operatorname*{mex}_{\substack{\gamma' < \gamma^n \\ \alpha' < \alpha}} (\gamma^n \otimes \alpha') \oplus \gamma'.$$

By induction and (a′), we can rewrite each excludent as $(\gamma^n \times \alpha') + \gamma'$, which is the typical ordinal less than $\gamma^n \times \alpha$.

(d) Let f be the lexicographically earliest polynomial over $\mathcal{P}_\gamma$ with no nim-root in $\mathcal{P}_\gamma$. Then f must be a monic polynomial, so we can write

$$f(x) = x^{\boxed{n}} \oplus g(x)$$

with $\deg(g) < n$. We will show that every $\alpha < g(\gamma)$ is an excludent for $\gamma^{\boxed{n}}$, but $g(\gamma)$ itself is not, so that necessarily $\gamma^{\boxed{n}} = g(\gamma)$. Write

$$g(x) = \bigoplus_{i=0}^{n-1} \big(x^{\boxed{i}} \otimes \alpha_i\big), \quad \text{each } \alpha_i \in \mathcal{P}_\gamma.$$

Then by (c′), we have

$$g(\gamma) = \bigoplus_{i=0}^{n-1} \big(\gamma^{\boxed{i}} \otimes \alpha_i\big) = \sum_{i=0}^{n-1} \big(\gamma^i \times \alpha_i\big).$$

Moreover, if g' is a polynomial earlier than g, then also by (c′)

$$g'(\gamma) = \bigoplus_{i=0}^{n-1} \big(\gamma^{\boxed{i}} \otimes \alpha'_i\big) = \sum_{i=0}^{n-1} \big(\gamma^i \times \alpha'_i\big).$$

Regarding these expressions as ordinal expansions in base γ, we see that as g' ranges over polynomials earlier than g, the values $g'(\gamma)$ range over ordinals less than $g(\gamma)$. Thus it suffices to show that every such $g'(\gamma)$ is an excludent for $\gamma^{\textcircled{n}}$, but $g(\gamma)$ itself is not.

But the excludents for $\gamma^{\textcircled{n}}$ have the form

$$\gamma^{\textcircled{n}} \oplus \bigotimes_{j=1}^{n} (\gamma \oplus \beta_j), \quad \text{each } \beta_j \in \mathcal{P}_\gamma.$$

(As in (c′), this can be seen by inductively expanding the definition of $\otimes$.) Equivalently, they have the form

$$\gamma^{\textcircled{n}} \oplus f'(\gamma),$$

where f' is a polynomial of degree n that splits in $\mathcal{P}_\gamma$. But the hypotheses of the theorem imply that all polynomials earlier than f split in $\mathcal{P}_\gamma$, but f itself does not.

(e) This is automatic, since γ lies outside the algebraically closed subfield $\mathcal{P}_\gamma$.

□

The Algebraic Closure of $\mathcal{P}_2$

The Simplest Extension Theorem can be used to investigate algebraic extensions of the field $\mathcal{P}_2 = \{0, 1\}$. First note that if $\mathcal{P}_\alpha$ is a group, then the next group is $\mathcal{P}_{\alpha\times 2}$. Since the union of groups is a group, the first few groups are

$$\mathcal{P}_2,\ \mathcal{P}_4,\ \mathcal{P}_8,\ \mathcal{P}_{16},\ \ldots,\ \mathcal{P}_\omega,\ \mathcal{P}_{\omega\times 2},\ \mathcal{P}_{\omega\times 4},\ \ldots,\ \mathcal{P}_{\omega^2},\ \mathcal{P}_{\omega^2\times 2},\ \ldots$$

where the subscripts range over ordinals of the form 2^β.

Now if $\mathcal{F}$ is an algebraic field extension of $\mathcal{P}_2$, then every element of $\mathcal{F}$ has finite multiplicative order, and it follows that every subring of $\mathcal{F}$ is a field. So below the first transcendental element, each ring extension of $\mathcal{P}_2$ will already be a field extension.

Moreover, if $\alpha \in \mathcal{F}$, then α generates a finite field $\mathcal{E} = \mathcal{P}_2(\alpha) < \mathcal{F}$, and since $\mathcal{E}$ is finite, it must contain a square nim-root for α. This shows that every algebraic extension of $\mathcal{P}_2$ is perfect (that is, contains square nim-roots for all its elements), so each such extension already has nim-roots for polynomials of the form $x^{\textcircled{2}} \oplus \alpha$. The next simplest candidates for irreducible polynomials are given by

$$f_\alpha(x) = x^{\textcircled{2}} \oplus x \oplus \alpha \quad (\alpha \geq 1).$$

Now $f_1(0) = f_1(1) = 1$, so in fact f_1 has no nim-root in $\mathcal{P}_2$. By the Simplest Extension Theorem, 2 must be a nim-root of f_1, and we have $2^{\textcircled{2}} = 2 \oplus 1 = 3$, giving the next field $\mathcal{P}_4$.

Then as α ranges over $\mathcal{P}_4$, the values $\alpha^{②} \oplus \alpha$ range over $\{0,1\}$ (by inspection), so f_2 has no nim-root in $\mathcal{P}_4$. By the Simplest Extension Theorem, 4 is a nim-root of f_2, so that $4^{②} = 4 \oplus 2 = 6$, generating the new field $\mathcal{P}_{16}$.

Continuing like this, suppose that $\mathcal{P}_n$ is a finite field, with $n = 2^{2^k}$. Then by Lemma IV.5.6(e) (page 217), $\mathcal{P}_n$ has nim-roots for f_α for all $\alpha < n/2$, but not for $f_{n/2}$. So by the Simplest Extension Theorem, $n^{②} \oplus n = n/2$, and we obtain the identity

$$n^{②} = \tfrac{3}{2}n \quad \text{for } n \text{ of the form } 2^{2^k}.$$

It follows that $\mathcal{P}_\omega$ has nim-roots for every f_α. It is therefore quadratically closed, since any polynomial of the form $x^{②} \oplus (\eta \otimes x) \oplus \alpha$ can be solved by the transformation $x = \eta \otimes y$.

Now the fact that $\mathcal{P}_\omega$ is a quadratically closed field is not new: we proved it back in Section IV.5, along with the identity $n^{②} = \frac{3}{2}n$ for Fermat 2-powers n, by a direct combinatorial argument. The Simplest Extension Theorem gives an alternate proof, one that is more elegantly algebraic and that can be extended far into the transfinite.

For example, consider the simplest extension of $\mathcal{P}_\omega$. The simplest cubic polynomials have the form

$$g_\alpha(x) = x^{③} \oplus \alpha \quad (\alpha \geq 1).$$

Obviously g_1 has a nim-root in $\mathcal{P}_\omega$. But 2 cannot have a cube nim-root in $\mathcal{P}_\omega$, since any such root would have multiplicative order 9, and 9 is not a divisor of any integer of the form $2^{2^k} - 1$. So g_2 has no nim-root in $\mathcal{P}_\omega$, and by the Simplest Extension Theorem it follows that $g_2(\omega) = 0$. We arrive at the remarkable identity

$$\omega^{③} = 2.$$

Certainly $\omega^{②} \otimes n = \omega^2 \times n$ for all $n < \omega$, by Lemma 4.4(c′), so the next field is $\mathcal{P}_{\omega^3}$. Now $\mathcal{P}_{\omega^3}$ is quadratically closed: if $\mathcal{F}$ is any quadratic extension of $\mathcal{P}_2(\omega)$, then $\mathcal{F}$ is a finite field extension of $\mathcal{P}_2$, so its Galois group over $\mathcal{P}_2$ is Abelian. Therefore $\mathcal{F} = \mathcal{E}(\omega)$ for some quadratic extension $\mathcal{E}$ of $\mathcal{P}_2$. But $\mathcal{E}$ is isomorphic to a subfield of $\mathcal{P}_\omega$ (since $\mathcal{P}_\omega$ is quadratically closed), so $\mathcal{F}$ is isomorphic to a subfield of $\mathcal{P}_{\omega^3}$.

Moreover $\mathcal{P}_{\omega^3}$ contains cube nim-roots for all $n < \omega$, so the simplest irreducible polynomial is

$$g_\omega(x) = x^{③} \oplus \omega,$$

so by the Simplest Extension Theorem

$$(\omega^3)^{③} = \omega.$$

By the same argument we obtain a sequence of cubic extensions at steps

$$2,\ \omega,\ \omega^3,\ \omega^{3^2},\ \omega^{3^3},\ \ldots,$$

in which each successive term is a cube nim-root of the previous one. The union of these extensions is $\mathcal{P}_{\omega^\omega}$, in which every element has a cube nim-root.

In fact $\mathcal{P}_{\omega^\omega}$ is cubically closed (we'll prove this in Theorem 4.5 below), so the next extensions are by fifth nim-roots. Now $2^{⑤} = 3$ and $3^{⑤} = 2$, so $x^{⑤} \oplus 2$ and $x^{⑤} \oplus 3$ already have solutions. But 4 has multiplicative order 15 (by inspection of $\mathcal{P}_{16}$, Figure IV.5.1 on page 217), so any fifth nim-root of 4 must have order 75, whereas 25 is not a divisor of any integer of the form $2^{2^k} - 1$. Therefore $x^{⑤} \oplus 4$ is irreducible, and by the Simplest Extension Theorem

$$(\omega^\omega)^{⑤} = 4.$$

Then in the sequence

$$4,\ \omega^\omega,\ \omega^{\omega\times 5},\ \omega^{\omega\times 5^2},\ \omega^{\omega\times 5^3},\ \ldots,$$

each successive term is a fifth nim-root of the previous one. The union $\mathcal{P}_{\omega^{\omega^2}}$ has nim-roots for all polynomials of degree 5, and the next sequence is

$$\alpha_7,\ \omega^{\omega^2},\ \omega^{\omega^2\times 7},\ \omega^{\omega^2\times 7^2},\ \omega^{\omega^2\times 7^3},\ \ldots,$$

where α_7 denotes the least element of $\mathcal{P}_{\omega^{\omega^2}}$ with no seventh nim-root. (It's not terribly difficult to compute $\alpha_7 = \omega + 1$; see Exercise 4.1.)

In general, we have the following theorem.

Theorem 4.5. *Let p be the $(k+1)^{st}$ odd prime. Then in the sequence*

$$\alpha_p,\ \omega^{\omega^k},\ \omega^{\omega^k\times p},\ \omega^{\omega^k\times p^2},\ \ldots,\ \omega^{\omega^k\times p^i},\ \ldots,$$

each successive term is a pth nim-root of the previous one. Here $\alpha_p < \omega^{\omega^k}$ denotes the least ordinal with no p^{th} nim-root in $\mathcal{P}_{\omega^{\omega^k}}$.

Moreover, each term of this sequence except for α_p is a field, and their union $\omega^{\omega^{k+1}}$ is a field with nim-roots for all polynomials of degree $\leq p$.

Proof. Let $\gamma = \omega^{\omega^k}$. For convenience, call a field $\mathcal{F}$ **p-ically closed** if $\mathcal{F}$ contains nim-roots for all polynomials of degree $\leq p$. Then supposing (by induction) that the theorem is true for all $p' < p$, we have that $\mathcal{P}_\gamma$ is p'-ically closed. (For the base case $p = 3$, we know that $\mathcal{P}_\omega$ is quadratically closed from the preceding analysis of finite nim arithmetic.)

Therefore $x^{ⓟ} \oplus \alpha_p$ is the simplest irreducible polynomial over $\mathcal{P}_\gamma$, so by the Simplest Extension Theorem,

$$\gamma^{ⓟ} = \alpha_p.$$

By Lemma 4.4(c′), it follows that the next field occurs at stage $\gamma^p = \omega^{\omega^k\times p}$.

We first claim that $\mathcal{P}_{\gamma^p}$ is also p'-ically closed for all $p' < p$. To see this, suppose that $\mathcal{F}$ is any algebraic extension of $\mathcal{P}_2(\gamma)$ by a polynomial of degree p'. Then $\mathcal{F}$ is a finite field extension of $\mathcal{P}_2$, so its Galois group over $\mathcal{P}_2$ is Abelian. Therefore $\mathcal{F} = \mathcal{E}(\gamma)$ for some algebraic extension $\mathcal{E}$ of $\mathcal{P}_2$ by a polynomial of degree p'. But $\mathcal{E}$ is isomorphic to a subfield of $\mathcal{P}_\gamma$ (since $\mathcal{P}_\gamma$ is p'-ically closed), so $\mathcal{F}$ is isomorphic to a subfield of $\mathcal{P}_{\gamma^p} = \mathcal{P}_\gamma(\gamma)$.

Moreover, $\mathcal{P}_{\gamma^p}$ contains p^{th} nim-roots for all elements of $\mathcal{P}_\gamma$, so $x^{\boxed{p}} \oplus \gamma$ is the simplest irreducible polynomial over $\mathcal{P}_{\gamma^p}$. By the Simplest Extension Theorem,

$$(\gamma^p)^{\boxed{p}} = \gamma.$$

Iterating this argument shows that in the sequence

$$\alpha_p,\ \gamma,\ \gamma^p,\ \gamma^{p^2},\ \ldots,\ \gamma^{p^i},\ \ldots,$$

each successive term is a pth nim-root of the previous one. This proves the first part of the theorem.

To complete the proof, we must show that $\mathcal{P}_{\gamma^\omega}$ is p-ically closed. Suppose (for contradiction) that it is not, and let f be the simplest irreducible polynomial over $\mathcal{P}_{\gamma^\omega}$ of degree p. Then the coefficients of f generate a finite field $\mathcal{E} < \mathcal{P}_{\gamma^\omega}$. Since f is irreducible over $\mathcal{P}_{\gamma^\omega}$, there exists an extension $\mathcal{F} > \mathcal{E}$ of degree p such that $\mathcal{F}$ is not isomorphic to any subfield of $\mathcal{P}_{\gamma^\omega}$.

But since p is odd and $\mathcal{E}$ is a finite field of characteristic 2, any degree-p extension of $\mathcal{E}$ can be obtained by a pth nim-root. Since $\mathcal{F}$ is not isomorphic to any subfield $\mathcal{P}_{\gamma^\omega}$, this implies that some $\beta < \gamma^\omega$ has no p^{th} nim-root in $\mathcal{P}_{\gamma^\omega}$, a contradiction. □

As an immediate corrollary, we have the following:

Theorem 4.6. *Let $\tau = \omega^{\omega^\omega}$. Then $\mathcal{P}_\tau$ is an algebraic closure of $\mathcal{P}_2$, and τ is transcendental over $\mathcal{P}_\tau$.*

Proof. By Theorem 4.5, each field $\mathcal{P}_{\omega^{\omega^k}}$ has nim-roots for all polynomials of degree $\leq p$, where p is the $(k+1)^{\text{st}}$ odd prime. Therefore their union $\mathcal{P}_\tau$ is algebraically closed, and the transcendence of τ follows from the Simplest Extension Theorem. □

Nim Arithmetic in $\mathcal{P}_\tau$

Throughout the rest of this section we will write $\tau = \omega^{\omega^\omega}$. We can use Theorem 4.5 to describe the arithmetic structure of $\mathcal{P}_\tau$, which is an elegant generalization of finite nim arithmetic (cf. Section IV.5). Note that for all k,

$$\omega^{\omega^k} = (2^\omega)^{\omega^k} = 2^{\omega^{k+1}}.$$

So every ordinal $\alpha < \tau$ can be written in the form

$$\alpha = 2^{\beta_1} + 2^{\beta_2} + \cdots + 2^{\beta_k}$$

with $\omega^\omega > \beta_1 > \beta_2 > \cdots > \beta_k$. We know how to add the terms of this expression:

$$2^\alpha \oplus 2^\beta = \begin{cases} 0 & \text{if } \alpha = \beta; \\ 2^\alpha + 2^\beta & \text{otherwise.} \end{cases}$$

Multiplication distributes over addition, so to describe the arithmetic of $\mathcal{P}_\tau$, it suffices to show how to multiply individual terms of the form 2^β.

Recall (from Section IV.5) that in the finite case, every integer of the form 2^k can be expressed as a product of Fermat 2-powers. The problem therefore reduces to multiplication of individual Fermat 2-powers. The approach in the transfinite case is similar, with an expanded universe of primitive multiplicands.

Normal form of 2^β. Denote by p_k the $(k+1)^{\text{st}}$ prime, so that $p_0 = 2$, $p_1 = 3$, etc. Then for each prime power $q = p_k^n$, we define

$$\kappa_q = 2^{\omega^k \times p_k^{n-1}}.$$

The ordinals κ_q are the primitive multiplicands in the normal form of 2^α. In particular, note that $\kappa_2 = 2^{2^0}$, $\kappa_4 = 2^{2^1}$, $\kappa_8 = 2^{2^2}$, ..., which is exactly the sequence of Fermat 2-powers. Likewise, when p is the $(k+1)^{\text{st}}$ odd prime, then $p = p_{k+1}$ and we have

$$\kappa_{p^n} = 2^{\omega^{k+1} \times p^{n-1}} = \omega^{\omega^k \times p^{n-1}},$$

which is exactly the sequence of successive p^{th} nim-roots in Theorem 4.5.

Now every $\beta < \omega^\omega$ can be expressed as a polynomial in ω,

$$\beta = \sum_{k=0}^{n} \omega^k \times c_k, \quad \text{each } c_k < \omega.$$

Therefore

$$2^\beta = \prod_{k=0}^{n} 2^{\omega^k \times c_k}.$$

Writing each c_k in base p_k, we can write 2^α in the normal form

$$2^\beta = \prod_q \kappa_q^{a_q}$$

with $a_q < p$ whenever q is a power of p. When an ordinal is so written, Lemma 4.4(c′) shows that nim-product is interchangeable with ordinary product:

$$\prod_q \kappa_q^{a_q} = \bigotimes_q \kappa_q^{\boxed{a_q}}.$$

Therefore to complete the picture of arithmetic in $\mathcal{P}_\tau$, it suffices to describe the nim-products for individual terms $\kappa_q^{\textcircled{a}}$.

Products of $\kappa_q^{\textcircled{a}}$. Let $q = p^n$. For all a and b, we certainly have

$$\kappa_q^{\textcircled{a}} \otimes \kappa_q^{\textcircled{b}} = \kappa_q^{\boxed{a+b}}$$

(by definition). When $a + b < p$ this expression resolves normally to κ_q^{a+b}. The remaining unresolved cases all have $a + b \geq p$ and so depend on the evaluation of terms $\kappa_q^{\textcircled{p}}$. This is given by the following rule.

Theorem 4.7. *Let $q = p^n$, with p prime. Then:*

$$\kappa_q^{\textcircled{p}} = \begin{cases} \kappa_q + \prod_{i=1}^{n-1} \kappa_{2^i} & \text{if } p = 2; \\ \kappa_{p^{n-1}} & \text{if } p \text{ is odd and } n > 1; \\ \alpha_p & \text{if } p \text{ is odd and } n = 1, \end{cases}$$

with α_p as in Theorem 4.5.

Proof. The case $p = 2$ is just a restatement of finite nim arithmetic (Theorem IV.5.3 on page 215). Likewise, the remaining cases are a restatement of Theorem 4.5, since (as noted above)

$$\omega^{\omega^k \times p^{n-1}} = 2^{\omega^{k+1} \times p^{n-1}} = \kappa_q$$

when $p = p_{k+1}$ is the $(k+1)^{\text{st}}$ odd prime. $\square$

The only mysterious values here are the ordinals α_p, which must be calculated individually. Lenstra has (miraculously!) developed an algorithm for computing α_p, by reducing the calculations to sequences of operations on finite fields. Figure 4.1 on the following page shows the values for all $p \leq 173$, and some additional details are given in the notes for this section on page 452.

Example. Let's compute the value of

$$\alpha = \left(\omega^{\omega^3}\right)^{\textcircled{66}}.$$

Here $p_4 = 11$, and by Figure 4.1 we see that $\alpha_{11} = \omega^\omega + 1$. Therefore

$$\left(\omega^{\omega^3}\right)^{\textcircled{11}} = \omega^\omega + 1 = \omega^\omega \oplus 1,$$

so

$$\alpha = (\omega^\omega \oplus 1)^{\textcircled{6}}.$$

Expanding the sixth nim-power in characteristic 2 gives

$$\alpha = (\omega^\omega)^{\textcircled{6}} \oplus (\omega^\omega)^{\textcircled{4}} \oplus (\omega^\omega)^{\textcircled{2}} \oplus 1.$$

p	α_p	p	α_p	p	α_p
3	2	47	$\omega^{\omega^7}+1$	107	$\omega^{\omega^{14}}+1$
5	4	53	$\omega^{\omega^4}+1$	109	ω^3+4
7	$\omega+1$	59	$\omega^{\omega^8}+1$	113	$\omega^{\omega^2}+4$
11	$\omega^\omega+1$	61	$\omega^\omega+\omega$	127	$\omega^{\omega^2}+1$
13	$\omega+4$	67	$\omega^{\omega^3}+\omega$	131	$\omega^{\omega^4}+\omega^\omega$
17	16	71	$\omega^{\omega^2}+\omega^\omega$	137	$\omega^{\omega^5}+1$
19	ω^3+4	73	ω^3+1	139	$\omega^{\omega^7}+\omega$
23	$\omega^{\omega^3}+1$	79	$\omega^{\omega^4}+1$	149	$\omega^{\omega^{10}}+1$
29	$\omega^{\omega^2}+4$	83	$\omega^{\omega^{11}}+1$	151	$\omega^\omega+\omega$
31	$\omega^\omega+1$	89	$\omega^{\omega^3}+1$	157	$\omega^{\omega^4}+1$
37	ω^3+4	97	$\omega+256$	163	$\omega^{27}+4$
41	$\omega^\omega+1$	101	$\omega^{\omega\times 5}+1$	167	$\omega^{\omega^{21}}+1$
43	$\omega^{\omega^2}+1$	103	$\omega^{\omega^5}+\omega$	173	$\omega^{\omega^{12}}+4$

Figure 4.1. The values α_p for odd primes $p \leq 173$.

Since $p_2 = 5$ and $\alpha_5 = 4$, we have $(\omega^\omega)^{\textcircled{6}} = \omega^\omega \otimes (\omega^\omega)^{\textcircled{5}} = \omega^\omega \otimes 4$, so that

$$\alpha = (\omega^\omega \otimes 4) \oplus (\omega^\omega)^{\textcircled{4}} \oplus (\omega^\omega)^{\textcircled{2}} \oplus 1$$

in normal form. Therefore

$$\alpha = \omega^{\omega\times 4} + \omega^{\omega\times 2} + \omega^\omega \times 4 + 1.$$

Transcendental Extensions of $\mathcal{P}_\tau$

Since τ is transcendental over $\mathcal{P}_\tau$, polynomials in τ are evaluated normally (by Lemma 4.4), so that the next ring occurs at step

$$\tau^\omega = \omega^{\omega^\omega \times \omega} = \omega^{\omega^{\omega+1}}.$$

But this ring is not a field, since it contains no inverse for τ, and by the Simplest Extension Theorem it follows that $\tau^\omega = 1 \oslash \tau$.

Then we find that expressions of the form $\big(\tau^\omega \otimes f(\tau)\big) \oplus g(\tau)$ evaluate normally, for polynomials f and g. That is,

$$\begin{aligned} &\tau^\omega \times \big((\tau^m \times \alpha_m) + \cdots + \alpha_0\big) + \big((\tau^n \times \beta_n) + \cdots + \beta_0\big) \\ &= \tau^{\boxed{-1}} \otimes \big((\tau^{\boxed{m}} \otimes \alpha_m) \oplus \cdots \oplus \alpha_0\big) \oplus \big((\tau^{\boxed{n}} \otimes \beta_n) \oplus \cdots \oplus \beta_0\big), \end{aligned}$$

showing that another ring occurs at $\tau^{\omega+\omega}$, so that $\tau^{\omega+\omega}$ must be an inverse for $\tau+1$. One can continue like this to show that successive inverses are adjoined at steps

$$\tau^{\omega},\ \tau^{\omega+\omega},\ \tau^{\omega+\omega\times 2},\ \ldots,\ \tau^{\omega+\omega\times\alpha},\ \ldots \quad (\alpha<\tau).$$

So the next field occurs at

$$\tau^{\omega+\omega\times\tau}=\tau^{\tau}=\omega^{\omega^{\omega^{\omega}}}=\omega^{\tau}.$$

Again the Simplest Extension Theorem shows that ω^{τ} must be a square nim-root for τ, so that $(\omega^{\tau})^{⓶}=\tau$. This argument can be iterated to show that in the sequence

$$\tau,\ \omega^{\tau},\ \omega^{\tau\times 2},\ \omega^{\tau\times 4},\ \ldots,\ \omega^{\tau\times 2^n},\ \ldots$$

each successive term is the square nim-root of the previous one, a sort of high-level analogue of Theorem 4.5 for $p=2$. It follows that

$$\mathcal{P}_{\omega^{\tau}\times\omega}$$

is a perfect field, and hence so are all its extensions up to the next transcendental (since its characteristic is 2). The simplest polynomial with no solution in $\mathcal{P}_{\omega^{\tau}\times\omega}$ is therefore

$$x^{⓶}\oplus x\oplus\tau$$

which, by the Simplest Extension Theorem, must have $\omega^{\tau\times\omega}$ as a nim-root. There are many more irreducible quadratics. It can be shown (though we won't prove it here) that the next quadratically closed field is in fact $\mathcal{P}_{\epsilon_0}$, where ϵ_0 is the first epsilon number (Definition B.3.16 on page 468):

$$\epsilon_0=\omega^{\omega^{\omega^{\omega^{\cdot^{\cdot^{\cdot}}}}}}=\sup\left\{\omega,\ \omega^{\omega},\ \omega^{\omega^{\omega}},\ \omega^{\omega^{\omega^{\omega}}},\ \ldots\right\}.$$

The next extension is by a cubic, yielding the remarkable identity

$$\epsilon_0^{⓷}=\tau.$$

But the extension by ϵ_0 introduces new unsolved quadratics, and in fact the next quadractically closed field occurs at the next epsilon number ϵ_1, which is therefore also the next cubic extension. It is unknown whether this pattern continues, and the following question, asked by Conway thirty-five years ago, is still open:

Question. What is the least ordinal $\alpha>\tau$ with $\mathcal{P}_{\alpha}$ algebraically closed?

Exercises

4.1 Prove the identity $\alpha_7 = \omega + 1$. (First show that no field of order 2^{2^k} can contain an element of order 7, and use this to prove that every integer has a seventh nim-root in $\mathcal{P}_\omega$. Then show that $\mathcal{P}_4(\omega)$ has order exactly 64 and that ω has a seventh nim-root in this field, but $\omega + 1$ does not.)

4.2 If $\mathcal{P}_\gamma < \mathcal{P}_\delta$ are fields, then $\delta = \gamma^\alpha$ for some α. Moreover if $\alpha = 4$, then $\mathcal{P}_{\gamma^2}$ is also a field.

4.3 If $p = 2^{2^k} + 1$ is a Fermat prime, then $\alpha_p = 2^{2^k}$.

4.4 For each integer $n \geq 2$, let $\kappa_n = \min\{\alpha \in \mathcal{P}_\tau : n \text{ divides } \deg(\alpha)\}$, where $\deg(\alpha)$ denotes the degree of the irreducible polynomial of α over $\mathcal{P}_2$.

(a) Show that this definition is consistent with the earlier definition of κ_q (when q is a prime power).

(b) Let $n \geq 2$, let p be the smallest prime dividing n, and let q be the largest power of p dividing n. Prove that

$$\kappa_n = \begin{cases} \kappa_{n/q} & \text{if } q \text{ divides } \deg(\kappa_{n/q}); \\ \kappa_{n/q} + \kappa_q & \text{otherwise.} \end{cases}$$

(c) Conclude that every κ_n can be written uniquely as a finite sum

$$\kappa_n = \kappa_{q_1} + \kappa_{q_2} + \cdots + \kappa_{q_k},$$

where $q_1, q_2, \ldots, q_k$ are distinct maximal prime-power divisors of n.

4.5 For any $\alpha \in \mathcal{P}_\tau$, the multiplicative group of $\mathcal{P}_\omega(\alpha)$ is generated by

$$\{\alpha \oplus m : m < \omega\}.$$

(Let f be the irreducible nim-polynomial of α over P_ω. Let g be any nim-polynomial with coefficients in P_ω; if $g(\alpha) \neq 0$, then f and g are relatively prime. Use the theorem—you don't have to prove it!—that for every sufficiently large integer n, there exists an irreducible monic polynomial h of degree n with $g \equiv h \pmod{f}$. Show that if $n = 2^k$, then h splits over ω and therefore $g(\alpha)$ can be written as a product of ordinals of the form $\alpha \oplus m$.)

4.6 Let $f(p) = \deg(\zeta_p)$, where ζ_p is a primitive p^{th} nim-root of unity. Let κ_n be as in Exercise 4.4. Prove that for all primes p:

(a) $\alpha_p \geq \kappa_{f(p)}$. (First show that $\zeta_p \in \ker(h)$, where $h : \mathcal{P}_2(\alpha_p) \to \mathcal{P}_2(\alpha_p)$ is given by $h(x) = x^{\textcircled{p}}$. Conclude that $f(p)$ divides $\deg(\alpha_p)$.)

(b) $\alpha_p \leq \kappa_{f(p)} \oplus m$ for some integer $m \geq 0$. (Show that some $\beta \in \mathcal{P}_2(\kappa_{f(p)})$ is not a pth nim-power in κ_p, and use Exercise 4.5.)

(c) $\alpha_p = \kappa_{f(p)} \oplus m$ for some integer $m \geq 0$.

(d) The value of α_p can be effectively determined.

Notes

The Field $\mathbf{ON}_2$ was introduced by Conway, who proved the Simplest Extension Theorem along with Lemma 4.4. The treatment here essentially follows ONAG, though with some changes of notation.

Remarkably, the notion of "algebraic simplicity" inherent in $\mathbf{ON}_2$ closely parallels the historical cognitive development of mathematics: the ranking established by Theorem 4.3 is essentially the order in which each algebraic operation came to be accepted as a legitimate way of constructing new numbers from old! It is rather extraordinary that such a hierarchy emerges from a few simple genetic definitions, which appear to make no mention of algebraic or transcendental numbers.

Hendrik Lenstra has developed methods for effective calculation of the ordinals α_p [**Len77**]. In particular, Lenstra showed that every such ordinal has the form

$$\alpha_p = \kappa_{q_1} \oplus \cdots \oplus \kappa_{q_k} \oplus m,$$

where $m < \omega$, the integers $q_1, \ldots, q_k$ range over various factors of $p-1$, and all the values $q_1, \ldots, q_k, m$ can be effectively determined. A proof of this result is given by Exercises 4.4, 4.5, and 4.6. Since Theorem 4.7 (also due to Lenstra) reduces the arithmetic of $\mathcal{P}_\tau$ to calculations involving individual ordinals α_p, Lenstra's methods show that the arithmetic of $\mathcal{P}_\tau$ is completely effective.

Lenstra used these techniques to compute α_p for all $p \leq 43$, and the table was later extended to $p \leq 67$ by Lieven Le Bruyn [**LB09**]. In preparing this textbook, I verified Le Bruyn's results and tabulated the remaining values in Figure 4.1, using a custom extension to *cgsuite*.

Proofs that $\mathcal{P}_{\epsilon_0}$ and $\mathcal{P}_{\epsilon_1}$ are the next quadratically closed fields after $\mathcal{P}_\tau$ are also due to Lenstra [**Len78**]. Much work remains to be done in this area.

A generalization in a different direction was recently studied by Joseph DiMuro [**DiM11**], who defined characteristic-p analogues of $\mathbf{ON}_2$, denoted by $\mathbf{ON}_p$. Di Muro showed that Lenstra's techniques for effective computation in $\mathcal{P}_\tau$ can be extended to algebraic subfields of $\mathbf{ON}_p$. However, it is unclear how one might extend Conway's genetic definition of multiplication to $\mathbf{ON}_p$ (but see Exercise IV.1.5 on page 182 for a genetic definition of addition for $\mathbf{ON}_3$).

Exercises 4.3 through 4.6: Lenstra.

Appendix A

Open Problems

Another extensive list of open problems, with an emphasis on specific games and examples, is maintained by Guy and Nowakowski [**GN13**].

Chapter II

Open Problem (page 93). Give an algorithm to determine the outcome of an arbitrary flower garden or show that the problem is NP-hard.

Conjecture (page 100). Let G be a positive infinitesimal. Then writing all uptimals in base G,

$$0.a_1a_2\cdots a_n = \left\{-G^R + 0.a_1^+a_2^+\cdots a_j^+,\ 0.a_1a_2\cdots a_k \;\middle|\; G^R + 0.a_1^-a_2^-\cdots a_j^-\right\}.$$

Question (page 150). Can the atomic weight theory be generalized to higher-order uptimals?

Open Problem (page 151). Extend the atomic weight calculus to a procedure for computing $\mathrm{aw}(G)$ for arbitrary atomic G.

Question (page 151). What can be said about the structure of $\mathbb{G}^+$? What is its isomorphism type?

Question (page 151). Are there natural representatives of arbitrary elements of $\mathbb{G}^+$ that help clarify its structure, in the same way atomic weight clarifies the structure of $\mathbb{G}^0/\mathcal{Z}$?

Chapter III

Open Problem (page 166). Improve upon the Fraser–Wolfe bounds for the cardinality of $\mathbb{G}_n$ (and in particular $\mathbb{G}_4$).

Open Problem (page 167). Investigate symmetries of the generalized lattices $\mathrm{ch}(\mathcal{A})$.

Open Problem (page 167). Describe the hereditary structure of $\mathcal{HT}_n$ (via join-irreducibles).

Conjecture (page 167). $\mathcal{HT}_n$ has exactly $2^{a_n} - 2$ antichains, where a_n is the number of antichains of the complete Boolean algebra on n variables.

Open Problem (page 178). Characterize the structure of $\langle \mathbb{G}_3 \rangle$ and $\langle \mathbb{G}_3^0 \rangle$ as Abelian groups.

Open Problem (page 178). Characterize the structure of $\mathbb{G}$ as a partially ordered Abelian group.

Chapter IV

Question (page 194). Is every finite octal game eventually periodic?

Question (page 196). Does there exist a complete theory of finite subtraction games?

Open Problem (page 197). Identify an octal game with period longer than **0.106**.

Open Problem (page 197). Extend the analysis of hexadecimal games.

Chapter V

Open Problem (page 249). Characterize the abstract structure of the monoid $\mathcal{M}$ of misère impartial games. (Conjecture: $\mathcal{M}$ is isomorphic to the free commutative monoid on a countable set of generators, modulo a factor of $\mathbb{Z}_2$ corresponding to $*$.)

Question (page 268). Let $(\mathcal{Q}, \mathcal{P}) = \mathcal{Q}(\mathscr{A})$, with quotient map $\Phi : \mathscr{A} \to \mathcal{Q}$, and let $G, H \in \mathscr{A}$. Does $\Phi(G) = \Phi(H)$ necessarily imply $\mathscr{G}(G) = \mathscr{G}(H)$?

Question (page 268). If $(\mathcal{Q}, \mathcal{P})$ is a misère quotient and $\mathcal{A}$ is a maximal subgroup of $\mathcal{Q}$, must $\mathcal{A} \cap \mathcal{P}$ be nonempty?

Open Problem (page 268). Specify an algorithm to determine whether or not $\mathcal{Q}(\mathscr{A})$ is infinite, given a finitely generated set of impartial games $\mathscr{A}$.

Open Problem (page 269). Specify an algorithm to calculate a presentation for $\mathcal{Q}(\mathscr{A})$, given a finitely generated set of impartial games $\mathscr{A}$.

Open Problem (page 276). Develop a structure theory for partizan misère quotients.

Open Problem (page 278). Extend the theory of partizan games in misère play.

Chapter VI

Open Problem (page 299). Extend the atomic weight calculus to (an appropriate subclass of) stoppers.

Open Problem (page 309). Characterize the cycle patterns that can occur in a canonical stopper. In particular, determine whether LLLRLRRRLR is a possible pattern.

Open Problem (page 309). Exhibit a ruleset (with reasonably "natural" rules) containing a position G, such that the canonical form of G has a wild cycle.

Question (page 317). Is there an alternative notion of simplest form that works for *all* finite loopy games (in particular, for the Carousel)?

Question (page 317). Is the sum of finite stoppers necessarily stopper-sided?

Question (page 317). Given stoppers S and T with $S \geq T$, must there necessarily exist a game G with $G = S \mathbin{\&} T$?

Open Problem (page 321). Give an algorithm to determine whether an arbitrary loopy game G is stopper-sided and to calculate its sides if it is.

Conjecture (Stability Conjecture, page 328). Every game of stable degree is stable.

Open Problem (page 332). Investigate the class and variety structure of various idempotents such as $\uparrow^{\mathbf{on}}$, $\mathbf{star}_m$, and $\spadesuit$.

Chapter VII

Open Problem (page 377). Develop a thermographic calculus that works for all loopy games.

Open Problem (page 377). Invent some new loopy games that play well under a koban restriction, as a source of further examples for the generalized temperature theory.

Open Problem (page 396). Develop a formal treatment of the dogmatic theory.

Open Problem (page 396). Extend the dogmatic theory to (any reasonable subclass of) complex loopy games.

Chapter VIII

Question (page 409). Can the atomic weight calculus be generalized to $\mathbf{PG}^0$?

Open Problem (page 410). Describe the partial-order structure of $\mathbb{G}^0_\alpha$, for $\alpha \geq \omega$.

Conjecture (page 437). $\mathrm{b}(x \times y) \leq \mathrm{b}(x) \times \mathrm{b}(y)$ for all $x, y \in \mathbf{SN}$.

Question (page 438). Does there exist a reasonable general definition of surreal integration?

Question (page 451). What is the least ordinal $\alpha > \tau$ with $\mathcal{P}_\alpha$ algebraically closed?

Appendix B

Mathematical Prerequisites

This appendix gives a cursory overview of some of the mathematical background for combinatorial game theory. It is provided primarily for reference and review. Some prior familiarity with the subject matter is assumed, and detailed proofs and examples are mostly omitted. In all cases, references are given for a more detailed treatment of each topic.

The material on Abelian groups in Section 1 is essential and is used throughout this book. Most of the remaining material is relevant to just a few chapters: lattices (Section 2) are important to the structure theory in Chapter III; ordinals (Section 3) are central to the study of transfinite games in Chapter VIII; and commutative semigroups (Section 4) are a key part of the misère theory in Chapter V.

1. Abelian Groups

Definition 1.1. An **Abelian group** is a set $\mathcal{A}$, together with a binary operation $+$ on $\mathcal{A}$ and an **identity element** $0 \in \mathcal{A}$, satisfying the following axioms:

(i) (associativity) $a + (b + c) = (a + b) + c$ for all $a, b, c \in \mathcal{A}$;

(ii) (commutativity) $a + b = b + a$ for all $a, b \in \mathcal{A}$;

(iii) (identity) $a + 0 = a$ for all $a \in \mathcal{A}$;

(iv) (negatives) for each $a \in \mathcal{A}$, there exists $b \in \mathcal{A}$ such that $a + b = 0$.

Throughout this section, the word "group" will mean "Abelian group." The negative of $a \in \mathcal{A}$ will be denoted by $-a$, and we'll write $a - b$ to mean $a + (-b)$. For $n \in \mathbb{N}$, we also write

$$n \cdot a = \overbrace{a + a + \cdots + a}^{n \text{ times}}$$

and for $n < 0$, we let $n \cdot a = |n| \cdot (-a)$.

Definition 1.2. A **subgroup** of $\mathcal{A}$ is a subset $\mathcal{B} \subset \mathcal{A}$ that is a group. If $\mathcal{B}$ is a subgroup of $\mathcal{A}$, then we write $\mathcal{B} < \mathcal{A}$.

If $\mathcal{B} < \mathcal{A}$, then we can define

$$a \equiv b \pmod{\mathcal{B}} \quad \text{if} \quad a - b \in \mathcal{B}.$$

This is an equivalence relation, and the set of equivalence classes forms a group under the induced operation. This is the **quotient group $\mathcal{A}/\mathcal{B}$**.

The set of integers $\mathbb{Z}$ (equipped with ordinary integer addition) is clearly a group. Denote by $n\mathbb{Z}$ the set

$$n\mathbb{Z} = \{na : a \in \mathbb{Z}\}.$$

Clearly $n\mathbb{Z} < \mathbb{Z}$, so there exists a quotient $\mathbb{Z}/n\mathbb{Z}$, the group of **integers modulo** n. We write $\mathbb{Z}_n$ as shorthand for this group. The groups $\mathbb{Z}$ and $\mathbb{Z}_n$ are collectively known as **cyclic groups**.

An element $a \in \mathcal{A}$ is a **torsion element** if $n \cdot a = 0$ for some n. Then $\mathcal{A}$ is a **torsion group** if every $a \in \mathcal{A}$ is a torsion element, and it is **torsion-free** if 0 is the only torsion element of $\mathcal{A}$.

Finitely Generated Abelian Groups

Definition 1.3. If $\mathcal{S} \subset \mathcal{A}$, then the subgroup of $\mathcal{A}$ **generated by** $\mathcal{S}$ is the smallest subgroup $\mathcal{B} < \mathcal{A}$ with $\mathcal{S} \subset \mathcal{B}$.

Intersections of subgroups are clearly subgroups, so a unique "smallest" $\mathcal{B}$ necessarily exists. We write $\langle \mathcal{S} \rangle$ for the subgroup of $\mathcal{A}$ generated by $\mathcal{S}$; the enclosing group $\mathcal{A}$ will always be clear from the context.

Definition 1.4. A group $\mathcal{A}$ is **finitely generated** if $\mathcal{A} = \langle \mathcal{S} \rangle$ for some finite subset $\mathcal{S} \subset \mathcal{A}$.

The finitely generated Abelian groups are remarkably easy to classify.

Definition 1.5. If $\mathcal{A}_1, \mathcal{A}_2, \ldots, \mathcal{A}_k$ are Abelian groups, then their **direct product** is given by

$$\mathcal{A}_1 \times \mathcal{A}_2 \times \cdots \times \mathcal{A}_k = \big\{(a_1, \ldots, a_k) : \text{each } a_i \in \mathcal{A}_i\big\}$$

with componentwise addition:

$$(a_1, \ldots, a_k) + (b_1, \ldots, b_k) = (a_1 + b_1, \ldots, a_k + b_k).$$

We can now state the *Structure Theorem for Finitely Generated Abelian Groups*:

Theorem 1.6. *Let $\mathcal{A}$ be a finitely generated Abelian group. Then there are cyclic groups $\mathcal{A}_1, \mathcal{A}_2, \ldots, \mathcal{A}_k$ such that*

$$\mathcal{A} \cong \mathcal{A}_1 \times \mathcal{A}_2 \times \cdots \times \mathcal{A}_k.$$

General Abelian Groups

If $\mathcal{A}$ is a group, then we write $\mathcal{A}^k$ as shorthand for the direct product of k copies of $\mathcal{A}$:

$$\mathcal{A}^k = \overbrace{\mathcal{A} \times \mathcal{A} \times \cdots \times \mathcal{A}}^{k \text{ times}}.$$

We also write $\mathcal{A}^\omega$ for the direct *sum* of a countable number of copies of $\mathcal{A}$. Formally, $\mathcal{A}^\omega$ is the set of all sequences $(a_0, a_1, a_2, \ldots)$ of countable length, with each $a_i \in \mathcal{A}$, such that the **support** of the sequence is finite:

$$a_i \neq 0 \text{ for just finitely many values of } i.$$

Typical examples include $\mathbb{Z}^\omega$ and $\mathbb{Z}_2^\omega$. These groups are obviously *not* finitely generated. Other important groups that are not finitely generated include:

- the group $\mathbb{Q}$ of rational numbers;
- the group $\mathbb{D}$ of **dyadic** rationals (rational numbers whose denominator is a power of 2);
- $\mathbb{Q}/\mathbb{Z}$ and $\mathbb{D}/\mathbb{Z}$.

These examples illustrate the following important property.

Definition 1.7. Let p be a prime number. A group $\mathcal{A}$ is **p-divisible** if for every $a \in \mathcal{A}$, there exists $b \in \mathcal{A}$ such that $p \cdot b = a$. We say $\mathcal{A}$ is **divisible** if it is p-divisible for all p.

Clearly $\mathbb{Q}$ and $\mathbb{Q}/\mathbb{Z}$ are divisible. Likewise, $\mathbb{D}$ and $\mathbb{D}/\mathbb{Z}$ are 2-divisible, but not p-divisible for any $p > 2$. No finitely generated group is p-divisible, for any p; nor is $\mathbb{Z}^\omega$ or $\mathbb{Z}_2^\omega$.

2. Partial Orders

A detailed treatment of the material in this section can be found in a standard combinatorics reference such as [**Sta86**].

Definition 2.1. A **partially ordered set** (or just **poset**) is a set $\mathcal{P}$, together with a relation $\geq$ on $\mathcal{P}$, that satisfies the following axioms:

(i) (reflexivity) $x \geq x$ for all $x \in \mathcal{P}$;

(ii) (antisymmetry) if $x \geq y$ and $y \geq x$, then $x = y$, for all $x, y \in \mathcal{P}$; and

(iii) (transitivity) if $x \geq y$ and $y \geq z$, then $x \geq z$, for all $x, y, z \in \mathcal{P}$.

The relation $\geq$ is a **partial order** of the set $\mathcal{P}$.

We write $x \leq y$ to mean $y \geq x$, and $x > y$ to mean $x \geq y$ and $x \neq y$ (likewise $x < y$). Similarly, if $x \not\geq y$ and $y \not\geq x$, then we say x and y are **incomparable** and write $x \not\gtrless y$.

A finite poset $\mathcal{P}$ is typically illustrated by drawing a **Hasse diagram**. Figure 2.1 shows an example for a typical seven-element poset:

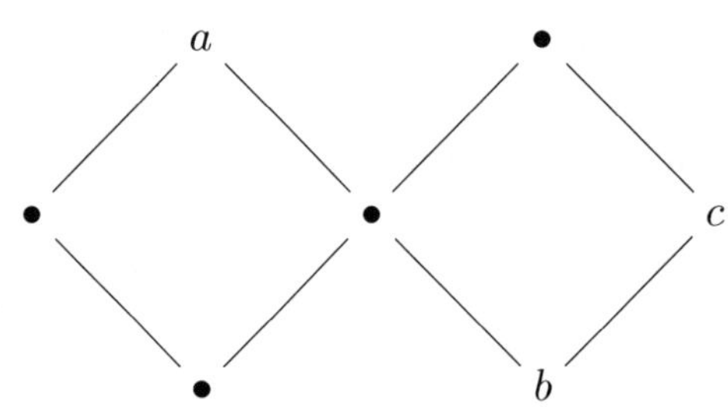

Figure 2.1. A typical seven-element poset.

Each vertex of the diagram in Figure 2.1 is an element of $\mathcal{P}$, and the edges represent partial-order relations. If $x, y \in \mathcal{P}$, then $x \geq y$ if and only if there is a path from x to y, traveling entirely along downward-sloping edges. In the diagram above, for example, we have $a > b$ and $c > b$, but $a \not\gtrless c$.

If $\mathcal{S}$ is a set, then the collection $\operatorname{Pow}(\mathcal{S})$ of all subsets of $\mathcal{S}$ is a poset, with $\mathcal{X} \geq \mathcal{Y}$ if and only if $\mathcal{X} \supset \mathcal{Y}$.

If $\mathcal{P}$ is a poset, then every subset $\mathcal{Q} \subset \mathcal{P}$ is also a poset (a **subposet** of $\mathcal{P}$), equipped with the partial-order induced by $\geq$.

If $\mathcal{P}$ and $\mathcal{Q}$ are posets, then a function $f : \mathcal{P} \to \mathcal{Q}$ is a **homomorphism** if

$$x \geq y \quad \Longleftrightarrow \quad f(x) \geq f(y)$$

for all $x, y \in \mathcal{P}$. If f is one-to-one, then it is a partial-order **isomorphism**.

A **totally ordered set** is a poset $\mathcal{P}$ in which no two elements are incomparable. If $\mathcal{P}$ is any poset, then a subposet $\mathcal{Q} \subset \mathcal{P}$ is a **chain** if $\mathcal{Q}$ is totally ordered, and an **antichain** if any two elements of $\mathcal{Q}$ are incomparable.

Lattices

Definition 2.2. A **lattice** is a poset $\mathcal{L}$ equipped with two binary operations $\vee$ and $\wedge$ (**join** and **meet**), satisfying the following additional axioms:

(iv) (Least upper bound) $x \vee y \geq x$ and $x \vee y \geq y$, for all $x, y \in \mathcal{L}$. Moreover, for any $z \in \mathcal{L}$ with $z \geq x$ and $z \geq y$, we have $z \geq x \vee y$.

(v) (Greatest lower bound) $x \wedge y \leq x$ and $x \wedge y \leq y$, for all $x, y \in \mathcal{L}$. Moreover, for any $z \in \mathcal{L}$ with $z \leq x$ and $z \leq y$, we have $z \leq x \wedge y$.

Every finite lattice $\mathcal{L}$ has a unique maximal element, obtained by successively joining every element of $\mathcal{L}$. In particular, the seven-element poset pictured in Figure 2.1 is not a lattice, since it has two distinct maximal elements. It can be made into a lattice by adjoining two further elements as in Figure 2.2.

Definition 2.3. A lattice $\mathcal{L}$ is **distributive** if $x \wedge (y \vee z) = (x \wedge y) \vee (x \wedge z)$ for all $x, y, z \in \mathcal{L}$.

The example in Figure 2.2 is distributive (we'll see this in a moment). The lattices $\mathcal{M}_3$ and $\mathcal{N}_5$, whose Hasse diagrams are shown in Figure 2.3 on the next page, are examples of nondistributive lattices.

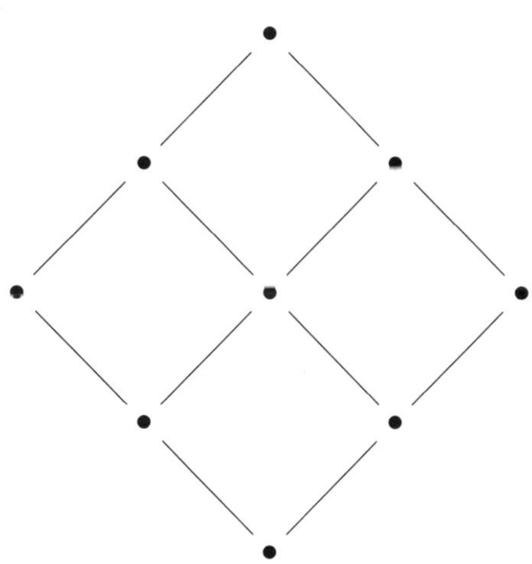

Figure 2.2. A nine-element distributive lattice.

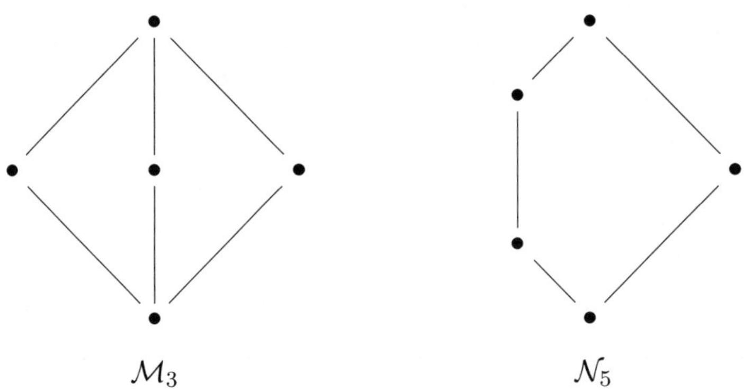

Figure 2.3. The archetypal nondistributive lattices $\mathcal{M}_3$ and $\mathcal{N}_5$.

Distributivity is a key property. Its name derives from the fact that meet "distributes" over join, just as multiplication (in a ring) distributes over addition. There are several other equivalent characterizations of distributive lattices that give more insight into their structure. Here's the first:

Theorem 2.4. *The following are equivalent, for a lattice $\mathcal{L}$:*

(i) *$\mathcal{L}$ is distributive.*

(ii) *$\mathcal{L}$ contains no sublattice isomorphic to $\mathcal{M}_3$ or $\mathcal{N}_5$.*

Here a **sublattice** is a subposet that is closed under join and meet.

For every set $\mathcal{S}$, the powerset lattice $\mathrm{Pow}(\mathcal{S})$ is distributive. It is the **Boolean lattice** (or **Boolean algebra**) over $\mathcal{S}$.

Join-Irreducibles

Another characterization of distributivity is given in terms of **join-irreducible** elements of a lattice $\mathcal{L}$.

Definition 2.5. $x \in \mathcal{L}$ is **join-irreducible** if no $y, z < x$ satisfy $y \vee z = x$. Likewise, x is **meet-irreducible** if no $y, z > x$ satisfy $y \wedge z = x$.

Remarkably, if $\mathcal{L}$ is a finite distributive lattice, then it is characterized by the induced poset of its join-irreducible elements. This deep relationship between finite posets and distributive lattices is given by the following theorem.

Definition 2.6. Let $\mathcal{P}$ be a poset. An **ideal** of $\mathcal{P}$ is a subset $\mathcal{Q} \subset \mathcal{P}$ that is downward-closed: if $x \in \mathcal{Q}$ and $y \leq x$, then also $y \in \mathcal{Q}$.

The set of ideals $\mathcal{I}(\mathcal{P})$ of a finite poset $\mathcal{P}$ forms a distributive lattice.

Theorem 2.7 (Birkhoff's Theorem). *Let $\mathcal{L}$ be any finite distributive lattice, and let $\mathcal{J}$ be the poset of its join-irreducible elements. Then $\mathcal{L} \cong \mathcal{I}(\mathcal{J})$.*

Correspondingly, let $\mathcal{P}$ be any finite poset. Then $\mathcal{I}(\mathcal{P})$ is a distributive lattice whose poset of join-irreducibles is isomorphic to $\mathcal{P}$.

(In the Birkhoff correspondence, the join-irreducibles of $\mathcal{I}(\mathcal{P})$ are the ideals that are generated by a single element of $\mathcal{P}$.)

3. Ordinals

The ordinals are a basic concept in set theory, and they are central to the theory of transfinite games in Chapter VIII. Proofs of most results in this section can be found in [**Jec02**]; other standard references include [**Kun83, Kan03**].

Definition 3.1. A **well-ordering** of a set $\mathcal{S}$ is a total ordering $\leq$, such that every subset $\mathcal{A} \subset \mathcal{S}$ has a $\leq$-minimal element.

Theorem 3.2. *The following are equivalent, for a total ordering $\leq$ of a set $\mathcal{S}$:*

(i) *$\leq$ is a well-ordering of $\mathcal{S}$.*

(ii) *There is no infinite, strictly decreasing sequence of elements of $\mathcal{S}$.*

A **well-ordered set** is a set $\mathcal{S}$ equipped with a well-ordering $\leq$.

Definition 3.3. Let $\mathcal{S}$ and $\mathcal{T}$ be well-ordered sets. We say $\mathcal{S}$ and $\mathcal{T}$ are **order-isomorphic** if there is an order-preserving bijection between them.

If $\mathcal{S}$ and $\mathcal{T}$ are *finite* well-ordered sets, then $\mathcal{S}$ and $\mathcal{T}$ are order-isomorphic if and only if they have the same cardinality. (This is easy to prove by induction.) However this is *not* true in the infinite case. For example, let $\mathcal{S} = \mathbb{N}$, and let $\mathcal{T} = \mathbb{N} \cup \{\omega\}$, where $\omega > n$ for all $n \in \mathbb{N}$. Then $|\mathcal{S}| = |\mathcal{T}| = \aleph_0$, but $\mathcal{S} \not\cong \mathcal{T}$, since $\mathcal{T}$ has a largest element and $\mathcal{S}$ does not.

The isomorphism type of a well-ordered set is known as its **ordertype**. This is made precise by defining canonical representatives for each isomorphism class. These canonical well-ordered sets are the **ordinals**; we'll give a formal definition after stating a few examples.

Every natural number is an ordinal, defined like so:

$$0 = \emptyset,$$
$$1 = \{0\},$$
$$2 = \{0, 1\},$$
$$3 = \{0, 1, 2\},$$

and so on, until we reach the first **transfinite ordinals**:

$$\begin{aligned}
\omega \quad &= \{0, 1, 2, 3, \ldots\} = \mathbb{N}, \\
\omega + 1 &= \{0, 1, 2, 3, \ldots, \omega\}, \\
\omega + 2 &= \{0, 1, 2, 3, \ldots, \omega, \omega + 1\}, \\
\omega \times 2 &= \{0, 1, 2, 3, \ldots, \omega, \omega + 1, \omega + 2, \ldots\}.
\end{aligned}$$

If α is an ordinal, then the set membership relation $\in$ is transitive on α: if $\gamma \in \beta \in \alpha$, then $\gamma \in \alpha$. (Equivalently, every element of α is a subset of α.) Any set with this property is called a **transitive set**. We can now state the usual definition of ordinal:

Definition 3.4. An **ordinal** is a transitive set well-ordered by $\in$. The Class of all ordinals is denoted by **ON**.

If α is an ordinal, then so is every $\beta \in \alpha$. Because every ordinal is a transitive set, $\in$ is a transitive Relation on the Class **ON**, and we customarily write $\beta < \alpha$ in place of $\beta \in \alpha$. Likewise, we write $\beta \leq \alpha$ if either $\beta < \alpha$ or $\beta = \alpha$.

Theorem 3.5. *$\leq$ is a Well-Ordering of* **ON**.

Successor and Limit Ordinals

If β is an ordinal, then so is $\alpha = \beta \cup \{\beta\}$. Clearly $\beta < \alpha$, and if γ is any ordinal with $\gamma < \alpha$, then it must be the case that $\gamma \leq \beta$. So α is the "next ordinal" after β.

Definition 3.6. If β is any ordinal, then the **successor** of β is the ordinal $\alpha = \beta \cup \{\beta\}$, and we write $\alpha = \beta + 1$.

Moreover, if $\mathcal{A}$ is any set of ordinals, then its union is also an ordinal.

Definition 3.7. Let $\mathcal{A} \subset$ **ON** be a set of ordinals. The **supremum** of $\mathcal{A}$ is the ordinal

$$\sup(\mathcal{A}) = \bigcup \mathcal{A} = \{\xi \in \mathbf{ON} : \xi < \alpha \text{ for some } \alpha \in \mathcal{A}\}.$$

We can now define the two main types of ordinals.

Definition 3.8. An ordinal $\alpha > 0$ is a **successor ordinal** if $\alpha = \beta + 1$ for some $\beta \in$ **ON**, and a **limit ordinal** if $\alpha = \sup(\alpha)$.

Theorem 3.9. *Every $\alpha > 0$ is either a successor ordinal or a limit ordinal, but not both.*

Theorem 3.10 (Transfinite Induction). *Let ϕ be a proposition. Suppose that for every ordinal α,*

$$\Big(\phi(\beta) \text{ for all } \beta < \alpha\Big) \quad \Longrightarrow \quad \phi(\alpha).$$

Then $\phi(\alpha)$ is true for all α.

Proofs by transfinite induction typically have two cases, depending on whether α is a successor or a limit ordinal.

Ordinal Arithmetic

Addition, multiplication, and exponentiation of ordinals are defined as follows.

Definition 3.11. $\alpha \dot{+} \beta$ is defined by induction on β:

$$\alpha \dot{+} \beta = \begin{cases} \alpha & \text{if } \beta = 0; \\ (\alpha \dot{+} \gamma) + 1 & \text{if } \beta = \gamma + 1 \text{ is a successor ordinal;} \\ \sup\{\alpha \dot{+} \gamma : \gamma < \beta\} & \text{if } \beta \text{ is a limit ordinal.} \end{cases}$$

Intuitively, we can think of $\alpha \dot{+} \beta$ as the ordertype obtained when α and β are placed end-to-end. Ordinal addition is associative, but *not* commutative: for example

$$\omega \dot{+} 1 = \omega \cup \{\omega\} > \omega,$$

whereas

$$1 \dot{+} \omega = \sup\{1 \dot{+} n : n < \omega\} = \omega.$$

Definition 3.12. $\alpha \dot{\times} \beta$ is defined by induction on β:

$$\alpha \dot{\times} \beta = \begin{cases} 0 & \text{if } \beta = 0; \\ (\alpha \dot{\times} \gamma) + \alpha & \text{if } \beta = \gamma + 1 \text{ is a successor ordinal;} \\ \sup\{\alpha \dot{\times} \gamma : \gamma < \beta\} & \text{if } \beta \text{ is a limit ordinal.} \end{cases}$$

We can think of $\alpha \dot{\times} \beta$ as the ordertype obtained when a full copy of α is substituted for each element of β. Ordinal multiplication is also noncommutative:

$$\omega \dot{\times} 2 = \omega \dot{+} \omega,$$

whereas

$$2 \dot{\times} \omega = \sup\{2 \dot{\times} n : n < \omega\} = \omega.$$

Definition 3.13. α^β is defined by induction on β:

$$\alpha^\beta = \begin{cases} 1 & \text{if } \beta = 0; \\ (\alpha^\gamma) \dot{\times} \alpha & \text{if } \beta = \gamma + 1 \text{ is a successor ordinal;} \\ \sup\{\alpha^\gamma : \gamma < \beta\} & \text{if } \beta \text{ is a limit ordinal.} \end{cases}$$

Ordinal exponentiation is *not* the same operation as cardinal exponentiation. For example, although $2^{\aleph_0} > \aleph_0$ (cardinal exponentation), we have

$$2^\omega = \sup\{2^n : n < \omega\} = \omega$$

(ordinal exponentiation).

A typical ordinal constructed using these operations is

$$\omega^{\omega^\omega} \times 6 + \omega^{\omega^{11}+\omega^2} + \omega^3 \times 4 + 18.$$

Addition and multiplication are associative, and multiplication distributes over addition as expected. Exponentiation satisfies

$$\alpha^{\beta+\gamma} = \alpha^\beta \times \alpha^\gamma.$$

Normal Form

Theorem 3.14 (Cantor Normal Form Theorem). *Every ordinal α can be expressed uniquely as a finite sum*

$$\omega^{\gamma_1} \dot{\times}\, n_1 \dot{+} \cdots \dot{+}\, \omega^{\gamma_k} \dot{\times}\, n_k$$

with $\gamma_1 > \gamma_2 > \cdots > \gamma_k$ and each $n_i < \omega$.

The Cantor Normal Form Theorem essentially gives a "base ω" expansion for α. There is a corresponding binary expansion:

Theorem 3.15 (Binary Normal Form Theorem). *Every ordinal α can be expressed uniquely as a finite sum*

$$2^{\gamma_1} + \cdots + 2^{\gamma_k}$$

with $\gamma_1 > \gamma_2 > \cdots > \gamma_k$.

Often, the exponents γ_i in the Cantor Normal Form Theorem will satisfy $\gamma_i < \alpha$, so that we obtain an expression for α in terms of smaller ordinals. This is not always the case, however. For example, let

$$\epsilon = \sup\left\{\omega, \omega^\omega, \omega^{\omega^\omega}, \omega^{\omega^{\omega^\omega}}, \dots\right\}.$$

Then ϵ satisfies $\epsilon = \omega^\epsilon$, so the normal form for ϵ involves ϵ itself. These fixed points are an important enough class of ordinals to deserve a special name.

Definition 3.16. An **epsilon number** is an ordinal ϵ that satisfies $\epsilon = \omega^\epsilon$.

We denote the αth epsilon number by ϵ_α. So the example given above is ϵ_0 ("epsilon nought"), and the rest can be described inductively:

$$\epsilon_\alpha = \begin{cases} \sup\left\{\omega^{\epsilon_\gamma+1}, \omega^{\omega^{\epsilon_\gamma+1}}, \omega^{\omega^{\omega^{\epsilon_\gamma+1}}}, \dots\right\} & \text{if } \alpha = \gamma+1; \\ \sup\left\{\epsilon_\gamma : \gamma < \alpha\right\} & \text{if } \alpha \text{ is a limit ordinal.} \end{cases}$$

ϵ_α is countable whenever α is countable, so there is a rich hierarchy of explictly constructible countable ordinals. Even the ϵ notation is insufficient to describe the full hierarchy; for example, let

$$\zeta = \sup\left\{\epsilon_0, \epsilon_{\epsilon_0}, \epsilon_{\epsilon_{\epsilon_0}}, \ldots\right\}.$$

Then ζ is an ϵ-fixed point, in the sense that $\epsilon_\zeta = \zeta$. We could proceed to define higher ζ_α and let η be a ζ-fixed point—then iterate the construction through the transfinite Greek "alphabet" and diagonalize ....

For the purposes of this text, such constructions are of little use, so the reader is directed to the references at the beginning of this section for investigations into the higher transfinite.

Natural Sum and Product

We've been writing $\dot{+}$ and $\dot{\times}$ for ordinal sum and product. This is to distinguish them from the **natural sum** and **product**, which *are* commutative and are the same operations as the Conway sum and product (which is defined over all surreal numbers and studied extensively in Chapter VIII). The natural sum and product of two ordinals can be succinctly defined in terms of their normal forms:

$$\begin{aligned} \text{if} \quad \alpha &= \omega^{\gamma_1} \dot{\times} m_1 \dot{+} \cdots \dot{+} \omega^{\gamma_k} \dot{\times} m_k \\ \text{and} \quad \beta &= \omega^{\gamma_1} \dot{\times} n_1 \dot{+} \cdots \dot{+} \omega^{\gamma_k} \dot{\times} n_k, \end{aligned}$$

then

$$\alpha + \beta = \omega^{\gamma_1} \dot{\times} (m_1 + n_1) \dot{+} \cdots \dot{+} \omega^{\gamma_k} \dot{\times} (m_k + n_k)$$

and

$$\alpha \times \beta = \sum_{i,j=1}^{k} (\omega^{\gamma_i + \gamma_j} \dot{\times} m_i n_j).$$

(The summation $\sum$ and the exponents $\gamma_i + \gamma_j$ are natural sums.) In other words, $\alpha + \beta$ and $\alpha \times \beta$ are obtained by adding and multiplying the normal forms for α and β like ordinary polynomials.

Then it's not hard to show that

$$\omega^\gamma \times n + \alpha = \omega^\gamma \dot{\times} n \dot{+} \alpha$$

for all γ, n, and $\alpha < \omega^\gamma$. It follows that Cantor normal form can be written in terms of natural sum and product, with exactly the same terms as before:

$$\omega^{\gamma_1} \times n_1 + \cdots + \omega^{\gamma_k} \times n_k = \omega^{\gamma_1} \dot{\times} n_1 \dot{+} \cdots \dot{+} \omega^{\gamma_k} \dot{\times} n_k.$$

Therefore ordinal and natural operations are interchangeable in expressing normal forms.

4. Commutative Semigroups

A detailed treatment of this material can be found in [**Gri01**].

Definition 4.1. A **commutative semigroup** is a set $\mathcal{S}$, together with a binary operation $\cdot$ on $\mathcal{S}$, that satisfies the following axioms:

(i) (associativity) $x \cdot (y \cdot z) = (x \cdot y) \cdot z$, for all $x, y, z \in \mathcal{S}$;

(ii) (commutativity) $x \cdot y = y \cdot x$, for all $x, y \in \mathcal{S}$.

A commutative semigroup $\mathcal{S}$ is a **commutative monoid** if there exists an **identity element** $1 \in \mathcal{S}$ such that:

(iii) (identity) $x \cdot 1 = x$ for all $x \in \mathcal{S}$.

We'll usually write xy as shorthand for $x \cdot y$.

Definition 4.2. An **idempotent** is an element $z \in \mathcal{S}$ such that $z \cdot z = z$.

Definition 4.3. A **subgroup** of $\mathcal{S}$ is a subset $\mathcal{A} \subset \mathcal{S}$ such that $\mathcal{A}$ is isomorphic to an Abelian group. We say $\mathcal{A}$ is a **maximal subgroup** if it is maximal among all subgroups of $\mathcal{S}$.

If $\mathcal{A}$ is a subgroup of $\mathcal{S}$, then $x \cdot x = x$ for a unique $x \in \mathcal{A}$, so that $\mathcal{A}$ contains exactly one idempotent. Conversely, given an idempotent $x \in \mathcal{S}$, there is a unique maximal subgroup of $\mathcal{S}$ containing x (obtained, say, by taking the union of all groups containing x). It follows that there is a one-to-one correspondence between idempotents and maximal subgroups of $\mathcal{S}$.

There is a natural partial order among idempotents of a semigroup:

$$x \leq y \quad \text{if} \quad xy = x.$$

If $xy = x$, then we say x **absorbs** y (or is **stronger** than y). If x and y are idempotents, then certainly $xy \leq x$ and $xy \leq y$. Moreover any idempotent z that absorbs both x and y certainly absorbs xy as well, so that xy is a greatest lower bound for x and y. It follows that the idempotents of $\mathcal{S}$ form a lower semilattice (i.e., a partial order with a meet operation, but not necessarily a join).

Let $\equiv$ be an equivalence relation on $\mathcal{S}$. If $x \equiv y$ implies $xz \equiv yz$, for all $x, y, z \in \mathcal{S}$, then $\equiv$ is a **congruence**. Given a semigroup $\mathcal{S}$ and a congruence $\equiv$, the set of equivalence classes forms a **quotient semigroup** $\mathcal{S}/{\equiv}$.

The **mutual divisibility** congruence is a fundamental example.

Definition 4.4. Let $x, y \in \mathcal{S}$. We say x **divides** y, and we write $x \mid y$, if $xz = y$ for some $z \in \mathcal{S}$. If $x \mid y$ and $y \mid x$, then we say x and y are **mutually divisible**.

Certainly $x \mid y$ implies $xz \mid yz$, so mutual divisibility is a congruence. Its congruence classes are the **mutual divisibility classes** of $\mathcal{S}$. Each maximal subgroup is a mutual divisibility class.

Generators and Relations

Let $\mathcal{S}$ be a commutative semigroup and let $\mathcal{X} \subset \mathcal{S}$ be a subset. Then the subsemigroup of $\mathcal{S}$ **generated by** $\mathcal{X}$, denoted by $\langle \mathcal{X} \rangle$, is the smallest subsemigroup of $\mathcal{S}$ containing $\mathcal{X}$. If $\mathcal{S} = \langle \mathcal{X} \rangle$ for some finite $\mathcal{X}$, then we say $\mathcal{S}$ is **finitely generated**.

Finitely generated Abelian groups are completely characterized by the structure theorem (Theorem 1.6 on page 461). The structure of finitely generated commutative semigroups is not quite so rigid, but can still be nicely described in the following way.

Definition 4.5. The **free commutative semigroup** on the generators $\mathcal{X}$ is the set of all monomials in $\mathcal{X}$, i.e., the set of all expressions of the form

$$x_1^{n_1} x_2^{n_2} \cdots x_k^{n_k}$$

with each $x_i \in \mathcal{X}$, each $n_i \geq 1$, and $k \geq 1$.

The **free commutative monoid** on the generators $\mathcal{X}$ is defined the same way, but allowing $k = 0$.

If $\mathcal{X}$ is a set of generators and $\mathcal{F}$ is the free commutative semigroup on $\mathcal{X}$, then a **relation** on $\mathcal{X}$ is a pair of elements (u, v), with $u, v \in \mathcal{F}$. A relation is normally written $u = v$.

A **semigroup** (or **monoid**) **presentation** is given by a generating set $\mathcal{X}$ together with a set of relations $\mathcal{R}$, denoted $\langle \mathcal{X} \mid \mathcal{R} \rangle$. A typical monoid presentation is

$$\langle a, b \mid a^2 = 1,\ b^3 = ab \rangle.$$

If $\langle \mathcal{X} \mid \mathcal{R} \rangle$ is a presentation, then we think of $\mathcal{R}$ as imposing certain equivalences on combinations of elements of $\mathcal{X}$. These equivalences include those explicitly mentioned by $\mathcal{R}$, but there might be others as well, possibly arising from the interaction between several different elements of $\mathcal{R}$. For example, the relation $b^3 = ab$ in the previous example implies $b^6 = a^2 b^2$, which together with $a^2 = 1$ implies $b^6 = b^2$.

There must always exist a certain *minimal* congruence generated by $\mathcal{R}$, in the following sense.

Theorem 4.6. *For any semigroup (or monoid) presentation $\langle \mathcal{X} \mid \mathcal{R} \rangle$, there exists a congruence $\equiv$ on $\mathcal{F}$ such that*

(i) *$u \equiv v$ for each relation $(u, v) \in \mathcal{R}$; and*

(ii) *if* $\sim$ *is any congruence satisfying (i), then* $x \sim y$ *implies* $x \equiv y$.

We can therefore identify each presentation $\langle \mathcal{X} \mid \mathcal{R} \rangle$ with the semigroup (or monoid) $\mathcal{F}/\equiv$, with $\equiv$ given in Theorem 4.6. This is the semigroup **presented by** $\langle \mathcal{X} \mid \mathcal{R} \rangle$, and it obeys a key universality property: any semigroup generated by $\mathcal{X}$ and satisfying the relations $\mathcal{R}$ must be a quotient of $\langle \mathcal{X} \mid \mathcal{R} \rangle$.

Definition 4.7. A semigroup (or monoid) $\mathcal{S}$ is said to be **finitely presented** if $\mathcal{S} \cong \langle \mathcal{X} \mid \mathcal{R} \rangle$, for some finite set $\mathcal{X}$ and finite set of relations $\mathcal{R}$.

Finitely presented semigroups are computationally effective. Given a finite presentation $\mathcal{S} \cong \langle \mathcal{X} \mid \mathcal{R} \rangle$, we first order the elements of $\mathcal{X}$ (in an arbitrary way). We then choose the relations $(u, v) \in \mathcal{R}$ so that $u < v$, under the induced lexicographic ordering of $\mathcal{F}$. (This is always possible, since the relations (u, v) and (v, u) are equivalent.) Then given an arbitrary element $x \in \mathcal{F}$ that represents an element of $\mathcal{S}$, we can find a canonical expression for x by successively replacing subexpressions v by u, for some $(u, v) \in \mathcal{R}$, until no further replacements are possible. Since the lexicographic ordering of $\mathcal{F}$ is well-founded, this procedure must eventually terminate, and so we obtain an algorithm for multiplying arbitrary elements of $\mathcal{S}$.

We can now state the main result in the structure theory of finitely generated commutative semigroups.

Theorem 4.8 (Rédei's Theorem). *Every finitely generated commutative semigroup is finitely presented.*

Finite Semigroups

When $\mathcal{S}$ is finite, we can say much more about its structure.

Lemma 4.9. *If* $\mathcal{S}$ *is finite and* $x \in \mathcal{S}$, *then* x^n *is an idempotent for some integer* $n \geq 1$.

Proof. The sequence of powers of x,

$$x,\ x^2,\ x^3,\ x^4,\ \ldots,$$

must contain a duplicate, say $x^r = x^{r+k}$ $(k > 0)$. Let n be the unique integer with $r \leq n < r + k$ and $n \equiv 0 \pmod{k}$. Then writing $n = tk$, we have

$$x^{2n} = x^{n+tk} = x^{r+tk} x^{n-r} = x^r x^{n-r} = x^n. \qquad \square$$

The idempotent given by Lemma 4.9 uniquely associates each $x \in \mathcal{S}$ to one of the maximal subgroups of $\mathcal{S}$, and this partitions $\mathcal{S}$ into **Archimedean components**. These components can be described in another way:

Definition 4.10. If $x, y \in \mathcal{S}$, then $x \sim y$ if $x \mid y^m$ and $y \mid x^n$ for some $m, n \geq 1$.

Certainly $x \sim y$ is a congruence. Moreover, if $x \in \mathcal{S}$ and y is the unique idempotent equal to some power of x, then $x \sim y$. So every congruence class modulo $\sim$ contains exactly one idempotent, and this characterizes the Archimedean components of $\mathcal{S}$.

The Kernel

If $\mathcal{S}$ is finite, then its poset of idempotents must also be finite. It therefore contains a minimal element z, obtained by multiplying together all idempotents of $\mathcal{S}$:

$$z = \prod \{x \in \mathcal{S} : x \text{ is an idempotent}\}.$$

Clearly $z \leq x$ for all idempotents $x \in \mathcal{S}$. The maximal subgroup $\mathcal{K}$ containing z is the **kernel** of $\mathcal{S}$ and has a number of quite nice properties:

Proposition 4.11. *Let $\mathcal{K}$ be the kernel of $\mathcal{S}$ and let $z \in \mathcal{K}$ be its idempotent.*

(a) $\mathcal{K} = z\mathcal{S} = \{zx : x \in \mathcal{S}\}$.

(b) *The map $x \mapsto zx$ is a surjective homomorphism from $\mathcal{S}$ onto $\mathcal{K}$.*

(c) *$\mathcal{K}$ is isomorphic to the group of fractions of $\mathcal{S}$.*

(d) *If $f : \mathcal{S} \to \mathcal{A}$ is a homomorphism onto an Abelian group $\mathcal{A}$, then f factors through the map $x \mapsto zx$. That is, there exists a map $g : \mathcal{K} \to \mathcal{A}$ making the following diagram commute:*

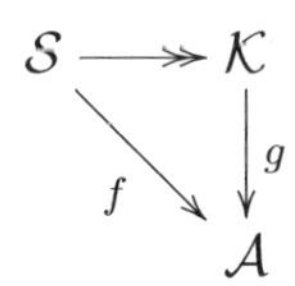

A similar idea shows that if $\mathcal{S}$ is a finite *monoid*, then its poset of idempotents is a lattice. Given $x, y \in \mathcal{S}$, put

$$z = \prod \{w \in \mathcal{S} : w \text{ is an idempotent and } w \geq x \text{ and } w \geq y\}.$$

Then z is a least upper bound for x and y. The fact that $\mathcal{S}$ is a monoid guarantees that the product set is nonempty, since $1 \geq x$ and $1 \geq y$ for all x and y.

We arrive at a remarkably clear picture of the structure of a finite commutative monoid $\mathcal{S}$. It is partitioned into a lattice of maximal subgroups, together with nilpotent extensions for each subgroup. Multiplication over $\mathcal{S}$ respects both the structure of the lattice and the structure of each maximal subgroup.

Appendix C

A Finite Loopfree History

Combinatorial game theory is a rich and fascinating subject, with a fittingly rich and fascinating history. What follows is just a part of its story, tracing the evolution of the disjunctive theory into a mature mathematical discipline, from ancient times down to about 1994. As with any history, many important topics are skimmed or omitted, and the choice of emphasis undoubtedly reflects the author's biases. One hopes that, nonetheless, even this brief survey will instill a sense of wonder: at the efforts of a great many people to craft a singular and beautiful theory, driven by dedication and inspiration and, above all, an overriding love of games.

Origins

Games have been played and studied by many civilizations throughout history. SENET, perhaps the oldest board game known to present-day scholars, has been discovered in predynastic Egyptian tombs dating to at least 3000 BCE. The ROYAL GAME OF UR originated in Mesopotamia around the same time; several of its variants are still played in modern-day Iraq. Mancala games have been played in Africa for centuries; in Mesoamerica, PATOLLI was popular as early as 600 AD. All of these are two-player perfect-information games, although several involve random elements such as dice.

Around 2,500 years ago (and perhaps even earlier), there arose in China a game of extraordinary depth, beauty, and complexity. Known as WEIQI in China, BADUK in Korea, and by its Japanese name GO most everywhere else, its popularity has never waned. Today it enjoys a huge following, with tens of millions of players throughout the world, including several thousand active professionals.

Go was also one of the earliest games to be studied in a systematic way. Texts devoted to history and strategy appeared in China as early as 1100 AD, and by the 14th century there was a rich literature of problem books, opening theory, and general strategy.

Similar developments occurred in parallel in the West. By 1600 a healthy debate had ensued on modern Chess strategy; a 1561 treatise by Ruy López challenged the contemporary opening orthodoxy. Endgame theory also entered the literature around this time, marking a conscious effort to identify classes of positions with exact solutions.

These early studies were largely pedagogical, and efforts to develop a *mathematical* theory of games did not take place until much later. A 1713 letter from James Waldegrave to the French probabilist Pierre-Remond de Montmort describes perhaps the first such attempt: a complete solution to two-player Le Her, a card game popular at the time. Waldegrave's solution is essentially a minimax strategy, but it appears he made no attempt to extend it into a more general theory.

By the late 19th century, recreational mathematics had evolved into a mature subject. It was popularized in the United States by Sam Loyd, and in Britain by Henry Dudeney; both authors wrote prolifically about games and puzzles of all sorts. W. W. Rouse Ball's classic *Mathematical Recreations and Essays*, later revised by Coxeter, first appeared in 1892. Contemporaneously in France, Édouard Lucas published a monumental four-volume treatise on recreational mathematics containing signficant material on combinatorial games, including a complete strategy for Hare and Hounds.

These developments occurred in parallel with the great formalization program that transformed mathematical thought at the turn of the 20th century. The German mathematician Ernst Zermelo was a central force in these efforts. Best known for his work on axiomatic set theory (he's the Z in ZFC), Zermelo was also an avid Chess player, and he was particularly fascinated by endgame problems in which the outcome could be completely determined. To Zermelo, who had spent years developing the foundations of set theory, it was natural to seek a formal setting for mathematical games as well. In a 1913 paper, *On an application of set theory to the theory of the game of Chess* [**Zer13**], he addressed the question:

> Can the value of an arbitrary position ... be determined, or at least defined, in a mathematically objective manner? [**Zer13**]

Zermelo proceeded to outline a remarkable chain of reasoning, decades ahead of its time. He observed that the set of possible Chess positions is finite and that any play can be represented as a sequence of consecutive positions. He then proved by induction that the outcome of every CHESS position is determined, and he generalized to arbitrary finite combinatorial games. This result is now known as the Fundamental Theorem (Theorem I.1.5 on page 9); elsewhere in the literature it is sometimes called *Zermelo's Theorem*. Zermelo's key observation, that positions and moves in combinatorial games can be framed in an abstract mathematical setting, set the stage for all future work in game theory.

Zermelo's result has been credited as "the first formal theorem in the theory of games" [**SW01**]—so that game theory itself was incubated in a combinatorial context, fifteen years prior to von Neumann's first paper on the subject. But even Zermelo was not really the first, for the theory of NIM preceded his work by more than a decade.

NIM and the Impartial Theory

The precise origins of NIM are uncertain. JIAN SHIZI, a similar take-away game, was played in ancient China; it is unclear whether NIM is derived from it or arose independently in Europe. Whatever the case, there is clear evidence of NIM being played in 16th-century Italy, and by the late 1800s it was a popular pub game in both Europe and America.

In the 1890s the game came to the attention of the American geometer Charles Leonard Bouton. Bouton spent two years in Leipzig, 1896–1898, completing his doctorate under the great Norwegian mathematician Sophus Lie. It is possible that he learned of NIM during this time (the name "NIM" resembles the German word *nimm*, meaning *take*); however Bouton himself never mentioned any such connection. Bouton quickly recognized that NIM admits a concise winning strategy in both normal and misère play. In 1901 he published a paper, *Nim, a game with a complete mathematical theory* [**Bou01**], in which he classified the $\mathscr{P}$-positions and proved the nim-addition rule.

At a stroke, combinatorial game theory was born, but it was slow to catch on, and the next 30 years saw scattered progress. The NIM variant WYTHOFF was studied by the Dutch number theorist Willem Abraham Wythoff, who published an analysis of its $\mathscr{P}$-positions (our Theorem IV.3.4) in a 1907 paper. Another variant, NIM_k, was solved by E. H. Moore in 1910 (see Exercise IV.1.3 on page 182).

KAYLES also arose around this time. It was introduced by Dudeney, for whom a "heap" was a row of bowling pins:

> It is assumed that the ancient players had become so expert that they could always knock down any single kayle-pin, or any two kayle-pins

> that stood close together. They therefore altered the game, and it was agreed that the player who knocked down the last pin was the winner. [**Dud19**]

In the late 1930s Michael Goldberg analyzed KAYLES up to heaps of size 20 but failed to arrive at the full solution.

Emanuel Lasker was perhaps the most sustained advocate of combinatorial games throughout the 1920s. Today Lasker is best known as history's longest-reigning CHESS champion (1894–1921) and as a research mathematician who made important contributions to commutative algebra. But he also had a passionate interest in games of all sorts, and mathematical games lay conveniently at the intersection of these interests. Among his inventions was LASKER'S NIM ($4.\overline{3}$ in octal notation); he published a solution in his 1931 book *Brettspiele der Völker* [**Las31**]. Lasker understood the Fundamental Theorem and the outcome classes $\mathscr{P}$ and $\mathscr{N}$ but stopped short of discovering the full Sprague–Grundy theory.

In addition to his other achievements, Lasker was an instrumental force in popularizing GO in the west. He had immense respect for GO and wrote in *Brettspiele der Völker* (quoted by Jörg Bewersdorff):

> GO has a much more penetrating logical structure than CHESS. It surpasses CHESS in simplicity and is its equal in its demands on the players' imagination. [**Bew04**]

No trifling words from a world CHESS champion!

The full impartial theory was finally isolated in the 1930s, independently by Roland Sprague and Patrick Michael Grundy. Sprague was first out of the gate, but unfortunately his results were published in an obscure Japanese journal, and wide distribution was hindered by the political conditions of the time. Sprague's work went largely unrecognized until much later, and today Grundy's name is most commonly associated with the theory.

Grundy was just 22 when his breakthrough was published, in 1939. Although Grundy clearly recognized its wide applicability, it appears he knew of few examples: he focused almost exclusively on NIM and GRUNDY'S GAME (cf. page 185), with no mention of KAYLES. After writing a dissertation on commutative algebra, Grundy continued on to a career in experimental statistics and education research, and his theory of impartial games fell into neglect. Systematic applications of the Sprague–Grundy theory would have to await Richard Guy, who burst onto the scene in the late 1940s with a deep reservoir of examples and has remained a central figure in the subject right up to the present day.

Richard Guy

Richard Kenneth Guy was born in 1916 in Warwickshire, England. The son of schoolteachers, Guy developed an early interest in mathematics and a lifelong love of CHESS. As a student at Caius College, Cambridge, Guy spent a great deal of time playing and studying CHESS, and he was particularly fascinated with the endgame, "where complete analysis was often possible." He became well known as a composer of endgame problems, winning several prizes and publishing in the *British Chess Magazine*.

During the war Guy served as a meteorologist in the Royal Air Force, in Iceland and Bermuda. Upon returning to England, he moved to southeast London to lecture at Goldsmiths College. There he found himself a bicycle ride away from the noted CHESS problemist Thomas Rayner Dawson. A rubber chemist by trade, Dawson had achieved fame as a prolific composer of fairy chess problems (involving exotic pieces not used in ordinary CHESS games), composing over 5,000 of them during his lifetime. At the time, Dawson was retired and served as editor of both the problems and endgames sections of the *British Chess Magazine*. His health was failing and he found endgames to be something of a distraction, so he suggested that Guy take over as endgames editor. Guy happily accepted, and the two of them interacted frequently over the next few years.

Dawson, a successful amateur mathematician, loved to discuss combinatorial problems, and in 1947 he showed Guy a problem he had invented, the game now known as DAWSON'S CHESS. Dawson posed it as a misère game and understandably found it difficult to make much progress. Guy was intrigued. Fortuitously, he forgot that Dawson had proposed misère play and instead began to study the normal-play version. Unaware of Sprague and Grundy's earlier work, but familiar with Goldberg's partial results on KAYLES, Guy soon (re-)discovered the full impartial theory. He used the theory to calculate many values of DAWSON'S CHESS, enough to make their periodicity apparent. Repeating the exercise with KAYLES, Guy quickly realized that his analysis generalizes to a wide class of games, and he converged on the definition of octal game and a proof of the periodicity theorem (Theorem IV.2.7 on page 190).

Excited by his results, Guy approached Theodor Estermann, a professor of number theory with whom he was studying at the time, to ask if the results had any interest. Estermann replied that he had no idea and instead directed Guy to speak with Cedric Austen Bardell Smith.

As it turns out, Smith had been an exact contemporary of Guy's at Cambridge—Guy recalls seeing him in lectures—although they had no interaction there. After graduating from Cambridge, Smith was hired as a statistician at the Galton Laboratory, a genetics research center at the University College London.

Smith was familiar with Grundy's earlier discoveries. He was delighted by Guy's results—for although Grundy had discovered the full impartial theory, he had never found many applications for it, apart from GRUNDY'S GAME and a few trivial examples. Guy's octal games introduced a vast new space of possibilities, and he and Smith extended his results and in 1949 wrote them up into a paper. Unfortunately it was rejected by the London Mathematical Society and didn't see publication until Guy and Smith resubmitted it to the Proceedings of the Cambridge Philosophical Society in 1956.

The resulting paper, *The G-values of various games* [**GS56**], marked a pivotal moment in the history of combinatorial game theory. While Sprague and Grundy first obtained the complete impartial theory, the Guy–Smith paper demonstrated that the theory has genuine substance: there are nontrivial problems, such as KAYLES, that become vastly simpler in the proper theoretical context.

The 1950s

Several other notable developments occurred in parallel to the work of Guy and Smith. In 1956 *Scientific American* recruited a little-known journalist named Martin Gardner to write a column on recreational mathematics. Over the next 25 years his column, *Mathematical Games*, evolved into an immensely popular and inspirational force in the subject. Many great young mathematicians, including Berlekamp and Conway, were inspired by and contributed to Gardner's columns over the years.

Although Gardner had little formal mathematical training, he became perhaps the most influential force in recreational mathematics of the 20th century. And though he made few direct theoretical contributions to combinatorial game theory, the importance of his role in popularizing and legitimizing the subject cannot be overstated. *Winning Ways* itself is dedicated:

> To Martin Gardner, who has brought more mathematics to more millions than anyone else. [**BCG01**]

Early investigations into partizan combinatorial games also took place during this period. In the 1950s, John Milnor studied games with real-valued payoffs assigned to terminal positions and proved a version of the mean value theorem [**Miln53**]. Milnor's work appears to be the first serious effort to develop a disjunctive theory of partizan games.

The Milnor theory was substantially refined by Olof Hanner in a 1959 paper, *Mean play of sums of positional games* [**Han59**]. Hanner understood the distinction between sente and gote and converged on the "correct" definition of mean value. His work anticipates many ideas in the later development of the temperature theory.

The advent of computers around this time opened new horizons in combinatorial games research. The EDSAC, one of the first computers in Britain, went online at Cambridge in 1949. Soon afterward Guy and Smith enlisted C. Brian Haselgrove, a fellow Cambridge graduate, to program it to calculate several hundred values of various octal games. Meanwhile Grundy, now working as a statistician at the Rothamsted Experimental Station, computed more than a thousand values of GRUNDY'S GAME. Thus computers and combinatorial games were inextricably bound together from the earliest days of both.

Sadly, Grundy's health began to fail around this time, and he died at age 42 in mysterious circumstances. According to Guy,

> Grundy had psychological troubles.... He probably committed suicide .... He fell off a cliff in south England, and it seems very likely he actually threw himself off.... He had done a number of quite nice things, but he thought he wasn't appreciated. [**Guy11**]

Today Grundy's influence on combinatorial game theory is inescapable, and it is a tragic irony that he did not live to see the full impact of his work. Yet just as one chapter in the history of combinatorial games came to a close, another would soon begin. Shortly after Grundy's death, Richard Guy made his next great contribution to combinatorial game theory: he attracted the interest of John Horton Conway, a bright graduate student who had just arrived at Cambridge.

John Conway

Conway was born in Liverpool in 1937. He showed a prodigious interest in mathematics from an early age. Conway recalls:

> My mother said that when I was three or four, I knew the powers of two. When I was about eleven, ... at an interview for my new school, I told the interviewer that I wanted to be a mathematician at Cambridge. [**Sei94**]

This dream stayed with him and after receiving his BA in 1959, Conway enrolled in the Cambridge mathematics department. He achieved early success by proving a difficult conjecture of Davenport: every integer can be written as the sum of 37 fifth powers. With his doctorate secure, Conway was free to study whatever he wished, and he soon became fascinated with transfinite numbers in set theory.

Conway's interest in games was also kindled around this time. In 1960, Richard Guy's son Michael entered Cambridge as an undergraduate, and he and Conway soon became friends. Richard had previously taught Michael much of what he knew about combinatorial games, and Michael passed this knowledge on to Conway, who became keenly interested. Soon Michael introduced Conway to his father, launching a decades-long friendship and collaboration. Almost immediately, Conway asked what happens in the partizan case, though ten years were to pass before he could formulate a satisfactory answer.

After obtaining his Ph.D., Conway remained at Cambridge as a lecturer in mathematical logic. He soon became disillusioned with set theory:

> In my late twenties, I became very depressed. I felt that I wasn't doing real mathematics; I hadn't published, and I was feeling very guilty because of that. [**Sei94**]

Then in 1965, John Leech discovered the *Leech lattice*, a dense sphere packing in 24-dimensional Euclidean space. Upon learning of the Leech lattice, Conway successfully isolated its symmetry group, leading in turn to several other surprising discoveries, including three previously unknown finite simple groups. This work catapulted Conway to fame and earned him an international reputation as a mathematical innovator.

At around this time, Richard Guy left for Canada, settling permanently in Calgary. He continued to return to Cambridge each summer to work with Conway. Then in 1967, Guy attended the Conference on Combinatorial Mathematics and Its Applications in Chapel Hill, NC, where he was approached by an enterprising young mathematician named Elwyn Ralph Berlekamp. Berlekamp had read the Guy–Smith paper and had applied the analysis of KAYLES in a nontrivial way to obtain new DOTS AND BOXES strategies. He was quite excited by these results, and he suggested to Guy that the two of them write a book on combinatorial games. Guy replied that he knew just the person who could help them, and *Winning Ways* was born.

Elwyn Berlekamp

Elwyn Berlekamp's interest in games began in childhood. He recalls being "hooked on games from at least the age of six" [**Ber12**]; in grade school, he would often sit in the back of the classroom and play DOTS AND BOXES. Growing up in postwar Ohio and Kentucky, Berlekamp became a fan of Martin Gardner's *Scientific American* column, through which he was exposed to many ideas in mathematical games.

In 1960, while still an undergraduate, Berlekamp interned at Bell Labs as research assistant for the pioneering investment theorist John Kelly. Kelly showed him Bouton's strategy for NIM, and not long afterward he encountered the work of Guy and Smith. Berlekamp later wrote:

> I was awestruck [by] Bouton's solution to the game of NIM, and awestruck again when I first read some of the work of Richard Guy and Cedric Smith. The fact that some games have mathematically based winning strategies made a big impression on me. [**Ber00a**]

As a student at MIT, Berlekamp became an avid BRIDGE player. He wrote a master's thesis on algorithms for solving double-dummy BRIDGE problems. He also contributed a regular BRIDGE column to *The Tech*, MIT's student newspaper, earning him a reputation as a games enthusiast.

One day he was approached by a group of students who had written a computer program to play 3×3 DOTS AND BOXES. They challenged Berlekamp to a demonstration game, and to his surprise, the computer won. With his interest piqued, Berlekamp spent the next several days studying the game, and soon discovered a perfect winning strategy for second player. He returned to challenge the computer to a rematch, and defeated it soundly, winning seven games out of ten. The theory of DOTS AND BOXES would continue to hold his interest for the next fifty years.

Berlekamp completed his doctorate in 1964 and was immediately recruited to join the engineering faculty at UC Berkeley. There his interest in games intensified. He discovered solutions to DUKEGO (a simpler variant of the famous Angel problem) and new strategies for HARE AND HOUNDS that applied to larger board sizes.

In 1967 Berlekamp left Berkeley for Bell Labs. It was there that he recognized the connection between DOTS AND BOXES and KAYLES. Soon afterward he met Guy at the Chapel Hill conference, and Guy invited him to give a talk on games at Calgary, which further stimulated his interest.

Berlekamp wrote up his results on DOTS AND BOXES in a technical report, which to his surprise became enmeshed in Bell Labs' internal politics: it was returned to Berlekamp's desk with a note asking him to explain how it pertained to the "needs of the Bell system." In a six-page response, Berlekamp argued that the study of DOTS AND BOXES contributes to an understanding of abstract graph theory, which is vital to Bell's long-term research interests. Fortunately the senior executives at Bell were sympathetic to Berlekamp's cause and encouraged him to continue to pursue his curiosity, but the incident left a lasting impression. Berlekamp emerged with increased conviction that it is important to study games and to educate the public about their value.

In 1969 Guy introduced him to Conway at an Oxford number theory conference. Berlekamp describes their meeting:

> We skipped the conference and spent one day at one of the pubs, [playing various games] for ten hours or so .... Neither of us ever beat the other person at one of his own games, but the other always had more games that *he* could always win! [**Ber12**]

It was clear that writing a book on games together was a good idea, and so began a journey that would occupy all three authors for more than a decade. "We imagined we'd finish it in a couple years," Berlekamp later said, "but it became an addiction." [**Ber12**]

Winning Ways

Berlekamp, Conway, and Guy initially envisioned a book collecting all the results then known: the Guy–Smith theory of octal games; Berlekamp's work on DUKEGO and HARE AND HOUNDS; and of course DOTS AND BOXES. But shortly after they began, there occurred a pivotal event that dramatically altered the scope and depth of the subject matter. In Conway's words:

> We had the British GO champion [Jon Diamond] in our math department at Cambridge, and I used to watch him play games in the vain hope that someday I would understand them. I never did. But I did see that in the end, a GO game decomposed into a sum of little games, and I thought it was a good idea to study this kind of sum. You could do the same thing with other games—CHECKERS, DOMINOES. And then I discovered that certain games behaved very much like numbers. [**Shu95**]

Conway had dreamed of a partizan analogue ever since Michael Guy first introduced him to the impartial theory a decade earlier. But it was GO that finally inspired him to discover the axiomatic theory of partizan games, and—to his astonishment—there was a novel theory of numbers embedded within. To Conway, with his background in mathematical logic and number theory, it was natural to consider transfinite games alongside finite ones, and almost immediately this led him to discover the surreal numbers:

> The theory was a real big shock to me. It was bizarre ... crazy, but it was true! It was like climbing to the top of the beanstalk, and there was the enchanted castle. I had no idea what to expect.... Exploring that world took me some time. [**Sei94**]

Berlekamp and Guy were also captivated by the new theory of partizan games, though they were mostly interested in finite games. Many of the examples introduced in *Winning Ways*, including HACKENBUSH and TOADS AND FROGS, arose around this time.

The early 1970s was a period of intense activity and collaboration. In early 1970 Berlekamp spent a semester at the University of London and made frequent trips to Cambridge to visit Conway. Guy too continued to visit Cambridge for several weeks each summer, and he organized semi-annual meetings in Calgary where all three could meet and coordinate, a tradition that would continue for more than a decade. Guy describes their collaboration:

> The book *Winning Ways* is a collection of essays. The ideas came from Berlekamp and Conway, but they were mainly written by me, almost entirely from Conway's dictation, either of his original ideas or of the ideas written out by Berlekamp .... [**AA93**]

In 1971 Berlekamp left Bell Labs and returned to Berkeley. In a fortunate coincidence, at almost exactly the same time, Conway began sabbatical at Caltech, and the two of them met almost weekly during this period.

Berlekamp, Conway, and Guy gradually pieced together the rudiments of the partizan theory. Early on, $\uparrow$ and $*$ were recognized as infinitesimals, but their structure was little understood and they were called simply ϵ and δ.

All three co-authors were assisted by a succession of excellent graduate students. As early as 1967 Jack Kenyon, a master's student of Richard Guy's at Calgary, published the first thesis on combinatorial games. Kenyon found new solutions to octal games with longer periods (including **0.156** of period 349) and carried out some early investigations into hexadecimal games (see Exercise IV.2.11 on page 196). Richard Austin, with the help of a new generation of computers, carried these efforts still further in his 1976 master's thesis.

In the mid-1970s Conway spent a semester at Calgary, and together he and Guy taught a full-year course on combinatorial games. Among their students was Richard Nowakowski, who at the time was studying graph theory with Guy. Nowakowski was hooked, and his interest in games stayed with him after he graduated. He went on to become (and still remains) one of the most active and influential contributors to the subject in recent times.

Starting immediately after his return to Berkeley, Berlekamp regularly taught a course on combinatorial games, through which he recruited many graduate students over the years.[1] One of his students, Shuo-Yen Robert Li, completed the first combinatorial games Ph.D. in 1974. Shortly afterwards Li published the first paper on loopy games (see Notes to Section VI.4), and he made several other contributions to combinatorial games before embarking on a decorated career in coding theory and information science. Dean Hickerson, another of Berlekamp's students, made important contributions

[1] Including this author, more than 30 years later!

to the temperature theory (more on this below) and was first to show that there are 1,474 games born by day 3. Hickerson is also noted for his many contributions to the theory of CONWAY'S GAME OF LIFE.

At Cambridge, Clive Bach was deeply involved in the theory of loopy games; in particular, Bach's Carousel (Figure VI.4.3 on page 316) is named for him. Bach also made several contributions to surreal numbers and transfinite nim arithmetic.

Out of all the students who were involved with *Winning Ways*, Simon Norton had the most profound impact. Today Norton is best known for his work on group theory: he and Conway proposed the famous *monstrous moonshine* conjecture (Richard Borcherds's proof of this conjecture earned him a Fields Medal in 1998). Though his work on combinatorial games is less publicized, he was a key contributor to much of the theory in *Winning Ways*. Among his inventions are Norton's Lemma, the atomic weight calculus, and the game SNORT; he also contributed heavily to the development of the temperature theory and to the structure theory of $\mathbb{G}$ and **PG** (see Notes to Sections III.3 and VIII.1).

Norton's influence on *Winning Ways* is perhaps best summarized by the last sentence of its preface:

> And thank you, Simon! [**BCG01**]

Toward Publication

Work on *Winning Ways* continued and the theory began to take shape. Berlekamp, Conway and Guy had very different working styles. Berlekamp's formal mathematical style clashed with Conway's more whimsical approach, often leading to intense disagreements about the structure and content of the text. Guy recalls that:

> Occasionally Conway and Berlekamp would almost come to blows .... [Berlekamp] wanted to get on with the work and we were making puns back and forth or going off into some complete irrelevance, nothing to do with mathematics.... But it was wasting time as far as he was concerned. [**AA93**]

Differences in vision also emerged. Conway's fascination with the surreal numbers deepened, whereas Berlekamp—though he remained keenly interested in the partizan theory—felt that emphasizing transfinite numbers would dramatically reduce the breadth of the book's appeal.

Things came to a head in 1975. Conway became increasingly frustrated over the direction of *Winning Ways*. He desperately wanted to publish the theory of surreal numbers, and as it became clear that *Winning Ways* would not be an appropriate forum, Conway hatched the idea for *On Numbers and Games*. He later wrote:

> A busy term was approaching, and it seemed that this "transfinite" material just had to be got out of the way before that term started if *Winning Ways* was ever to be published. So I sat down for that week and wrote [*On Numbers and Games*], and then confessed the fact to my co-authors. [**Con01**]

Conway's "confession" caught his co-authors completely by surprise. It was ill-received by Berlekamp, for in addition to the surreal numbers, ONAG included a substantial amount of joint work on games. Fortunately Guy was a patient mediator and despite ONAG's publication in 1976, work on *Winning Ways* resumed with renewed energy.

The scope of the project continued to increase, growing to two volumes and involving the use of color printing and unusual symbols and typefaces. The publisher, Academic Press, proved willing to accommodate the authors' requests, helping to give *Winning Ways* its unique flair and authority. Unfortunately, as publication neared, executives at Academic Press became concerned with its growing expense—and responded by eliminating its advertising budget! So in 1982 *Winning Ways* finally went to press, but with little official promotion. Compounding the problem, Martin Gardner, who had been eager to promote *Winning Ways* in his *Scientific American* column, retired in 1981, just before publication.

Despite these setbacks, *Winning Ways* was a resounding critical success, and for the first time combinatorial game theory began to gain currency in a serious way. It was immediately recognized as a cornerstone of the recreational mathematics literature: in his review for *Winning Ways*, Gardner himself wrote that

> *Winning Ways* is the greatest contribution of this century to the burgeoning field of recreational mathematics. No other work has been so packed with completely new and significant material, or presented with so much wit, depth, and clarity. [**Gar82**]

What Gardner left unsaid was that *Winning Ways* also heralded the birth of an entirely new mathematical discipline. Combinatorial game theory, formerly a motley collection of fascinating but scattered results, was now presented as a unified and coherent subject all its own. Its recognition as a "serious" (rather than "recreational") discipline took some time to achieve, but fortunately this process was helped along immensely by a parallel series of developments, halfway around the world, led chiefly by the Israeli mathematician Aviezri Fraenkel.

Aviezri Fraenkel

Fraenkel was born in Munich in 1929. His family presciently moved to Basel, Switzerland, three years later, and again to Jerusalem in early 1939. In 1947 he entered the Technion to study electrical engineering, but in his first year the Israeli War of Independence intervened, and the university temporarily closed.

After the war, Fraenkel completed his degree in electrical engineering, followed by several years in the signal corps to finish his military service. Soon after, he joined the team that built the WEIZAC, the first computer in Israel, at the Weizmann Institute of Science. The WEIZAC team was led by UCLA Professor Gerald Estrin, and once the project was completed, Estrin invited Fraenkel to the Ph.D. program in computer engineering at UCLA.

After three years of study at UCLA, Fraenkel was ready for his qualifying exams, and he selected as the subjects of examination Computer Engineering, Modern Algebra, and Applied Mathematics. To his surprise, the engineering department declared that the latter two could not be considered distinct subjects for the exam! His wife Shaula later wrote:

> To the rescue came the late Prof. Ernst Straus .... He told Aviezri that he could take those three subjects for the Qualifying exams for a Ph.D. in Mathematics. The relieved Aviezri came home and told his wife that she would get a mathematician husband, not an engineer. Aviezri passed his exam with flying colors, became a mathematician, and has been one ever since. [**Fra01**]

In 1962, following a postdoctoral appointment at the University of Oregon, Fraenkel returned to the Weizmann Institute, where he would remain for the duration of his career. It was there that he began to study combinatorial games. After discovering NIM and WYTHOFF in a problems book by Yaglom and Yaglom [**YY67**], Fraenkel became curious about some related problems in the theory of complementary sequences and soon published a series of papers on the subject [**Fra69, Fra73**]. During this period, he grew increasingly intrigued by the curious dichotomy between the simplicity of NIM and the hardness of CHESS and GO:

> I was fascinated by the fact that NIM and CHESS are both in the same family, yet there seems to be such a huge gap between the easy strategy for the former and the seeming lack of an efficient strategy for the latter. In an attempt to make some progress, I thought that the natural way was to identify the mathematical differences between NIM-like and CHESS-like games, and attack these separately in a divide and conquer fashion. [**Fra12**]

These ideas were to appear as a motivating theme throughout Fraenkel's early papers on game theory. He first experimented with ways of making NIM-like games more CHESS-like, primarily by studying examples of impartial games with cycles. This quickly led to an independent (re-)discovery of the generalized Sprague–Grundy theory (see Section IV.4). He then investigated a class of *annihilation games*: impartial games played on graphs, with tokens that "annihilate" each other when they come into contact, a sort of simplified variant of the capture rule in CHESS. (See the Notes to Section IV.4 on page 214 for discussion.) But further progress proved difficult, and to Fraenkel's frustration, CHESS remained hopelessly out of reach.

Now he began to suspect that there was something really *different* about CHESS that precludes an efficient solution, and it is around this time that the idea of a true dichotomy became explicit in his writings [**Fra78, Fra80**]. Addressing the problem from the other direction, Fraenkel turned to the fledgling computational complexity theory. To his great relief, this change of tactics led quickly to a milestone result, published in a 1981 paper coauthored with David Lichtenstein [**FL81**]: CHESS, suitably generalized to an $n \times n$ board, is EXPTIME-complete.

This seemed to give an answer, of sorts, to the question of what distinguishes NIM from CHESS. Other hardness results trickled in around this time: Lichtenstein and Sipser showed that GO is PSPACE-hard; *Winning Ways* includes a proof that HACKENBUSH is NP-hard. Building on these results, Fraenkel advocated computational complexity as a unifying framework for separating "hard" games from "easy." Today this subject is known as *algorithmic combinatorial game theory*; Figure I.4.10 on page 45 provides a sampling of its many recent successes.

Despite such progress, much work remains to be done. Fraenkel's comments, penned in 1996, still ring true today:

> It appears that ... there is a very large set of games whose complexities are hard to determine. The set of these games is termed *Wonderland*, because we are wondering about the complexity classification of its members.... Today, most games belong to Wonderland, and despite recent impressive progress, the tools for reducing Wonderland are still few and inadequate. [**Fra96a**]

Examples of games still in "Wonderland" include DOMINEERING, FLOWERS, and misère DAWSON'S KAYLES. Only with a better grasp of such examples can Fraenkel's original ambition—a true *understanding* of the differences between NIM and CHESS and of where the line between them is drawn—be fully realized. In the meantime, Fraenkel's achievements stand as a major contribution to the foundation of combinatorial game theory as a mathematically rigorous subject.

Mathematical Go

While the 1970s was a formative period for combinatorial game theory throughout the world, the next decade saw further developments that signaled its transition to a mature subject.

Soon after *Winning Ways* was published, Berlekamp undertook a detailed study of DOMINEERING. To help understand its structure, he invented a simplified form of DOMINEERING called BLOCKBUSTING. Berlekamp found that BLOCKBUSTING has a striking property: *warming inverts chilling* (see Notes to Section II.5 on page 122), so that results about "chilled BLOCKBUSTING" can be translated directly to the ordinary version. Berlekamp used similar techniques to obtain solutions for infinitely many positions of $2 \times n$ and $3 \times n$ DOMINEERING.

In 1988 the American Mathematical Society staged a symposium in honor of von Neumann, and Berlekamp was invited to give a talk on combinatorial games. After the talk, a member of the audience asked if there were any real-world games to which the theory could be applied. Berlekamp replied that DOTS AND BOXES is one example and that GO *should* be another, but he added that no one had yet succeeded at the latter.

The question motivated Berlekamp to renew a search for applications to GO. He constructed GO problems on $6 \times n$ boards that are essentially isomorphic to certain BLOCKBUSTING positions, so that the theory of BLOCKBUSTING can be applied to them. However, the problems were highly contrived and required large values of n in order to work.

In the fall of 1989 David Wolfe, a graduate student at Berkeley, took Berlekamp's class. Wolfe was an expert GO player and president of the local GO club, and he was captivated by Berlekamp's constructions. Together they developed a rich theory of temperature-1 endgames, culminating in the "9-dan Stumping Problem" (Figure I.4.2 on page 29), which has stumped every professional player who has attempted to solve it. Their work was published in Wolfe's 1991 thesis [**Wol91**] and a subsequent book, *Mathematical Go* [**BW94**]. It remains among the most successful applications of the partizan theory to date.

The theory of partizan games had reached maturity. It could at last be applied to say something profound about GO, the game that first inspired it two decades before. Conway's original vision had finally been realized, and in response to *Mathematical Go* he offered these remarks:

> The addition theory for partizan games grew out of my pitiful attempts to understand the game of GO. Now the failure of those attempts no longer matters, because Berlekamp and Wolfe have succeeded beyond my wildest dreams, and I heartily congratulate them. [**Con93**]

Games Past and Future

Combinatorial game theory owes a great debt to recreational mathematics, and many of its motivating themes and examples derive from puzzles, amusements, and children's games. This association gives the subject a unique flair but unfortunately has probably hindered its acceptance by a wider community. "People ... if they hear 'games,'" says Guy, "they think it must just be a lot of trivialities." [**Guy11**]

The great G. H. Hardy, who included Bouton's Theorem in his classic *Introduction to the Theory of Numbers*, nonetheless later wrote:

> Chess problems are *unimportant*. The best mathematics is *serious* as well as beautiful.... The "seriousness" of a mathematical theorem lies, not in its practical consequences, ... but in the *significance* of the mathematical ideas which it connects.... The inferiority of the chess problem ... lies not in its consequences but in its content. [**Har92**]

Of course we now know that there is a great deal of "serious" mathematics in the study of combinatorial games as well. Guy comments on the evolution of combinatorial game theory and combinatorics generally:

> Until we published *Winning Ways* no one paid much attention to combinatorial games—it's really only just taking off now. Combinatorics itself took a long time too. When I first came to this continent in 1965 and I started going to combinatorics conferences, there were coding theorists and experimental design people and people working in projective planes and people working on Steiner systems and so on, and some aspects of group theory. These people were all doing the same mathematics, in a different language. It was only in the 1960s and 1970s that they all realized they were all doing the same thing, and looking at it from a different point of view. And the subject of combinatorics sort of took off. Combinatorics is now pretty well respectable, and hopefully combinatorial game theory will become respectable within a finite time now. [**Guy11**]

Fortunately perceptions have indeed begun to change, thanks to progress on a number of fronts over the past several decades. First, and most obviously, combinatorial game theory has successfully been applied to say something meaningful about games people actually play—GO and DOTS AND BOXES are prime examples.

There are also increasing connections between combinatorial games and other mathematical domains. Examples of natural PSPACE-complete problems are rare in computational complexity; combinatorial games provide a wealth of them. Misère quotients are novel examples of commutative monoids that arise "in nature," providing a new application for techniques in abstract commutative algebra. Lexicodes, popularized by Conway and

Sloane, reveal interesting connections between nim arithmetic and classical coding theory. The surreal numbers have obvious connections to nonstandard analysis and the theory of valued fields. The list goes on.

The significance of a CHESS problem lies in its power to inspire. From Zermelo to Dawson to Berlekamp and Conway and Guy, games people play have inspired new kinds of mathematical thought, leading to elegant theory—and ultimately, yes, deep and surprising connections to other domains. Far from trivial, games are an enduring, and essentially *mathematical*, aspect of human experience, and there is much we can learn from them. Perhaps Conway said it best:

> I used to feel guilty in Cambridge that I spent all day playing games, while I was supposed to be doing mathematics. Then, when I discovered surreal numbers, I realized that playing games *is* mathematics. [**Con99**]

Games are important because people play them, and combinatorial game theory has flowered because it is both fun *and* serious. In the years since *Mathematical Go*, interest in the subject has rapidly increased, and the growing list of open problems hints at how much work remains to be done. It is an exciting time for combinatorial game theory, and we can look forward to more discoveries and more surprises in the years ahead.

Notes

David Parlett's encyclopedic survey of the history of board games [**Par99**] is an excellent reference for material on the early games mentioned here (and many others). Loyd [**Loy14**], Dudeney [**Dud17, Dud19**], Ball [**Bal05**] and Lucas [**Luc94**] were prolific popularizers of recreational mathematics.

See Dimand and Dimand [**DD96**] for the early history of game *theory*. An in-depth analysis of Zermelo's contributions appears in a paper by Schwalbe and Walker [**SW01**], including an English translation of Zermelo's original analysis [**Zer13**].

Bouton's defining paper [**Bou01**] is readily available online; the origins of the name NIM are discussed in [**Ros53, Wal53**]. Wythoff [**Wyt07**] and Moore [**Moo10**] both published their early results, and Goldberg's early work on KAYLES appears in some prior editions of Ball. Sprague [**Spr36, Spr37**] and Grundy [**Gru39**] each published their results in the late 1930s; see also [**Spr47**].

The material on Berlekamp, Conway, Fraenkel and Guy is drawn from a variety of biographical sources [**Guy91b, AA93, Sei94, Shu95, Fra01**], as well as taped interviews with Berlekamp [**Ber12**] and Guy [**Guy11**] conducted by the author, and personal correspondence with Fraenkel [**Fra12**]. Kenyon [**Ken67**], Austin [**Aus76**], Li [**Li74**] and Wolfe [**Wol91**] wrote historically significant theses that are mentioned in the text.

Other perspectives on the history of combinatorial game theory may be found in [**Now08, CSNS11**]; the latter includes significant material on early games.

Bibliography

This bibliography includes material that is directly relevant to the subject matter covered in this book. Each reference includes a list of page numbers indicating where in the text it is cited; these appear in square brackets at the far right of the entry.

A far more comprehensive combinatorial games bibliography, now listing more than 1,700 references, is maintained (and regularly updated) by Aviezri Fraenkel [**Fra09**].

[AA93] Donald J. Albers and Gerald L. Alexanderson, *A conversation with Richard K. Guy*, The College Math. J. **24** (1993), no. 2, 122–148. [485, 486, 492]

[AGNW05] Michael H. Albert, J.P. Grossman, Richard J. Nowakowski, and David Wolfe, *An introduction to Clobber*, INTEGERS: The Electr. J. Combin. Number Theory **5** (2005), no. 2, #A01. [151]

[AN09] Michael H. Albert and Richard J. Nowakowski (eds.), *Games of no chance 3*, MSRI Publications, no. 56, Cambridge University Press, Cambridge, 2009. [25]

[AN11] ———, *Lattices of games*, Order (2011), 1–10. [166]

[ANW07] Michael H. Albert, Richard J. Nowakowski, and David Wolfe, *Lessons in play: An introduction to combinatorial game theory*, A K Peters, Ltd. / CRC Press, Natick, MA, 2007. [25]

[All84] Dean T. Allemang, *Machine computation with finite games*, Master's thesis, Trinity College, Cambridge, 1984, `http://miseregames.org/allemang/`. [242, 248, 258, 269]

[All01] ———, *Generalized genus sequences for misère octal games*, Internat. J. Game Theory **30** (2001), no. 4, 539–556. [242, 258]

[All09] Meghan R. Allen, *An investigation of partizan misère games*, Ph.D. thesis, Dalhousie University, 2009. [278]

[All13] ———, *Peeking at partizan misère quotients*, in Nowakowski and Wolfe [**NW13**]. [278]

[All87] Norman L. Alling, *Foundations of analysis over surreal number fields*, Mathematics Studies, no. 141, North–Holland, 1987. [423]

[Aus76] Richard B. Austin, *Impartial and partisan games*, Master's thesis, University of Calgary, 1976. [197, 492]

[Bal05] W. W. Rouse Ball, *Mathematical recreations and essays*, fourth ed., Macmillan and Co., Ltd., 1905. [492]

[Bec08] József Beck, *Combinatorial games: Tic-tac-toe theory*, Cambridge University Press, Cambridge, 2008. [43]

[Ber88] Elwyn R. Berlekamp, *Blockbusting and Domineering*, J. Combin. Theory, Ser. A **49** (1988), 67–116. [49, 122, 123]

[Ber96] ———, *The economist's view of combinatorial games*, in Nowakowski [**Now96**], pp. 365–405. [49, 341, 351, 377, 396]

[Ber00a] ———, *The Dots and Boxes game: Sophisticated child's play*, A K Peters, Ltd. / CRC Press, Natick, MA, 2000. [50, 483]

[Ber00b] ———, *Sums of $N \times 2$ Amazons*, Game Theory, Optimal Stopping, Probability and Statistics (F. Thomas Bruss and Lucien Le Cam, eds.), Lecture Notes–Monograph Series, no. 35, Institute of Mathematical Statistics, 2000, pp. 1–34. [123, 341]

[Ber02] ———, *Idempotents among partisan games*, in Nowakowski [**Now02**], pp. 3–23. [351]

[Ber09] ———, *Yellow–Brown Hackenbush*, in Albert and Nowakowski [**AN09**], pp. 413–418. [123]

[Ber12] ———, 2012, personal communication (taped interview). [482, 484, 492]

[BCG01] Elwyn R. Berlekamp, John H. Conway, and Richard K. Guy, *Winning ways for your mathematical plays*, second ed., A K Peters, Ltd. / CRC Press, Natick, MA, 2001. [25, 45, 259, 480, 486]

[BK96] Elwyn R. Berlekamp and Yonghoan Kim, *Where is the "Thousand-Dollar Ko"?*, in Nowakowski [**Now96**], pp. 203–226. [49]

[BMS96] Elwyn R. Berlekamp, Martin Müller, and William L. Spight, *Generalized thermography: Algorithms, implementation, and application to Go endgames*, Technical Report TR-96-030, Internat. Comp. Sci. Inst., Berkeley, CA, 1996. [49, 377]

[BN03] Elwyn R. Berlekamp and Teigo Nakamura, *Analysis of composite corridors*, in Schaeffer et al. [**SMB03**], pp. 213–229. [49]

[BP03] Elwyn R. Berlekamp and Matt Pearson, *Entrepreneurial Chess*, unpublished manuscript, 2003. [49]

[BS02] Elwyn R. Berlekamp and Katherine W. Scott, *Forcing your opponent to stay in control of a loony Dots-and-Boxes endgame*, in Nowakowski [**Now02**], pp. 317–330. [50]

[BW94] Elwyn R. Berlekamp and David Wolfe, *Mathematical Go: Chilling gets the last point*, A K Peters, Ltd. / CRC Press, Natick, MA, 1994. [30, 100, 123, 490]

[Bew04] Jörg Bewersdorff, *Luck, logic and white lies: The mathematics of games*, A K Peters, Ltd. / CRC Press, Natick, MA, 2004. [341, 478]

[BF90] Uri Blass and Aviezri S. Fraenkel, *The Sprague–Grundy function of Wythoff's game*, Theoret. Comput. Sci. **75** (1990), 311–333. [206]

[Bou01] Charles L. Bouton, *Nim, a game with a complete mathematical theory*, Ann. of Math. **3** (1901), no. 2, 35–39. [25, 45, 231, 477, 492]

[BUvdH00] D. M. Breuker, Jos W. H. M. Uiterwijk, and H. Jaap van den Herik, *Solving 8×8 Domineering*, Theoret. Comput. Sci. **230** (2000), no. 1–2, 195–206. [46]

[Bul02] Nathan Bullock, *Domineering: Solving large combinatorial search spaces*, ICGA J. **25** (2002), no. 2, 67–84. [46]

[Cal96] Dan Calistrate, *The reduced canonical form of a game*, in Nowakowski [**Now96**], pp. 409–416. [135]

[Cal98] ———, *Combinatorial games of Ramsey and Conway types*, Ph.D. thesis, University of Calgary, 1998. [135]

[CPW02] Dan Calistrate, Marc Paulhus, and David Wolfe, *On the lattice structure of finite games*, in Nowakowski [**Now02**], pp. 25–30. [159, 166]

[CSNS11] Alda Carvalho, Carlos Santos, João Neto, and Jorge N. Silva, *History of combinatorial games*, Proceedings of Board Game Studies Colloquium XIII, 2011. [492]

[CT02] Alice Chan and Alice Tsai, $1 \times n$ *Konane: A summary of results*, in Nowakowski [**Now02**], pp. 331–339. [50]

[Cin10] Alessandro Cincotti, *n-player partizan games*, Theoret. Comput. Sci. **411** (2010), no. 34–36, 3224–3234. [44]

[Con78] John H. Conway, *Loopy games*, Advances in Graph Theory (Bèla Bollobás, ed.), Ann. Discrete Math., no. 3, 1978, pp. 55–74. [300, 309, 321]

[Con90] ———, *Integral lexicographic codes*, Discrete Math. **83** (1990), no. 2-3, 219–235. [183]

[Con93] ———, *Letter to A K Peters, Ltd. in reference to* Mathematical Go [**BW94**], 1993. [490]

[Con99] ———, *Public lecture, Princeton, NJ,* Oct. 27, 1999. [492]

[Con01] ———, *On numbers and games*, second ed., A K Peters, Ltd. / CRC Press, Natick, MA, 2001. [25, 50, 150, 227, 231, 248, 258, 487]

[CS92] John H. Conway and William L. Sibert, *Mathematical Kayles*, Internat. J. Game Theory **20** (1992), no. 3, 237–246. [45, 259]

[CS] John H. Conway and Aaron N. Siegel, *notes on the algebraic structure of misère impartial games*, unpublished. [248]

[CS86] John H. Conway and Neil J. A. Sloane, *Lexicographic codes: Error-correcting codes from game theory*, IEEE Transactions on Information Theory **32** (1986), no. 3, 337–348. [183]

[Daw73] Thomas R. Dawson, *Caissa's wild roses, 1935*, Five Classics of Fairy Chess, Dover Publications, Inc., 1973. [15]

[DDE02] Erik D. Demaine, Martin L. Demaine, and David Eppstein, *Phutball endgames are hard*, in Nowakowski [**Now02**], pp. 351–360. [50]

[DH09] Erik D. Demaine and Robert A. Hearn, *Playing games with algorithms: Algorithmic combinatorial game theory*, in Albert and Nowakowski [**AN09**], pp. 1–42. [46]

[Der10] Dariusz Dereniowski, *Phutball is PSPACE-hard*, Theoret. Comput. Sci. **411** (2010), no. 44–46, 3971–3978. [45]

[DD96] Mary Ann Dimand and Robert W. Dimand, *The history of game theory, Volume 1: From the beginnings to 1945*, Routledge Studies in the History of Economics, no. 8, Routledge, 1996. [492]

[DiM11] Joseph DiMuro, *On* $\mathbf{On}_p$, 2011, preprint. `http://arxiv.org/abs/1108.0962`. [453]

[DFP99] Andreas W. M. Dress, Achim Flammenkamp, and Norbert Pink, *Additive periodicity of the Sprague–Grundy function of certain Nim games*, Adv. in Appl. Math. **22** (1999), 249–270. [206]

[Dud17] Henry E. Dudeney, *Amusements in mathematics*, Thomas Nelson and Sons, Ltd., 1917. [492]

[Dud19] ———, *The Canterbury puzzles*, second ed., Thomas Nelson and Sons, Ltd., 1919. [478, 492]

[Ehr94] Philip Ehrlich, *Real numbers, generalizations of the reals, and theories of continua*, Synthese Library, no. 242, Kluwer Academic Publishers, 1994. [438]

[Ehr12] ———, *The absolute arithmetic continuum and the unification of all numbers great and small*, Bulletin of Symbolic Logic **18** (2012), no. 1, 1–45. [438]

[Eis95] David Eisenbud, *Commutative algebra: With a view toward algebraic geometry*, Graduate Texts in Mathematics, no. 150, Springer, 1995. [436]

[Elk96] Noam D. Elkies, *On numbers and endgames: Combinatorial game theory in Chess endgames*, in Nowakowski [**Now96**], pp. 135–150. [31]

[Elk02] ———, *Higher nimbers in pawn endgames on large chessboards*, in Nowakowski [**Now02**], pp. 61–78. [31]

[ES73] Paul Erdős and John L. Selfridge, *On a combinatorial game*, J. Combin. Theory Ser. A **14** (1973), no. 3, 298–301. [42]

[Eri96] Jeff Erickson, *New Toads and Frogs results*, in Nowakowski [**Now96**], pp. 299–310. [50]

[Ern95] Michael D. Ernst, *Playing Konane mathematically: A combinatorial game-theoretic analysis*, UMAP J. **16** (1995), no. 2, 95–121. [50]

[EH60] Max Euwe and David Hooper, *A guide to Chess endings*, David McKay, 1960. [31]

[Fer84] Thomas S. Ferguson, *Misère annihilation games*, J. Combin. Theory Ser. A **37** (1984), no. 3, 205–230. [214]

[Fin12] Alex Fink, *Lattice games without rational strategies*, J. Combin. Theory Ser. A **119** (2012), 450–459. [269]

[Fla] Achim Flammenkamp, *Sprague–Grundy values of octal games*, `http://wwwhomes.uni-bielefeld.de/achim/octal.html`. [193, 197]

[Fla79] James A. Flanigan, *An analysis of some take-away and loopy partizan graph games*, Ph.D. thesis, University of California, Los Angeles, 1979. [51]

[Fla81] ———, *Selective sums of loopy partizan graph games*, Internat. J. Game Theory **10** (1981), 1–10. [51]

[Fla83] ———, *Slow joins of loopy games*, J. Combin. Theory, Ser. A **34** (1983), no. 1, 46–59. [51]

[Fra69] Aviezri S. Fraenkel, *The bracket function and complementary sets of integers*, Canadian J. Math. **21** (1969), 6–27. [488]

[Fra73] ———, *Complementing and exactly covering sequences*, J. Combin. Theory, Ser. A **14** (1973), 8–20. [488]

[Fra78] ———, *AMS book reviews for* [**Con01**] *and* [**Knu74**], Bull. Amer. Math. Soc. **84** (1978), no. 6, 1328–1336. [489]

[Fra80] ———, *From Nim to Go*, Combinatorial Mathematics, Optimal Designs and Their Applications (Jaya Srivastava, ed.), Annals of Discrete Mathematics, vol. 6, North-Holland, 1980, pp. 137–156. [489]

[Fra82] ———, *How to beat your Wythoff games' opponent on three fronts*, Amer. Math. Monthly **89** (1982), 353–361. [206]

[Fra84] ———, *Wythoff games, continued fractions, cedar trees and Fibonacci searches*, Theoret. Comput. Sci. **29** (1984), 49–73. [206]

[Fra96a] ———, *Combinatorial games: Selected bibliography with a succinct gourmet introduction*, in Nowakowski [**Now96**], pp. 493–537. [489]

[Fra96b] ———, *Scenic trails ascending from sea-level Nim to alpine Chess*, in Nowakowski [**Now96**], pp. 13–42. [46, 183]

[Fra04] ———, *Complexity, appeal and challenges of combinatorial games*, Theoret. Comput. Sci. **313** (2004), 393–415. [207]

[Fra09] ———, *Combinatorial games (dynamic survey)*, Electr. J. Combin. (2009), #DS2,
`http://www.combinatorics.org/ojs/index.php/eljc/article/view/ds2`. [25]

[Fra12] ———, October 2012, personal communication. [488, 492]

[FL81] Aviezri S. Fraenkel and David Lichtenstein, *Computing a perfect strategy for* $n \times n$ *Chess requires time exponential in* n, J. Combin. Theory, Ser. A **31** (1981), no. 2, 199–214. [45, 489]

[FLN88] Aviezri S. Fraenkel, Martin Loebl, and Jaroslav Nešetřil, *Epidemiography II. Games with a dozing yet winning player*, J. Combin. Theory, Ser. A **49** (1988), no. 1, 129–144. [38]

[FL91] Aviezri S. Fraenkel and Mordechai Lorberbom, *Nimhoff games*, J. Combin. Theory, Ser. A **58** (1991), no. 1, 1–25. [206]

[FN85] Aviezri S. Fraenkel and Jaroslav Nešetřil, *Epidemiography*, Pacific J. Math. **118** (1985), no. 2, 369–381. [38]

[FO98] Aviezri S. Fraenkel and Michal Ozery, *Adjoining to Wythoff's game its P-positions as moves*, Theoret. Comput. Sci. **205** (1998), 283–296. [206]

[FP73] Aviezri S. Fraenkel and Yehoshua Perl, *Constructions in combinatorial games with cycles*, Infinite and Finite Sets, Vol. 2 (A. Hajnal, R. Rado, and V. T. Sós, eds.), Colloq. Math. Soc. János Bolyai, no. 10, North-Holland, 1973, pp. 667–699. [214]

[FR03] Aviezri S. Fraenkel and Ofer Rahat, *Complexity of error-correcting codes derived from combinatorial games*, in Schaeffer et al. [**SMB03**], pp. 201–212. [183]

[FT75] Aviezri S. Fraenkel and Uzi Tassa, *Strategy for a class of games with dynamic ties*, Comput. Math. Appl. **1** (1975), 237–254. [214]

[FT82] ———, *Strategies for compounds of partizan games*, Math. Proc. Cambridge Philos. Soc. **92** (1982), 193–204. [289]

[FY76] Aviezri S. Fraenkel and Yaacov Yesha, *Theory of annihilation games*, Bull. Amer. Math. Soc. **82** (1976), 775–777. [214]

[FY82] ———, *Theory of annihilation games—I*, J. Combin. Theory, Ser. B **33** (1982), 60–82. [214]

[FY86] ———, *The generalized Sprague–Grundy function and its invariance under certain mappings*, J. Combin. Theory, Ser. A **43** (1986), 165–177. [214]

[Fra01] Shaula Fraenkel, *Aviezri Fraenkel: A brief biography*, Electr. J. Combin. **8** (2001), no. 2, #I1. [488, 492]

[Fras02] William E. Fraser, *Computer-assisted thermographic analysis of Go endgames*, Ph.D. thesis, University of California, Berkeley, 2002. [49, 377, 396]

[FHW05] William E. Fraser, Susan Hirshberg, and David Wolfe, *The structure of the distributive lattice of games born by day n*, INTEGERS: The Electr. J. Combin. Number Theory **5** (2005), no. 2, #A06. [166]

[FW04] William E. Fraser and David Wolfe, *Counting the number of games*, Theoret. Comput. Sci. **313** (2004), 527–532. [166]

[FL09] Eric J. Friedman and Adam S. Landsberg, *On the geometry of combinatorial games: A renormalization approach*, in Albert and Nowakowski [**AN09**], pp. 349–376. [206]

[FKUB05] Timothy Furtak, Masashi Kiyomi, Takeaki Uno, and Michael Buro, *Generalized Amazons is PSPACE-complete*, IJCAI'05: Proceedings of the 19th International Joint Conference on Artificial Intelligence, 2005, pp. 132–137. [45, 123]

[Gal74] David Gale, *A curious Nim-type game*, Amer. Math. Monthly **81** (1974), 876–879. [50]

[GP89] Anil Gangolli and Thane E. Plambeck, *A note on periodicity in some octal games*, Internat. J. Game Theory **18** (1989), no. 3, 311–320. [197]

[Gar82] Martin Gardner, *Review of* Winning Ways [**BCG01**], 1982. [487]

[Gas96] Ralph Gasser, *Solving Nine Men's Morris*, in Nowakowski [**Now96**], pp. 101–113. [46]

[Gon86] Harry Gonshor, *An introduction to the theory of surreal numbers*, London Math. Soc. Lecture Note Series, no. 110, Cambridge University Press, 1986. [423, 437]

[Gri01] Pierre A. Grillet, *Commutative semigroups*, Advances in Mathematics, no. 2, Springer, 2001. [470]

[GN13] J.P. Grossman and Richard J. Nowakowski, *A ruler regularity in hexadecimal games*, in Nowakowski and Wolfe [**NW13**]. [197]

[GS09] J.P. Grossman and Aaron N. Siegel, *Reductions of partizan games*, in Albert and Nowakowski [**AN09**], pp. 427–445. [136]

[Gru39] Patrick M. Grundy, *Mathematics and games*, Eureka **2** (1939), 6–8. [492]

[GrS56] Patrick M. Grundy and Cedric A. B. Smith, *Disjunctive games with the last player losing*, Proc. Cambridge Philos. Soc. **52** (1956), 527–533. [231, 248]

[GM11] Alan Guo and Ezra Miller, *Lattice point methods for combinatorial games*, Adv. in Applied Math. **46** (2011), 363–378. [269]

[GM12] ———, *Algorithms for lattice games*, Internat. J. Game Theory (2012), 1–12. [269]

[Guy91a] Richard K. Guy (ed.), *Combinatorial games*, Proceedings of Symposia in Applied Mathematics, no. 43, American Mathematical Society, Providence, RI, 1991.

[Guy91b] ———, *Mathematics from fun & fun from mathematics: An informal autobiographical history of combinatorial games*, Paul Halmos: Celebrating 50 Years of Mathematics (John H. Ewing and F. W. Gehring, eds.), Springer–Verlag, New York, NY, 1991, pp. 287–295. [492]

[Guy91c] ———, *Unsolved problems in combinatorial games*, in *Combinatorial games* [**Guy91a**], pp. 183–189. [49]

[Guy91d] ———, *What is a game?*, in *Combinatorial games* [**Guy91a**], pp. 1–21. [159]

[Guy11] ———, January 2011, personal communication (taped interview). [481, 491, 492]

[GN13] Richard K. Guy and Richard J. Nowakowski, *Unsolved problems in combinatorial games*, in Nowakowski and Wolfe [**NW13**]. [25, 455]

[GS56] Richard K. Guy and Cedric A. B. Smith, *The G-values of various games*, Proc. Cambridge Philos. Soc. **52** (1956), 514–526. [45, 196, 197, 480]

[HJ63] Alfred Hales and Robert Jewett, *Regularity and positional games*, Trans. Amer. Math. Soc. **106** (1963), 222–229. [42]

[Han59] Olof Hanner, *Mean play of sums of positional games*, Pacific J. Math. **9** (1959), no. 1, 81–99. [82, 122, 480]

[Har92] G. H. Hardy, *A mathematician's apology*, Cambridge University Press, 1992. [491]

[Hea06] Robert A. Hearn, *Games, puzzles, and computation*, Ph.D. thesis, Massachusetts Institute of Technology, 2006. [123]

[Hea09] ———, *Amazons, Konane, and Cross Purposes are PSPACE-complete*, in Albert and Nowakowski [**AN09**], pp. 287–306. [45, 123]

[HD09] Robert A. Hearn and Erik D. Demaine, *Games, puzzles, and computation*, A K Peters, Ltd. / CRC Press, Natick, MA, 2009. [46]

[Ho] Nhan Bao Ho, *Subtraction games with three element subtraction sets*, `http://arxiv.org/abs/1202.2986`. [196]

[HN04] Sam Howse and Richard J. Nowakowski, *Periodicity and arithmetic-periodicity in hexadecimal games*, Theoret. Comput. Sci. **313** (2004), no. 3, 463–472. [197]

[HS02] Scott Huddleston and Jerry Shurman, *Transfinite Chomp*, in Nowakowski [**Now02**], pp. 183–212. [50]

[Jec02] Thomas Jech, *Set theory*, third millennium ed., Springer Monographs in Mathematics, Springer–Verlag, 2002. [399, 465]

[JN09] Lim Y. Jin and Jurg Nievergelt, *Tigers and Goats is a draw*, in Albert and Nowakowski [**AN09**], pp. 163–176. [46]

[KM] Thomas Kahle and Ezra Miller, *Decompositions of commutative monoid congruences and binomial ideals*, `http://arxiv.org/abs/1107.4699`. [269]

[Kan03] Akihiro Kanamori, *The higher infinite: Large cardinals in set theory from their beginnings*, second ed., Springer Monographs in Mathematics, Springer–Verlag, 2003. [465]

[Kao97] Kuo Yuan Kao, *Sums of hot and tepid combinatorial games*, Ph.D. thesis, University of North Carolina at Charlotte, 1997. [49]

[Ken67] Jack C. Kenyon, *Nim-like games and the Sprague–Grundy theory*, Master's thesis, University of Calgary, 1967. [197, 492]

[Kim95] Yonghoan Kim, *New values in Domineering and loopy games in Go*, Ph.D. thesis, University of California, Berkeley, 1995. [49]

[Knu74] Donald E. Knuth, *Surreal numbers*, Addison–Wesley, 1974. [423]

[Kun83] Kenneth Kunen, *Set theory: An introduction to independence proofs*, Studies in Logic and the Foundations of Mathematics, no. 102, North Holland, 1983. [399, 465]

[Lan96] Howard A. Landman, *Eyespace values in Go*, in Nowakowski [**Now96**], pp. 227–257. [49]

[Lan02] ———, *A simple FSM-based proof of the additive periodicity of the Sprague–Grundy function of Wythoff's game*, in Nowakowski [**Now02**], pp. 383–386. [206]

[Lar] Urban Larsson, *Star and permutation games*, to appear. [207]

[Las31] Emanuel Lasker, *Brettspiele der völker*, A. Scherl g.m.b.h., 1931. [478]

[Lau99] François Laubie, *A recursive definition of p-ary addition without carry*, J. de Théorie des Nombres de Bordeaux **11** (1999), 307–315. [183]

[LB09] Lieven Le Bruyn, *On2: Extending Lenstra's list*, 2009, `http://www.neverendingbooks.org/on2-extending-lenstras-list` [453]

[Len77] Hendrik W. Lenstra, *On the algebraic closure of two*, Proc. Kon. Ned. Akad. Wet. Series A **80** (1977), 389–396, `http://hdl.handle.net/1887/3796`. [453]

[Len78] ———, *Nim multiplication*, 1978, I.H.E.S., Bures-sur-Yvette, `http://hdl.handle.net/1887/2125`. [221, 453]

[LP97] Harry R. Lewis and Christos H. Papadimitriou, *Elements of the theory of computation*, second ed., Prentice–Hall, 1997. [36]

[Li74] S.-Y. Robert Li, *Generalized impartial games*, Ph.D. thesis, University of California, Berkeley, 1974. [492]

[Li76] ———, *Sums of zuchswang games*, J. Combin. Theory, Ser. A **21** (1976), 52–67. [289, 321]

[Li78] ———, *n-person Nim and n-person Moore's Games*, Internat. J. Game Theory **7** (1978), no. 1, 31–36. [43]

[LS80] David Lichtenstein and Michael Sipser, *Go is polynomial-space hard*, J. of the Association for Computing Machinery **27** (1980), no. 2, 393–401. [45]

[Loe96] Daniel E. Loeb, *Stable winning coalitions*, in Nowakowski [**Now96**], pp. 451–471. [43]

[Loy14] Sam Loyd, *Sam Loyd's cyclopedia of 5000 puzzles, tricks, and conundrums*, The Lamb Publishing Company, 1914. [492]

[Luc94] Édouard Lucas, *Récréations mathématiques*, Gauthier–Villars, 1882–1894, four volumes. [492]

[Lur02] Jacob Lurie, *On a conjecture of Conway*, Illinois J. Math. **46** (2002), no. 2, 497–506. [412]

[McK11] Neil A. McKay, *Canonical forms of uptimals*, Theoret. Comput. Sci. **412** (2011), no. 52, 7122–7132. [100]

[MMN13] Neil A. McKay, Rebecca Milley, and Richard J. Nowakowski, *Misère-play Hackenbush sprigs*, to appear in Internat. J. Game Theory, 2013. [278]

[MO07] G. A. Mesdal and Paul Ottaway, *Simplification of partizan games in misère play*, INTEGERS: The Electr. J. Combin. Number Theory **7** (2007), no. 1, #G06. [277]

[Mil13] Ezra Miller, *Affine stratifications from finite misère quotients*, J. Algebraic Combin. **37** (2013), no. 1, 1–9. [269]

[MS05] Ezra Miller and Bernd Sturmfels, *Combinatorial commutative algebra*, Graduate Texts in Mathematics, no. 227, Springer, 2005. [269]

[Mill13] Rebecca Milley, *Restricted universes of partizan misère games*, Ph.D. thesis, Dalhousie University, 2013. [278]

[MNO12] Rebecca Milley, Richard J. Nowakowski, and Paul Ottaway, *The misère monoid of one-handed alternating games*, INTEGERS: The Electr. J. Combin. Number Theory **12B** (2012), #A1. [278]

[MR13] Rebecca Milley and Gabriel Renault, *Dead ends in misère play: The misère monoid of canonical numbers*, to appear in INTEGERS: The Electr. J. Combin. Number Theory, 2013. [278]

[Miln53] John Milnor, *Sums of positional games*, Contributions to the Theory of Games, Volume II (H. W. Kuhn and A. W. Tucker, eds.), Annals of Math. Studies, no. 28, Princeton Univ. Press, Princeton, 1953. [82, 122, 480]

[Moe91] David J. Moews, *Sums of games born on days 2 and 3*, Theoret. Comput. Sci. **91** (1991), no. 1, 119–128. [135, 178]

[Moe93] ———, *On some combinatorial games connected with Go*, Ph.D. thesis, University of California, Berkeley, 1993. [321]

[Moe96a] ———, *Infinitesimals and coin-sliding*, in Nowakowski [**Now96**], pp. 315–327. [100]

[Moe96b] ———, *Loopy games and Go*, in Nowakowski [**Now96**], pp. 259–272. [321]

[Moe02] ———, *The abstract structure of the group of games*, in Nowakowski [**Now02**], pp. 49–57. [178, 411]

[Moo10] E. H. Moore, *A generalization of the game called Nim*, Ann. Math., 2nd Ser. **11** (1910), no. 3, 93–94. [492]

[MT02] Martin Müller and Theodore Tegos, *Experiments in computer Amazons*, in Nowakowski [**Now02**], pp. 243–260. [123]

[Nak09] Tiego Nakamura, *Counting liberties in Go capturing races*, in Albert and Nowakowski [**AN09**], pp. 177–196. [49]

[Niv09] Gabriel Nivasch, *More on the Sprague–Grundy function for Wythoff's game*, in Albert and Nowakowski [**AN09**], pp. 377–410. [206]

[Now96] Richard J. Nowakowski (ed.), *Games of no chance*, MSRI Publications, no. 29, Cambridge University Press, Cambridge, 1996. [25]

[Now02] Richard J. Nowakowski (ed.), *More games of no chance*, MSRI Publications, no. 42, Cambridge University Press, Cambridge, 2002. [25]

[Now08] ———, *The history of combinatorial game theory*, Proceedings of the Board Game Studies Colloquium XI, 2008. [492]

[NO11] Richard J. Nowakowski and Paul Ottaway, *Option-closed games*, Contributions to Disc. Math. **6** (2011), no. 1, 142–153. [136]

[NW13] Richard J. Nowakowski and David Wolfe (eds.), *Games of no chance 4*, MSRI Publications, Cambridge University Press, Cambridge, 2013. [25]

[Ott09] Paul Ottaway, *Combinatorial games with restricted options under normal and misère play*, Ph.D. thesis, Dalhousie University, 2009. [136]

[Pap93] Christos H. Papadimitriou, *Computational complexity*, Addison Wesley, 1993. [36]

[Par99] David Parlett, *The Oxford history of board games*, Oxford University Press, 1999. [492]

[Pin93] Norbert Pink, *Über die grundyfunktionen des wythoffspiels und verwandter spiele*, Ph.D. thesis, Universität Heidelberg, 1993. [206]

[Pla92] Thane E. Plambeck, *Daisies, Kayles, and the Sibert–Conway decomposition in misère octal games*, Theoret. Comput. Sci. **96** (1992), no. 2, 361–388. [259]

[Pla05] ———, *Taming the wild in impartial combinatorial games*, INTEGERS: The Electr. J. Combin. Number Theory **5** (2005), no. 1, #G05. [259]

[Pla09] ———, *Advances in losing*, in Albert and Nowakowski [**AN09**], pp. 57–89.

[PS08a] Thane E. Plambeck and Aaron N. Siegel, *Misère quotients for impartial games*, J. Combin. Theory Ser. A **115** (2008), no. 4, 593–622. [259, 265, 268]

[PS08b] ———, *Misère quotients for impartial games: Supplementary material*, 2008, `http://arxiv.org/abs/math/0612616`, pp. 1–17. [268, 269]

[Ple91] Vera S. Pless, *Games and codes*, in Guy [**Guy91a**], pp. 101–110. [183]

[Pro00] James G. Propp, *Three-player impartial games*, Theoret. Comput. Sci. **233** (2000), no. 1-2, 263–278. [43]

[Rob83] J. M. Robson, *The complexity of Go*, Information Processing 83, Proceedings of the IFIP 9th World Computer Congress, North-Holland/IFIP, 1983, pp. 413–417. [45]

[Ros53] Alan S. C. Ross, *The name of the game of Nim*, The Mathematical Gazette **37** (1953), no. 320, 119–120. [492]

[SS08] Carlos P. Santos and Jorge N. Silva, *Konane has infinite nim-dimension*, INTEGERS: The Electr. J. Combin. Number Theory **8** (2008), no. 1, #G02. [50]

[Sch96] Jonathan H. Schaeffer, *Solving the game of Checkers*, in Nowakowski [**Now96**], pp. 119–133. [46]

[SBB+07] Jonathan H. Schaeffer, Neil Burch, Yngvi Björnsson, Akihiro Kishimoto, Martin Müller, Robert Lake, Paul Lu, and Steve Sutphen, *Checkers is solved*, Science **317** (2007), no. 5844, 1518–1522. [46]

[SMB03] Jonathan H. Schaeffer, Martin Müller, and Yngvi Björnsson (eds.), *Computers and games: Third international conference, CG'02*, Lecture Notes in Comp. Sci., no. 2883, Berlin, University of Alberta, Springer–Verlag, 2003.

[SS06] Dierk Schleicher and Michael Stoll, *An introduction to Conway's games and numbers*, Moscow Math. J. **6** (2006), no. 2, 359–388. [423]

[Sch52] Frederik Schuh, *Spel van delers (The game of Divisors)*, Nieuw Tijdschrift voor Wiskunde **39** (1952), 299–304. [50]

[SW01] Ulrich Schwalbe and Paul Walker, *Zermelo and the early history of game theory*, Games and Economic Behavior **34** (2001), no. 1, 123–137. [477, 492]

[Sei94] Charles Seife, *Mathemagician*, The Sciences (1994), 12–15, `http://www.users.cloud9.net/~cgseife/conway.html`. [481, 482, 484, 492]

[Shu95] Polly Shulman, *Infinity plus one, and other surreal numbers*, Discover (1995), 96–105, `http://discovermagazine.com/1995/dec/infinityplusonea599`. [484, 492]

[Siea] Aaron N. Siegel, *Lattices of reduced game values*, unpublished. [167]

[Sieb] ———, *Loopy and loopfree canonical values in Hare and Hounds*, unpublished. [321]

[Sie05] ———, *Loopy games and computation*, Ph.D. thesis, University of California, Berkeley, 2005. [166, 309, 321]

[Sie09a] ———, *Backsliding Toads and Frogs*, in Albert and Nowakowski [**AN09**], pp. 197–214. [321]

[Sie09b] ———, *Coping with cycles*, in Albert and Nowakowski [**AN09**], pp. 91–123. [50, 309]

[Sie09c] ———, *New results in loopy games*, in Albert and Nowakowski [**AN09**], pp. 215–232. [309, 321]

[Sie13a] ———, *Partizan misère canonical form*, in Nowakowski and Wolfe [**NW13**]. [277]

[Sie13b] ———, *The structure and classification of misère quotients*, in Nowakowski and Wolfe [**NW13**]. [268]

[Sie11] Angela A. Siegel, *On the structure of games and their posets*, Ph.D. thesis, Dalhousie University, 2011. [136, 159, 167]

[Smi66] Cedric A. B. Smith, *Graphs and composite games*, J. Combin. Theory, Ser. A **1** (1966), 51–81. [50, 214]

[Sna02] Raymond G. Snatzke, *Exhaustive search in Amazons*, in Nowakowski [**Now02**], pp. 261–278. [123]

[Sna03] ———, *Exhaustive search and databases in the application of combinatorial game theory to the game Amazons*, Ph.D. thesis, Universität Augsburg, Augsburg, Germany, 2003. [123]

[Sna04] ———, *New results of exhaustive search in the game Amazons*, Theoret. Comput. Sci. **313** (2004), 499–509. [123]

[Spi99] William L. Spight, *Extended thermography for multiple kos in Go*, Computers and Games: First International Conference, CG'98 (H. Jaap van den Herik and Hiroyuki Iida, eds.), Lecture Notes in Comp. Sci., no. 1558, Springer–Verlag, Berlin, 1999, pp. 232–251. [49, 377]

[Spi02] ———, *Go thermography: The 4/21/98 Jiang–Rui endgame*, in Nowakowski [**Now02**], pp. 89–105. [49, 377]

[Spi03] ———, *Evaluating kos in a neutral threat environment: Preliminary results*, in Schaeffer et al. [**SMB03**], pp. 413–428. [49, 396]

[Spr36] Roland P. Sprague, *Über mathematische Kampfspiele*, Tôhoku Math. J. **41** (1935–36), 438–444. [492]

[Spr37] ———, *Über zwei Abarten von Nim*, Tôhoku Math. J. **43** (1937), 351–359. [492]

[Spr47] ———, *Bemerkungen über eine spezielle Abelsche Gruppe*, Math. Z. **51** (1947), 82–84. [492]

[Sta86] Richard P. Stanley, *Enumerative combinatorics, Volume I*, Cambridge Studies in Advanced Mathematics, no. 49, Cambridge University Press, Cambridge, 1986. [462]

[Ste07] Fraser Stewart, *The sequential join of combinatorial games*, INTEGERS: The Electr. J. Combin. Number Theory **7** (2007), no. 1, #G03. [51]

[Sti96] Lewis Stiller, *Multilinear algebra and Chess endgames*, in Nowakowski [**Now96**], pp. 151–192. [46]

[Str85] Philip D. Straffin, *Three-person winner-take-all games with McCarthy's revenge rule*, College J. Math. **16** (1985), no. 5, 386–394. [43]

[SU93] Walter Stromquist and Daniel Ullman, *Sequential compounds of combinatorial games*, Theoret. Comput. Sci. **119** (1993), no. 2, 311–321. [51]

[Sun05] Xinyu Sun, *Wythoff's sequence and N-heap Wythoff's conjectures*, Disc. Math. **300** (2005), 180–195. [207]

[SZ04] Xinyu Sun and Doron Zeilberger, *On Fraenkel's n-heap Wythoff conjecture*, Annals of Combin. **8** (2004), 225–238. [207]

[Tak02] Takenobu Takizawa, *An application of mathematical game theory to Go endgames: Some width-two-entrance rooms with and without kos*, in Nowakowski [**Now02**], pp. 107–124. [49]

[Tha] Thotsaporn Thanatipanonda, *Three results of combinatorial game Toads and Frogs*, to appear. [50]

[Tha11] ———, *Further hopping with Toads and Frogs*, Electr. J. Combin. **18** (2011), no. 1, #P67. [50]

[TZ09] Thotsaporn Thanatipanonda and Doron Zeilberger, *A symbolic finite-state approach for automated proving of theorems in combinatorial game theory*, J. Difference Equations and Appl. **15** (2009), 111–118. [50]

[Tho99] Chris Thompson, *Count of day 6 misere-inequivalent impartial games*, 1999, posted to usenet `rec.games.abstract` on February 19, 1999. [248]

[vdDE01] Lou van den Dries and Philip Ehrlich, *Fields of surreal numbers and exponentiation*, Fundamenta Mathematicae **167** (2001), no. 2, 173–188. [437]

[Wal53] J. L. Walsh, *The name of the game of Nim*, The Mathematical Gazette **37** (1953), no. 322, 290, reader commentary on an article by Ross [**Ros53**]. [492]

[Wei07] Michael P. R. Weimerskirch, *On infinite indistinguishability quotient monoids in misère impartial combinatorial games*, Ph.D. thesis, University of Minnesota, 2007. [269]

[Wol91] David Wolfe, *Mathematics of Go: Chilling corridors*, Ph.D. thesis, University of California, Berkeley, 1991. [49, 490, 492]

[Wyt07] Willem A. Wythoff, *A modification of the game of Nim*, Niew Archief voor Wiskunde **7** (1907), 199–202. [206, 492]

[YY67] Akiva M. Yaglom and Isaak M. Yaglom, *Challenging mathematical problems with elementary solutions, Volume II*, English ed., University of Chicago, 1967. [488]

[Zer13] Ernst F. F. Zermelo, *Über eine anwendung der mengenlehre auf die theorie des schachspiels*, Proc. Fifth Congress Mathematicians, Cambridge University Press, Cambridge, 1913, pp. 501–504. [476, 492]

Glossary of Notation

This glossary is a list of various notation used throughout *Combinatorial Game Theory*, ordered by first appearance in the text. Each symbol is given together with a brief description, as well as the page (or pages) of the book on which its definition may be found.

Notation	Meaning or Description	Page(s)
$a \oplus b$	the nim-sum of a and b	2, 438
$\mathscr{N}, \mathscr{P}, \mathscr{L}, \mathscr{R}$	the four (loopfree) outcome classes	6
G^L, G^R	a typical Left or Right option of G	8
$o(G)$	the outcome class of G	10, 55
$G + H$	the disjunctive sum of G and H	11, 54
$G = H$	G is equal to H	11, 55
$\mathbb{G}$	the group of short partizan game values	12, 55
$G \cong H$	G is isomorphic to H	12
$\lfloor x \rfloor, \lceil x \rceil$	the floor and ceiling of x	24
$\mathrm{lb}(n)$	the base-2 logarithm of n, rounded down	24
$\lvert \mathcal{S} \rvert$	the cardinality of the set $\mathcal{S}$	24
$\mathrm{Pow}(\mathcal{S})$	the powerset of $\mathcal{S}$	24
$f[\mathcal{S}], f^{-1}[\mathcal{S}]$	the image and preimage of $\mathcal{S}$ under f	24
$\mathcal{A}^\omega$	the direct sum of countably many copies of $\mathcal{A}$	24
$\{G^L \mid G^R\}$	the game with typical Left option G^L and typical right option G^R	53
0	zero, the empty game	54
$\tilde{\mathbb{G}}_n$	the set of games born by day n	54
$\tilde{\mathbb{G}}$	the set of short partizan games	54

Notation	Meaning or Description	Page(s)
$-G$	the negative of G	54
$G - H$	the disjunctive sum of G and $-H$	54
$n \cdot G$	the disjunctive sum of n copies of G	54
$\{G \parallel H \mid J\}$	the game $\{G \mid \{H \mid J\}\}$	
$G \geq H$	G is greater than or equal to H	57
$G \mathrel{\Vert\!\!\rhd} H$	G is greater than or confused with H	58
$G \not\gtrless H$	G is confused with H	58
$1, 2, -1, -2, \ldots$	positive and negative integers	58
$*$	star	58
$\uparrow, \downarrow, \Uparrow, \Downarrow$	up, down, double-up, double-down	59
$\mathbb{G}^0$	the group of dicotic game values	60
$\mathbb{G}_n$	the set of game values born by day n	60
$\mathrm{b}(G)$	the birthday of G	60, 400
$\mathbb{G}_n^0$	the set of dicotic game values born by day n	61
$\tilde{\mathrm{b}}(G)$	the formal birthday of G	61
$\mathrm{ch}(\mathcal{A})$	the set of children of games in $\mathcal{A}$	61
$\Delta^L(G), \Delta^R(G)$	a typical Left or Right incentive of G	62
$\mathbb{D}$	the group of dyadic rationals	69
$\frac{1}{2}, \frac{1}{4}, -\frac{1}{2}, -\frac{1}{4}, \ldots$	dyadic rational numbers	69
$[a, b]$ or $]a, b[$	an open or closed interval with endpoints a, b	71, 71
$[a, b[$ or $]a, b]$	a half-open interval with endpoints a, b	71, 71
$\pm x$	the switch $\{x \mid -x\}$	73
$\mathcal{C}(G)$	the confusion interval of G	75
$L(G), R(G)$	the Left and Right stops of G	75
$m(G)$	the mean value of G	79
$G \ll H$	G is infinitesimal relative to H	84
$*m$	star-m, the nimber of order m	84
$\Uparrow\!\uparrow, \Downarrow\!\downarrow, \Uparrow\!\Uparrow, \Downarrow\!\Downarrow$	triple- and quadruple-up and down	85
$\uparrow n$	up-n (shorthand for $n \cdot \uparrow$)	85
$\uparrow n{*}m$	up-n-star-m (shorthand for $\uparrow n + {*m}$)	85
$+_G, -_G$	tiny-G and miny-G	88
$G : H$	the ordinal sum of G and H	89
$\uparrow^n, \downarrow_n, \uparrow^{[n]}, \downarrow_{[n]}$	up-n^{th} and relatives	94
$0.a_1a_2a_3\ldots$	the uptimal with coefficients $a_1, a_2, a_3, \ldots$	95
$\mathcal{UC}(G)$	the uptimal confusion interval of G	95

Notation	Meaning or Description	Page(s)
$\{0^k \mid G\}$	shorthand for $\{0 \parallel \{0^{k-1} \mid G\}\}$	96
$G^n, G^{[n]}$	generalized uptimals	99
G_t	G cooled by t	102
$t(G)$	the temperature of G	102
$L_t(G), R_t(G)$	the Left and Right scores of G at t	103
$\tilde{L}_t(G), \tilde{R}_t(G)$	the Left and Right scaffolds of G at t	108
$\int^T G$	G heated by T	112
$\oint^T G$	G overheated by T	114
$\oint_S^T G$	G overheated from S to T	115
$G \equiv H$	G is infinitesimally close to H	124
$G \geqq H$	G is greater than or equal to H modulo an infinitesimal difference	124
$\mathrm{rcf}(G)$	the reduced canonical form of G	129
G_*	G reduced by $*$	131
☆	a remote star	138
$G \sim H$	G and H are equivalent modulo ☆	138
$G \gtrsim H$	G is greater than or equal to H modulo ☆	138
$G \cdot \uparrow$	the Norton product of G by $\uparrow$	141
$\mathrm{aw}(G)$	the atomic weight of G	142
$\lfloor G \rfloor, \lceil G \rceil$	the floor and ceiling sets for G in $\mathbb{G}_n$	159, 159
$G \vee H, G \wedge H$	the meet and join of G and H in $\mathbb{G}_n$	159, 159
G^c	the companion of G	163
$\mathcal{HT}_n$	the set of hereditarily transitive game values born by day n	167
$\langle \mathcal{S} \rangle$	the subgroup generated by $\mathcal{S}$	167
$\mathrm{cl}(\mathcal{A})$	the closure of $\mathcal{A}$	168, 253
G'	a typical option of the impartial game G	179
$G' \in G$	G' is an option of the impartial game G	180, 208
$\mathscr{G}(G)$	the nim value of G	180
$\mathrm{mex}(\mathcal{S})$	the minimal excluded value of $\mathcal{S}$	180
$\mathbb{G}^{\mathrm{I}}$	the group of impartial game values	181
$\mathscr{G}(n)$	the nim value of a heap of size n	184
$\mathbf{d}_0.\mathbf{d}_1\mathbf{d}_2\mathbf{d}_3\ldots$	the octal game with code digits $\mathbf{d}_0$, $\mathbf{d}_1$, $\mathbf{d}_2$, $\ldots$	188
$\underset{\geq m}{\mathrm{mex}}(\mathcal{S})$	the minimal excluded value of $\mathcal{S}$ among integers $\geq m$	203

Notation	Meaning or Description	Page(s)
$\mathscr{D}$	the outcome class of drawn positions	209
$\operatorname{rank}(G)$	the rank of G as a loopy impartial game	212
$a \otimes b$	the nim-product of a and b	215, 439
$a \oslash b$	the nim-quotient of a and b	219, 440
$a^{\boxed{n}}$	the nth nim-power of a	219, 440
$o^+(G), o^-(G)$	the normal-play and misère-play outcomes of G	223
$\mathscr{G}^+(G), \mathscr{G}^-(G)$	the normal and misère nim values of G	229
$\mathscr{G}^\pm(G)$	the genus of G	231
$\delta(G)$	the discriminant of G	233
$*G_\#, *GHJ$	shorthand notation for misère impartial games	235
$\mathscr{G}^*(G)$	the extended genus of G	239
G^-	the mate of G	243
$G \bowtie H$	G is linked to H	244
$\mathcal{Q}(\mathscr{A})$	the misère quotient of the set $\mathscr{A}$	250
$\mathcal{T}_n$	the tame misère quotient on n generators	254
G°	the adjoint of G	270
$G \ltimes H, G \rtimes H$	G is downlinked or uplinked to H	274
$\overline{G}$	the negative of G (synonymous with $-G$)	282
$\hat{\mathscr{P}}, \check{\mathscr{P}}, \hat{\mathscr{N}}, \check{\mathscr{N}}$	the four partizan loopy outcome classes	282
on, **off**, **dud**	the loopy games: on, off, dud	284
over, **under**	the loopy games: over, under	291
$\Sigma_n^X(G)$	the nth sidling approximation to G (from X)	292
$G \sim_n H$	G and H are similar for n moves	295
$G \simeq H$	G and H are similar	295
$\uparrow^{[\mathbf{on}]}, \uparrow^{\mathbf{on}}$	the loopy games: upon, uponth	296
ace, $\mathbf{pip}_n$	the pip games	298
$\hat{o}(G), \check{o}(G)$	the biased outcomes of G	309
$G \mathrel{\hat{=}} H, G \mathrel{\check{=}} H$	G is equal to H (biased)	310
$G \mathrel{\hat{\geq}} H, G \mathrel{\check{\geq}} H$	G is greater than or equal to H (biased)	310
$G = S \,\&\, T$	G has onside S and offside T	310
$G \mathbin{+\!\!\uparrow} H, G \mathbin{+\!\!\downarrow} H$	the upsum and downsum of G and H	318
G°	the degree (of loopiness) of G	322
$V(G)$	the variety of G	327
$\mathscr{E}_t^\delta$	the coupon stack of temperature t and granularity δ	336

Notation	Meaning or Description	Page(s)
$L_t^\delta(G), R_t^\delta(G)$	the enriched scores of G	336, 355
$L_{\text{full}}(G), R_{\text{full}}(G)$	the Left and Right fullstops of G	339, 355
$\mathscr{E}_t$	a thick coupon stack of temperature t	341
$t_{\text{b}}(G)$	the board temperature of G	344
$t_{\text{a}}(G)$	the ambient temperature of G	345
$\text{s}(G)$	the subposition count of G	372
θ_x	the standard threat of size x	381
Θ_n	the nth threat environment	381
$L_t^\sharp(G), R_t^\sharp(G)$	the Left komaster scores of G	381
$L_t^\flat(G), R_t^\flat(G)$	the Right komaster scores of G	381
$m^\sharp(G), m^\flat(G)$	the mast value with Left or Right as komaster	382
$t^\sharp(G), t^\flat(G)$	the temperature with Left or Right as komaster	382
$L_t^{\sharp\sharp}(G), R_t^{\sharp\sharp}(G)$	the Left komonster scores of G	392
$L_t^{\flat\flat}(G), R_t^{\flat\flat}(G)$	the Right komonster scores of G	392
ω	omega, the least infinite ordinal	398
PG	the Group of long partizan game values	399
IG	the Group of long impartial game values	400
$\mathbf{PG}^0$	the Group of long dicotic game values	400
SN	the Field of surreal numbers	401
$\pm\mathbb{N}$	the game $\pm(0, 1, 2, \dots)$	404
$\mathbf{PG}^\infty$	the Group of all-moderate game values	405
∞	the game $\{\mathbb{N} \mid \pm\mathbb{N}\}$	406
∞^α	the largest moderate games	407
$x \times y$	the Conway product of long games x and y	412
x_β	the βth approximation to the surreal number x	417
$\sigma \mid \beta$	the restriction of the sign sequence σ to length β	418
$\text{supp}(x)$	the support of the surreal number x	428
$\mathbf{ON}_2$	the simplest Field of characteristic 2	439
$\mathcal{P}_\gamma$	the set of predecessors of the ordinal γ	439

Author Index

Index of Games

Index

In this index, references to the definitions of terminology are printed using **boldface** page numbers. References to primary statements or proofs of theorems are printed using *italic* page numbers.

Selected Published Titles in This Series

146 **Aaron N. Siegel,** Combinatorial Game Theory, 2013
144 **Shun-Jen Cheng and Weiqiang Wang,** Dualities and Representations of Lie Superalgebras, 2012
143 **Alberto Bressan,** Lecture Notes on Functional Analysis, 2013
142 **Terence Tao,** Higher Order Fourier Analysis, 2012
141 **John B. Conway,** A Course in Abstract Analysis, 2012
140 **Gerald Teschl,** Ordinary Differential Equations and Dynamical Systems, 2012
139 **John B. Walsh,** Knowing the Odds, 2012
138 **Maciej Zworski,** Semiclassical Analysis, 2012
137 **Luis Barreira and Claudia Valls,** Ordinary Differential Equations, 2012
136 **Arshak Petrosyan, Henrik Shahgholian, and Nina Uraltseva,** Regularity of Free Boundaries in Obstacle-Type Problems, 2012
135 **Pascal Cherrier and Albert Milani,** Linear and Quasi-linear Evolution Equations in Hilbert Spaces, 2012
134 **Jean-Marie De Koninck and Florian Luca,** Analytic Number Theory, 2012
133 **Jeffrey Rauch,** Hyperbolic Partial Differential Equations and Geometric Optics, 2012
132 **Terence Tao,** Topics in Random Matrix Theory, 2012
131 **Ian M. Musson,** Lie Superalgebras and Enveloping Algebras, 2012
130 **Viviana Ene and Jürgen Herzog,** Gröbner Bases in Commutative Algebra, 2011
129 **Stuart P. Hastings and J. Bryce McLeod,** Classical Methods in Ordinary Differential Equations, 2012
128 **J. M. Landsberg,** Tensors: Geometry and Applications, 2012
127 **Jeffrey Strom,** Modern Classical Homotopy Theory, 2011
126 **Terence Tao,** An Introduction to Measure Theory, 2011
125 **Dror Varolin,** Riemann Surfaces by Way of Complex Analytic Geometry, 2011
124 **David A. Cox, John B. Little, and Henry K. Schenck,** Toric Varieties, 2011
123 **Gregory Eskin,** Lectures on Linear Partial Differential Equations, 2011
122 **Teresa Crespo and Zbigniew Hajto,** Algebraic Groups and Differential Galois Theory, 2011
121 **Tobias Holck Colding and William P. Minicozzi II,** A Course in Minimal Surfaces, 2011
120 **Qing Han,** A Basic Course in Partial Differential Equations, 2011
119 **Alexander Korostelev and Olga Korosteleva,** Mathematical Statistics, 2011
118 **Hal L. Smith and Horst R. Thieme,** Dynamical Systems and Population Persistence, 2011
117 **Terence Tao,** An Epsilon of Room, I: Real Analysis, 2010
116 **Joan Cerdà,** Linear Functional Analysis, 2010
115 **Julio González-Díaz, Ignacio García-Jurado, and M. Gloria Fiestras-Janeiro,** An Introductory Course on Mathematical Game Theory, 2010
114 **Joseph J. Rotman,** Advanced Modern Algebra, Second Edition, 2010
113 **Thomas M. Liggett,** Continuous Time Markov Processes, 2010
112 **Fredi Tröltzsch,** Optimal Control of Partial Differential Equations, 2010
111 **Simon Brendle,** Ricci Flow and the Sphere Theorem, 2010
110 **Matthias Kreck,** Differential Algebraic Topology, 2010
109 **John C. Neu,** Training Manual on Transport and Fluids, 2010
108 **Enrique Outerelo and Jesús M. Ruiz,** Mapping Degree Theory, 2009

For a complete list of titles in this series, visit the
AMS Bookstore at **www.ams.org/bookstore/gsmseries/**.